Differential Geometry and Its Application, 2nd Edition

Differential Geometry and Its Application, 2nd Edition

Editor

Mića Stanković

MDPI

Basel • Beijing • Wuhan • Barcelona • Belgrade • Novi Sad • Cluj • Manchester

Editor
Mića Stanković
University of Nis
Niš
Serbia

Editorial Office
MDPI AG
Grosspeteranlage 5
4052 Basel, Switzerland

This is a reprint of articles from the Special Issue published online in the open access journal *Axioms* (ISSN 2075-1680) (available at: https://www.mdpi.com/journal/axioms/special_issues/KLBCBVV42T).

For citation purposes, cite each article independently as indicated on the article page online and as indicated below:

Lastname, A.A.; Lastname, B.B. Article Title. *Journal Name* **Year**, *Volume Number*, Page Range.

ISBN 978-3-7258-2301-7 (Hbk)
ISBN 978-3-7258-2302-4 (PDF)
doi.org/10.3390/books978-3-7258-2302-4

Contents

axioms

MDPI

Editorial

Differential Geometry and Its Application, 2nd edition

Mića S. Stanković

Faculty of Sciences and Mathematics, University of Niš, Višegradska 33, 18000 Niš, Serbia; mica.stankovic@pmf.edu.rs

1. Introduction

With this Editorial, we present a Special Issue of *Axioms* entitled 'Differential Geometry and Its Application, 2nd edition'. We launched this Special Edition as a continuation of 'Differential Geometry and Its Application'. This Special Issue provides a platform showcasing the latest achievements in many branches of theoretical and practical mathematical studies. These relate to Riemannian theories, generalized Riemannian spaces and their mappings. The scope of this Special Issue also includes Finsler geometry, Kenmotsu manifolds, Kaehler manifolds, manifolds with non-symmetric linear connections, cosymplectic manifolds, contact manifolds, statistical manifolds, Minkowski spaces, geodesic mappings, almost geodesic mappings, holomorphically projective mappings, warped products of manifolds, complex space forms, quaternionic space forms, golden manifolds, inequalities, invariants, immersions, etc. Potential authors are encouraged to submit papers that present new ideas in the field of differential geometry, in addition to the above topics. Given the broad scope and widespread interest in this topic, more works should be published in this area.

2. Overview of the Published Papers

This Special Issue contains 17 papers which were accepted for publication after a rigorous reviewing process.

The authors of the first contribution consider dual representations of Bertrand offsets. Surfaces are specified and several new results are gained in terms of their integral invariants. A new description of Bertrand offsets for developable surfaces is given. Furthermore, the authors obtained several relationships through the striction curves of Bertrand offsets of ruled surfaces and their integral invariants.

In the second contribution, the authors find some conditions under which the tangent bundle TM has a dualistic structure. Then, they introduce infinitesimal affine transformations on statistical manifolds and investigate these structures on a special statistical distribution, as well as on a tangent bundle of a statistical manifold. Moreover, they also study the mutual curvatures of a statistical manifold M and its tangent bundle TM, investigating their relations. More precisely, the authors obtain the mutual curvatures of well-known connections on the tangent bundle TM (the complete, horizontal, and Sasaki connections) and study their vanishing.

In the third contribution, the authors study several non-Riemannian quantities in Finsler geometry. These non-Riemannian quantities play an important role in understanding the geometric properties of Finsler metrics. In particular, the authors find differential equations of Finsler warped product metrics with vanishing ξ-curvature or vanishing H-curvature. Furthermore, they show that, for Finsler warped product metrics, the ξ-curvature vanishes if and only if the H-curvature vanishes.

The authors of the fourth contribution investigate the geometrical axioms of Riemannian submersions in the context of the η-Ricci-Yamabe soliton ($\eta - RY$ soliton) with a potential field. They give the categorization of each fiber of Riemannian submersion as an $\eta - RY$ soliton, an η-Ricci soliton, and an η-Yamabe soliton. Additionally, the authors

Citation: Stanković, M.S. Differential Geometry and Its Application, 2nd edition. *Axioms* **2024**, *13*, 561. https://doi.org/10.3390/axioms13080561

Received: 15 August 2024
Revised: 16 August 2024
Accepted: 16 August 2024
Published: 18 August 2024

consider the many circumstances under which a target manifold of Riemannian submersion is an $\eta - RY$ soliton, an η-Ricci soliton, an η-Yamabe soliton, or a quasi-Yamabe soliton. They deduce a Poisson equation on a Riemannian submersion in a specific scenario if the potential vector field ω of the soliton is of a gradient type $=: grad(\eta)$, providing some examples of an $\eta - RY$ soliton to illustrate their findings. Finally, the authors explore a number theoretic approach to Riemannian submersion with totally geodesic fibers.

The aim of the authors of the fifth contribution is to characterize a Riemannian 3-manifold M^3 equipped with a semi-symmetric metric ξ-connection $\widetilde{\nabla}$ with ρ-Einstein and gradient ρ-Einstein solitons. The existence of a gradient ρ-Einstein soliton in an M^3 admitting $\widetilde{\nabla}$ is ensured by constructing a non-trivial example; in this way, some of the authors' results are verified. By using the standard tensorial technique, the authors prove that the scalar curvature of $(M^3, \widetilde{\nabla})$ satisfies the Poisson equation $\Delta R = \frac{4(2-\sigma-6\rho)}{\rho}$.

In the sixth contribution, the authors utilize the axode invariants to derive novel hyperbolic proofs of the Euler–Savary and Disteli formulae. The widely recognized inflection circle is situated on the hyperbolic dual unit sphere, in accordance with the principles of the kinematic theory of spherical locomotions. Subsequently, a time-like line congruence is defined and its spatial equivalence is thoroughly studied. The formulated assertions degenerate into a quadratic form, which facilitates a comprehensive understanding of the geometric features of the inflection line congruence.

A principal curve on a surface plays a paramount role in reasonable implementations (contribution seven). A curve on a surface is a principal curve if its tangents are principal directions. Using the Serret–Frenet frame, the surface pencil couple can be expressed as linear combinations of the components of the local frames in Galilean 3-space $\mathbb{G}_3$. With these parametric representations, a family of surfaces using principal curves (curvature lines), the authors construct the necessary and sufficient conditions for the given Bertrand couple to be the principal curves on these surfaces. Moreover, they also analyze the necessary and sufficient conditions for the given Bertrand couple to satisfy the principal curves and the geodesic requirements. As implementations of their main conclusions, the authors expand some models to confirm the method.

In the eighth contribution, the authors examine the behavior of the simplest realistic Oregonator model of the BZ-reaction from the perspective of KCC theory. In order to reduce the complexity of the model, the authors initially transformed the first-order differential equation of the Oregonator model into a system of second-order differential equations. Using this approach, the authors describe the evolution of the Oregonator model in geometric terms by considering it as a geodesic in a Finsler space. The authors found five KCC invariants using the general expression of the nonlinear and Berwald connections. To understand the chaotic behavior of the Oregonator model, the deviation vector and its curvature around equilibrium points are studied. The authors then obtain the necessary and sufficient conditions for the parameters of the system in order to achieve Jacobi stability near the equilibrium points. Furthermore, a comprehensive examination was conducted to compare the linear and Jacobi stabilities of the Oregonator model at its equilibrium points, the authors then highlight these instances with a few illustrative examples.

The author of the ninth contribution gives an expository account of differential cohomology and the classification of higher line bundles (also known as S^1-banded gerbes) with a connection. He begins by examining how Čech cohomology is used to classify principal bundles and defines their characteristic classes, introducing differential cohomology a la Cheeger and Simons and S^1-banded gerbes with a connection.

In the tenth contribution, the authors study isotropic submanifolds in locally metallic product space forms. Firstly, they establish the Chen–Ricci inequality for such submanifolds and determine the conditions under which the inequality becomes equal. Additionally, the authors explore the minimality of Lagrangian submanifolds in locally metallic product space forms, applying the result to create a classification theorem for isotropic submanifolds whose mean curvature is constant. More specifically, they demonstrate that the submanifolds are either a product of two Einstein manifolds with Einstein constants, or they are isometric to a totally geodesic submanifold. The authors provide several examples to support their findings.

In the eleventh contribution, the authors study submanifolds tangent to the Reeb vector field in trans-Sasakian manifolds. They prove Chen's first inequality and the Chen–Ricci inequality, respectively, for submanifolds in trans-Sasakian manifolds which admit a semi-symmetric, non-metric connection. Moreover, the authors obtain a generalized Euler inequality for special contact slant submanifolds in trans-Sasakian manifolds endowed with a semi-symmetric non-metric connection.

In the twelfth contribution, the authors study and classify left-invariant cross-curvature solitons on Lorentzian three-dimensional Lie groups.

In the thirteenth contribution, the authors explores the Ricci tensor of slant submanifolds within locally metallic product space forms equipped with a semi-symmetric metric connection (SSMC). The authors' investigation includes a derivation of the Chen–Ricci inequality and an in-depth analysis of its equality. More precisely, if the mean curvature vector at a point vanishes, then the equality case of this inequality is achieved by a unit tangent vector at the point when, and only when, the vector belongs to the normal space. Finally, they have shown that when a point is a totally geodesic point or is totally umbilical with $n = 2$, the equality of this inequality holds true for all unit tangent vectors at that point, and vice versa.

In the fourteenth contribution, the authors' focus revolves around the establishment of a geometric inequality, commonly referred to as Chen's inequality. They specifically apply this inequality to assess the square norm of the mean curvature vector and the warping function of warped product slant submanifolds. Their investigation takes place within the context of locally metallic product space forms with quarter-symmetric metric connections. Additionally, they delve into the condition that determines when equality is achieved within the inequality. Furthermore, the authors also explore a number of implications of their findings.

The author of contribution fifteen proves that a 2-gerbe has a torsion Dixmier–Douady class if, and only if, the gerbe has locally constant cocycle data. As an example application, the author gives an alternative description of flat twisted vector bundles in terms of locally constant transition maps. These results generalize to n-gerbes for $n = 1$ and $n = 3$, providing insights into the structure of higher gerbes and their applications in the geometry of twisted vector bundles.

In the sixteenth contribution, the authors define quasi-canonical biholomorphically projective and equitorsion quasi-canonical biholomorphically projective mappings. Some relations between the corresponding curvature tensors of the generalized Riemannian spaces GR_N and $G\overline{R}_N$ are obtained. At the end, they found the invariant geometric object of an equitorsion quasi-canonical biholomorphically projective mapping.

Finally, in the seventeenth contribution, the authors formulate a data-independent latent space regularization constraint for general unsupervised autoencoders. The regularization relies on sampling the autoencoder Jacobian at Legendre nodes, which are the centers of the Gauss–Legendre quadrature. Revisiting this classic allows the authors to prove that regularized autoencoders ensure a one-to-one re-embedding of the initial data manifold into their latent representation. Demonstrations show that previously proposed regularization strategies, such as contractive autoencoding, cause topological defects even in simple examples, as do convolutional-based (variational) autoencoders. In contrast, topological preservation is ensured by standard multilayer perceptron neural networks when regularized using this approach. This observation extends from the classic FashionMNIST dataset to (low-resolution) MRI brain scans, suggesting that reliable low-dimensional representations of complex high-dimensional datasets can be achieved using this regularization technique.

3. Conclusions

A total of 17 papers were published in this Special Issue, 'Differential Geometry and Its Application, 2nd edition'. In these works, researchers interested in various aspects of Riemannian space theory and related topics will find interesting insights and inspiring results.

Funding: This research was funded by the grant from the Ministry of Science, Technological Development and Innovation of the Republic of Serbia 451-03-65/2024-03/200124.

Conflicts of Interest: The author declares no conflicts of interest.

List of Contributions

1. Nazra, S.H.; Abdel-Baky, R.A. Bertrand Offsets of Ruled Surfaces with Blaschke Frame in Euclidean 3-Space. *Axioms* **2023**, *12*, 649.
2. Peyghan, E.; Seifipour, D.; Mihai, I. Infinitesimal Affine Transformations and Mutual Curvatures on Statistical Manifolds and Their Tangent Bundle. *Axioms* **2023**, *12*, 667.
3. Sun, L.; Zhang, X.; Wu, M. Finsler Warped Product Metrics with Special Curvature Properties. *Axioms* **2023**, *12*, 784.
4. Siddiqi, M.D.; Mofarreh, F.; Akyol, M.A.; Hakami, A.H. η-Ricci-Yamabe Solitons along Riemannian Submersions. *Axioms* **2023**, *12*, 796.
5. Haseeb, A.; Chaubey, S.K.; Mofarreh, F.; Ahmadini, A.A.H. A Solitonic Study of Riemannian Manifolds Equipped with a Semi-Symmetric Metric ξ-Connection. *Axioms* **2023**, *12*, 809.
6. Almoneef, A.A.; Abdel-Baky, R.A. Kinematic Geometry of a Timelike Line Trajectory in Hyperbolic Locomotions. *Axioms* **2023**, *12*, 915.
7. Alluhaibi, N.; Abdel-Baky, R.A. Surface Pencil Couple with Bertrand Couple as Joint Principal Curves in Galilean 3-Space. *Axioms* **2023**, *12*, 1022.
8. Gupta, M.K.; Sahu, A.; Yadav, C.K.; Goswami, A.; Swarup, C. KCC Theory of the Oregonator Model for Belousov-Zhabotinsky Reaction. *Axioms* **2023**, *12*, 1133.
9. Park, B. Differential Cohomology and Gerbes: An Introduction to Higher Differential Geometry. *Axioms* **2024**, *13*, 60.
10. Li, Y.; Aquib, M.; Khan, M.A.; Al-Dayel, I.; Youssef, M.Z. Chen-Ricci Inequality for Isotropic Submanifolds in Locally Metallic Product Space Forms. *Axioms* **2024**, *13*, 183.
11. Mohammed, M.; Massamba, F.; Mihai, I.; Elamin, A.E.A.M.A.; Aldien, M.S. Some Chen Inequalities for Submanifolds in Trans-Sasakian Manifolds Admitting a SemiSymmetric Non-Metric Connection. *Axioms* **2024**, *13*, 195.
12. Azami, S.; Jafari, M.; Haseeb, A.; Ahmadini, A.A.H. Cross Curvature Solitons of Lorentzian Three-Dimensional Lie Groups. *Axioms* **2024**, *13*, 211.
13. Li, Y.; Aquib, M.; Khan, M.A.; Al-Dayel, I.; Masood, K. Analyzing the Ricci Tensor for Slant Submanifolds in Locally Metallic Product Space Forms with a Semi-Symmetric Metric Connection. *Axioms* **2024**, *13*, 454.
14. Li, Y.; Aquib, M.; Khan, M.A.; Al-Dayel, I.; Masood, K. Analyzing the Ricci Tensor for Slant Submanifolds in Locally Metallic Product Space Forms with a Semi-Symmetric Metric Connection. *Axioms* **2024**, *13*, 486.
15. Park, B. Geometry of Torsion Gerbes and Flat Twisted Vector Bundles. *Axioms* **2024**, *13*, 504.
16. Milenković, V.M.; Stanković, M.S. Quasi-Canonical Biholomorphically Projective Mappings of Generalized Riemannian Space in the Eisenhart Sense. *Axioms* **2024**, *13*, 528.
17. Ramanaik, C.K.; Willmann, A.; Suarez Cardona, J.-E.; Hanfeld, P.; Hoffmann, N.; Hecht, M. Ensuring Topological Data-Structure Preservation under Autoencoder Compression Due to Latent Space Regularization in Gauss-Legendre Nodes. *Axioms* **2024**, *13*, 535.

axioms

MDPI

Article

Bertrand Offsets of Ruled Surfaces with Blaschke Frame in Euclidean 3-Space

Sahar H. Nazra [1] and Rashad A. Abdel-Baky [2,*]

[1] Department of Mathematical Sciences, College of Applied Sciences, Umm Al-Qura University, lMecca 24382, Saudi Arabia; shnazra@uqu.edu.sa
[2] Department of Mathematics, Faculty of Science, University of Assiut, Assiut 71516, Egypt
* Correspondence: baky1960@aun.edu.eg

Abstract: Dual representations of the Bertrand offset-surfaces are specified and several new results are gained in terms of their integral invariants. A new description of Bertrand offsets of developable surfaces is given. Furthermore, several relationships through the striction curves of Bertrand offsets of ruled surfaces and their integral invariants are obtained.

Keywords: Bertrand offsets; height dual functions; striction curve

MSC: (2010) 53A04; 53A05; 53A17

Citation: Nazra, S.H.; Abdel-Baky, R.A. Bertrand Offsets of Ruled Surfaces with Blaschke Frame in Euclidean 3-Space. *Axioms* **2023**, *12*, 649. https://doi.org/10.3390/axioms12070649

Academic Editor: Mića Stanković

Received: 2 June 2023
Revised: 21 June 2023
Accepted: 25 June 2023
Published: 29 June 2023

1. Introduction

The approach of Bertrand offsets for ruled surfaces is an important and effective tool in model-based manufacturing of mechanical products, and geometric modelling. Offsets of these sort surfaces can be utilized to create geometric models of shell-type sorts and thick surfaces [1–4]. So, many engineers and geometers have inspected and attained numerous geometrical-kinematic properties of the ruled surfaces in Euclidean and non-Euclidean spaces; for instance Ravani and Ku adapted the theory of Bertrand curves for ruled surfaces based on line geometry [5]. They showed that a ruled surface can have an infinity of Bertrand offsets, in the same approach as a plane curve can have an infinity of Bertrand mates. Based on the E. Study map, Küçük and Gürsoy gave various adjectives of Bertrand offsets of trajectory ruled surfaces in terms of the interrelationships through the projection areas for the spherical images of Bertrand offsets and their integral invariants [6]. In [7], Kasap and Kuruoglu acquired the connections through integral invariants of the couple of the Bertrand ruled surfaces in Euclidean 3-space $\mathbb{E}^3$. In [8] Kasap and Kuruoglu initiated the address of Bertrand offsets of ruled surfaces in Minkowski 3-space. The involute-evolute offsets of ruled surface is offered by Kasap et al. in [9]. Orbay et al. [10] started the investigation of Mannheim offsets of the ruled surface. Onder and Ugurlu gained the relationships through the invariants of Mannheim offsets of timelike ruled surfaces, and they gave the conditions for these surface offsets to be developable [11]. Aldossary and Abdel-Baky explained the theory of Bertrand curves for ruled surfaces, based on the E. Study map [12]. Senturk and Yuce have considered the integral invariants of the offsets by the geodesic Frenet frame [13]. Important contributions to the Bertrand offsets of these ruled surfaces have been studied in [14–16].

In this, a generalization of the theory of Bertrand curves is offered for ruled and developable surfaces in Euclidean 3-space $\mathbb{E}^3$. Using the E. Study map, two ruled surfaces which are offset in the sense of Bertrand are defined. It is shown that, generally, any ruled surface can have a binary infinity of Bertrand offsets; however for a developable ruled surface to have a developable Bertrand offset, a linear equation should be specified among the curvature and torsion of its edge of regression. Furthermore, it is shown that the developable offsets of a developable surface are parallel offsets. The results, in addition to being of theoretical interest, have applications in geometric modelling and the manufacturing of products.

2. Basic Concepts

Dual numbers are the set of all pairs of real numbers written as

$$\mathbb{D} = \{\widehat{a} = a + \varepsilon a^*,\ a,\ a^* \in \mathbb{R}\},$$

where the dual unit ε satisfies the relationships $\varepsilon \neq 0,\ \varepsilon 1 = 1\varepsilon,\ \varepsilon^2 = 0$. The application of line geometry and dual number representation of line trajectories can be found in the works [1–5,17] , the dual number is used to recast the point displacement relationship into relationships of lines. As stated, the dual numbers were first introduced by W. Clifford after him E. Study used it as a tool for his research on the differential line geometry. Given dual numbers $\widehat{a} = a + \varepsilon a^*$, and $\widehat{b} = b + \varepsilon b^*$ the rules for combination can be defined as:

$$\left.\begin{array}{r} Equality : \widehat{a} = \widehat{b} \Longleftrightarrow a = b,\ a^* = b^*, \\ Addition : \widehat{a} + \widehat{b} = (a+b) + \varepsilon(a^* + b^*), \\ Multiplication : \widehat{ab} = ab + \varepsilon(a^* b + a b^*). \end{array}\right\}$$

The set of dual numbers, denoted as $\mathbb{D}$, forms a commutative group under addition. The associative laws hold for multiplication and dual numbers are distributive. As a result, the division of dual numbers is defined as:

$$\frac{\widehat{a}}{\widehat{b}} = \frac{a}{b} + \varepsilon\left(\frac{a^* b - a b^*}{b^2}\right),\ b \neq 0.$$

A dual number is called a pure dual when $\widehat{a} = \varepsilon a^*$. Division by a pure dual number is not defined. An example of dual number is the dual angle between two skew lines in space defined as: $\widehat{\theta} = \theta + \varepsilon\theta^*$, where θ is projected angle between the lines and θ^* is the minimal distance between the lines along their common perpendicular line. A differentiable function $f(x)$ can be defined for a dual variable $f(x + \varepsilon x^*)$ by expanding the function using a Taylor series:

$$f(x + \varepsilon x^*) = f(x) + \varepsilon x^* \frac{df(x)}{dx}.$$

So, we can give the followings:

$$\left.\begin{array}{r} \sin^{-1}(\theta + \varepsilon\theta^*) = \sin^{-1}\theta + \varepsilon\frac{\theta^*}{\sqrt{1-\theta^2}}, \\ \cos^{-1}(\theta + \varepsilon\theta^*) = \cos^{-1}\theta - \varepsilon\frac{\theta^*}{\sqrt{1-\theta^2}}, \\ \tan^{-1}(\theta + \varepsilon\theta^*) = \tan^{-1}\theta + \varepsilon\frac{\theta^*}{1+\theta^2}. \end{array}\right\}$$

Other functions may also be defined in this manner. It may also shown that, for an positive integer n,

$$\widehat{a}^n = a^n + \varepsilon n a^* a^{n-1} = a^n\left(1 + \varepsilon n \frac{a^*}{a}\right).$$

E. Study's Map

An oriented line L in the Euclidean 3-space $\mathbb{E}^3$ can be determined by a point $\mathbf{p} \in L$ and a normalized direction vector $\mathbf{a}$ of L, i.e., $\|\mathbf{a}\| = 1$. To obtain components for L, one forms the moment vector $\mathbf{a}^* = \mathbf{p} \times \mathbf{a}$,with respect to the origin point in $\mathbb{E}^3$. If $\mathbf{p}$ is substituted by any point $\mathbf{q} = \mathbf{p} + t\mathbf{a},\ t \in \mathbb{R}$,on L, this suggest that $\mathbf{a}^*$ is independent of $\mathbf{p}$ on L. The two vectors $\mathbf{a}$, and $\mathbf{a}^*$ are not independent of one another; they fulfil the following two equations:

$$< \mathbf{a}, \mathbf{a} > = 1,\quad < \mathbf{a}^*, \mathbf{a} > = 0.$$

The six components $a_i,\ a_i^*(i=1,2,3)$ of $\mathbf{a}$, and $\mathbf{a}^*$ are called the normalized Plűcker coordinates of the line L. Hence, the two vectors $\mathbf{a}$, and $\mathbf{a}^*$ determine the oriented line L.

Conversely, any six-tuple $a_i,\ a_i^*(i=1,2,3)$ with

$$a_1^2+a_2^2+a_3^2=1,\ \ a_1a_1^*+a_2a_2^*+a_3a_3^*=0.$$

represent a line in $\mathbb{E}^3$. Thus, the set of all oriented lines in $\mathbb{E}^3$ is in one-to-one correspondence with pairs of vectors in $\mathbb{E}^3$.

For vectors $(\mathbf{a}^*,\mathbf{a})\in\mathbb{E}^3\times\mathbb{E}^3$ we define the set

$$\mathbb{D}^3=\mathbb{D}\times\mathbb{D}\times\mathbb{D}=\{\widehat{\mathbf{a}}=\mathbf{a}+\varepsilon\mathbf{a}^*;\ \varepsilon\neq0,\ \varepsilon1=1\varepsilon,\ \varepsilon^2=0.\}.$$

Then for any two vectors $\widehat{\mathbf{a}},\widehat{\mathbf{b}}\in\mathbb{D}^3$, the scalar product is defined by:

$$<\widehat{\mathbf{a}},\widehat{\mathbf{b}}>=<\mathbf{a},\mathbf{b}>+\varepsilon(<\mathbf{a}^*,\mathbf{b}>+<\mathbf{a},\mathbf{b}^*>),$$

and the norm of $\widehat{\mathbf{a}}$ is defined by:

$$\|\widehat{\mathbf{a}}\|=\|\mathbf{a}\|+\varepsilon\frac{<\mathbf{a}^*,\mathbf{a}>}{\|\mathbf{a}\|},\ \ \|\mathbf{a}\|\neq0.$$

Hence, we may write the dual vector $\widehat{\mathbf{a}}$ as a dual multiplier of a dual vector in the form

$$\widehat{\mathbf{a}}=\|\widehat{\mathbf{a}}\|\widehat{\mathbf{e}},$$

where $\widehat{\mathbf{e}}$ is referred to as the axis. The ratio

$$h=\frac{<\mathbf{a}^*,\mathbf{a}>}{\|\mathbf{a}\|^2},$$

is called the pitch along the axis $\widehat{\mathbf{e}}$; If $h=0$ and $\|\mathbf{a}\|=1$, $\widehat{\mathbf{a}}$ is an oriented line, and when h is finite, $\widehat{\mathbf{a}}$ is a proper screw; and when h is infinite, $\widehat{\mathbf{a}}$ is called a couple. A dual vector with norm equal to unit is called a dual unit vector. Hence, each oriented line $L=(\mathbf{a},\mathbf{a}^*)\in\mathbb{E}^3$ is represented by dual unit vector

$$\widehat{\mathbf{a}}=\mathbf{a}+\varepsilon\mathbf{a}^*(<\mathbf{a},\mathbf{a}>=1,\ \ <\mathbf{a}^*,\mathbf{a}>=0).$$

The dual unit sphere in $\mathbb{D}^3$ is specified as

$$\mathbb{K}=\{\widehat{\mathbf{a}}\in\mathbb{D}^3\mid\|\widehat{\mathbf{a}}\|^2=\widehat{a}_1^2+\widehat{a}_2^2+\widehat{a}_3^2=1\}.$$

Via this we have the E. Study's map: The set of all oriented lines in the Euclidean 3-space $\mathbb{E}^3$ is in one-to-one correspondence with the set of points on dual unit sphere in the dual 3-space $\mathbb{D}^3$.

This dualized form of line representation along with the E. Study's map leads to a new interpretation of the scalar and vectorial products of two lines. For two directed lines $\widehat{\mathbf{a}}$, and $\widehat{\mathbf{b}}$ the dual angle $\widehat{\theta}=\theta+\varepsilon\theta^*$combines the angle θ and the minimal distance θ^*. This gives rise to geometric interpretations of the following products of the dual unit vectors [1–4]:

$$<\widehat{\mathbf{a}},\widehat{\mathbf{b}}>=\cos\widehat{\theta}=\cos\theta-\varepsilon\theta^*\sin\theta.$$

The following special cases can be given:

1. If $<\widehat{\mathbf{a}},\widehat{\mathbf{b}}>=0$, then $\theta=\frac{\pi}{2}$ and $\theta^*=0$; this means that the two lines $\widehat{\mathbf{a}}$, and $\widehat{\mathbf{b}}$ meet at right angle,
2. If $<\widehat{\mathbf{a}},\widehat{\mathbf{b}}>=$ pure dual, then $\theta=\frac{\pi}{2}$ and $\theta^*\neq0$; the lines $\widehat{\mathbf{a}}$, and $\widehat{\mathbf{b}}$ are orthogonal skew lines,
3. If $<\widehat{\mathbf{a}},\widehat{\mathbf{b}}>=$ pure real, then $\theta\neq\frac{\pi}{2}$ and $\theta^*=0$; the lines $\widehat{\mathbf{a}}$, and $\widehat{\mathbf{b}}$ are intersect,

4. If $< \widehat{\mathbf{a}}, \widehat{\mathbf{b}} > = 1$, then $\theta = 0$ and $\theta^* = 0$; the lines $\widehat{\mathbf{a}}$, and $\widehat{\mathbf{b}}$ are coincident (their directions are the same or opposite).

3. The Blaschke Approach

In this section, we consider the Blaschke approach for ruled surfaces by bearing in mind the E. Study map. Therefore, based on the notations in Section 2, a regular dual curve

$$t \in \mathbb{R} \mapsto \widehat{\mathbf{x}}(t) \in \mathbb{K},$$

is a ruled surface $(\widehat{x})$ in Euclidean 3-space $\mathbb{E}^3$. The lines $\widehat{\mathbf{x}}(t)$ are the generators of $(\widehat{x})$. Hence, ruled surfaces and dual curves are synonymous in this paper. The dual unit vector

$$\widehat{\mathbf{t}}(t) = \mathbf{t} + \varepsilon \mathbf{t}^* = \frac{d\widehat{\mathbf{x}}(t)}{dt} \left\| \frac{d\widehat{\mathbf{x}}(t)}{dt} \right\|^{-1}$$

is central normal of $(\widehat{x})$. The dual unit vector $\widehat{\mathbf{g}}(t) = \mathbf{g}(t) + \varepsilon \mathbf{g}^*(t) = \widehat{\mathbf{x}} \times \widehat{\mathbf{t}}$ is central tangent of $(\widehat{x})$. So, we have the moving Blaschke frame $\{\widehat{\mathbf{x}}(t), \widehat{\mathbf{t}}(t), \widehat{\mathbf{g}}(t)\}$ on $\widehat{\mathbf{x}}(t)$. Then, the Blaschke formulae read:

$$\frac{d}{dt}\begin{pmatrix} \widehat{\mathbf{x}} \\ \widehat{\mathbf{t}} \\ \widehat{\mathbf{g}} \end{pmatrix} = \begin{pmatrix} 0 & \widehat{p} & 0 \\ -\widehat{p} & 0 & \widehat{q} \\ 0 & -\widehat{q} & 0 \end{pmatrix} \begin{pmatrix} \widehat{\mathbf{x}} \\ \widehat{\mathbf{t}} \\ \widehat{\mathbf{g}} \end{pmatrix}, \tag{1}$$

where

$$\widehat{p}(t) = p(t) + \varepsilon p^*(t) = \left\| \frac{d\widehat{\mathbf{x}}(t)}{dt} \right\|, \ \widehat{q} = q + \varepsilon q^* = \det(\widehat{\mathbf{x}}, \frac{d\widehat{\mathbf{x}}(t)}{dt}, \frac{d^2\widehat{\mathbf{x}}(t)}{dt^2}),$$

are the Blaschke invariants of the dual curve $\widehat{\mathbf{x}}(t) \in \mathbb{K}$. The dual unit vectors $\widehat{\mathbf{x}}$, $\widehat{\mathbf{t}}$, and $\widehat{\mathbf{g}}$ corresponding to three concurrent mutually orthogonal oriented lines in $\mathbb{E}^3$ and they intersected at a point $\mathbf{c}$ on $\widehat{\mathbf{x}}$ named central point. The locus of the central points is the striction curve $\mathbf{c}(t)$ on $(\widehat{x})$. The dual arc-length $\widehat{s}$ of $\widehat{\mathbf{x}}(t)$ is specified by

$$d\widehat{s} = ds + \varepsilon ds^* = \left\| \frac{d\widehat{\mathbf{x}}(t)}{dt} \right\| dt = \widehat{p}(t)dt. \tag{2}$$

Then, we may have

$$\begin{pmatrix} \widehat{\mathbf{x}}' \\ \widehat{\mathbf{t}}' \\ \widehat{\mathbf{g}}' \end{pmatrix} = \begin{pmatrix} 0 & 1 & 0 \\ -1 & 0 & \widehat{\gamma}(\widehat{s}) \\ 0 & -\widehat{\gamma}(\widehat{s}) & 0 \end{pmatrix} \begin{pmatrix} \widehat{\mathbf{x}} \\ \widehat{\mathbf{t}} \\ \widehat{\mathbf{g}} \end{pmatrix} = \widehat{\omega}(\widehat{s}) \times \begin{pmatrix} \widehat{\mathbf{x}} \\ \widehat{\mathbf{t}} \\ \widehat{\mathbf{g}} \end{pmatrix}; \ ('= \frac{d}{d\widehat{s}}), \tag{3}$$

where $\widehat{\omega} = \omega + \varepsilon \omega^* = \widehat{\text{fl}}\widehat{\mathbf{x}} + \widehat{\mathbf{g}}$ is the Darboux vector, and $\widehat{\gamma}(\widehat{s}) := \gamma + \varepsilon \gamma^*$ is the dual geodesic curvature of $\widehat{\mathbf{x}}(\widehat{s}) \in \mathbb{K}$. The tangent of $\mathbf{c}(s)$ is specified by [16]:

$$\frac{d\mathbf{c}}{ds} = F(s)\mathbf{x} + \mu(s)\mathbf{g}. \tag{4}$$

The functions $\gamma(s)$, $F(s)$ and $\mu(s)$ are the curvature (construction) parameters of $(\widehat{x})$. These functions described as follows: γ is the geodesic curvature of the spherical image curve $\mathbf{x} = \mathbf{x}(s)$; $F(s)$ explains the angle among the ruling of $(\widehat{x})$ and the tangent to the striction curve; and $\mu(s)$ is the distribution parameter at the ruling. These parameters prepare an approach for constructing ruled surface by

$$(\widehat{x}) : \mathbf{y}(s, v) = \int_0^s (F(s)\mathbf{x}(s) + \mu(s)\mathbf{g}(s))ds + v\mathbf{x}(s), \ v \in \mathbb{R}. \tag{5}$$

The unit normal vector field at any point is

$$\xi(s,v) = \frac{\frac{\partial \mathbf{y}(s,v)}{\partial s} \times \frac{\partial \mathbf{y}(s,v)}{\partial v}}{\left\| \frac{\partial \mathbf{y}(s,v)}{\partial s} \times \frac{\partial \mathbf{y}(s,v)}{\partial v} \right\|} = \pm \frac{\mu \mathbf{t} - v\mathbf{g}}{\sqrt{\mu^2 + v^2}}, \tag{6}$$

which is the central normal at the striction point ($v = 0$). Let φ be the angle among the unit normal vector $\mathbf{n}$ and the central normal $\mathbf{t}$, then

$$\mathbf{n}(s,v) = \cos\varphi \mathbf{t} - \sin\varphi \mathbf{g}.$$

It is evident that:

$$\tan\varphi = \frac{v}{\mu}.$$

Hence, we have the following [4,17]:

Corollary 1. *The tangent plane of the non-developable ruled surface turns clearly among π along a ruling.*

Furthermore, the dual unit vector with the same sense as $\widehat{\omega}$ is also specified by

$$\widehat{\mathbf{b}}(\widehat{s}) := \mathbf{b} + \varepsilon \mathbf{b}^* = \frac{\widehat{\omega}}{\|\widehat{\omega}\|} = \frac{\widehat{\gamma}}{\sqrt{\widehat{\gamma}^2 + 1}} \widehat{\mathbf{x}} + \frac{1}{\sqrt{\widehat{\gamma}^2 + 1}} \widehat{\mathbf{g}}.$$

It is obvious that $\widehat{\mathbf{b}}(\widehat{s})$ is the Disteli-axis of $(\widehat{x})$. Let $\widehat{\psi} = \psi + \varepsilon\psi^*$ be the dual radius of curvature among $\widehat{\mathbf{b}}$ and $\widehat{\mathbf{x}}$. Then,

$$\widehat{\mathbf{b}}(\widehat{s}) = \cos\widehat{\psi}\widehat{\mathbf{x}} + \sin\widehat{\psi}\widehat{\mathbf{g}}, \text{ with } \cot\widehat{\psi} = \frac{\widehat{q}}{\widehat{p}}. \tag{7}$$

In fact, it is necessary to have the dual curvature $\widehat{\kappa}(\widehat{s})$, and the dual torsion $\widehat{\tau}(\widehat{s})$. Therefore, the Serret-Frenet frame of $\widehat{\mathbf{x}}(\widehat{s}) \in \mathbb{K}$ is made up of the set $\{\widehat{\mathbf{t}}(\widehat{s}), \widehat{\mathbf{n}}(\widehat{s}), \widehat{\mathbf{b}}(\widehat{s})\}$. Then, the relative orientation is given by

$$\begin{pmatrix} \widehat{\mathbf{t}} \\ \widehat{\mathbf{n}} \\ \widehat{\mathbf{b}} \end{pmatrix} = \begin{pmatrix} 0 & 1 & 0 \\ -\sin\widehat{\psi} & 0 & \cos\widehat{\psi} \\ \cos\widehat{\psi} & 0 & \sin\widehat{\psi} \end{pmatrix} \begin{pmatrix} \widehat{\mathbf{x}} \\ \widehat{\mathbf{t}} \\ \widehat{\mathbf{g}} \end{pmatrix}.$$

Similarly, we can describe the dual Serret-Frenet formulae

$$\begin{pmatrix} \widehat{\mathbf{t}}' \\ \widehat{\mathbf{n}}' \\ \widehat{\mathbf{b}}' \end{pmatrix} = \begin{pmatrix} 0 & \widehat{\kappa} & 0 \\ -\widehat{\kappa} & 0 & \widehat{\tau} \\ 0 & -\widehat{\tau} & 0 \end{pmatrix} \begin{pmatrix} \widehat{\mathbf{t}} \\ \widehat{\mathbf{n}} \\ \widehat{\mathbf{b}} \end{pmatrix},$$

where

$$\left.\begin{array}{l} \widehat{\gamma}(\widehat{s}) = \gamma + \varepsilon(\Gamma - \gamma\mu) = \cot\psi - \varepsilon\psi^*(1 + \cot^2\psi), \\ \widehat{\kappa}(\widehat{s}) := \kappa + \varepsilon\kappa^* = \sqrt{1 + \widehat{\gamma}^2} = \frac{1}{\sin\widehat{\psi}} = \frac{1}{\widehat{\rho}(\widehat{s})}, \\ \widehat{\tau}(\widehat{s}) := \tau + \varepsilon\tau^* = \pm\widehat{\psi}' = \pm\frac{\widehat{\gamma}'}{1+\widehat{\gamma}^2}. \end{array}\right\} \tag{8}$$

Height Dual Functions

In correspondence with [18], a dual point $\widehat{\mathbf{b}}_0 \in \mathbb{K}$ will be named a $\widehat{\mathbf{b}}_k$ evolute of the dual curve $\widehat{\mathbf{x}}(\widehat{s}) \in \mathbb{K}$; for all $\widehat{s}$ such that $< \widehat{\mathbf{b}}_0, \widehat{\mathbf{x}}(\widehat{s}) >= 0$, but $< \widehat{\mathbf{b}}_0, \widehat{\mathbf{x}}_1^{k+1}(\widehat{s}) >\neq 0$. Here $\widehat{\mathbf{x}}^{k+1}$ signalizes the k-th derivatives of $\widehat{\mathbf{x}}(\widehat{s})$ with respect to $\widehat{s}$. For the 1st evolute $\widehat{\mathbf{b}}$ of $\widehat{\mathbf{x}}(\widehat{s})$, we have $< \widehat{\mathbf{b}}, \widehat{\mathbf{x}}' >= \pm < \widehat{\mathbf{b}}, \widehat{\mathbf{t}} >= 0$, and $< \widehat{\mathbf{b}}, \widehat{\mathbf{x}}'' >= \pm < \widehat{\mathbf{b}}, -\widehat{\mathbf{x}} + \widehat{\mathbf{flg}} >\neq 0$. So, $\widehat{\mathbf{b}}$ is at least a $\widehat{\mathbf{b}}_2$ evolute of $\widehat{\mathbf{x}}(\widehat{s}) \in \mathbb{K}$.

We now address a dual function $\widehat{a} : I \times \mathbb{K} \to \mathbb{D}$, by $\widehat{a}(\widehat{s}, \widehat{\mathbf{b}}_0) = < \widehat{\mathbf{b}}_0, \widehat{\mathbf{x}} >$. We call $\widehat{a}$ a height dual function on $\widehat{\mathbf{x}}(\widehat{s}) \in \mathbb{K}$. We use the notation $\widehat{a}(\widehat{s}) = \widehat{a}(\widehat{s}, \widehat{\mathbf{b}}_0)$ for any stationary point $\widehat{\mathbf{b}}_0 \in \mathbb{K}$. Hence, we state the following:

Proposition 1. *Under the above hypotheses, the following holds:*

i $\widehat{a}$ *will be stationary in the 1st approximation iff* $\widehat{\mathbf{b}}_0 \in Sp\{\widehat{\mathbf{x}},\widehat{\mathbf{g}}\}$, *that is,*

$$\widehat{a}' = 0 \Leftrightarrow < \widehat{\mathbf{x}}, \widehat{\mathbf{b}}_0 > = 0 \Leftrightarrow < \widehat{\mathbf{t}}, \widehat{\mathbf{b}}_0 > = 0 \Leftrightarrow \widehat{\mathbf{b}}_0 = \widehat{a}_1 \widehat{\mathbf{x}} + \widehat{a}_3 \widehat{\mathbf{g}};$$

for some dual numbers $\widehat{a}_1, \widehat{a}_3 \in \mathbb{D}$, *and* $\widehat{a}_1^2 + \widehat{a}_3^2 = 1$.

ii $\widehat{a}$ *will be stationary in the 2nd approximation iff* $\widehat{\mathbf{b}}_0$ *is* $\widehat{\mathbf{b}}_2$ *evolute of* $\widehat{\mathbf{b}}_0 \in \mathbb{K}$, *that is,*

$$\widehat{a}' = \widehat{a}'' = 0 \Leftrightarrow \widehat{\mathbf{b}}_0 = \pm \widehat{\mathbf{b}}.$$

iii $\widehat{a}$ *will be stationary in the 3rd approximation iff* $\widehat{\mathbf{b}}_0$ *is* $\widehat{\mathbf{b}}_3$ *evolute of* $\widehat{\mathbf{b}}_0 \in \mathbb{K}$, *that is,*

$$\widehat{a}' = \widehat{a}'' = \widehat{a}''' = 0 \Leftrightarrow \widehat{\mathbf{b}}_0 = \pm \widehat{\mathbf{b}}, \text{ and } \widehat{\gamma}' \neq 0.$$

iv $\widehat{a}$ *will be stationary in the 4th approximation iff* $\widehat{\mathbf{b}}_0$ *is* $\widehat{\mathbf{b}}_4$ *evolute of* $\widehat{\mathbf{b}}_0 \in \mathbb{K}$, *that is,*

$$\widehat{a}' = \widehat{a}'' = \widehat{a}''' = \widehat{a}^{iv} = 0 \Leftrightarrow \widehat{\mathbf{b}}_0 = \pm \widehat{\mathbf{b}}, \widehat{\gamma}' = 0, \text{ and } \widehat{\gamma}'' \neq 0.$$

Proof. For the 1st differentiation of $\widehat{a}$ we gain

$$\widehat{a}' = < \widehat{\mathbf{x}}', \widehat{\mathbf{b}}_0 >. \tag{9}$$

So, we gain

$$\widehat{a}' = 0 \Leftrightarrow < \widehat{\mathbf{t}}, \widehat{\mathbf{b}}_0 > = 0 \Leftrightarrow \widehat{\mathbf{b}}_0 = \widehat{a}_1 \widehat{\mathbf{x}} + \widehat{a}_3 \widehat{\mathbf{g}}; \tag{10}$$

for some dual numbers $\widehat{a}_1, \widehat{a}_3 \in \mathbb{D}$, and $\widehat{a}_1^2 + \widehat{a}_3^2 = 1$, the result is clear. 2- Differentiation of Equation (10) leads to:

$$\widehat{a}'' = < \widehat{\mathbf{x}}'', \widehat{\mathbf{b}}_0 > = < -\widehat{\mathbf{x}} + \widehat{\gamma} \widehat{\mathbf{g}}, \widehat{\mathbf{b}}_0 > . \tag{11}$$

By the Equations (10) and (11) we have:

$$\widehat{a}' = \widehat{a}'' = 0 \Leftrightarrow < \widehat{\mathbf{x}}, \widehat{\mathbf{b}}_0 > = < \widehat{\mathbf{x}}, \widehat{\mathbf{b}}_0 > = 0 \Leftrightarrow \widehat{\mathbf{b}}_0 = \pm \frac{\widehat{\mathbf{x}} \times \widehat{\mathbf{x}}}{\|\widehat{\mathbf{x}} \times \widehat{\mathbf{x}}\|} = \pm \widehat{\mathbf{b}}.$$

3- Differentiation of Equation (11) leads to:

$$\widehat{a}''' = < \widehat{\mathbf{x}}''', \widehat{\mathbf{b}}_0 > = -\left(1 + \widehat{\gamma}^2\right) < \widehat{\mathbf{t}}, \widehat{\mathbf{b}}_0 > + \widehat{\gamma}' < \widehat{\mathbf{g}}, \widehat{\mathbf{b}}_0 >$$

Hence, we have:

$$\widehat{a}' = \widehat{a}'' = \widehat{a}''' = 0 \Leftrightarrow \widehat{\mathbf{b}}_0 = \pm \widehat{\mathbf{b}}, \text{ and } \widehat{\gamma}' \neq 0.$$

4- By the analogous arguments, we can also have:

$$\widehat{a}' = \widehat{a}'' = \widehat{a}''' = \widehat{a}'''' = 0 \Leftrightarrow \widehat{\mathbf{b}}_0 = \pm \widehat{\mathbf{b}}, \widehat{\gamma}' = 0, \text{ and } \widehat{\gamma}'' \neq 0.$$

This completed the proof. □

In view of the Proposition 1, we have the following:

(a) The osculating circle $\mathbb{S}(\widehat{\rho}, \widehat{\mathbf{b}}_0)$ of $\widehat{\mathbf{x}}(\widehat{s}) \in \mathbb{K}$ is displayed by

$$< \widehat{\mathbf{b}}_0, \widehat{\mathbf{x}} > = \widehat{\rho}(\widehat{s}), \ < \widehat{\mathbf{x}}, \widehat{\mathbf{b}}_0 > = 0, < \widehat{\mathbf{x}}, \widehat{\mathbf{b}}_0 > = 0,$$

which are specified via the condition that the osculating circle must have osculate of at least 3rd order at $\widehat{\mathbf{x}}(\widehat{s}_0)$ iff $\widehat{\gamma}^{'} \neq 0$.

(b) The osculating circle $\mathbb{S}(\widehat{\rho}, \widehat{\mathbf{b}}_0)$ and the curve $\widehat{\mathbf{x}}(\widehat{s}) \in \mathbb{K}$ have at least 4-th order at $\widehat{\mathbf{x}}(s_0)$ iff $\widehat{\gamma}^{'} = 0$, and $\widehat{\gamma}^{''} \neq 0$.

In this manner, by catching into contemplation the evolutes of $\widehat{\mathbf{x}}(\widehat{s}) \in \mathbb{K}$, we can gain a sequence of evolutes $\widehat{\mathbf{b}}_2, \widehat{\mathbf{b}}_3,..., \widehat{\mathbf{b}}_n$. The proprietorships and the mutual connections through these evolutes and their involutes are very enjoyable problems. For instance, it is not difficult to address that when $\widehat{\mathbf{b}}_0 = \pm\widehat{\mathbf{b}}$, and $\widehat{\gamma}^{'} = 0$, $\widehat{\mathbf{x}}(\widehat{s})$ is existing at $\widehat{\psi}$ is stationary relative to $\widehat{\mathbf{b}}_0$. In this case, the Disteli-axis is stationary up to 2nd order, and the line $\widehat{\mathbf{x}}$ moves over it with stationary pitch. Thus, the ruled surface $(\widehat{x})$ with stationary Disteli-axis is formed by line $\widehat{\mathbf{x}}$ situated at an stationary distance ψ^* and stationary angle ψ with respect to the Disteli-axis $\widehat{\mathbf{b}}$, that is,

$$\widehat{\gamma}(\widehat{s}) := \gamma + \varepsilon(F - \gamma\mu) = \cot\widehat{\psi} = \widehat{c},$$

where $\widehat{c} = c + \varepsilon c^* \in \mathbb{D}$. By separating the real and dual parts, the following theorem can be stated:

Theorem 1. *A non-developable ruled surface $(\widehat{x})$ is a stationary Disteli-axis iff $\gamma(s)$ = constant, and $(F - \gamma\mu)$ = constant.*

Furthermore, in the case of

$$\widehat{\gamma}(\widehat{s}) := \gamma + \varepsilon(F - \gamma\mu) = 0 = \cot\psi - \varepsilon\psi^*(1 + \cot^2\psi),$$

then $\widehat{\mathbf{x}}(\widehat{s})$ is a dual great circle on $\mathbb{K}$, that is,

$$\widehat{c} = \{\widehat{\mathbf{x}} \in \mathbb{K} \mid <\widehat{\mathbf{x}}, \widehat{\mathbf{b}}> = 0; \text{ with } \left\|\widehat{\mathbf{b}}\right\|^2 = 1\}.$$

In this case, all the rulings of $(\widehat{x})$ intersected orthogonally with the stationary Disteli-axis $\widehat{\mathbf{b}}$, that is, $\psi = \frac{\pi}{2}$, and $\psi^* = 0$. Thus, we have $\widehat{\gamma}(\widehat{s}) := \gamma + \varepsilon(F - \gamma\mu) = 0 \Leftrightarrow (\widehat{x})$ is a helicoidal ruled surface.

Corollary 2. *A non-developable ruled surface $(\widehat{x})$ is a helicoidal ruled surface iff $\gamma(s) = 0$, and $F(s) = 0$.*

For $\widehat{\gamma}(\widehat{s})$ is a constant dual number, from the Equations (3) and (8), we have the ODE, $\widehat{\mathbf{x}}^{'''} + \widehat{\kappa}^2\widehat{\mathbf{x}}^{'} = \mathbf{0}$. After several algebraic manipulations, the general solution of this equation is:

$$\widehat{\mathbf{x}}(\widehat{\theta}) = \left(\sin\widehat{\psi}\sin\widehat{\theta}, -\sin\widehat{\psi}\cos\widehat{\theta}, \cos\widehat{\psi}\right).$$

Here $\widehat{\kappa}\widehat{s} := \widehat{\theta} = \theta + \varepsilon\theta^*$; where $0 \leq \theta \leq 2\pi$, and $\theta^* \in \mathbb{R}$. If we set $\widehat{\theta} = \theta(1 + \varepsilon h)$, h being the pitch of the screw movement, then we have

$$\begin{pmatrix} \widehat{\mathbf{x}} \\ \widehat{\mathbf{t}} \\ \widehat{\mathbf{g}} \end{pmatrix} = \begin{pmatrix} \sin\widehat{\psi}\sin\widehat{\theta} & -\sin\widehat{\psi}\cos\widehat{\theta} & \cos\widehat{\psi} \\ \cos\widehat{\theta} & \sin\widehat{\theta} & 0 \\ -\sin\widehat{\theta}\cos\widehat{\psi} & \cos\widehat{\theta}\cos\widehat{\psi} & \sin\widehat{\psi} \end{pmatrix} \begin{pmatrix} \widehat{\mathbf{i}} \\ \widehat{\mathbf{j}} \\ \widehat{\mathbf{k}} \end{pmatrix}, \tag{12}$$

and

$$d\widehat{s} = \left\|\frac{d\widehat{\mathbf{x}}}{d\theta}\right\| d\theta = (1 + \varepsilon h)\sin\widehat{\psi} d\theta, \tag{13}$$

Then,

$$\mu = \psi^*\cot\psi + h, \text{ and } F = h\cot\psi - \psi^*. \tag{14}$$

Further, the Disteli-axis $\widehat{\mathbf{b}}$ is:

$$\widehat{\mathbf{b}} = \frac{\widehat{\gamma}\widehat{\mathbf{r}}+\widehat{\mathbf{g}}}{\sqrt{\widehat{\gamma}^2+1}} = \widehat{\mathbf{k}}. \tag{15}$$

This means that the stationary axis of the helical movement is the Disteli-axis $\widehat{\mathbf{b}}$. From real and dual parts, respectively, we have:

$$\mathbf{x}(\theta) = (\sin\psi\sin\theta, -\sin\psi\cos\theta, \cos\psi),$$

and

$$\mathbf{x}^*(\theta) = \begin{pmatrix} r_1^* \\ r_2^* \\ r_3^* \end{pmatrix} = \begin{pmatrix} \theta^*\cos\theta\sin\psi + \psi^*\cos\psi\sin\theta \\ \theta^*\sin\theta\sin\psi - \psi^*\cos\psi\cos\theta \\ -\psi^*\sin\psi \end{pmatrix}.$$

Let $m(m_1, m_2, m_3)$ be a point on $\widehat{\mathbf{x}}$. Since $m\times x = x^*$ we have the system of linear equations in m_1, m_2, and m_3:

$$\left.\begin{aligned} m_2\cos\psi + m_3\sin\psi\cos\theta &= x_1^*, \\ -m_1\cos\psi + m_3\sin\psi\sin\theta &= x_2^*, \\ -m_1\sin\psi\cos\theta - m_2\sin\psi\sin\theta &= x_3^*. \end{aligned}\right\}$$

The matrix of coefficients of unknowns m_1, m_2, and m_3 is:

$$\begin{pmatrix} 0 & \cos\psi & \sin\psi\cos\theta \\ -\cos\psi & 0 & \sin\psi\sin\theta \\ -\sin\psi\cos\theta & -\sin\psi\sin\theta & 0 \end{pmatrix},$$

and thus its rank is 2 with $\psi \neq p\pi$ (p is an integer), and $\psi \neq 0$. The rank of the augmented matrix:

$$\begin{pmatrix} 0 & \cos\psi & \sin\psi\cos\theta & x_1^* \\ -\cos\psi & 0 & \sin\psi\sin\theta & x_2^* \\ -\sin\psi\cos\theta & -\sin\psi\sin\theta & 0 & x_3^* \end{pmatrix},$$

is 2. Then this set has infinitely numerous solutions given with

$$\begin{aligned} m_1 &= \psi^*\cos\theta - (\theta^* - m_3)\tan\psi\sin\theta, \\ m_2 &= \psi^*\sin\theta + (\theta^* - m_3)\tan\psi\sin\theta, \\ & m_1\cos\theta + m_2\sin\theta = \psi^*. \end{aligned} \tag{16}$$

Since m_3 is selected at random, then we may occupy $\theta^* - m_3 = 0$. In this case, Equation (16) reads

$$m_1 = \psi^*\cos\theta,\ m_2 = \psi^*\sin\theta,\ m_3 = \theta^*.$$

We now just find the base curve as;

$$\mathbf{m}(\theta) = (\psi^*\cos\theta, \psi^*\sin\theta, h\theta).$$

It can be show that $<\mathbf{m}', \mathbf{x}'>= 0$; $(' = \frac{d}{d\theta})$ so the base curve of $(\widehat{x})$ is its striction curve. The curvature $\kappa_c(\theta)$, and the torsion $\tau_c(\theta)$, respectively, are

$$\kappa_c(\theta) = \frac{\psi^*}{\psi^{*2}+h^2}, \text{ and } \tau_c(\theta) = \frac{h}{\psi^{*2}+h^2}.$$

Then $\mathbf{c}(\theta)$ is a cylindrical helix, and the ruled surface $(\widehat{x})$ is:

$$(\widehat{x}) : \mathbf{y}(\theta, v) = \begin{pmatrix} \psi^*\cos\theta + v\sin\psi\sin\theta \\ \psi^*\sin\theta - v\sin\psi\cos\theta \\ h\theta + v\cos\psi \end{pmatrix}, \tag{17}$$

where ψ, ψ^*, and h can control the shape of $(\widehat{x})$. The stationary Disteli-axis ruled surface $(\widehat{x})$ can be classified into four types according to the shapes of their striction curves:

(1) Archimedes helicoid with the striction curve is a helix: for $h = \psi^* = 1$, $\psi = \frac{\pi}{4}$, $0 \leq \theta \leq 2\pi$ and $-7 \leq v \leq 7$ (Figure 1),
(2) Right helicoid with the striction curve is a helix,: for $h = \psi^* = 1$, $\psi = \frac{\pi}{2}$, $0 \leq \theta \leq 2\pi$ and $-3 \leq v \leq 3$ (Figure 2),
(3) Hyperboloid of one-sheet with the striction curve is a circle: for $h = 0$, $\psi^* = 1$, $\psi = \frac{\pi}{4}$, $0 \leq \theta \leq 2\pi$ and $-7 \leq v \leq 7$ (Figure 3),
(4) A cone with the striction curve is a point: for $h = \psi^* = 0$, $\psi = \frac{\pi}{4}$, $0 \leq \theta \leq 2\pi$ and $-7 \leq v \leq 7$ (Figure 4).

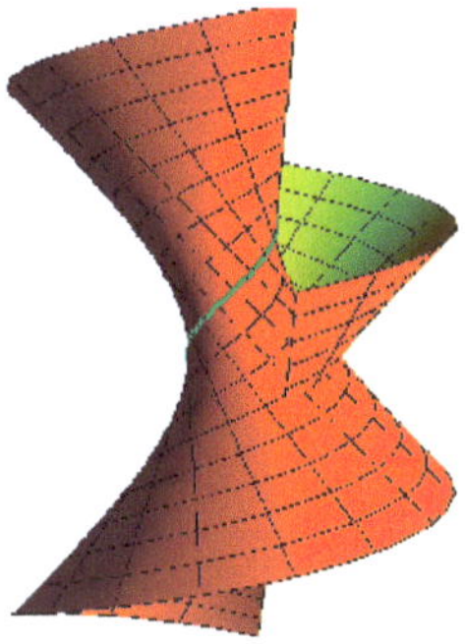

Figure 1. Archimedecs helicoid.

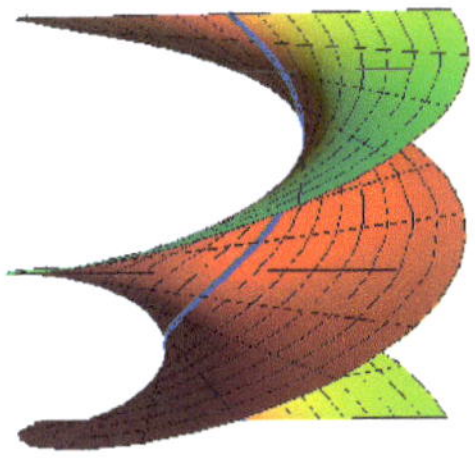

Figure 2. Right helicoid.

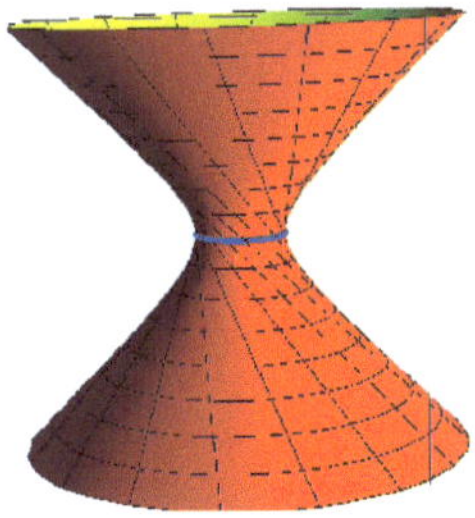

Figure 3. Hyperboloid of one-sheet.

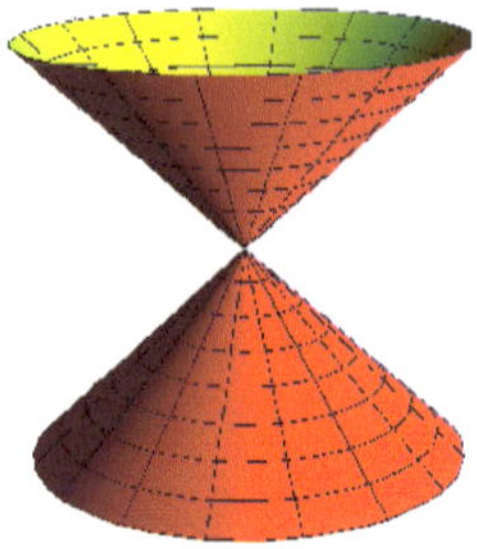

Figure 4. A cone.

4. Bertrand Offsets of Ruled Surfaces

In this section, we consider the Bertrand offsets of ruled and developable surfaces, then a theory comparable to the theory of Bertrand curves can be raised for such surfaces.

Definition 1. *Let $(\widehat{x})$ and $(\widehat{\overline{x}})$ be two non-developable ruled surfaces in $\mathbb{E}^3$. The surface $(\widehat{\overline{x}})$ is said to be Bertrand offsets of $(\widehat{x})$ if there exists a one-to-one correspondence among their rulings such that both surfaces have a mutual central normal at the corresponding striction points.*

Let $(\widehat{\overline{x}})$ be a Bertrand offset of $(\widehat{x})$ with the Blaschke frame $\{\widehat{\overline{\mathbf{x}}}(\widehat{\overline{s}}), \widehat{\overline{\mathbf{t}}}(\widehat{\overline{s}}), \widehat{\overline{\mathbf{g}}}(\widehat{\overline{s}})\}$, it can be computed as mentioned in the above equations. Consider $\widehat{\phi} = \phi + \varepsilon\phi^*$ be the dual angle among the rulings of $(\widehat{x})$ and $(\widehat{\overline{x}})$ at the corresponding points, that is,

$$< \widehat{\overline{\mathbf{x}}}, \widehat{\mathbf{x}} >= \cos\widehat{\phi}. \tag{18}$$

By differentiating of Equation (18) with respect to $\widehat{s}$, we find

$$< \widehat{\overline{\mathbf{t}}}, \widehat{\mathbf{x}} > \widehat{\overline{s}}' + < \widehat{\overline{\mathbf{x}}}, \widehat{\mathbf{t}} >= -\widehat{\phi}' \sin\widehat{\phi}. \tag{19}$$

Since $(\widehat{x})$ and $(\widehat{\overline{x}})$ are Bertrand offsets $(\widehat{\mathbf{t}} = \widehat{\overline{\mathbf{t}}})$, then we have $\widehat{\phi}' = 0$, so that $\widehat{\phi} = \phi + \varepsilon\phi^*$ is an stationary dual number. Thus, the following Theorem can be given:

Theorem 2. *The offset angle ϕ and the offset distance ϕ^* among the rulings of a non-developable ruled surface and its Bertrand offset are constants.*

It is evident from Theorem 2 that a ruled surface, generally, has a double infinity of Bertrand offsets. Each Bertrand offset can be formed by an stationary linear offset $\phi^* \in \mathbb{R}$ and an stationary angular offset $\phi \in [0, 2\pi]$. Any two surfaces of this pencil of ruled surfaces are reciprocal of one another; if $(\widehat{\overline{x}})$ is a Bertrand offset of $(\widehat{x})$, then $(\widehat{x})$ is likewise a Bertrand offset of $(\widehat{\overline{x}})$. So, we can write

$$\widehat{\mathbf{x}}(\widehat{\overline{s}}) = \cos\widehat{\phi}\widehat{\mathbf{x}}(\widehat{s}) + \sin\widehat{\phi}\widehat{\mathbf{g}}(\widehat{s}). \tag{20}$$

In the view of the fact that for a ruled surface and its Bertrand offset the central normals coincide, it follows from Theorem 1 that the central tangents of the two ruled surfaces also construct the same stationary dual angle at the matching striction points. Thus, the relationship among their Blaschke frames can be written as:

$$\begin{pmatrix} \widehat{\overline{\mathbf{x}}}(\widehat{s}) \\ \widehat{\overline{\mathbf{t}}}(\widehat{s}) \\ \widehat{\overline{\mathbf{g}}}(\widehat{s}) \end{pmatrix} = \begin{pmatrix} \cos\widehat{\phi} & 0 & \sin\widehat{\phi} \\ 0 & 1 & 0 \\ -\sin\widehat{\phi} & 0 & \cos\widehat{\phi} \end{pmatrix} \begin{pmatrix} \widehat{\mathbf{x}}(\widehat{s}) \\ \widehat{\mathbf{t}}(\widehat{s}) \\ \widehat{\mathbf{g}}(\widehat{s}) \end{pmatrix}. \tag{21}$$

The major point to note here is the technique we have used (compared with [5,12]). The equation of the striction curve of the offset surface $(\widehat{\overline{x}})$, in terms of its base surface $(\widehat{x})$, can therefore be written as

$$\overline{\mathbf{c}}(\overline{s}) = \mathbf{c}(s) + \phi^{*}\mathbf{t}(s). \tag{22}$$

Hence, the equation of $(\widehat{\overline{x}})$ in terms of $(\widehat{x})$ can be written as

$$(\widehat{\overline{x}}) : \overline{\mathbf{y}}(s, v) = \mathbf{c}(s) + \phi^{*}\mathbf{t}(s) + v(\cos\phi\mathbf{x}(s) + \sin\phi\mathbf{g}(s)),\ v \in \mathbb{R}. \tag{23}$$

Let $\overline{\xi}(\overline{s}, v)$ be the unit normal of an arbitrary point on $(\widehat{\overline{x}})$. Then, as in Equation (6), we have:

$$\overline{\xi}(\overline{s}, v) = \frac{\overline{\mu}\overline{\mathbf{t}} - v\overline{\mathbf{g}}}{\sqrt{\overline{\mu}^{2} + v^{2}}}, \tag{24}$$

where $\overline{\mu}$ is the distribution parameter of $(\widehat{\overline{x}})$. It is evident from Equations (6) and (24) that the normal to a ruled surface and its Bertrand offsets are not the same. This signifies that the Bertrand offsets of a ruled surface are, generally, not parallel offsets. Therefore, the parallel conditions among $(\widehat{\overline{x}})$ in terms of $(\widehat{x})$ can be specified by the following theorem:

Theorem 3. *A non-developable ruled surface $(\widehat{x})$ and its Bertrand offset $(\widehat{\overline{x}})$ are parallel offsets if and only if (i) $\mu = \overline{\mu}$, (ii) each edge of the Blaschke frame of $(\widehat{x})$ is collinear with the conformable edge for $(\widehat{\overline{x}})$.*

Proof. Suppose a non-developable ruled surface $(\widehat{x})$ and its Bertrand offset $(\widehat{\overline{x}})$ are parallel offsets, or $\overline{\xi}(\overline{s}, v) \times \xi(s, v) = \mathbf{0}$, we have the following expression by Equations (6) and (24)

$$v(\mu\cos\phi - \overline{\mu})\mathbf{x} + v^{2}\sin\phi\mathbf{t} + v\mu\sin\phi\mathbf{g} = \mathbf{0}.$$

The above equation should be hold true for any value $v \neq 0$, which leads to $\phi = 0$ and $\mu = \overline{\mu}$.

Suppose that the two conditions of Theorem 2 hold true, that is, $\phi = 0$, $\mu = \overline{\mu}$, and then substitute them into $\overline{\xi}(\overline{s}, v) \times \xi(s, v)$, we have

$$\overline{\xi}(\overline{s}, v) \times \mathbf{n}(s, v) = \frac{\overline{\mu}\mathbf{t} - v\mathbf{g}}{\sqrt{\overline{\mu}^{2} + v^{2}}} \times \frac{\mu\mathbf{t} - v\mathbf{g}}{\sqrt{\mu^{2} + v^{2}}}$$

the result of the above equation is zero vector, which implies that $(\widehat{x})$ and $(\widehat{\overline{x}})$ are parallel offsets. □

Deriving again in the same manner, but now for developable surface $\mu = 0$, we have:

Corollary 3. *A developable ruled surface $(\widehat{x})$ and its developable Bertrand offset $(\widehat{\overline{x}})$ are parallel offsets if and only if each edge of the Blaschke frame of $(\widehat{x})$ is collinear with the conformable edge for $(\widehat{\overline{x}})$.*

Corollary 4. *A developable ruled surface $(\widehat{x})$ and its non-developable Bertrand offset $(\widehat{\overline{x}})$ can not be parallel offsets.*

On the other hand, we also have

$$\frac{d}{d\widehat{\overline{s}}}\begin{pmatrix} \widehat{\overline{\mathbf{x}}}(\widehat{\overline{s}}) \\ \widehat{\overline{\mathbf{t}}}(\widehat{\overline{s}}) \\ \widehat{\overline{\mathbf{g}}}(\widehat{\overline{s}}) \end{pmatrix} = \begin{pmatrix} 0 & 1 & 0 \\ -1 & 0 & \widehat{\overline{\gamma}} \\ 0 & -\widehat{\overline{\gamma}} & 0 \end{pmatrix}\begin{pmatrix} \widehat{\overline{\mathbf{x}}}(\widehat{\overline{s}}) \\ \widehat{\overline{\mathbf{t}}}(\widehat{\overline{s}}) \\ \widehat{\overline{\mathbf{g}}}(\widehat{\overline{s}}) \end{pmatrix}, \tag{25}$$

where

$$d\widehat{\overline{s}} = (\cos\widehat{\phi} + \widehat{\gamma}\sin\widehat{\phi})d\widehat{s},\ \widehat{\overline{\gamma}}d\widehat{\overline{s}} = (\widehat{\gamma}\cos\widehat{\phi} - \sin\widehat{\phi})d\widehat{s}.$$

By eliminating $\frac{d\widehat{\overline{s}}}{d\widehat{s}}$, we attain

$$(\widehat{\overline{\gamma}} - \widehat{\gamma})\cos\widehat{\phi} + (1 + \widehat{\overline{\gamma}}\widehat{\gamma})\sin\widehat{\phi} = 0. \tag{26}$$

This is a dual version of Bertrand offsets of ruled surfaces in terms of their dual geodesic curvatures.

Theorem 4. *The non-developable ruled surfaces $(\widehat{x})$ and $(\widehat{\overline{x}})$ form a Bertrand offsets if and only if the Equation (23) is satisfied.*

Corollary 5. *The Bertrand offset $(\widehat{\overline{x}})$ of a helicoidal surface, generally, does not have to be a helicoidal.*

Corollary 6. *The Bertrand offset of an stationary Disteli-axis ruled surface is also an stationary Disteli-axis ruled surface.*

Furthermore, the striction curve of $(\widehat{\overline{x}})$, in terms of $\mathbf{c}(\theta)$, can be written as:

$$\overline{\mathbf{c}}(\theta) := \mathbf{c}(\theta) + \phi^*\mathbf{t}(\theta) = (\psi^*\cos\theta, \psi^*\sin\theta, h\theta) + \phi^*(\cos\theta, \sin\theta, 0). \tag{27}$$

With the help of the Equations (17), (21) and (27), we obtain

$$(\widehat{\overline{x}}) : \overline{\mathbf{y}}(\theta, v) = \begin{pmatrix} (\psi^* + \phi^*)\cos\theta + v\sin(\psi - \phi)\sin\theta \\ (\psi^* + \phi^*)\sin\theta - v\sin(\psi - \phi)\cos\theta \\ h\theta + v\cos(\psi - \phi) \end{pmatrix}. \tag{28}$$

Example 1. *In this example, we verify the idea of Corollary 5. In view of Theorem 1, and Equation (26) we have that: $\widehat{\gamma} = \cot\widehat{\psi} = 0$ ($\psi = \frac{\pi}{2}$, $\psi^* = 0$)$\Leftrightarrow$ $\widehat{\overline{\gamma}} + \cot\widehat{\phi} = 0$. Then, in view of Equations (17) and (28), the ruled surface*

$$(\widehat{\overline{x}}) : \overline{\mathbf{y}}(\theta, v) = \begin{pmatrix} \phi^*\cos\theta + v\cos\phi\sin\theta \\ \phi^*\sin\theta - v\cos\phi\cos\theta \\ h\theta + v\sin\phi \end{pmatrix},$$

is the Bertrand offset of the helicoidal surface (Figure 5)

$$(\widehat{x}) : \mathbf{y}(\theta, v) = (v\sin\theta, -v\cos\theta, h\theta).$$

Take $\phi^* = 1$, $\phi = \frac{\pi}{4}$ and $h = 1$ for example the Bertrand offset is shown Figure 6; where $0 \leq \theta \leq 2\pi$, $-3 \leq v \leq 3$. The graph of the helicoidal surface $(\widehat{x})$ with its Bertrand offset $(\widehat{\overline{x}})$ is shown in Figure 7.

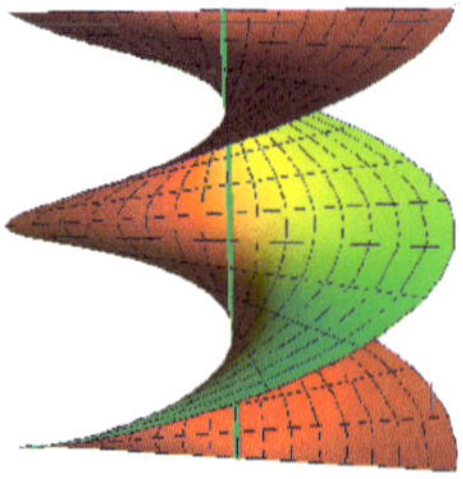

Figure 5. Helicoidal surface.

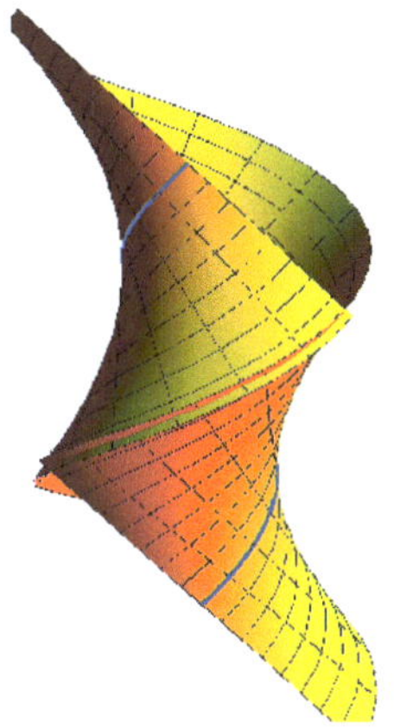

Figure 6. Bertrand offset of the helicoidal surface.

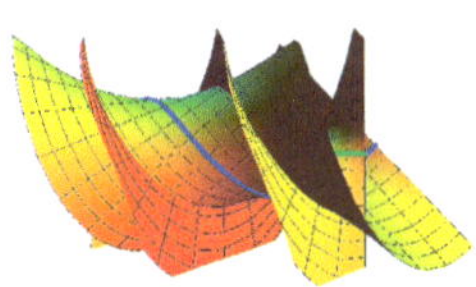

Figure 7. Helicoidal surface with its Bertrand offset.

The Striction Curves

In this subsection, we consider the striction curves of the Bertrand offsets. With the aid of Equation (19), the tangent of the striction curve $\overline{\mathbf{c}}(\overline{s})$ of $(\widehat{\overline{x}})$ is

$$\frac{d\overline{\mathbf{c}}(\overline{s})}{d\overline{s}} = [(F - \phi^*)\mathbf{x} + (\mu + \gamma\phi^*)\mathbf{g}]\frac{ds}{d\overline{s}}, \tag{29}$$

whereas, as in Equation (4), is:

$$\frac{d\overline{\mathbf{c}}(\overline{s})}{d\overline{s}} = \overline{F}(\overline{s})\overline{\mathbf{x}}(\overline{s}) + \overline{\mu}(\overline{s})\overline{\mathbf{g}}(\overline{s}). \tag{30}$$

From Equations (29) and (30) we attain

$$\frac{d\overline{s}}{ds} = \frac{F - \phi^*}{\overline{F}\cos\phi + \overline{\mu}\sin\phi} = \frac{\mu + \gamma\phi^*}{-\overline{F}\sin\phi + \overline{\mu}\cos\phi}. \tag{31}$$

Hence, we have two main different cases:

(a) In the case of $(\widehat{x})$ is tangential surface, that is, $\mu = 0$. In this case the Blaschke frame $\{\mathbf{x}(s), \mathbf{t}(s), \mathbf{g}(s)\}$ turn into the classical Serret-Frenet frame $\{\mathbf{t}(s), \mathbf{n}(s), \mathbf{b}(s)\}$ and the striction curve $\mathbf{c}(s)$ turns out to be the edge of regression of $(\widehat{x})$. Therefore, we find

$$\begin{pmatrix} \overline{\mathbf{t}} \\ \overline{\mathbf{n}} \\ \overline{\mathbf{b}} \end{pmatrix} = \begin{pmatrix} \cos\phi & 0 & -\sin\phi \\ 0 & 1 & 0 \\ \sin\phi & 0 & \cos\phi \end{pmatrix} \begin{pmatrix} \mathbf{t} \\ \mathbf{n} \\ \mathbf{b} \end{pmatrix}, \tag{32}$$

and

$$\overline{\mathbf{c}}(\overline{s}) = \mathbf{c}(s) + \phi^*\mathbf{n}(s). \tag{33}$$

Moreover, the curvature $\kappa(s)$ and the torsion $\tau(s)$ of $\mathbf{c}(s)$ can be specified by

$$\kappa(s) = \frac{1}{F(s)}, \tau(s) = \frac{\gamma(s)}{F(s)}.$$

Therefore, from Equation (31) we have

$$\overline{\mu} = \overline{F}\frac{(1-\phi^*\kappa)\sin\phi + \tau\phi^*\cos\phi}{(1-\phi^*\kappa)\cos\phi - \tau\phi^*\sin\phi}. \tag{34}$$

Thus the Bertrand offset of a developable surface is not developable, that is, $\overline{\mu}(s) \neq 0$ Furthermore, if the ruled surface $(\widehat{\overline{x}})$ is also a tangential developable, that is, $\overline{\mu}(s) = 0$. Then, from Equation (34), we gain

$$(1-\phi^*\kappa)\sin\phi + \tau\phi^*\cos\phi = 0. \tag{35}$$

Theorem 5. *$(\widehat{x})$ and $(\widehat{\overline{x}})$ are tangential Bertrand offsets if and only if their striction curves are Bertrand curves.*

Corollary 7.

(i) *If $\phi = 0$, then $\phi^* = 0$ or $\tau = 0$,*
(ii) *If $\phi^* = 0$, then $\phi = 0$, that is, the rulings are identical,*
(iii) *If $\tau = 0$, and $\phi^* \neq 0$, then $\kappa(v) = 1/\phi^*$ is stationary or $\phi = 0$,*
(iv) *$\phi = \pi/2$, and $\phi^* \neq 0$, then $\kappa(v) = 1/\phi^*$ is stationary.*

From Equation (28), we also have

$$\cos\phi d\overline{s} = (1-\phi^*\kappa)ds, \text{ and } \sin\phi d\overline{s} + \phi^*\tau ds = 0. \tag{36}$$

If $\mathbf{c}(s) = \mathbf{c}(s+2\pi)$ $(\overline{\mathbf{c}}(\overline{s}) = \overline{\mathbf{c}}(\overline{s}+2\pi))$, then $(\widehat{x})$ (resp. (x)) is a closed tangential surface. Let $\mathcal{L}$ (resp. $\overline{\mathcal{L}}$) be the length of the edge of regression of $(\widehat{x})$ (resp. $(\widehat{\overline{x}})$) $\mathbf{c}(v)$. Since both ϕ^*, and ϕ are stationary, integration of the first expression in Equation (36) yields

$$\overline{\mathcal{L}}\cos\phi = \mathcal{L} - \phi^* \oint_{\mathbf{c}(s)} \kappa ds.$$

If $\overline{\mathbf{n}} = -\mathbf{n}$, then $\mathbf{c}(s) = \overline{\mathbf{c}}(\overline{s}) + \phi^*\overline{\mathbf{n}}(\overline{s})$, so by the symmetry of the mating relationship,

$$\mathcal{L}\cos\overline{\phi} = \overline{\mathcal{L}} - \phi^* \oint_{\overline{\mathbf{c}}(\overline{s})} \overline{\kappa} d\overline{s}.$$

Hence $\cos\phi = \cos\overline{\phi}$, and adding the previous two equations

$$(\overline{\mathcal{L}} + \mathcal{L})(1-\cos\phi) = \phi^*\left(\oint_{\mathbf{c}(s)} \kappa ds + \oint_{\overline{\mathbf{c}}(\overline{s})} \overline{\kappa} d\overline{s}\right), \tag{37}$$

or

$$(\overline{\mathcal{L}} + \mathcal{L})\sin^2\frac{\phi}{2} = \frac{\phi^*}{2}\left(\oint_{\mathbf{c}(s)} \kappa ds + \oint_{\overline{\mathbf{c}}(\overline{s})} \overline{\kappa} d\overline{s}\right). \tag{38}$$

In view of the Fenchel's inequality [17], we have

$$\overline{\mathcal{L}} + \mathcal{L} \geq 2\pi\phi^*.$$

Corollary 8. *The sum of the lengths of the striction curves of the tangential offsets is never inferior to 2π times the distance through their corresponding points.*

If the striction curve $\mathbf{c}(s)$ is self-mated curve, that is, $\overline{\mathbf{c}}(s) = \mathbf{c}(s)$, the Formula (37) turn into

$$\mathcal{L} = \frac{\phi^*}{1 - \cos\phi} \oint_{\mathbf{c}(s)} \kappa(s)ds, \text{ with } 1 - \cos\phi \neq 0. \tag{39}$$

If $\phi = (2p+1)\pi$ (p is an integer) then $\cos\phi = -1$, and therefore Equation (39) become

$$\mathcal{L} = \frac{\phi^*}{2} \oint_{\mathbf{c}(s)} \kappa(s)ds. \tag{40}$$

Furthermore, if the striction curve $\mathbf{c}(s)$ is a planar curve, then

$$\oint_{\mathbf{c}(s)} \kappa(s)ds = 2\pi. \tag{41}$$

From Equations (40) and (41) we get $\mathcal{L} = \pi\phi^*$. Thus we have the following theorem:

Theorem 6. *The length of a self-mated striction curve* $\mathbf{c}(s)$ *of* $(\widehat{x})$ *is* π *times its width (breadth).*

(b) In the case of $(\widehat{x})$ is a binormal surface, that is, $\digamma = 0$. In this case, from Equation (31), we have:

$$\overline{\digamma} = \overline{\mu} \frac{\phi^* \cos\phi - (\gamma\phi^* + \mu)\sin\phi}{\phi^* \sin\phi + (\gamma\phi^* + \mu)\cos\phi}. \tag{42}$$

Thus the Bertrand offset of a binormal is not binormal, that is, $\overline{\digamma}(s) \neq 0$. Further, if the Bertrand offset $(\widehat{\overline{x}})$ is also a binormal, then we can find:

$$(1 + \phi^*\kappa)\sin\phi - \phi^*\tau\cos\phi = 0.$$

Theorem 7. $(\widehat{x})$ *and* $(\widehat{\overline{x}})$ *are binormal Bertrand offsets if and only if their striction curves are Bertrand curves.*

In a similar manner, all the results of the tangential surface may be given for the binormal ruled surface.

5. Conclusions

In this study, an extension of Bertrand offsets of curves for ruled, and developable surfaces has been improved. Noteworthy, there are numerous similarities through the Bertrand curves and the Bertrand offsets for ruled surfaces. For example, a ruled surface can have an infinity of Bertrand offsets in analogy with a plane curve can have an infinity of Bertrand mates. From this result the proofs of the theorems of Fenchel and Barbier are reproved by an elegant method. Meanwhile, the derivations of some useful geometric relations, examples and instructive figures of the ruled surfaces are included. For future study, we will catch with the novel ideas that Gaussian and mean curvatures of these Bertrand offsets can be gained, when the Weingarten map for the Bertrand offsets ruled surfaces is realized. We will also address integrating the work of singularity theory and submanifold theory and so forth, given in [19–22], with the results of this paper to explore new methods to find more theorems related to symmetric properties on this subject.

Author Contributions: Conceptualization, S.H.N. and R.A.A.-B. methodology, S.H.N. and R.A.A.-B. investigation, S.H.N. and R.A.A.-B.; writing—original draft preparation, S.H.N. and R.A.A.-B.; writing—review and editing, S.H.N. and R.A.A.-B. All authors have read and agreed to the published version of the manuscript.

Funding: This research received no external funding.

Data Availability Statement: Our manuscript has no associate data.

Conflicts of Interest: The authors declare no conflict of interest.

References

1. Veldkamp, G.R. On the use of Dual numbers, Vectors, and Matrices in Instantaneous Spatial Kinematics. *Mech. Mach. Theory* **1976**, *11*, 141–156. [CrossRef]
2. Bottema, O.; Roth, B. *Theoretical Kinematics*; North-Holland Press: New York, NY, USA, 1979.
3. Karger, A.; Novak, J. *Space Kinematics and Lie Groups*; Gordon and Breach Science Publishers: New York, NY, USA, 1985.
4. Pottman, H.; Wallner, J. *Computational Line Geometry*; Springer: Berlin/Heidelberg, Germany, 2001.
5. Ravani, B.; Ku, T.S. Bertrand offsets of ruled and developable surfaces. *Comput. Aided Des.* **1991**, *23*, 145–152. [CrossRef]
6. Küçük, A. ; Gürsoy, O. On the invariants of Bertrand trajectory surface offsets. *AMC* **2003**, 11–23 . [CrossRef]
7. Kasap, E.; Kuruoglu, N. Integral invariants of the pairs of the Bertrand ruled surface. *Bull. Pure Appl. Sci. Sect. E Math.* **2002**, *21*, 37–44.
8. Kasap, E.; Kuruoglu, N. The Bertrand offsets of ruled surfaces in $\mathbb{R}^3_1$. *Acta Math Vietnam.* **2006**, *31*, 39–48.
9. Kasap, E.; Yuce, S.; Kuruoglu, N. The involute-evolute offsets of ruled surfaces. *Iran. Sci Tech Trans. A* **2009**, *33*, 195–201.
10. Orbay, K.; Kasap, E.; Aydemir, I. Mannheim offsets of ruled surfaces. *Math. Probl. Eng.* **2009**, *2009*, 160917. [CrossRef]
11. Onder, M.; Ugurlu, H.H. Frenet frames and invariants of timelike ruled surfaces. *Ain. Shams Eng. J.* **2013**, *4*, 507–513. [CrossRef]
12. Aldossary, M.T.; Abdel-Baky, R.A. On the Bertrand offsets for ruled and developable surfaces. *Boll. Unione Mat. Ital.* **2015**, *8*,53–64. [CrossRef]
13. Sentrk, G.Y.; Yuce, S. Properties of integral invariants of the involute-evolute offsets of ruled surfaces. *Int. J. Pure Appl. Math.* **2015**, *102*, 75. [CrossRef]
14. Sentrk, G.Y.; Yuce, S.; Kasap, E. Integral Invariants of Mannheim offsets of ruled surfaces. *Appl. Math. E-Notes* **2016**, *16*, 198–209.
15. Sentrk, G.Y.; Yuce, S. Bertrand offsets of ruled surfaces with Darboux frame. *Results Math.* **2017**, *72*, 1151–1159. [CrossRef]
16. Sentrk, G.Y.; Yuce, S. On the evolute offsets of ruled surfaces usin the Darboux frame. *Commun. Fac. Sci. Univ. Ank. Ser. A1 Math. Stat. Vo.* **2019**, *68*, 1256–1264. [CrossRef]
17. Gugenheimer, H.W. *Differential Geometry*; Graw-Hill: New York, NY, USA, 1956.
18. Bruce, J.W.; Giblin, P.J. *Curves and Singularities*, 2nd ed.; Cambridge University Press: Cambridge, UK, 1992.
19. Li, Y.L.; Zhu, Y.S.; Sun, Q.Y. Singularities and dualities of pedal curves in pseudo-hyperbolic and de Sitter space. *Int. J. Geom. Methods Mod. Phys.* **2021**, *18*, 1–31. [CrossRef]
20. Li, Y.L.; Nazra, S.; Abdel-Baky, R.A. Singularities Properties of timelike sweeping surface in Minkowski 3-Space. *Symmetry* **2022**, *14*, 1996. [CrossRef]
21. Li, Y.L.; Chen, Z.; Nazra, S.; Abdel-Baky, R.A. Singularities for timelike developable surfaces in Minkowski 3-Space. *Symmetry* **2023**, *15*, 277. [CrossRef]
22. Almoneef, A.A.; Abdel-Baky, R.A. Singularity properties of spacelike circular surfaces. *Symmetry* **2023**, *15*, 842. [CrossRef]

axioms

Article

Infinitesimal Affine Transformations and Mutual Curvatures on Statistical Manifolds and Their Tangent Bundles

Esmaeil Peyghan [1], Davood Seifipour [2] and Ion Mihai [3,*]

[1] Department of Mathematics, Faculty of Science, Arak University, Arak 38156-8-8349, Iran; e-peyghan@araku.ac.ir
[2] Department of Mathematics, Abadan Branch, Islamic Azad University, Abadan 63178-36531, Iran; d-seifipour@phd.araku.ac.ir
[3] Department of Mathematics, Faculty of Mathematics and Computer Science, University of Bucharest, 010014 Bucharest, Romania
* Correspondence: imihai@fmi.unibuc.ro

Abstract: The purpose of this paper is to find some conditions under which the tangent bundle TM has a dualistic structure. Then, we introduce infinitesimal affine transformations on statistical manifolds and investigate these structures on a special statistical distribution and the tangent bundle of a statistical manifold too. Moreover, we also study the mutual curvatures of a statistical manifold M and its tangent bundle TM and we investigate their relations. More precisely, we obtain the mutual curvatures of well-known connections on the tangent bundle TM (the complete, horizontal, and Sasaki connections) and we study the vanishing of them.

Keywords: infinitesimal affine transformations; mutual curvature; statistical manifold; tangent bundle

MSC: 53B12; 53B20; 53C15

Citation: Peyghan, E.; Seifipour, D.; Mihai, I. Infinitesimal Affine Transformations and Mutual Curvatures on Statistical Manifolds and Their Tangent Bundle. *Axioms* **2023**, *12*, 667. https://doi.org/10.3390/axioms12070667

Academic Editor: Mića Stanković

Received: 24 May 2023
Revised: 22 June 2023
Accepted: 4 July 2023
Published: 6 July 2023

1. Introduction

Concerning the necessity of the study of tangent bundles, it can be stated that the concept of the tangent bundle is used widely in classical mechanics and especially in Lagrangian formalism. The tangent bundle can also describe the motion of objects in all classical mechanics scenarios, with a coordinate system such that the first n tuples of it represent the position of an object and the second n tuples of it represent the velocity of the object. This coordinate system is an effective tool in the study of geometric structures of the TM. So, we can regard the TM as the state space in classical mechanics. In fact, the tangent bundle of a differential manifold M assembles all of the tangent vectors in M. On the other hand, vector fields on tangent bundles belong to basic concepts of pure and applied differential geometry, global analysis, and mathematical physics. Semisprays, sprays, and geodesic sprays are important classes of vector fields on tangent bundles. For instance, the semispray theory has been used in the calculus of variations on manifolds to characterize extremal curves of a variational functional as integral curves of the Hamilton or Euler–Lagrange vector fields. Sprays and semisprays also provide a natural framework for the extension of classical results of analytical mechanics to contemporary mechanical problems and stimulate a broad research field in the global theory of nonconservative systems, symmetries, and constraint theory (see [1] for more details).

The geometry of tangent bundles with Riemannian lift metrics has been studied very extensively in recent years (see [2–8], for instance). In this paper, we consider one of the famous Riemannian lift metrics, the Sasaki metric. Then, using this metric and its Levi-Civita connection, we investigate some of the geometric structures on the TM. The focus of this paper is concerned with the equations of a dualistic structure on the TM. Then, we investigate the mutual curvature of the TM and its relation with the curvature of M and the mutual curvature of M.

Information geometry is the combination and interaction of differential geometry and statistics [9]. In addition, it is an important and useful bridge between applicable and pure sciences (see [9,10], for instance). In this area, we use and extend the methods of differential geometry in probability theory. The mathematical point of view of information geometry started with C. R. Rao. He showed that a statistical model should be a differentiable Riemannian manifold, via the Fisher information matrix. This means that we can define a Riemannian metric in the space of probability distributions. In fact, information geometry is the study of natural geometric structures using families of probability distributions. Two of the main objects in this area are statistical connections and statistical manifolds. They have applications in fields such as computer science and physics. In fact, a statistical manifold is a manifold whose points are probability distributions (see [9–12]).

A statistical structure on a manifold M is a pair (g, ∇) such that g is a Riemannian (semi-Riemannian) metric and ∇ is a torsion-free linear connection such that ∇g is totally symmetric. A Riemannian (semi-Riemannian) manifold (M, g) together with the Levi-Civita connection ∇ of g is a typical example of a statistical manifold. In other words, statistical manifolds can be regarded as generalizations of Riemannian (semi-Riemannian) manifolds. Statistical manifolds provide geometric models of probability distributions. The geometries of statistical manifolds have been applied to various fields of information science, information theory, neural networks, machine learning, image processing, statistical mechanics, etc. (see [9,10], for instance).

The organization of this paper is as follows: In the first part we introduce the concept of the infinitesimal affine transformation of the Riemannian manifold (M, g) with respect to the affine connection ∇. Then, using an explicit example we find conditions such that a vector field X be an infinitesimal affine transformation of the 2-dimensional statistical manifold M_2. In fact, we solve a system of partial differential equations and this solution gives us the general form of any infinitesimal affine transformation of M_2. To continue, we also find conditions under which the vector fields X^V and X^C are an infinitesimal affine transformation of the TM with respect to the (α)-connection $\nabla^{(\alpha)}$.

In the second part, we study the geometry of the dualistic structure of M and the mutual curvature on the statistical manifolds. Then, we prove that under which conditions R and R^* are parallel with respect to the (α)-connection $\nabla^{(\alpha)}$, where R and R^* are curvature tensors of the dualistic structure (∇, ∇^*) on M. Then, we extend this problem to the mutual curvature R_{∇,∇^*} as mentioned in [13]. Moreover, we find conditions such that the (α)-curvature $R^{(\alpha)}$ is parallel with respect to the (β)-connection $\nabla^{(\beta)}$. In the next part, we provide conditions such that the TM equipped with them has a dualistic structure, and then, we find equations in which the TM equipped with them is a conjugate symmetric space. We also study the mutual curvature $\widetilde{R}_{\overset{H}{\nabla},\overset{C}{\nabla}}$ on the TM and its relation with the curvature of M, where ∇ is an affine connection on M and $\overset{C}{\nabla}$ and $\overset{H}{\nabla}$ are the complete lift and horizontal lift connections on the TM, respectively. At the end, we investigate the mutual curvature $\widetilde{R}_{\overset{1}{\nabla}^S,\overset{2}{\nabla}^S}$, where $\{\overset{1}{\nabla}, \overset{2}{\nabla}\}$ is a pair of the Levi-Civita connections of two non-isometric Riemannian metrics g_1 and g_2, and $\overset{1}{\nabla}^S$ and $\overset{2}{\nabla}^S$ are the Levi-Civita connections of the Sasaki lift metrics g_1^S and g_2^S on the TM, respectively. Moreover, we prove that the mutual curvature $\widetilde{R}_{\overset{S}{\nabla},\overset{C}{\nabla}}$ vanishes if and only if M is a flat space with respect to the ∇, where ∇ is the Levi-Civita connection of the metric g and ∇^S is the Levi-Civita connection of the Sasaki lift metric g^S. Moreover, in this case we prove that the mutual curvature $\widetilde{R}_{\overset{S}{\nabla},\overset{C}{\nabla}}$ reduces to the Riemannian curvature tensor of the Levi-Civita connection ∇^S. At the end of this paper, we give an explicit example of the mutual curvature.

2. Preliminaries

In this section, we introduce some basic facts that we use throughout the paper.

Let (x^i), $i = 1, \cdots, n$, be a coordinate system on M, (x^i, y^i) be the induced coordinate system on the TM, and $\{\frac{\partial}{\partial x^i}|_{(x,y)}, \frac{\partial}{\partial y^i}|_{(x,y)}\}$ be the natural basis of $T_{(x,y)}TM$. Then, the various lifts of a vector field $X = X^i\partial_i$ on M (complete lift, horizontal lift, and vertical lift, respectively) are defined as follows:

$$X^C = X^i\frac{\partial}{\partial x^i} + y^a(\partial_a X^i)\frac{\partial}{\partial y^i}, \quad X^H = X^i\frac{\partial}{\partial x^i} - y^a\Gamma^k_{ai}X^i\frac{\partial}{\partial y^k}, \quad X^V = X^i\frac{\partial}{\partial y^i}.$$

It is known that $T_{(x,y)}TM$ can be decomposed to $H_{(x,y)}TM \oplus V_{(x,y)}TM$, where $H_{(x,y)}TM$ is spanned by $\{\frac{\delta}{\delta x^i}|_{(x,y)} := (\frac{\partial}{\partial x^i})^h = \frac{\partial}{\partial x^i}|_{(x,y)} - y^k\Gamma^j_{ki}(x)\frac{\partial}{\partial y^j}|_{(x,y)}\}$ and $V_{(x,y)}$ is spanned by $\{\frac{\partial}{\partial y^i}|_{(x,y)} := (\frac{\partial}{\partial x^i})^v\}$. For simplicity, we write ∂_i, δ_i, and $\partial_{\bar{i}}$ instead of $\frac{\partial}{\partial x^i}$, $\frac{\delta}{\delta x^i}$, and $\frac{\partial}{\partial y^i}$.

Let (M, g) be a Riemannian manifold. Similar to the lifts of vector fields, we can construct the Sasaki lift metric g^S on the TM as follows:

$$g^S_{(x,y)}(X^H, Y^H) = g_x(X, Y), \qquad g^S_{(x,y)}(X^V, Y^H) = 0, \qquad g^S_{(x,y)}(X^V, Y^V) = g_x(X, Y). \quad (1)$$

The Levi-Civita connection of the Sasaki metric g^S is as follows:

$$(\nabla^S_{X^H}Y^H)_{(x,y)} = (\nabla_X Y)^H_{(x,y)} - \frac{1}{2}(R_x(X, Y)y)^V, \qquad (\nabla^S_{X^V}Y^H)_{(x,y)} = \frac{1}{2}(R_x(y, X)Y)^H, \quad (2)$$

$$(\nabla^S_{X^H}Y^V)_{(x,y)} = (\nabla_X Y)^V_{(x,y)} + \frac{1}{2}(R_x(y, Y)X)^H, \qquad (\nabla^S_{X^V}Y^V)_{(x,y)} = 0, \quad (3)$$

for all vector fields X, Y on M, and $(x, y) \in TM$.

If ∇ is a linear connection, then the horizontal lift connection $\overset{H}{\nabla}$ and complete lift connection $\overset{C}{\nabla}$ of ∇ are, respectively, defined by [14]:

$$\begin{aligned} &\overset{H}{\nabla}_{X^H}Y^H = (\nabla_X Y)^H, \quad \overset{H}{\nabla}_{X^H}Y^V = (\nabla_X Y)^V, \quad \overset{H}{\nabla}_{X^V}Y^H = \overset{H}{\nabla}_{X^V}Y^V = 0, \\ &\overset{C}{\nabla}_{X^H}Y^H = (\nabla_X Y)^H + (R(y, X)Y)^V, \quad \overset{C}{\nabla}_{X^V}Y^H = \overset{C}{\nabla}_{X^V}Y^V = 0, \\ &\overset{C}{\nabla}_{X^H}Y^V = (\nabla_X Y)^V, \quad \overset{C}{\nabla}_{X^C}Y^C = (\nabla_X Y)^C, \quad \overset{C}{\nabla}_{X^C}Y^V = \overset{C}{\nabla}_{X^V}Y^C = (\nabla_X Y)^V. \end{aligned} \quad (4)$$

According to [14], the Lie brackets of the horizontal lift and vertical lift of vector fields are as follows:

$$[X^H, Y^H] = [X, Y]^H - (R(X, Y)y)^V, \; [X^H, Y^V] = (\nabla_X Y)^V - T(X, Y)^V, \; [X^V, Y^V] = 0. \quad (5)$$

Let (M, g) be an n-dimensional Riemannian manifold and ∇ be an affine connection on M. A Codazzi couple on M is a pair (g, ∇) such that the cubic tensor field $C = \nabla g$ is totally symmetric, namely, the Codazzi equations hold:

$$(\nabla_X g)(Y, Z) = (\nabla_Y g)(Z, X) = (\nabla_Z g)(X, Y), \qquad \forall X, Y, Z \in \chi(M).$$

In this case, the triplet (M, g, ∇) is called a Codazzi manifold and ∇ is called a Codazzi connection. Furthermore, if ∇ is torsion free, then (M, g, ∇) is a statistical manifold, (g, ∇) is a statistical couple, and ∇ is a statistical connection. In local coordinates, C has the following form:

$$C(\partial_i, \partial_j, \partial_k) = \partial_i g(\partial_j, \partial_k) - g(\nabla_{\partial_i}\partial_j, \partial_k) - g(\partial_j, \nabla_{\partial_i}\partial_k),$$

and so $C_{ijk} = \partial_i g_{jk} - \Gamma^r_{ij}g_{rk} - \Gamma^r_{ik}g_{jr}$, $C_{ijk} = C_{jki} = C_{kij}$, where Γ^r_{ij} are the connection coefficients of ∇.

We also recall that if (M, g) is a Riemannian (pseudo-Riemannian) manifold, two affine torsion-free connections ∇ and ∇^* on M are said to be dual connections with respect to g if the following equation is satisfied:

$$Xg(Y,Z) = g(\nabla_X Y, Z) + g(Y, \nabla^*_X Z), \qquad \forall X, Y, Z \in \chi(M), \tag{6}$$

and in this case, we call (g, ∇, ∇^*) a dualistic structure on M. Furthermore, if we denote by R and R^* the curvature tensors of ∇ and ∇^*, then we say that (M, g) is a conjugate symmetric space if $R = R^*$. Moreover, if (g, ∇, ∇^*) is a dualistic structure on M, then $\{\nabla^{(\alpha)}\}_{\alpha \in R}$ given by

$$\nabla^{(\alpha)} = \frac{1+\alpha}{2}\nabla + \frac{1-\alpha}{2}\nabla^*, \qquad \forall \alpha \in \mathbb{R}, \tag{7}$$

is a family of affine connections, which is called an α-connection. It is known that if ∇ and ∇^* are statistical connections, then $\nabla^{(\alpha)}$ is a statistical connection for any $\alpha \in \mathbb{R}$ [15,16].

3. Infinitesimal Affine Transformations on Statistical Manifolds

Definition 1. *Let (M, g) be a Riemannian manifold, X be a vector field, and ∇ be an affine connection on M. Then, X is said to be an infinitesimal affine transformation of M with respect to ∇ if $L_X \nabla = 0$, where $L_X \nabla$ is the Lie derivative of ∇ with respect to X given by*

$$(L_X \nabla)(Y, Z) = L_X(\nabla_Y Z) - \nabla_Y(L_X Z) - \nabla_{[X,Y]} Z. \tag{8}$$

Setting $X = X^i \partial_i$, $Y = \partial_j$, and $Z = \partial_k$, then the local expression of (8) is as follows:

$$(L_{X^i\partial_i}\nabla)(\partial_j, \partial_k) = \{X^i \partial_i(\Gamma^t_{jk}) - \Gamma^r_{jk}\partial_r(X^t) + \partial_j\partial_k(X^t) + \partial_k(X^i)\Gamma^t_{ji} + \partial_j(X^i)\Gamma^t_{ik}\}\partial_t, \tag{9}$$

Example 1. *Here, we consider a p-dimensional statistical manifold. The importance of the distribution family introduced below lies in the fact that its member is a non-Gaussian multivariate distribution while the marginal distribution is Gaussian, which implies that a set of marginal distributions does not uniquely determine the multivariate normal distribution [17]. A p-dimensional statistical manifold is defined by*

$$M = \left\{ f(\mathbf{x};\lambda) | f(\mathbf{x};\lambda) = 2\Pi^p_{i=1}\frac{\sqrt{\lambda_i}}{\sqrt{2\pi}}e^{-\frac{\lambda_i x_i^2}{2}}, \mathbf{x} \in \Omega_p,\ \lambda \in \mathbb{R}^p_+ \right\},$$

where

$$\Omega_p = \{x = (x_1, \cdots, x_p) \in \mathbb{R}^p | \Pi^p_{i=1} x_i > 0\},$$
$$\mathbb{R}^p_+ = \{x = (x_1, \cdots, x_p) \in \mathbb{R}^p | x_i > 0, i = 1, \cdots, p\}.$$

The distribution in M can be rewritten as

$$f(\mathbf{x};\lambda) = e^{\frac{1}{2}}\sum_{i=1}^{p}\log(-\theta_i) + \sum_{i=1}^{p}\theta_i x_i^2 + \frac{p}{2}\log 2 - \log\sqrt{2\pi},$$

where $\theta_i = -\frac{1}{2}\lambda_i$. This is one member of the exponential family with the natural coordinates $(\theta_1, \cdots, \theta_p)$ and the potential function $\psi(\theta) = -\frac{1}{2}\sum^p_{i=1}\log(-\theta_i)$. It is known that for the exponential family, the Fisher information is just the second derivative of the potential function

$$g_{ij} = \frac{\partial^2 \psi}{\partial\theta_i \partial\theta_j} = -\frac{1}{2}\frac{1}{\theta_i\theta_j}\delta_{ij}, \tag{10}$$

and the α-connection is the third derivative of the potential function

$$\Gamma^{(\alpha)}_{ijk} = \frac{1-\alpha}{2}\frac{\partial^3\psi}{\partial\theta_i\partial\theta_j\partial\theta_k} = -\frac{1-\alpha}{2}\frac{1}{\theta_i\theta_j\theta_k}\delta_{ijk}, \tag{11}$$

where $\delta_{ii} = 1$ for $i = 1, \cdots, p$, $\delta_{ij} = 0$ for $i \neq j$, and $\delta_{iii} = 1$ for $i = 1, \cdots, p$, $\delta_{ijk} = 0$ for unequal i, j, k (see [18] for more details). For $p = 2$, the matrix expression of the metric g given by (10) and its inverse matrix are as follows:

$$g = \begin{bmatrix} -\frac{1}{2\theta_1^2} & 0 \\ 0 & -\frac{1}{2\theta_2^2} \end{bmatrix}, \quad g^{-1} = \begin{bmatrix} -2\theta_1^2 & 0 \\ 0 & -2\theta_2^2 \end{bmatrix}. \tag{12}$$

Combining (11) and (10) we get

$$\Gamma_{11}^{(\alpha)1} = \frac{1-\alpha}{\theta_1}, \quad \Gamma_{22}^{(\alpha)2} = \frac{1-\alpha}{\theta_2}, \quad \Gamma_{ij}^{(\alpha)k} = 0, \ \text{for unequal } i, j, k. \tag{13}$$

Now, let $X = X^1\partial_1 + X^2\partial_2$ be an infinitesimal affine transformation of M with respect to ∇, where $\partial_1 = \frac{\partial}{\partial\theta_1}$ and $\partial_2 = \frac{\partial}{\partial\theta_2}$. Then, from (9) we obtain

$$\{X^i\partial_i(\Gamma_{jk}^t) - \Gamma_{jk}^r\partial_r(X^t) + \partial_j\partial_k(X^t) + \partial_k(X^i)\Gamma_{ji}^t + \partial_j(X^i)\Gamma_{ik}^t\} = 0. \tag{14}$$

If we take $j, k, t \in \{1, 2\}$, then we obtain the following system of equations

$$\frac{X^1(\alpha-1)}{\theta_1^2} + \partial_1^2(X^1) + \partial_1(X^1)(\frac{1-\alpha}{\theta_1}) = 0, \tag{15}$$

$$\frac{X^2(\alpha-1)}{\theta_2^2} + \partial_2^2(X^2) + \partial_2(X^2)(\frac{1-\alpha}{\theta_2}) = 0, \tag{16}$$

$$(\frac{\alpha-1}{\theta_1})\partial_1(X^2) + \partial_1^2(X^2) = 0, \tag{17}$$

$$(\frac{\alpha-1}{\theta_2})\partial_2(X^1) + \partial_2^2(X^1) = 0, \tag{18}$$

$$\partial_1\partial_2(X^1) + \partial_2(X^1)(\frac{1-\alpha}{\theta_1}) = 0, \tag{19}$$

$$\partial_1\partial_2(X^2) + \partial_1(X^2)(\frac{1-\alpha}{\theta_2}) = 0. \tag{20}$$

From (19), we have $\partial_2(\partial_1 X^1 + X^1(\frac{1-\alpha}{\theta_1})) = 0$. So, $(\partial_1 X^1 + X^1(\frac{1-\alpha}{\theta_1}))$ is a function with respect to θ_1 only, i.e., $\partial_1 X^1 + X^1(\frac{1-\alpha}{\theta_1}) = f(\theta_1)$. Thus,

$$X^1(\frac{1-\alpha}{\theta_1}) = f(\theta_1) - \partial_1 X^1. \tag{21}$$

From (15), we find

$$\partial_1(X^1(\frac{1-\alpha}{\theta_1})) + \partial_1^2 X^1 = 0. \tag{22}$$

Substituting (21) in (22), we obtain $\partial_1 f = 0$, which gives $f(\theta_1) = A$, where A is a constant. So, we have

$$\partial_1 X^1 + X^1(\frac{1-\alpha}{\theta_1}) = A, \tag{23}$$

which is a linear differential equation with respect to θ_1. Thus, we have the following solution:

$$X^1 = e^{-\int \frac{1-\alpha}{\theta_1} d\theta_1}[\int A e^{\int \frac{1-\alpha}{\theta_1} d\theta_1} d\theta_1 + C(\theta_2)] \tag{24}$$

$$= \theta_1^{\alpha-1}(A\frac{\theta_1^{2-\alpha}}{2-\alpha} + C(\theta_2)) = \frac{A\theta_1}{2-\alpha} + C(\theta_2)\theta_1^{\alpha-1}. \tag{25}$$

It is easy to check that $X^1 = \frac{A\theta_1}{2-\alpha} + C(\theta_2)\theta_1^{\alpha-1}$ *satisfies (15) and (19). Setting* X^1 *in (18) we obtain*

$$\theta_1^{\alpha-1}(C''(\theta_2) + \frac{\alpha-1}{\theta_2}C'(\theta_2)) = 0. \tag{26}$$

So, we have the following ordinary differential equation

$$C''(\theta_2) + \frac{\alpha-1}{\theta_2}C'(\theta_2) = 0. \tag{27}$$

It is easily seen that the above ODE has the solution

$$C(\theta_2) = \frac{E}{\alpha}\theta_2^{\alpha} + F, \tag{28}$$

where E and F are constants. Therefore, we have

$$X^1 = \frac{A}{2-\alpha}\theta_1 + (\frac{E}{\alpha}\theta_2^{\alpha} + F)\theta_1^{\alpha-1}. \tag{29}$$

Similarly, we obtain

$$X^2 = \frac{B}{2\alpha}\theta_2 + (\frac{G}{\alpha}\theta_1^{\alpha} + H)\theta_2^{\alpha-1}, \tag{30}$$

where B, G, and H are constants.

Let ∇ be an affine connection on the Riemannian manifold M and let X be a vector field on M. Then, we have (see [14])

$$L_{X^V}\overset{C}{\nabla} = (L_X\nabla)^V, \qquad L_{X^C}\overset{C}{\nabla} = (L_X\nabla)^C. \tag{31}$$

Now, we assume that (g, ∇, ∇^*) is a dualistic structure on the Riemannian manifold M and let X be an infinitesimal affine transformation of M with respect to ∇ and ∇^*. Then, (7) gives us

$$(L_X\nabla^{(\alpha)})(Y,Z) = \frac{1+\alpha}{2}(L_X\nabla)(Y,Z) + \frac{1-\alpha}{2}(L_X\nabla^*)(Y,Z). \tag{32}$$

The above equation means that X is an infinitesimal affine transformation of M with respect to the α-connection $\nabla^{(\alpha)}$. Now, if we replace ∇ by $\nabla^{(\alpha)}$ in (31), we obtain

$$L_{X^V}(\nabla^{(\alpha)})^C = (L_X\nabla^{(\alpha)})^V, \qquad L_{X^C}(\nabla^{(\alpha)})^C = (L_X\nabla^{(\alpha)})^C. \tag{33}$$

From the hypothesis and (33), we deduce that X^V and X^C are infinitesimal affine transformations of the TM with respect to $(\nabla^{(\alpha)})^C$. So, we conclude the following theorem.

Theorem 1. *Let* (g, ∇, ∇^*) *be a dualistic structure on the Riemannian manifold M. If X is an infinitesimal affine transformation of M with respect to* ∇ *and* ∇^**, then X is an infinitesimal affine transformation of M with respect to the* α*-connection* $\nabla^{(\alpha)}$*. Moreover,* X^V *and* X^C *are infinitesimal affine transformations of the TM with respect to* $(\nabla^{(\alpha)})^C$*.*

4. Mutual Curvature on Statistical Manifolds

In this section, we introduce the concept of mutual curvature for the Riemannian manifold (M, g), and to continue, we consider the dualistic structure (∇, ∇^*) on M. Then, we show under which conditions the mutual curvature R_{∇,∇^*} is parallel with respect to the α-connection $\nabla^{(\alpha)}$. Finally, we find conditions under which the mutual curvature $R_{\nabla^{(\alpha)},\nabla^{(\beta)}}$ is parallel with respect to the γ-connection $\nabla^{(\gamma)}$.

Definition 2 ([13])**.** *Let (M,g) be a Riemannian manifold and let $(\overset{1}{\nabla},\overset{2}{\nabla})$ be a pair of connections. Then, their mutual curvature is the $(1,3)$-tensor $R_{\overset{1}{\nabla},\overset{2}{\nabla}}$, which is defined by the following formula:*

$$R_{\overset{1}{\nabla},\overset{2}{\nabla}}(X,Y)Z=\frac{1}{2}\{\overset{1}{\nabla}_X\overset{2}{\nabla}_YZ-\overset{1}{\nabla}_Y\overset{2}{\nabla}_XZ-\overset{1}{\nabla}_{[X,Y]}Z+\overset{2}{\nabla}_X\overset{1}{\nabla}_YZ-\overset{2}{\nabla}_Y\overset{1}{\nabla}_XZ-\overset{2}{\nabla}_{[X,Y]}Z\},\tag{34}$$

for all $X,Y,Z\in\chi(M)$.

It should be noted that the definition of mutual (or relative) curvature was previously presented by the authors in [16,19] in two different ways. However, as D. Iosifidis showed in [13], none of these are tensors.

Theorem 2. *Let (∇,∇^*) be a dualistic structure on the Riemannian manifold (M,g). Then, the following relation holds:*

$$R_{\nabla,\nabla^*}(X,Y)Z=R_{\nabla}(X,Y)Z+(\nabla_YK)(X,Z)-(\nabla_XK)(Y,Z),\qquad \forall X,Y,Z\in\chi(M),\tag{35}$$

where K is the difference tensor of ∇. Moreover, the difference tensor K is Codazzi-coupled (i.e., $(\nabla_XK)(Y,Z)=(\nabla_YK)(X,Z)$) if and only if R_{∇,∇^*} and R_∇ coincide. Furthermore, the mutual curvature R_{∇,∇^*} reduces to R_∇ whenever ∇ is the Levi-Civita connection.

Proof. Using $\nabla^*=\nabla-2K$ and direct computations, we obtain

$$\begin{aligned}
&R_{\nabla,\nabla^*}(X,Y)Z\\
&=\frac{1}{2}\{\nabla_X\nabla^*_YZ-\nabla_Y\nabla^*_XZ-\nabla_{[X,Y]}Z+\nabla^*_X\nabla_YZ-\nabla^*_Y\nabla_XZ-\nabla^*_{[X,Y]}Z\}\\
&=\frac{1}{2}\{\nabla_X(\nabla_YZ-2K_YZ)-\nabla_Y(\nabla_XZ-2K_XZ)-\nabla_{[X,Y]}Z\\
&+(\nabla_X\nabla_YZ-2K_X\nabla_YZ)-(\nabla_Y\nabla_XZ-2K_Y\nabla_XZ)-(\nabla_{[X,Y]}Z-2K_{[X,Y]}Z)\}\\
&=\frac{1}{2}\{2R_\nabla(X,Y)Z+2(\nabla_Y(K_XZ)-K_X(\nabla_YZ)+K_Y(\nabla_XZ)+K_{[X,Y]}Z-\nabla_X(K_YZ))\}\\
&=R_\nabla(X,Y)Z+(\nabla_YK)(X,Z)-(\nabla_XK)(Y,Z).
\end{aligned}$$

□

If we put $X=\partial_i$, $Y=\partial_j$, and $Z=\partial_k$ in (35) and we denote the connection coefficients of ∇ and $\nabla^{(0)}$ by Γ^r_{ij} and $\Gamma^{(0)r}_{ij}$, respectively, then we obtain the local expression of (35) as follows:

$$\begin{aligned}
&R_{\nabla,\nabla^*}(\partial_i,\partial_j)\partial_k=R^m_{\nabla ijk}\partial_m\\
&+\{\partial_j(\Gamma^m_{ik}-\Gamma^{(0)m}_{ik})+(\Gamma^r_{ik}-\Gamma^{(0)r}_{ik})\Gamma^m_{jr}-\Gamma^r_{ji}(\Gamma^m_{rk}-\Gamma^{(0)m}_{rk})-\Gamma^r_{jk}(\Gamma^m_{ir}-\Gamma^{(0)m}_{ir})\}\partial_m\\
&-\{\partial_j(\Gamma^m_{jk}-\Gamma^{(0)m}_{jk})+(\Gamma^r_{jk}-\Gamma^{(0)r}_{jk})\Gamma^m_{jr}-\Gamma^r_{ij}(\Gamma^m_{rk}-\Gamma^{(0)m}_{rk})-\Gamma^r_{ik}(\Gamma^m_{jr}-\Gamma^{(0)m}_{jr})\}\partial_m.
\end{aligned}\tag{36}$$

From (34) and by direct computations we obtain the following:

Lemma 1. *Let (M,g) be a Riemannian manifold and let $(\overset{1}{\nabla},\overset{2}{\nabla})$ be a pair of connections such that Γ^{1r}_{ij} and Γ^{2r}_{ij} are the connection coefficients of $\overset{1}{\nabla}$ and $\overset{2}{\nabla}$, respectively. Then, the following assertions hold:*

(1) $R_{\overset{1}{\nabla},\overset{2}{\nabla}}(X,Y)Z=R_{\overset{2}{\nabla},\overset{1}{\nabla}}(X,Y)Z$ *and* $R_{\overset{1}{\nabla},\overset{2}{\nabla}}(X,Y)Z=-R_{\overset{1}{\nabla},\overset{2}{\nabla}}(Y,X)Z$ *for all* $X,Y,Z\in\chi(M)$.

(2) $R_{\overset{1}{\nabla},\overset{2}{\nabla}}(X,Y)Z+R_{\overset{1}{\nabla},\overset{2}{\nabla}}(Y,Z)X+R_{\overset{1}{\nabla},\overset{2}{\nabla}}(Z,X)Y=0$, *whenever* $\overset{1}{\nabla},\overset{2}{\nabla}$ *are torsion-free connections.*

(3) The local expression of the mutual curvature $R_{\overset{1}{\nabla},\overset{2}{\nabla}}$ *is in the following form:* $R_{\overset{1}{\nabla},\overset{2}{\nabla}}(\partial_i,\partial_j)\partial_k=\frac{1}{2}\{(\partial_i(\Gamma^{2t}_{jk})+\Gamma^{2r}_{jk}\Gamma^{1t}_{ir}-\partial_j(\Gamma^{2t}_{ik})-\Gamma^{2r}_{ik}\Gamma^{1t}_{jr}+\partial_i(\Gamma^{1t}_{jk})+\Gamma^{1r}_{jk}\Gamma^{2t}_{ir}-\partial_j(\Gamma^{1t}_{ik})-\Gamma^{1r}_{ik}\Gamma^{2t}_{jr})\}\partial_t$.

Example 2. *Now, we compute the mutual curvature $R_{\nabla^{(\alpha)},\nabla^{(\beta)}}$ in Example 1. Direct computations give us the following*

$$R_{\nabla^{(\alpha)},\nabla^{(\beta)}}(\partial_1,\partial_2)\partial_1 = \frac{1}{2}\{\nabla^{(\alpha)}_{\partial_1}\nabla^{(\beta)}_{\partial_2}\partial_1 - \nabla^{(\alpha)}_{\partial_2}\nabla^{(\beta)}_{\partial_1}\partial_1 - \nabla^{(\alpha)}_{[\partial_1,\partial_2]}\partial_1 + \nabla^{(\beta)}_{\partial_1}\nabla^{(\alpha)}_{\partial_2}\partial_1 - \nabla^{(\beta)}_{\partial_2}\nabla^{(\alpha)}_{\partial_1}\partial_1 - \nabla^{(\beta)}_{[\partial_1,\partial_2]}\partial_1\}. \quad (37)$$

Using (13), we obtain

$$\nabla^{(\beta)}_{\partial_2}\partial_1 = 0, \quad \nabla^{(\beta)}_{\partial_1}\partial_1 = \frac{1-\beta}{\theta_1}\partial_1, \quad \nabla^{(\alpha)}_{\partial_2}(\frac{1-\beta}{\theta_1})\partial_1 = 0, \quad (38)$$

$$\nabla^{(\alpha)}_{\partial_2}\partial_1 = 0, \quad \nabla^{(\alpha)}_{\partial_1}\partial_1 = \frac{1-\alpha}{\theta_1}\partial_1, \quad \nabla^{(\beta)}_{\partial_2}(\frac{1-\alpha}{\theta_1})\partial_1 = 0. \quad (39)$$

Putting (38) and (39) into (37) gives us $R_{\nabla^{(\alpha)},\nabla^{(\beta)}}(\partial_1,\partial_2)\partial_1 = 0$. Similar computations imply

$$\nabla^{(\beta)}_{\partial_1}\partial_2 = 0, \quad \nabla^{(\beta)}_{\partial_2}\partial_2 = \frac{1-\beta}{\theta_2}\partial_2, \quad \nabla^{(\alpha)}_{\partial_1}(\frac{1-\beta}{\theta_2})\partial_2 = 0, \quad (40)$$

$$\nabla^{(\alpha)}_{\partial_1}\partial_2 = 0, \quad \nabla^{(\alpha)}_{\partial_2}\partial_2 = \frac{1-\alpha}{\theta_2}\partial_2, \quad \nabla^{(\beta)}_{\partial_1}(\frac{1-\alpha}{\theta_2})\partial_2 = 0. \quad (41)$$

Substituting (40) and (41) into (34), we derive that $R_{\nabla^{(\alpha)},\nabla^{(\beta)}}(\partial_2,\partial_1)\partial_2 = 0$. From the first item of Lemma 1, we deduce that the other components of the mutual curvature are zero.

Definition 3. *Let (M,g) be a Riemannian manifold and let ∇ be an affine connection with the curvature tensor R. Then, we say that M is a locally symmetric space if $\nabla R = 0$, i.e., $\nabla_X R = 0$, for each $X \in \chi(M)$, this means that R is parallel with respect to ∇.*

Now, we assume that (g,∇,∇^*) is a dualistic structure on the Riemannian manifold M and let R and R^* be curvature tensors of ∇ and ∇^*, respectively. If R is parallel with respect to ∇ and ∇^*, then by direct computations we have

$$\begin{aligned}
&(\nabla^{(\alpha)}_X R)(Y,Z)W \qquad (42)\\
&= \nabla^{(\alpha)}_X R(Y,Z)W - R(\nabla^{(\alpha)}_X Y,Z)W - R(Y,\nabla^{(\alpha)}_X Z)W - R(Y,Z)\nabla^{(\alpha)}_X W\\
&= \{\frac{1+\alpha}{2}\nabla_X R(Y,Z)W + \frac{1-\alpha}{2}\nabla^*_X R(Y,Z)W\}\\
&- R(\frac{1+\alpha}{2}\nabla_X Y + \frac{1-\alpha}{2}\nabla^*_X Y,Z)W\\
&- R(Y,\frac{1+\alpha}{2}\nabla_X Z + \frac{1-\alpha}{2}\nabla^*_X Z)W\\
&- R(Y,Z)(\frac{1+\alpha}{2}\nabla_X W + \frac{1-\alpha}{2}\nabla^*_X W)\\
&= \frac{1+\alpha}{2}\{\nabla_X R(Y,Z)W - R(\nabla_X Y,Z)W - R(Y,\nabla_X Z)W - R(Y,Z)\nabla_X W\}\\
&+ \frac{1-\alpha}{2}\{\nabla^*_X R(Y,Z)W - R(\nabla^*_X Y,Z)W - R(Y,\nabla^*_X Z)W - R(Y,Z)\nabla^*_X W\}\\
&= \frac{1+\alpha}{2}(\nabla_X R)(Y,Z)W + \frac{1-\alpha}{2}(\nabla^*_X R)(Y,Z)W.
\end{aligned}$$

Since R is parallel with respect to ∇ and ∇^*, then we derive that R is parallel with respect to the α-connection $\nabla^{(\alpha)}$. Moreover, if we replace R by R^* in the above relations and assume that R^* is parallel to ∇ and ∇^*, then we obtain that R^* is parallel to the

α-connection $\nabla^{(\alpha)}$. Now, we consider a pair $(\overset{1}{\nabla}, \overset{2}{\nabla})$ of connections on M. If in (42), we replace the mutual curvature $R_{\overset{1}{\nabla},\overset{2}{\nabla}}$ instead of R, then, the same as in (42), we obtain

$$(\nabla^{(\alpha)}_X R_{\overset{1}{\nabla},\overset{2}{\nabla}})(Y,Z)W = \frac{1+\alpha}{2}(\nabla_X R_{\overset{1}{\nabla},\overset{2}{\nabla}})(Y,Z)W + \frac{1-\alpha}{2}(\nabla^*_X R_{\overset{1}{\nabla},\overset{2}{\nabla}})(Y,Z)W. \tag{43}$$

If the mutual curvature $R_{\overset{1}{\nabla},\overset{2}{\nabla}}$ is parallel with respect to ∇ and ∇^*, then from (43), we obtain that $R_{\overset{1}{\nabla},\overset{2}{\nabla}}$ is parallel with respect to $\nabla^{(\alpha)}$. According to the above discussion we obtain the following:

Proposition 1. *Let (g, ∇, ∇^*) be a dualistic structure on the Riemannian manifold M, and R and R^* be curvature tensors of ∇ and ∇^*. Then, the following statements hold:*

(1) *If R (respectively, R^*) is parallel with respect to ∇ and ∇^*, then R (respectively, R^*) is parallel with respect to the α-connection $\nabla^{(\alpha)}$.*

(2) *The mutual curvature R_{∇,∇^*} is parallel with respect to $\nabla^{(\alpha)}$ whenever R_{∇,∇^*} is parallel with respect to ∇ and ∇^*.*

Now, we consider the dualistic structure (g, ∇, ∇^*) on the Riemannian manifold M and let $\{\nabla^{(\alpha)}\}_{\alpha \in R}$ be a family of α-connections. Equations (7), (34), (43), and direct computations give us the following:

Lemma 2. *Let (g, ∇, ∇^*) be a dualistic structure on the Riemannian manifold M, and let $\{\nabla^{(\alpha)}\}_{\alpha \in R}$ be a family of α-connections on M. Then, the following statements hold:*

(1)

$$\begin{aligned} R_{\nabla^{(\alpha)},\nabla^{(\beta)}}(X,Y)Z = {} & \frac{(1+\alpha)(1+\beta)}{4} R_{\nabla}(X,Y)Z + \frac{(1-\alpha\beta)}{2} R_{\nabla,\nabla^*}(X,Y)Z \\ & + \frac{(1-\alpha)(1-\beta)}{4} R_{\nabla^*}(X,Y)Z, \qquad \forall X,Y,Z \in \chi(M). \end{aligned}$$

(2) *If M is a flat space with respect to ∇ and ∇^* and the mutual curvature R_{∇,∇^*} vanishes, then the mutual curvature $R_{\nabla^{(\alpha)},\nabla^{(\beta)}}$ vanishes.*

(3) *The mutual curvature $R_{\nabla^{(\alpha)},\nabla^{(\beta)}}$ reduces to R_∇ (respectively, R_{∇^*}) whenever $\alpha = \beta = 1$ (respectively, $\alpha = \beta = -1$).*

(4) *The mutual curvature $R_{\nabla^{(\alpha)},\nabla^{(\beta)}}$ reduces to the mutual curvature R_{∇,∇^*} whenever $\alpha = -\beta = 1$.*

(5) *The mutual curvature $R_{\nabla^{(\alpha)},\nabla^{(\beta)}}$ is parallel with respect to the γ-connection $\nabla^{(\gamma)}$ whenever R_∇, R_{∇^*}, and the mutual curvatures R_{∇,∇^*} are parallel with respect to ∇ and ∇^*.*

As a consequence of the above lemma, if we consider $\alpha = \beta$, then we obtain the (α)-curvature $R^{(\alpha)}$ of the (α)-connection $\nabla^{(\alpha)}$ as follows:

$$R^{(\alpha)}(X,Y)Z = \frac{(1+\alpha)^2}{4} R_\nabla(X,Y)Z + \frac{(1-\alpha^2)}{2} R_{\nabla,\nabla^*}(X,Y)Z + \frac{(1-\alpha)^2}{4} R_{\nabla^*}(X,Y)Z. \tag{44}$$

To continue, we investigate under which conditions the (α)-curvature $R^{(\alpha)}$ of the (α)-connection $\nabla^{(\alpha)}$ is parallel with respect to the (β)-connection $\nabla^{(\beta)}$. Direct computations and (44) give us the following

$$\begin{aligned} (\nabla^{(\beta)}_X R^{(\alpha)})(Y,Z)W = {} & \frac{(1+\alpha)^2}{4} (\nabla^{(\beta)}_X R_\nabla)(Y,Z)W + \frac{(1-\alpha^2)}{2} (\nabla^{(\beta)}_X R_{\nabla,\nabla^*})(Y,Z)W \\ & + \frac{(1-\alpha)^2}{4} (\nabla^{(\beta)}_X R_{\nabla^*})(Y,Z)W. \end{aligned} \tag{45}$$

Let (∇, ∇^*) be a dualistic structure on the Riemannian manifold (M, g) with the curvature tensors R_∇ and R_{∇^*}, respectively, and let X be a vector field on M. Then, (44) implies the following;

$$(L_X R^{(\alpha)})(Y,Z,W) = \frac{(1+\alpha)^2}{4}(L_X R_\nabla)(Y,Z,W) + \frac{(1-\alpha^2)}{2}(L_X R_{\nabla,\nabla^*})(Y,Z,W) + \frac{(1-\alpha)^2}{4}(L_X R_{\nabla^*})(Y,Z,W). \tag{46}$$

Corollary 1. *Let (g, ∇, ∇^*) be a dualistic structure on the Riemannian manifold M. Then, the following statements hold:*

(1) *The (α)-curvature $R^{(\alpha)}$ is parallel with respect to the (β)-connection $\nabla^{(\beta)}$ whenever R_∇, R_{∇^*} and the mutual curvature R_{∇,∇^*} are parallel with respect to ∇ and ∇^*.*

(2) *The Lie derivative of the (α)-curvature $R^{(\alpha)}$ along X vanishes if the Lie derivatives of R_∇, R_{∇^*}, and R_{∇,∇^*} along X vanish.*

5. Dualistic Structure on the Tangent Bundle

In this section, we consider an arbitrary Riemannian metric $\bar{g}$ and two affine torsion-free connections $\widetilde{\nabla}$ and $\widetilde{\nabla}^*$ on the TM. Then, we investigate under which conditions $(\bar{g}, \widetilde{\nabla}, \widetilde{\nabla}^*)$ is a dualistic structure on the tangent bundle TM.

Let $\{\delta_i, \partial_{\bar{i}}\}$ be a basis for $T_v(TM)$, where $v \in TM$. According to this basis, we consider the Riemannian metric $\bar{g} = \alpha_{ij}dx^i dx^j + 2\beta_{ij}dx^i \delta y^j + \gamma_{ij}\delta y^i \delta y^j$ on the TM, where $\alpha_{ij}, \beta_{ij}, \gamma_{ij} \in C^\infty(TM)$. Since $\widetilde{\nabla}$ and $\widetilde{\nabla}^*$ are affine torsion-free connections on the TM and $\{\delta_i, \partial_{\bar{i}}\}$ is a basis, so the following identities hold:

$$\widetilde{\nabla}_{\delta_i}\delta_j = \Gamma^r_{ij}\delta_r + \Gamma^{\bar{r}}_{ij}\partial_{\bar{r}}, \qquad \widetilde{\nabla}_{\partial_{\bar{i}}}\partial_{\bar{j}} = \Gamma^r_{\bar{i}\bar{j}}\delta_r + \Gamma^{\bar{r}}_{\bar{i}\bar{j}}\partial_{\bar{r}}, \tag{47}$$

$$\widetilde{\nabla}_{\delta_i}\partial_{\bar{j}} = \Gamma^r_{i\bar{j}}\delta_r + \Gamma^{\bar{r}}_{i\bar{j}}\partial_{\bar{r}}, \qquad \widetilde{\nabla}_{\partial_{\bar{i}}}\delta_j = \Gamma^r_{\bar{i}j}\delta_r + \Gamma^{\bar{r}}_{\bar{i}j}\partial_{\bar{r}}, \tag{48}$$

$$\widetilde{\nabla}^*_{\delta_i}\delta_j = \Gamma^{*r}_{ij}\delta_r + \Gamma^{*\bar{r}}_{ij}\partial_{\bar{r}}, \qquad \widetilde{\nabla}^*_{\partial_{\bar{i}}}\partial_{\bar{j}} = \Gamma^{*r}_{\bar{i}\bar{j}}\delta_r + \Gamma^{*\bar{r}}_{\bar{i}\bar{j}}\partial_{\bar{r}}, \tag{49}$$

$$\widetilde{\nabla}^*_{\delta_i}\partial_{\bar{j}} = \Gamma^{*r}_{i\bar{j}}\delta_r + \Gamma^{*\bar{r}}_{i\bar{j}}\partial_{\bar{r}}, \qquad \widetilde{\nabla}^*_{\partial_{\bar{i}}}\delta_j = \Gamma^{*r}_{\bar{i}j}\delta_r + \Gamma^{*\bar{r}}_{\bar{i}j}\partial_{\bar{r}}, \tag{50}$$

where $\Gamma^A_{BC}, \Gamma^{*A}_{BC} \in C^\infty(TM)$, and $A, B, C \in \{1, \cdots n, \bar{1}, \cdots \bar{n}\}$. If we use torsion-freeness of $\widetilde{\nabla}$ and $\widetilde{\nabla}^*$ and applying (47)–(50), then we obtain the following:

Lemma 3. *Let (M, g) be a Riemannian manifold and let $\overset{g}{\nabla}$ be an affine connection on it. Let $\overset{g}{\Gamma}{}^r_{ij}$, $\overset{g}{T}{}^r_{ij}$, and $\overset{g}{R}{}^r_{ijs}$ be the connection coefficients, torsion components, and curvature components of $\overset{g}{\nabla}$, respectively. If $\widetilde{\nabla}$ and $\widetilde{\nabla}^*$ are two affine torsion-free connections on the TM, then the following equations hold:*

$$\Gamma^r_{i\bar{j}} = \Gamma^r_{\bar{j}i}, \qquad \Gamma^{\bar{r}}_{i\bar{j}} - \Gamma^{\bar{r}}_{\bar{j}i} = \overset{g}{\Gamma}{}^r_{ij} - \overset{g}{T}{}^r_{ij}, \qquad \Gamma^r_{ij} = \Gamma^r_{ji}, \tag{51}$$

$$\Gamma^{\bar{r}}_{ij} - \Gamma^{\bar{r}}_{ji} = -y^s \overset{g}{R}{}^r_{ijs}, \qquad \Gamma^r_{\bar{i}\bar{j}} = \Gamma^r_{\bar{j}\bar{i}}, \qquad \Gamma^{\bar{r}}_{\bar{i}\bar{j}} = \Gamma^{\bar{r}}_{\bar{j}\bar{i}}, \tag{52}$$

$$\Gamma^{*r}_{i\bar{j}} = \Gamma^{*r}_{\bar{j}i}, \qquad \Gamma^{*\bar{r}}_{i\bar{j}} - \Gamma^{*\bar{r}}_{\bar{j}i} = \overset{g}{\Gamma}{}^r_{ij} - \overset{g}{T}{}^r_{ij}, \qquad \Gamma^{*r}_{ij} = \Gamma^{*r}_{ji}, \tag{53}$$

$$\Gamma^{*\bar{r}}_{ij} - \Gamma^{*\bar{r}}_{ji} = -y^s \overset{g}{R}{}^r_{ijs}, \qquad \Gamma^{*r}_{\bar{i}\bar{j}} = \Gamma^{*r}_{\bar{j}\bar{i}}, \qquad \Gamma^{*\bar{r}}_{\bar{i}\bar{j}} = \Gamma^{*\bar{r}}_{\bar{j}\bar{i}}. \tag{54}$$

Proof. Using (5) and (47)–(50) and the torsion-freeness of $\widetilde{\nabla}$ and $\widetilde{\nabla}^*$ completes the proof. □

Now, if we put elements of $\{\delta_i, \partial_{\bar{i}}\}$ in (6), and use the above relations, we derive the following:

Proposition 2. *Let (M, g) be a Riemannian manifold and let $(TM, \bar{g})$ be its tangent bundle equipped with the Riemannian metric $\bar{g}$ (defined as above). If $\widetilde{\nabla}$ and $\widetilde{\nabla}^*$ are two affine torsion-free connections on the TM, then $(\bar{g}, \widetilde{\nabla}, \widetilde{\nabla}^*)$ is a dualistic structure on the TM if and only if (51)–(54) and the following equations hold:*

(1) $\partial_{\bar{i}}\gamma_{jk} = 2\Gamma^{r}_{\bar{i}\bar{j}}\beta_{rk} + \Gamma^{\bar{r}}_{\bar{i}\bar{j}}\gamma_{rk} + 2\Gamma^{*r}_{\bar{i}k}\beta_{jr} + \Gamma^{*\bar{r}}_{\bar{i}k}\gamma_{jr};$

(2) $\delta_i\alpha_{jk} = \Gamma^{r}_{ij}\alpha_{rk} + 2\Gamma^{\bar{r}}_{ij}\beta_{rk} + \Gamma^{*r}_{ik}\alpha_{jr} + 2\Gamma^{*\bar{r}}\beta_{jr};$

(3) $2\delta_i\beta_{jk} = 2\Gamma^{r}_{ij}\beta_{rk} + \Gamma^{\bar{r}}_{ij}\gamma_{rk} + \Gamma^{*r}_{i\bar{k}}\alpha_{jr} + 2\Gamma^{*\bar{r}}_{i\bar{k}}\beta_{jr};$

(4) $\partial_{\bar{k}}\alpha_{ij} = \Gamma^{r}_{\bar{k}i}\alpha_{rj} + 2\Gamma^{\bar{r}}_{\bar{k}i}\beta_{rj} + \Gamma^{*r}_{\bar{k}j}\alpha_{ir} + 2\Gamma^{*\bar{r}}_{\bar{k}j}\beta_{ir};$

(5) $2\partial_{\bar{i}}\beta_{jk} = \Gamma^{r}_{\bar{i}\bar{j}}\alpha_{rk} + 2\Gamma^{\bar{r}}_{\bar{i}\bar{j}}\beta_{rk} + 2\Gamma^{*r}_{\bar{i}k}\beta_{jr} + \Gamma^{*\bar{r}}_{\bar{i}k}\gamma_{jr};$

(6) $\delta_k\gamma_{ij} = 2\Gamma^{r}_{k\bar{i}}\beta_{rj} + \Gamma^{\bar{r}}_{k\bar{i}}\gamma_{rj} + 2\Gamma^{*r}_{k\bar{j}}\beta_{ir} + \Gamma^{*\bar{r}}_{k\bar{j}}\gamma_{ir}.$

Now, we assume that $(\bar{g}, \widetilde{\nabla}, \widetilde{\nabla}^*)$ is a dualistic structure on the TM, then we show under which conditions the TM is a conjugate symmetric space whenever $\widetilde{R}$ and $\widetilde{R}^*$ are curvature tensors of $\widetilde{\nabla}$ and $\widetilde{\nabla}^*$. We do this using a direct computation of the curvature components of $\widetilde{R}$ and $\widetilde{R}^*$ on the TM. According to the basis $\{\delta_i, \partial_{\bar{i}}\}$ for $T_v(TM)$, we imply that the curvature components of $\widetilde{R}$ and $\widetilde{R}^*$ are as follows:

$$\widetilde{R}(\delta_i, \delta_j)\delta_k = \widetilde{R}^{l}_{ijk}\delta_l + \widetilde{R}^{\bar{l}}_{ijk}\partial_{\bar{l}}, \qquad \widetilde{R}^*(\delta_i, \delta_j)\delta_k = \widetilde{R}^{*l}_{ijk}\delta_l + \widetilde{R}^{*\bar{l}}_{ijk}\partial_{\bar{l}}, \tag{55}$$

$$\widetilde{R}(\partial_{\bar{i}}, \partial_{\bar{j}})\partial_{\bar{k}} = \widetilde{R}^{l}_{\bar{i}\bar{j}\bar{k}}\delta_l + \widetilde{R}^{\bar{l}}_{\bar{i}\bar{j}\bar{k}}\partial_{\bar{l}}, \qquad \widetilde{R}^*(\partial_{\bar{i}}, \partial_{\bar{j}})\partial_{\bar{k}} = \widetilde{R}^{*l}_{\bar{i}\bar{j}\bar{k}}\delta_l + \widetilde{R}^{*\bar{l}}_{\bar{i}\bar{j}\bar{k}}\partial_{\bar{l}}, \tag{56}$$

$$\widetilde{R}(\delta_i, \delta_j)\partial_{\bar{k}} = \widetilde{R}^{l}_{ij\bar{k}}\delta_l + \widetilde{R}^{\bar{l}}_{ij\bar{k}}\partial_{\bar{l}}, \qquad \widetilde{R}^*(\delta_i, \delta_j)\partial_{\bar{k}} = \widetilde{R}^{*l}_{ij\bar{k}}\delta_l + \widetilde{R}^{*\bar{l}}_{ij\bar{k}}\partial_{\bar{l}}, \tag{57}$$

$$\widetilde{R}(\delta_i, \partial_{\bar{k}})\delta_j = \widetilde{R}^{l}_{i\bar{k}j}\delta_l + \widetilde{R}^{\bar{l}}_{i\bar{k}j}\partial_{\bar{l}}, \qquad \widetilde{R}^*(\delta_i, \partial_{\bar{k}})\delta_j = \widetilde{R}^{*l}_{i\bar{k}j}\delta_l + \widetilde{R}^{*\bar{l}}_{i\bar{k}j}\partial_{\bar{l}}, \tag{58}$$

$$\widetilde{R}(\partial_{\bar{k}}, \delta_i)\delta_j = \widetilde{R}^{l}_{\bar{k}ij}\delta_l + \widetilde{R}^{\bar{l}}_{\bar{k}ij}\partial_{\bar{l}}, \qquad \widetilde{R}^*(\partial_{\bar{k}}, \delta_i)\delta_j = \widetilde{R}^{*l}_{\bar{k}ij}\delta_l + \widetilde{R}^{*\bar{l}}_{\bar{k}ij}\partial_{\bar{l}}, \tag{59}$$

$$\widetilde{R}(\partial_{\bar{i}}, \partial_{\bar{j}})\delta_k = \widetilde{R}^{l}_{\bar{i}\bar{j}k}\delta_l + \widetilde{R}^{\bar{l}}_{\bar{i}\bar{j}k}\partial_{\bar{l}}, \qquad \widetilde{R}^*(\partial_{\bar{i}}, \partial_{\bar{j}})\delta_k = \widetilde{R}^{*l}_{\bar{i}\bar{j}k}\delta_l + \widetilde{R}^{*\bar{l}}_{\bar{i}\bar{j}k}\partial_{\bar{l}}, \tag{60}$$

$$\widetilde{R}(\partial_{\bar{i}}, \delta_k)\partial_{\bar{j}} = \widetilde{R}^{l}_{\bar{i}k\bar{j}}\delta_l + \widetilde{R}^{\bar{l}}_{\bar{i}k\bar{j}}\partial_{\bar{l}}, \qquad \widetilde{R}^*(\partial_{\bar{i}}, \delta_k)\partial_{\bar{j}} = \widetilde{R}^{*l}_{\bar{i}k\bar{j}}\delta_l + \widetilde{R}^{*\bar{l}}_{\bar{i}k\bar{j}}\partial_{\bar{l}}, \tag{61}$$

$$\widetilde{R}(\delta_k, \partial_{\bar{i}})\partial_{\bar{j}} = \widetilde{R}^{l}_{k\bar{i}\bar{j}}\delta_l + \widetilde{R}^{\bar{l}}_{k\bar{i}\bar{j}}\partial_{\bar{l}}, \qquad \widetilde{R}^*(\delta_k, \partial_{\bar{i}})\partial_{\bar{j}} = \widetilde{R}^{*l}_{k\bar{i}\bar{j}}\delta_l + \widetilde{R}^{*\bar{l}}_{k\bar{i}\bar{j}}\partial_{\bar{l}}. \tag{62}$$

According to (55)–(62), we obtain the following:

Proposition 3. *Let $(\bar{g}, \widetilde{\nabla}, \widetilde{\nabla}^*)$ be a dualistic structure on the TM and let $\widetilde{R}$ and $\widetilde{R}^*$ be the curvature tensors of $\widetilde{\nabla}$ and $\widetilde{\nabla}^*$ on the TM, respectively. Then, the TM is a conjugate symmetric space if and only if the following identities hold:*

$$\widetilde{R}^{l}_{ijk} = \widetilde{R}^{*l}_{ijk}, \qquad \widetilde{R}^{\bar{l}}_{ijk} = \widetilde{R}^{*\bar{l}}_{ijk}, \qquad \widetilde{R}^{l}_{\bar{i}\bar{j}\bar{k}} = \widetilde{R}^{*l}_{\bar{i}\bar{j}\bar{k}}, \qquad \widetilde{R}^{\bar{l}}_{\bar{i}\bar{j}\bar{k}} = \widetilde{R}^{*\bar{l}}_{\bar{i}\bar{j}\bar{k}},$$

$$\widetilde{R}^{l}_{ij\bar{k}} = \widetilde{R}^{*l}_{ij\bar{k}}, \qquad \widetilde{R}^{\bar{l}}_{ij\bar{k}} = \widetilde{R}^{*\bar{l}}_{ij\bar{k}}, \qquad \widetilde{R}^{l}_{i\bar{k}j} = \widetilde{R}^{*l}_{i\bar{k}j}, \qquad \widetilde{R}^{\bar{l}}_{i\bar{k}j} = \widetilde{R}^{*\bar{l}}_{i\bar{k}j},$$

$$\widetilde{R}^{l}_{\bar{k}ij} = \widetilde{R}^{*l}_{\bar{k}ij}, \qquad \widetilde{R}^{\bar{l}}_{\bar{k}ij} = \widetilde{R}^{*\bar{l}}_{\bar{k}ij}, \qquad \widetilde{R}^{l}_{\bar{i}\bar{j}k} = \widetilde{R}^{*l}_{\bar{i}\bar{j}k}, \qquad \widetilde{R}^{\bar{l}}_{\bar{i}\bar{j}k} = \widetilde{R}^{*\bar{l}}_{\bar{i}\bar{j}k},$$

$$\widetilde{R}^{l}_{\bar{i}k\bar{j}} = \widetilde{R}^{*l}_{\bar{i}k\bar{j}}, \qquad \widetilde{R}^{\bar{l}}_{\bar{i}k\bar{j}} = \widetilde{R}^{*\bar{l}}_{\bar{i}k\bar{j}}, \qquad \widetilde{R}^{l}_{k\bar{i}\bar{j}} = \widetilde{R}^{*l}_{k\bar{i}\bar{j}}, \qquad \widetilde{R}^{\bar{l}}_{k\bar{i}\bar{j}} = \widetilde{R}^{*\bar{l}}_{k\bar{i}\bar{j}}.$$

6. Mutual Curvatures of the Tangent Bundle

This section is concerned with the mutual curvatures of the TM. In this part we study the components of the mutual curvature $\widetilde{R}_{\widetilde{\nabla},\widetilde{\nabla}^*}$ of the TM whenever $(\bar{g}, \widetilde{\nabla}, \widetilde{\nabla}^*)$ is a dualistic structure on the TM.

We assume that (M, g) is a Riemannian manifold with affine connection $\overset{g}{\nabla}$ and $\overset{g}{\Gamma}{}^r_{ij}$, $\overset{g}{T}{}^r_{ij}$ and $\overset{g}{R}{}^r_{ijs}$ are the connection coefficients, torsion components, and curvature components of $\overset{g}{\nabla}$. According to (47)–(50) and using (34), we have

Theorem 3. *Let (M, g) be a Riemannian manifold and let $(TM, \bar{g})$ be its tangent bundle equipped with the Riemannian metric $\bar{g}$ (defined as above), and let $(\bar{g}, \widetilde{\nabla}, \widetilde{\nabla}^*)$ be a dualistic structure on the TM. If Γ^r_{ij} and Γ^{*r}_{ij} are the connection coefficients of $\widetilde{\nabla}$ and $\widetilde{\nabla}^*$ on the TM, respectively, then the components of the mutual curvature $\widetilde{R}_{\widetilde{\nabla},\widetilde{\nabla}^*}$ on the TM are as follows:*

$$
\begin{aligned}
\widetilde{R}_{\widetilde{\nabla},\widetilde{\nabla}^*}(\delta_i,\delta_j)\delta_k = \frac{1}{2}\{(&\Gamma^{*r}_{jk}\Gamma^t_{ir} + \delta_i(\Gamma^{*t}_{jk}) + \Gamma^{*\bar{r}}_{jk}\Gamma^t_{i\bar{r}} - \Gamma^{*r}_{ik}\Gamma^t_{jr} - \delta_j(\Gamma^{*t}_{ik}) - \Gamma^{*\bar{r}}_{ik}\Gamma^t_{j\bar{r}} + y^r\overset{g}{R}{}^l_{ijr}\Gamma^t_{\bar{l}k} \\
&+ \delta_i(\Gamma^t_{jk}) + \Gamma^r_{jk}\Gamma^{*t}_{ir} + \Gamma^{\bar{r}}_{jk}\Gamma^{*t}_{i\bar{r}} - \delta_j(\Gamma^t_{ik}) - \Gamma^r_{ik}\Gamma^{*t}_{jr} - \Gamma^{\bar{r}}_{ik}\Gamma^{*t}_{j\bar{r}} + y^l\overset{g}{R}{}^s_{ijl}\Gamma^{*t}_{\bar{s}k})\}\delta_t \\
&+ \frac{1}{2}\{(\Gamma^{*r}_{jk}\Gamma^{\bar{t}}_{ir} + \Gamma^{*\bar{r}}_{jk}\Gamma^{\bar{t}}_{i\bar{r}} + \delta_i(\Gamma^{*\bar{t}}_{jk}) - \Gamma^{*r}_{ik}\Gamma^{\bar{t}}_{jr} - \Gamma^{*\bar{r}}_{ik}\Gamma^{\bar{t}}_{j\bar{r}} - \delta_j(\Gamma^{*\bar{t}}_{ik}) + y^r\overset{g}{R}{}^l_{ijr}\Gamma^{\bar{t}}_{\bar{l}k} \\
&+ \Gamma^r_{jk}\Gamma^{*\bar{t}}_{ir} + \delta_i(\Gamma^{\bar{t}}_{jk}) + \Gamma^{\bar{r}}_{jk}\Gamma^{*\bar{t}}_{i\bar{r}} - \Gamma^r_{ik}\Gamma^{*\bar{t}}_{jr} - \delta_j(\Gamma^{\bar{t}}_{ik}) - \Gamma^{\bar{r}}_{ik}\Gamma^{*\bar{t}}_{j\bar{r}} + y^l\overset{g}{R}{}^s_{ijl}\Gamma^{*\bar{t}}_{\bar{s}k})\}\partial_{\bar{t}},
\end{aligned}
$$

$$
\begin{aligned}
\widetilde{R}_{\widetilde{\nabla},\widetilde{\nabla}^*}(\partial_{\bar{i}},\partial_{\bar{j}})\partial_{\bar{k}} = \frac{1}{2}\{(&\Gamma^{*r}_{\bar{j}\bar{k}}\Gamma^t_{\bar{i}r} + \partial_{\bar{i}}(\Gamma^{*t}_{\bar{j}\bar{k}}) + \Gamma^{*\bar{r}}_{\bar{j}\bar{k}}\Gamma^t_{\bar{i}\bar{r}} - \Gamma^{*r}_{\bar{i}\bar{k}}\Gamma^t_{\bar{j}r} - \partial_{\bar{j}}(\Gamma^{*t}_{\bar{i}\bar{k}}) - \Gamma^{*\bar{r}}_{\bar{i}\bar{k}}\Gamma^t_{\bar{j}\bar{r}} \\
&+ \partial_{\bar{i}}(\Gamma^t_{\bar{j}\bar{k}}) + \Gamma^r_{\bar{j}\bar{k}}\Gamma^{*t}_{\bar{i}r} + \Gamma^{\bar{r}}_{\bar{j}\bar{k}}\Gamma^{*t}_{\bar{i}\bar{r}} - \partial_{\bar{j}}(\Gamma^t_{\bar{i}\bar{k}}) - \Gamma^r_{\bar{i}\bar{k}}\Gamma^{*t}_{\bar{j}r} - \Gamma^{\bar{r}}_{\bar{i}\bar{k}}\Gamma^{*t}_{\bar{j}\bar{r}})\}\delta_t \\
&+ \frac{1}{2}\{(\Gamma^{*r}_{\bar{j}\bar{k}}\Gamma^{\bar{t}}_{\bar{i}r} + \Gamma^{*\bar{r}}_{\bar{j}\bar{k}}\Gamma^{\bar{t}}_{\bar{i}\bar{r}} + \partial_{\bar{i}}(\Gamma^{*\bar{t}}_{\bar{j}\bar{k}}) - \Gamma^{*r}_{\bar{i}\bar{k}}\Gamma^{\bar{t}}_{\bar{j}r} - \Gamma^{*\bar{r}}_{\bar{i}\bar{k}}\Gamma^{\bar{t}}_{\bar{j}\bar{r}} - \partial_{\bar{j}}(\Gamma^{*\bar{t}}_{\bar{i}\bar{k}}) \\
&+ \Gamma^r_{\bar{j}\bar{k}}\Gamma^{*\bar{t}}_{\bar{i}r} + \partial_{\bar{i}}(\Gamma^{\bar{t}}_{\bar{j}\bar{k}}) + \Gamma^{\bar{r}}_{\bar{j}\bar{k}}\Gamma^{*\bar{t}}_{\bar{i}\bar{r}} - \Gamma^r_{\bar{i}\bar{k}}\Gamma^{*\bar{t}}_{\bar{j}r} - \partial_{\bar{j}}(\Gamma^{\bar{t}}_{\bar{i}\bar{k}}) - \Gamma^{\bar{r}}_{\bar{i}\bar{k}}\Gamma^{*\bar{t}}_{\bar{j}\bar{r}})\}\partial_{\bar{t}},
\end{aligned}
$$

$$
\begin{aligned}
\widetilde{R}_{\widetilde{\nabla},\widetilde{\nabla}^*}(\delta_i,\delta_j)\partial_{\bar{k}} = \frac{1}{2}\{(&\Gamma^{*r}_{j\bar{k}}\Gamma^t_{ir} + \delta_i(\Gamma^{*t}_{j\bar{k}}) + \Gamma^{*\bar{r}}_{j\bar{k}}\Gamma^t_{i\bar{r}} - \Gamma^{*r}_{i\bar{k}}\Gamma^t_{jr} - \delta_j(\Gamma^{*t}_{i\bar{k}}) - \Gamma^{*\bar{r}}_{i\bar{k}}\Gamma^t_{j\bar{r}} + y^r\overset{g}{R}{}^l_{ijr}\Gamma^t_{\bar{l}\bar{k}} \\
&+ \delta_i(\Gamma^t_{j\bar{k}}) + \Gamma^r_{j\bar{k}}\Gamma^{*t}_{ir} + \Gamma^{\bar{r}}_{j\bar{k}}\Gamma^{*t}_{i\bar{r}} - \delta_j(\Gamma^t_{i\bar{k}}) - \Gamma^r_{i\bar{k}}\Gamma^{*t}_{jr} - \Gamma^{\bar{r}}_{i\bar{k}}\Gamma^{*t}_{j\bar{r}} + y^r\overset{g}{R}{}^s_{ijr}\Gamma^{*t}_{\bar{s}\bar{k}})\}\delta_t \\
&+ \frac{1}{2}\{(\Gamma^{*r}_{j\bar{k}}\Gamma^{\bar{t}}_{ir} + \Gamma^{*\bar{r}}_{j\bar{k}}\Gamma^{\bar{t}}_{i\bar{r}} + \delta_i(\Gamma^{*\bar{t}}_{j\bar{k}}) - \Gamma^{*r}_{i\bar{k}}\Gamma^{\bar{t}}_{jr} - \Gamma^{*\bar{r}}_{i\bar{k}}\Gamma^{\bar{t}}_{j\bar{r}} - \delta_j(\Gamma^{*\bar{t}}_{i\bar{k}}) + y^r\overset{g}{R}{}^l_{ijr}\Gamma^{\bar{t}}_{\bar{l}\bar{k}} \\
&+ \Gamma^r_{j\bar{k}}\Gamma^{*\bar{t}}_{ir} + \delta_i(\Gamma^{\bar{t}}_{j\bar{k}}) + \Gamma^{\bar{r}}_{j\bar{k}}\Gamma^{*\bar{t}}_{i\bar{r}} - \Gamma^r_{i\bar{k}}\Gamma^{*\bar{t}}_{jr} - \delta_j(\Gamma^{\bar{t}}_{i\bar{k}}) - \Gamma^{\bar{r}}_{i\bar{k}}\Gamma^{*\bar{t}}_{j\bar{r}} + +y^r\overset{g}{R}{}^l_{ijr}\Gamma^{*\bar{t}}_{\bar{l}\bar{k}})\}\partial_{\bar{t}},
\end{aligned}
$$

$$
\begin{aligned}
\widetilde{R}_{\widetilde{\nabla},\widetilde{\nabla}^*}(\delta_i,\partial_{\bar{k}})\delta_j = \frac{1}{2}\{(&\Gamma^{*r}_{\bar{k}j}\Gamma^t_{ir} + \delta_i(\Gamma^{*t}_{\bar{k}j}) + \Gamma^{*\bar{r}}_{\bar{k}j}\Gamma^t_{i\bar{r}} - \Gamma^{*r}_{ij}\Gamma^t_{\bar{k}r} - \partial_{\bar{k}}(\Gamma^{*t}_{ij}) - \Gamma^{*\bar{r}}_{ij}\Gamma^t_{\bar{k}\bar{r}} - (\overset{g}{\Gamma}{}^r_{ik} - \overset{g}{T}{}^r_{ik})\Gamma^t_{\bar{r}j} \\
&+ \delta_i(\Gamma^t_{\bar{k}j}) + \Gamma^r_{\bar{k}j}\Gamma^{*t}_{ir} + \Gamma^{\bar{r}}_{\bar{k}j}\Gamma^{*t}_{i\bar{r}} - \partial_{\bar{k}}(\Gamma^t_{ij}) - \Gamma^r_{ij}\Gamma^{*t}_{\bar{k}r} - \Gamma^{\bar{r}}_{ij}\Gamma^{*t}_{\bar{k}\bar{r}} - (\overset{g}{\Gamma}{}^r_{ik} - \overset{g}{T}{}^r_{ik})\Gamma^{*t}_{\bar{r}j})\}\delta_t \\
&+ \frac{1}{2}\{(\Gamma^{*r}_{\bar{k}j}\Gamma^{\bar{t}}_{ir} + \Gamma^{*\bar{r}}_{\bar{k}j}\Gamma^{\bar{t}}_{i\bar{r}} + \delta_i(\Gamma^{*\bar{t}}_{\bar{k}j}) - \Gamma^{*r}_{ij}\Gamma^{\bar{t}}_{\bar{k}r} - \Gamma^{*\bar{r}}_{ij}\Gamma^{\bar{t}}_{\bar{k}\bar{r}} - \partial_{\bar{k}}(\Gamma^{*\bar{t}}_{ij}) - (\overset{g}{\Gamma}{}^r_{ik} - \overset{g}{T}{}^r_{ik})\Gamma^{\bar{t}}_{\bar{r}j} \\
&+ \Gamma^r_{\bar{k}j}\Gamma^{*\bar{t}}_{ir} + \delta_i(\Gamma^{\bar{t}}_{\bar{k}j}) + \Gamma^{\bar{r}}_{\bar{k}j}\Gamma^{*\bar{t}}_{i\bar{r}} - \Gamma^r_{ij}\Gamma^{*\bar{t}}_{\bar{k}r} - \partial_{\bar{k}}(\Gamma^{\bar{t}}_{ij}) - \Gamma^{\bar{r}}_{ij}\Gamma^{*\bar{t}}_{\bar{k}\bar{r}} - (\overset{g}{\Gamma}{}^r_{ik} - \overset{g}{T}{}^r_{ik})\Gamma^{*\bar{t}}_{\bar{r}j})\}\partial_{\bar{t}},
\end{aligned}
$$

$$
\begin{aligned}
\widetilde{R}_{\widetilde{\nabla},\widetilde{\nabla}^*}(\partial_{\bar i},\partial_{\bar j})\delta_k = \frac{1}{2}\{(&\Gamma^{*r}_{\bar j k}\Gamma^t_{\bar i r} + \partial_{\bar i}(\Gamma^{*t}_{\bar j k}) + \Gamma^{*\bar r}_{\bar j k}\Gamma^t_{\bar i \bar r} - \Gamma^{*r}_{\bar i k}\Gamma^t_{\bar j r} - \partial_{\bar j}(\Gamma^{*t}_{\bar i k}) - \Gamma^{*\bar r}_{\bar i k}\Gamma^t_{\bar j \bar r} \\
&+ \partial_{\bar i}(\Gamma^t_{\bar j k}) + \Gamma^r_{\bar j k}\Gamma^{*t}_{\bar i r} + \Gamma^{\bar r}_{\bar j k}\Gamma^{*t}_{\bar i \bar r} - \partial_{\bar j}(\Gamma^t_{\bar i k}) - \Gamma^r_{\bar i k}\Gamma^{*t}_{\bar j r} - \Gamma^{\bar r}_{\bar i k}\Gamma^{*t}_{\bar j \bar r})\}\delta_t \\
&+ \frac{1}{2}\{(\Gamma^{*r}_{\bar j k}\Gamma^{\bar t}_{\bar i r} + \Gamma^{*\bar r}_{\bar j k}\Gamma^{\bar t}_{\bar i \bar r} + \partial_{\bar i}(\Gamma^{*\bar t}_{\bar j k}) - \Gamma^{*r}_{\bar i k}\Gamma^{\bar t}_{\bar j r} - \Gamma^{*\bar r}_{\bar i k}\Gamma^{\bar t}_{\bar j \bar r} - \partial_{\bar j}(\Gamma^{*\bar t}_{\bar i k}) \\
&+ \Gamma^r_{\bar j k}\Gamma^{*\bar t}_{\bar i r} + \partial_{\bar i}(\Gamma^{\bar t}_{\bar j k}) + \Gamma^{\bar r}_{\bar j k}\Gamma^{*\bar t}_{\bar i \bar r} - \Gamma^r_{\bar i k}\Gamma^{*\bar t}_{\bar j r} - \partial_{\bar j}(\Gamma^{\bar t}_{\bar i k}) - \Gamma^{\bar r}_{\bar i k}\Gamma^{*\bar t}_{\bar j \bar r})\}\partial_{\bar t},
\end{aligned}
$$

$$
\begin{aligned}
\widetilde{R}_{\widetilde{\nabla},\widetilde{\nabla}^*}(\partial_{\bar i},\delta_k)\partial_{\bar j} = \frac{1}{2}\{(&\Gamma^{*r}_{k\bar j}\Gamma^t_{\bar i r} + \partial_{\bar i}(\Gamma^{*t}_{k\bar j}) + \Gamma^{*\bar r}_{k\bar j}\Gamma^t_{\bar i \bar r} - \Gamma^{*r}_{\bar i\bar j}\Gamma^t_{kr} - \delta_k(\Gamma^{*t}_{\bar i\bar j}) - \Gamma^{*\bar r}_{\bar i\bar j}\Gamma^t_{k\bar r} + (\overset{g}{\Gamma}{}^r_{ki} - \overset{g}{T}{}^r_{ki})\Gamma^t_{\bar r\bar j} \\
&+ \partial_{\bar i}(\Gamma^t_{k\bar j}) + \Gamma^r_{k\bar j}\Gamma^{*t}_{\bar i r} + \Gamma^{\bar r}_{k\bar j}\Gamma^t_{\bar i\bar r} - \delta_k(\Gamma^t_{\bar i\bar j}) - \Gamma^r_{\bar i\bar j}\Gamma^t_{kr} - \Gamma^{\bar r}_{\bar i\bar j}\Gamma^{*t}_{k\bar r} + (\overset{g}{\Gamma}{}^r_{ki} - \overset{g}{T}{}^r_{ki})\Gamma^{*t}_{\bar r\bar j})\}\delta_t \\
&+ \frac{1}{2}\{(\Gamma^{*r}_{k\bar j}\Gamma^{\bar t}_{\bar i r} + \Gamma^{*\bar r}_{k\bar j}\Gamma^{\bar t}_{\bar i\bar r} + \partial_{\bar i}(\Gamma^{*\bar t}_{k\bar j}) - \Gamma^{*r}_{\bar i\bar j}\Gamma^{\bar t}_{kr} - \Gamma^{*\bar r}_{\bar i\bar j}\Gamma^{\bar t}_{k\bar r} - \delta_k(\Gamma^{*\bar t}_{\bar i\bar j}) + (\overset{g}{\Gamma}{}^r_{ki} - \overset{g}{T}{}^r_{ki})\Gamma^{\bar t}_{\bar r\bar j} \\
&+ \Gamma^r_{k\bar j}\Gamma^{*\bar t}_{\bar i r} + \partial_{\bar i}(\Gamma^{\bar t}_{k\bar j}) + \Gamma^{\bar r}_{k\bar j}\Gamma^{\bar t}_{\bar i\bar r} - \Gamma^r_{\bar i\bar j}\Gamma^{\bar t}_{kr} - \delta_k(\Gamma^{\bar t}_{\bar i\bar j}) - \Gamma^{\bar r}_{\bar i\bar j}\Gamma^{*\bar t}_{k\bar r} + (\overset{g}{\Gamma}{}^r_{ki} - \overset{g}{T}{}^r_{ki})\Gamma^{*\bar t}_{\bar r\bar j})\}\partial_{\bar t}.
\end{aligned}
$$

Now, we consider the pair $(\overset{H}{\nabla},\overset{C}{\nabla})$ of connections on the TM and study the mutual curvature $\widetilde{R}_{\overset{H}{\nabla},\overset{C}{\nabla}}$ on the TM and its relation with the curvature of M. Equations (34) and (4) imply the following:

$$\widetilde{R}_{\overset{H}{\nabla},\overset{C}{\nabla}}(\partial_{\bar i},\partial_{\bar j})\partial_{\bar k} = \widetilde{R}_{\overset{H}{\nabla},\overset{C}{\nabla}}(\delta_i,\partial_{\bar k})\delta_j = \widetilde{R}_{\overset{H}{\nabla},\overset{C}{\nabla}}(\partial_{\bar i},\partial_{\bar j})\delta_k = \widetilde{R}_{\overset{H}{\nabla},\overset{C}{\nabla}}(\partial_{\bar i},\delta_k)\partial_{\bar j} = 0, \tag{63}$$

$$
\begin{aligned}
\widetilde{R}_{\overset{H}{\nabla},\overset{C}{\nabla}}(\delta_i,\delta_j)\delta_k = {} & (R(\partial_i,\partial_j)\partial_k)^H + \frac{1}{2}(\nabla_{\partial_i}R(y,\partial_j)\partial_k - \nabla_{\partial_j}R(y,\partial_i)\partial_k)^V \\
&+ \frac{1}{2}(R(y,\partial_i)\nabla_{\partial_j}\partial_k - R(y,\partial_j)\nabla_{\partial_i}\partial_k)^V,
\end{aligned} \tag{64}
$$

$$\widetilde{R}_{\overset{H}{\nabla},\overset{C}{\nabla}}(\delta_i,\delta_j)\partial_{\bar k} = (R(\partial_i,\partial_j)\partial_k)^V. \tag{65}$$

If $\widetilde{R}_{\overset{H}{\nabla},\overset{C}{\nabla}}$ vanishes, then (64) and (65) imply that M is a flat space and $\nabla_{\partial_i}R(y,\partial_j)\partial_k = \nabla_{\partial_j}R(y,\partial_i)\partial_k$, for all i,j,k. Conversely, if M is flat, then $\widetilde{R}_{\overset{H}{\nabla},\overset{C}{\nabla}}$ vanishes.

Here, we investigate the mutual curvature $\widetilde{R}_{\overset{C}{\nabla^1},\overset{C}{\nabla^2}}$ on the TM and its relation with the mutual curvature R_{∇^1,∇^2}, where (∇^1,∇^2) is a pair of connections on M such that R_1 and R_2 are curvature tensors of ∇^1 and ∇^2, respectively. From (34) and (4) we conclude the following:

$$\widetilde{R}_{\overset{C}{\nabla^1},\overset{C}{\nabla^2}}(\partial_{\bar i},\partial_{\bar j})\partial_{\bar k} = \widetilde{R}_{\overset{C}{\nabla^1},\overset{C}{\nabla^2}}(\delta_i,\partial_{\bar k})\delta_j = \widetilde{R}_{\overset{C}{\nabla^1},\overset{C}{\nabla^2}}(\partial_{\bar i},\partial_{\bar j})\delta_k = \widetilde{R}_{\overset{C}{\nabla^1},\overset{C}{\nabla^2}}(\partial_{\bar i},\delta_k)\partial_{\bar j} = 0, \tag{66}$$

$$
\begin{aligned}
&\widetilde{R}_{\overset{C}{\nabla^1},\overset{C}{\nabla^2}}(\delta_i,\delta_j)\delta_k \\
&= (R_{\nabla^1,\nabla^2}(\partial_i,\partial_j)\partial_k)^H + \frac{1}{2}(R_1(y,\partial_i)\nabla^2_{\partial_j}\partial_k - R_1(y,\partial_j)\nabla^2_{\partial_i}\partial_k)^V \\
&+ \frac{1}{2}(\nabla^1_{\partial_i}R_2(y,\partial_j)\partial_k - \nabla^1_{\partial_j}R_2(y,\partial_i)\partial_k)^V + \frac{1}{2}(R_2(y,\partial_i)\nabla^1_{\partial_j}\partial_k - R_2(y,\partial_j)\nabla^1_{\partial_i}\partial_k)^V \\
&+ \frac{1}{2}(\nabla^2_{\partial_i}R_1(y,\partial_j)\partial_k - \nabla^2_{\partial_j}R_1(y,\partial_i)\partial_k)^V,
\end{aligned} \tag{67}
$$

$$\widetilde{R}_{\overset{C}{\nabla^1},\overset{C}{\nabla^2}}(\delta_i,\delta_j)\partial_{\bar k} = (R_{\nabla^1,\nabla^2}(\partial_i,\partial_j)\partial_k)^V. \tag{68}$$

According to (66)–(68), we derive that if the mutual curvature R_{∇^1,∇^2} vanishes and M is a flat space with respect to ∇^1 and ∇^2, then the mutual curvature $\widetilde{R}_{\overset{C}{\nabla^1},\overset{C}{\nabla^2}}$ vanishes. Conversely, if the mutual curvature $\widetilde{R}_{\overset{C}{\nabla^1},\overset{C}{\nabla^2}}$ vanishes, then from (68) we see that the mutual curvature R_{∇^1,∇^2} vanishes. Furthermore, if we consider the pair $\{\overset{H}{\nabla^1},\overset{H}{\nabla^2}\}$ of connections on the TM, then, the same as in the above discussion, we obtain that all of the components of the mutual curvature $\widetilde{R}_{\overset{H}{\nabla^1},\overset{H}{\nabla^2}}$ are zero except for

$$\widetilde{R}_{\overset{H}{\nabla^1},\overset{H}{\nabla^2}}(\delta_i,\delta_j)\delta_k = (R_{\nabla^1,\nabla^2}(\partial_i,\partial_j)\partial_k)^H, \qquad \widetilde{R}_{\overset{H}{\nabla^1},\overset{H}{\nabla^2}}(\delta_i,\delta_j)\partial_{\bar{k}} = (R_{\nabla^1,\nabla^2}(\partial_i,\partial_j)\partial_k)^V. \quad (69)$$

Therefore, from (69) we deduce that $\widetilde{R}_{\overset{H}{\nabla^1},\overset{H}{\nabla^2}}$ vanishes if and only if R_{∇^1,∇^2} vanishes. As mentioned in [14], if ∇ is a flat and torsion-free connection, then $\overset{C}{\nabla} = \overset{H}{\nabla}$. It follows that, if M is a flat space and ∇ is the Levi-Civita connection, then the mutual curvature $\widetilde{R}_{\overset{H}{\nabla},\overset{C}{\nabla}}$ reduces to the Riemannian curvature of the Levi-Civita connection ∇^S, because $\overset{C}{\nabla}$ and $\overset{H}{\nabla}$ reduce to the ∇^S. Now, we consider a pair of Levi-Civita connections (∇^1,∇^2) such that M is a flat space with respect to the ∇^i for $i = 1,2$. Since $\overset{C}{\nabla^i} = \overset{H}{\nabla^i}$ for $i = 1,2$, thus, the mutual curvatures $\widetilde{R}_{\overset{C}{\nabla^1},\overset{C}{\nabla^2}}$ and $\widetilde{R}_{\overset{H}{\nabla^1},\overset{H}{\nabla^2}}$ coincide and they reduce to the mutual curvature $\widetilde{R}_{\overset{1}{\nabla^S},\overset{2}{\nabla^S}}$. So, according to above discussion we have the following:

Lemma 4. *Let (M,g) be a Riemannian manifold with an affine connection ∇ and curvature tensor R and let (∇^1,∇^2) be a pair of connections on M such that R_1 and R_2 are curvature tensors of ∇^1 and ∇^2, respectively. Then, the following statements hold:*

(1) *If M is a flat space, then the mutual curvature $\widetilde{R}_{\overset{H}{\nabla},\overset{C}{\nabla}}$ is zero.*

(2) *If M is a flat space and ∇ is the Levi-Civita connection, then the mutual curvature $\widetilde{R}_{\overset{H}{\nabla},\overset{C}{\nabla}}$ reduces to the Riemannian curvature of the Levi-Civita connection ∇^S.*

(3) *If the mutual curvature $\widetilde{R}_{\overset{H}{\nabla},\overset{C}{\nabla}}$ vanishes, then M is flat.*

(4) *If the mutual curvature R_{∇^1,∇^2} vanishes and M is a flat space with respect to ∇^1 and ∇^2, then the mutual curvature $\widetilde{R}_{\overset{C}{\nabla^1},\overset{C}{\nabla^2}}$ vanishes.*

(5) *If the mutual curvature $\widetilde{R}_{\overset{C}{\nabla^1},\overset{C}{\nabla^2}}$ vanishes, then the mutual curvature R_{∇^1,∇^2} vanishes. Moreover, the mutual curvature $\widetilde{R}_{\overset{H}{\nabla^1},\overset{H}{\nabla^2}}$ vanishes if and only if the mutual curvature R_{∇^1,∇^2} vanishes.*

(6) *If ∇^1 and ∇^2 are Levi-Civita connections and M is a flat space with respect to ∇^i for $i = 1,2$, then $\widetilde{R}_{\overset{C}{\nabla^1},\overset{C}{\nabla^2}}$ and $\widetilde{R}_{\overset{H}{\nabla^1},\overset{H}{\nabla^2}}$ are equal and reduce to the mutual curvature $\widetilde{R}_{\overset{1}{\nabla^S},\overset{2}{\nabla^S}}$.*

6.1. Mutual Curvature with Respect to a Pair of Levi-Civita Connections in the Tangent Bundle

Let M be a smooth manifold and g_1 and g_2 be two non-isometric Riemannian metrics with the Levi-Civita connections ∇^1 and ∇^2 and the Riemannian curvature tensors R_1 and R_2, respectively. We consider the pair $(\overset{1}{\nabla^S},\overset{2}{\nabla^S})$ on the TM such that $\overset{1}{\nabla^S}$ and $\overset{2}{\nabla^S}$ are the Levi-Civita connections of the Sasaki lift metrics g_1^S and g_2^S, respectively. Now, we study on the components of the mutual curvature $\widetilde{R}_{\overset{1}{\nabla^S},\overset{2}{\nabla^S}}$ on the TM and its relation with the mutual curvature R_{∇^1,∇^2} on M. Equations (34), (5), (2), and (3) give us the following:

$$
\begin{aligned}
\widetilde{R}_{\overset{1}{\nabla}^S,\overset{2}{\nabla}^S}(\delta_i,\delta_j)\delta_k &= (R_{\nabla^1,\nabla^2}(\partial_i,\partial_j)\partial_k)^H \\
&+ \{\frac{1}{4}\overset{1}{\nabla}_{\partial_j}R_2(\partial_i,\partial_k)y + \frac{1}{4}R_1(\partial_j,\overset{2}{\nabla}_{\partial_i}\partial_k)y - \frac{1}{4}\overset{1}{\nabla}_{\partial_i}R_2(\partial_j,\partial_k)y - \frac{1}{4}R_1(\partial_i,\overset{2}{\nabla}_{\partial_j}\partial_k)y\}^V \\
&+ \{\frac{1}{4}R_1(y,R_1(\partial_i,\partial_j)y)\partial_k + \frac{1}{8}R_1(y,R_2(\partial_i,\partial_k)y)\partial_j - \frac{1}{8}R_1(y,R_2(\partial_j,\partial_k)y)\partial_i\}^H \\
&+ \{\frac{1}{4}\overset{2}{\nabla}_{\partial_j}R_1(\partial_i,\partial_k)y + \frac{1}{4}R_2(\partial_j,\overset{1}{\nabla}_{\partial_i}\partial_k)y - \frac{1}{4}\overset{2}{\nabla}_{\partial_i}R_1(\partial_j,\partial_k)y - \frac{1}{4}R_2(\partial_i,\overset{1}{\nabla}_{\partial_j}\partial_k)y\}^V \\
&+ \{\frac{1}{4}R_2(y,R_2(\partial_i,\partial_j)y)\partial_k + \frac{1}{8}R_2(y,R_1(\partial_i,\partial_k)y)\partial_j - \frac{1}{8}R_2(y,R_1(\partial_j,\partial_k)y)\partial_i\}^H.
\end{aligned} \tag{70}
$$

If we consider $y = 0$ in (70), then, except the first term, all of the terms on the right-hand side of the equation are zero. This implies that if $\widetilde{R}_{\overset{1}{\nabla}^S,\overset{2}{\nabla}^S}$ vanishes, then R_{∇^1,∇^2} vanishes. To continue, we also have

$$
\begin{aligned}
\widetilde{R}_{\overset{1}{\nabla}^S,\overset{2}{\nabla}^S}(\delta_i,\delta_j)\partial_{\bar{k}} &= (R_{\nabla^1,\nabla^2}(\partial_i,\partial_j)\partial_k)^V \\
&+ \{\frac{1}{4}R_1(y,\overset{2}{\nabla}_{\partial_j}\partial_k)\partial_i + \frac{1}{4}\overset{1}{\nabla}_{\partial_i}R_2(y,\partial_k)\partial_j - \frac{1}{4}R_1(y,\overset{2}{\nabla}_{\partial_i}\partial_k)\partial_j - \frac{1}{4}\overset{1}{\nabla}_{\partial_j}R_2(y,\partial_k)\partial_i\}^H \\
&- \{\frac{1}{8}R_1(\partial_j,R_2(y,\partial_k)\partial_i)y + \frac{1}{8}R_1(\partial_i,R_2(y,\partial_k)\partial_j)y\}^V, \\
&+ \{\frac{1}{4}R_2(y,\overset{1}{\nabla}_{\partial_j}\partial_k)\partial_i + \frac{1}{4}\overset{2}{\nabla}_{\partial_i}R_1(y,\partial_k)\partial_j - \frac{1}{4}R_2(y,\overset{1}{\nabla}_{\partial_i}\partial_k)\partial_j - \frac{1}{4}\overset{2}{\nabla}_{\partial_j}R_1(y,\partial_k)\partial_i\}^H \\
&- \{\frac{1}{8}R_2(\partial_j,R_1(y,\partial_k)\partial_i)y + \frac{1}{8}R_2(\partial_i,R_1(y,\partial_k)\partial_j)y\}^V.
\end{aligned} \tag{71}
$$

Furthermore, we have

$$
\begin{aligned}
\widetilde{R}_{\overset{1}{\nabla}^S,\overset{2}{\nabla}^S}(\delta_i,\partial_{\bar{k}})\delta_j &= \frac{1}{4}\{\overset{1}{\nabla}_{\partial_i}R_2(y,\partial_k)\partial_j - R_1(y,\partial_k)\overset{2}{\nabla}_{\partial_i}\partial_j - R_1(y,\overset{1}{\nabla}_{\partial_i}\partial_k)\partial_j\}^H \\
&- \frac{1}{8}(R_1(\partial_i,R_2(y,\partial_k)\partial_j)y)^V, \\
&+ \frac{1}{4}\{\overset{2}{\nabla}_{\partial_i}R_1(y,\partial_k)\partial_j - R_2(y,\partial_k)\overset{1}{\nabla}_{\partial_i}\partial_j - R_2(y,\overset{1}{\nabla}_{\partial_i}\partial_k)\partial_j\}^H \\
&- \frac{1}{8}(R_2(\partial_i,R_1(y,\partial_k)\partial_j)y)^V,
\end{aligned} \tag{72}
$$

$$
\begin{aligned}
\widetilde{R}_{\overset{1}{\nabla}^S,\overset{2}{\nabla}^S}(\partial_{\bar{i}},\partial_{\bar{j}})\delta_k &= \frac{1}{8}\{R_1(y,\partial_i)R_2(y,\partial_j)\partial_k - R_1(y,\partial_j)R_2(y,\partial_i)\partial_k\}^H \\
&+ \frac{1}{8}\{R_2(y,\partial_i)R_1(y,\partial_j)\partial_k - R_2(y,\partial_j)R_1(y,\partial_i)\partial_k\}^H,
\end{aligned} \tag{73}
$$

$$
\widetilde{R}_{\overset{1}{\nabla}^S,\overset{2}{\nabla}^S}(\partial_{\bar{i}},\delta_k)\partial_{\bar{j}} = \frac{1}{8}\{R_1(y,\partial_i)R_2(y,\partial_j)\partial_k + R_2(y,\partial_i)R_1(y,\partial_j)\partial_k\}^H, \tag{74}
$$

$$
\widetilde{R}_{\overset{1}{\nabla}^S,\overset{2}{\nabla}^S}(\partial_{\bar{i}},\partial_{\bar{j}})\partial_{\bar{k}} = 0.
$$

Using (70)–(74), we obtain the following:

Theorem 4. *Let M be a smooth manifold and g_1 and g_2 be two non-isometric Riemannian metrics with the Levi-Civita connections ∇^1 and ∇^2 and the Riemannian curvature tensors R_1 and R_2 such that $\overset{1}{\nabla}^S$ and $\overset{2}{\nabla}^S$ are the Levi-Civita connections of the Sasaki lift metrics g_1^S and g_2^S, respectively. Then, the following statements hold:*

(1) The mutual curvature $\widetilde{R}_{\overset{1}{\nabla}^S,\overset{2}{\nabla}^S}$ vanishes if the mutual curvature R_{∇^1,∇^2} vanishes and M is a flat space with respect to ∇^1 and ∇^2.

(2) *If the mutual curvature* $\widetilde{R}_{\overset{1}{\nabla}{}^S,\overset{2}{\nabla}{}^S}$ *vanishes, then the mutual curvature* R_{∇^1,∇^2} *vanishes.*

6.2. Mutual Curvatures in the Tangent Bundle with Different Connections

In this part, we study the mutual curvatures $\widetilde{R}_{\overset{1}{\nabla}{}^S,\overset{2}{\nabla}{}^H}$ and $\widetilde{R}_{\overset{1}{\nabla}{}^S,\overset{2}{\nabla}{}^C}$ on the TM and their geometric consequences, where ∇^1 is the Levi-Civita connection of the metric g.

Now, we consider Riemannian manifold (M,g) with the Levi-Civita connection $\overset{1}{\nabla}$ and Riemannian curvature R_1. Let $\overset{2}{\nabla}$ be an arbitrary affine connection on M. If we denote the Levi-Civita connection of the Sasaki lift metric g^S by $\overset{1}{\nabla}{}^S$ and the horizontal lift connection of $\overset{2}{\nabla}$ by $\overset{2}{\nabla}{}^H$, then from (34) and (4), we deduce that the components of the mutual curvature $\widetilde{R}_{\overset{1}{\nabla}{}^S,\overset{2}{\nabla}{}^H}$ are as follows:

$$\begin{aligned}\widetilde{R}_{\overset{1}{\nabla}{}^S,\overset{2}{\nabla}{}^H}(\delta_i,\delta_j)\delta_k &= (R_{\overset{1}{\nabla},\overset{2}{\nabla}}(\partial_i,\partial_j)\partial_k)^H + \frac{1}{4}\{R_1(\partial_j,\overset{2}{\nabla}_{\partial_i}\partial_k)y - R_1(\partial_i,\overset{2}{\nabla}_{\partial_j}\partial_k)y\}^V \\ &+ \frac{1}{4}\{R_1(y,R_1(\partial_i,\partial_j)y)\partial_k\}^H + \frac{1}{4}\{\overset{2}{\nabla}_{\partial_j}R_1(\partial_i,\partial_k)y - \overset{2}{\nabla}_{\partial_i}R_1(\partial_j,\partial_k)y\}^V.\end{aligned} \tag{75}$$

Setting $y = 0$ in (75), the second, third, and fourth terms on the right-hand side of Equation (75) are zero. In this case, if we assume that the mutual curvature $\widetilde{R}_{\overset{1}{\nabla}{}^S,\overset{2}{\nabla}{}^H}$ vanishes, then the mutual curvature $R_{\overset{1}{\nabla},\overset{2}{\nabla}}$ vanishes. We have

$$\begin{aligned}\widetilde{R}_{\overset{1}{\nabla}{}^S,\overset{2}{\nabla}{}^H}(\delta_i,\delta_j)\partial_{\bar{k}} &= (R_{\overset{1}{\nabla},\overset{2}{\nabla}}(\partial_i,\partial_j)\partial_k)^V + \frac{1}{4}\{R_1(y,\overset{2}{\nabla}_{\partial_j}\partial_k)\partial_i - R_1(y,\overset{2}{\nabla}_{\partial_i}\partial_k)\partial_j\}^H \\ &+ \frac{1}{4}\{\overset{2}{\nabla}_{\partial_i}R_1(y,\partial_k)\partial_j - \overset{2}{\nabla}_{\partial_j}R_1(y,\partial_k)\partial_i\}^H,\end{aligned} \tag{76}$$

$$\widetilde{R}_{\overset{1}{\nabla}{}^S,\overset{2}{\nabla}{}^H}(\delta_i,\partial_{\bar{k}})\delta_j = \frac{1}{4}\{\overset{2}{\nabla}_{\partial_i}R_1(y,\partial_k)\partial_j - R_1(y,\partial_k)\overset{2}{\nabla}_{\partial_i}\partial_j - R_1(y,\overset{1}{\nabla}_{\partial_i}\partial_k)\partial_j\}^H, \tag{77}$$

$$\widetilde{R}_{\overset{1}{\nabla}{}^S,\overset{2}{\nabla}{}^H}(\partial_{\bar{i}},\partial_{\bar{j}})\partial_{\bar{k}} = \widetilde{R}_{\overset{1}{\nabla}{}^S,\overset{2}{\nabla}{}^H}(\partial_{\bar{i}},\partial_{\bar{j}})\delta_k = \widetilde{R}_{\overset{1}{\nabla}{}^S,\overset{2}{\nabla}{}^H}(\partial_{\bar{i}},\delta_k)\partial_{\bar{j}} = 0. \tag{78}$$

According to (75)–(78) we derive that if $R_{\overset{1}{\nabla},\overset{2}{\nabla}}$ vanishes and M is a flat space with respect to $\overset{1}{\nabla}$, then $\widetilde{R}_{\overset{1}{\nabla}{}^S,\overset{2}{\nabla}{}^H}$ vanishes. As a special case, if $\overset{1}{\nabla} = \overset{2}{\nabla} = \nabla$, where ∇ is the Levi-Civita connection of g, then from (75)–(78) we derive that the mutual curvature $\widetilde{R}_{\nabla^S,\nabla^H}$ vanishes if M is a flat space with respect to ∇. Moreover, in this case, $\widetilde{R}_{\nabla^S,\nabla^H}$ reduces to the Riemannian curvature of the Levi-Civita connection ∇^S. Furthermore, if $\widetilde{R}_{\nabla^S,\nabla^H}$ vanishes, then M is a flat space. So, as a consequence of the above discussion we have the following:

Theorem 5. *Let (M,g) be a Riemannian manifold with the Levi-Civita connection $\overset{1}{\nabla}$ and let $\overset{2}{\nabla}$ be an affine connection on M. If we denote the Levi-Civita connection of the Sasaki lift metric g^S by $\overset{1}{\nabla}{}^S$ and the horizontal lift connection of $\overset{2}{\nabla}$ by $\overset{2}{\nabla}{}^H$, then the following assertions hold:*

(1) *If the mutual curvature* $\widetilde{R}_{\overset{1}{\nabla}{}^S,\overset{2}{\nabla}{}^H}$ *vanishes, then the mutual curvature* $R_{\overset{1}{\nabla},\overset{2}{\nabla}}$ *vanishes. Moreover, if the mutual curvature* $R_{\overset{1}{\nabla},\overset{2}{\nabla}}$ *vanishes and M is a flat space with respect to* $\overset{1}{\nabla}$*, then the mutual curvature* $\widetilde{R}_{\overset{1}{\nabla}{}^S,\overset{2}{\nabla}{}^H}$ *vanishes.*

(2) *The mutual curvature $\widetilde{R}_{\overset{S}{\nabla},\overset{H}{\nabla}}$ vanishes if and only if M is a flat space with respect to ∇, where ∇ is the Levi-Civita connection of metric g. Furthermore, in this case, the mutual curvature $\widetilde{R}_{\overset{S}{\nabla},\overset{H}{\nabla}}$ reduces to the Riemannian curvature of the Levi-Civita connection ∇^S.*

Now, we consider a Riemannian manifold (M,g) with the Levi-Civita connection $\overset{1}{\nabla}$. Let $\overset{2}{\nabla}$ be an affine connection on M, where R_1 and R_2 are curvature tensors of $\overset{1}{\nabla}$ and $\overset{2}{\nabla}$, respectively. If we denote the Levi-Civita connection of the Sasaki lift metric g^S by $\overset{1}{\nabla}{}^S$ and the complete lift connection of $\overset{2}{\nabla}$ by $\overset{2}{\nabla}{}^C$, then from (34) and (4) we conclude that the components of the mutual curvature $\widetilde{R}_{\overset{1}{\nabla}{}^S,\overset{2}{\nabla}{}^C}$ are as follows:

$$\begin{aligned}\widetilde{R}_{\overset{1}{\nabla}{}^S,\overset{2}{\nabla}{}^C}(\delta_i,\delta_j)\delta_k &= (R_{\overset{1}{\nabla},\overset{2}{\nabla}}(\partial_i,\partial_j)\partial_k)^H \\ &+\frac{1}{4}\{R_1(y,R_1(\partial_i,\partial_j)y)\partial_k+R_1(y,R_2(y,\partial_j)\partial_k)\partial_i-R_1(y,R_2(y,\partial_i)\partial_k)\partial_j\}^H \\ &+\{\frac{1}{4}R_1(\partial_j,\overset{2}{\nabla}_{\partial_i}\partial_k)y-\frac{1}{4}R_1(\partial_i,\overset{2}{\nabla}_{\partial_j}\partial_k)y+\frac{1}{2}\overset{1}{\nabla}_{\partial_i}R_2(y,\partial_j)\partial_k-\frac{1}{2}\overset{1}{\nabla}_{\partial_j}R_2(y,\partial_i)\partial_k\}^V \\ &+\{\frac{1}{2}R_2(y,\partial_i)\overset{1}{\nabla}_{\partial_j}\partial_k-\frac{1}{2}R_2(y,\partial_j)\overset{1}{\nabla}_{\partial_i}\partial_k+\frac{1}{4}\overset{2}{\nabla}_{\partial_j}R_1(\partial_i,\partial_k)y-\frac{1}{4}\overset{2}{\nabla}_{\partial_i}R_1(\partial_j,\partial_k)y\}^V.\end{aligned} \tag{79}$$

If we put $y=0$ in (79), then we derive that the second, third and fourth terms on the right-hand side of Equation (79) are zero. Thus, if $\widetilde{R}_{\overset{1}{\nabla}{}^S,\overset{2}{\nabla}{}^C}=0$, then from (79) we obtain $R_{\overset{1}{\nabla},\overset{2}{\nabla}}=0$. Furthermore, we have

$$\begin{aligned}\widetilde{R}_{\overset{1}{\nabla}{}^S,\overset{2}{\nabla}{}^C}(\delta_i,\delta_j)\partial_{\bar{k}} &= (R_{\overset{1}{\nabla},\overset{2}{\nabla}}(\partial_i,\partial_j)\partial_k)^V \\ &+\frac{1}{4}\{R_2(y,\partial_i)R_1(y,\partial_k)\partial_j-R_2(y,\partial_j)R_1(y,\partial_k)\partial_i\}^V \\ &+\frac{1}{4}\{R_1(y,\overset{2}{\nabla}_{\partial_j}\partial_k)\partial_i-R_1(y,\overset{2}{\nabla}_{\partial_i}\partial_k)\partial_j+\overset{2}{\nabla}_{\partial_i}R_1(y,\partial_k)\partial_j-\overset{2}{\nabla}_{\partial_j}R_1(y,\partial_k)\partial_i\}^H,\end{aligned} \tag{80}$$

$$\begin{aligned}\widetilde{R}_{\overset{1}{\nabla}{}^S,\overset{2}{\nabla}{}^C}(\delta_i,\partial_{\bar{k}})\delta_j &= \frac{1}{4}\{\overset{2}{\nabla}_{\partial_i}R_1(y,\partial_k)\partial_j-R_1(y,\partial_k)\overset{2}{\nabla}_{\partial_i}\partial_j-R_1(y,\overset{1}{\nabla}_{\partial_i}\partial_k)\partial_j\}^H \\ &+\frac{1}{4}\{R_2(y,\partial_i)R_1(y,\partial_k)\partial_j\}^V,\end{aligned} \tag{81}$$

$$\widetilde{R}_{\overset{1}{\nabla}{}^S,\overset{2}{\nabla}{}^C}(\partial_{\bar{i}},\partial_{\bar{j}})\delta_k=\widetilde{R}_{\overset{1}{\nabla}{}^S,\overset{2}{\nabla}{}^C}(\partial_{\bar{i}},\delta_k)\partial_{\bar{j}}=\widetilde{R}_{\overset{1}{\nabla}{}^S,\overset{2}{\nabla}{}^C}(\partial_{\bar{i}},\partial_{\bar{j}})\partial_{\bar{k}}=0. \tag{82}$$

From (79)–(82), we deduce that if $R_{\overset{1}{\nabla},\overset{2}{\nabla}}=0$ and M is a flat space with respect to $\overset{1}{\nabla}$ and $\overset{2}{\nabla}$, then $\widetilde{R}_{\overset{1}{\nabla}{}^S,\overset{2}{\nabla}{}^C}=0$. In the special case where $\overset{1}{\nabla}=\overset{2}{\nabla}=\nabla$, where ∇ is the Levi-Civita connection of metric g, then from (79)–(82) we obtain that if the mutual curvature $\widetilde{R}_{\nabla^S,\nabla^C}$ vanishes, then M is a flat space. Moreover, if M is a flat space, then $\widetilde{R}_{\nabla^S,\nabla^C}$ vanishes and in this case, $\widetilde{R}_{\nabla^S,\nabla^C}$ reduces to the Riemannian curvature of the Levi-Civita connection ∇^S. Thus, as a result of the above discussion we derive the gollowing:

Theorem 6. *Let (M,g) be a Riemannian manifold with the Levi-Civita connection $\overset{1}{\nabla}$ and let $\overset{2}{\nabla}$ be an affine connection on M with the curvature tensors R_1 and R_2, respectively. If $\overset{1}{\nabla}{}^S$ is the*

Levi-Civita connection of the Sasaki lift metric g^S *and* $\overset{2}{\nabla}{}^C$ *is the complete lift connection of* $\overset{2}{\nabla}$*, then the following assertions hold:*

(1) If the mutual curvature $\widetilde{R}_{\overset{1}{\nabla}{}^S,\overset{2}{\nabla}{}^C}$ *vanishes, then the mutual curvature* $R_{\overset{1}{\nabla},\overset{2}{\nabla}}$ *vanishes. Moreover, if the mutual curvature* $R_{\overset{1}{\nabla},\overset{2}{\nabla}}$ *vanishes and M is a flat space with respect to* $\overset{1}{\nabla}$ *and* $\overset{2}{\nabla}$*, then the mutual curvature* $\widetilde{R}_{\overset{1}{\nabla}{}^S,\overset{2}{\nabla}{}^C}$ *vanishes.*

(2) The mutual curvature $\widetilde{R}_{\nabla^S,\nabla^C}$ *vanishes if and only if M is a flat space with respect to* ∇*, where* ∇ *is the Levi-Civita connection of metric g. Moreover, in this case the mutual curvature* $\widetilde{R}_{\nabla^S,\nabla^C}$ *reduces to the Riemannian curvature of the Levi-Civita connection* ∇^S.

Author Contributions: Conceptualization, E.P. and D.S.; methodology, E.P.; software, D.S.; validation, E.P. and I.M.; formal analysis, I.M.; investigation, E.P. and I.M.; writing—original draft preparation, E.P.; writing—review and editing, E.P.; D.S. and I.M.; visualization, D.S..; supervision, I.M.; project administration, E.P. All authors have read and agreed to the published version of the manuscript.

Funding: This research received no external funding.

Data Availability Statement: Not applicable.

Conflicts of Interest: The authors declare no conflict of interest.

References

1. Li, T.; Krupka, D. The geometry of tangent bundles: Canonical vector fields. *Geometry* **2013**, *2013*, 364301. [CrossRef]
2. Abbasi, M.T.K.; Sarih, M. On some hereditary properties of Riemannian g-natural metrics on tangent bundles of Riemannian manifolds. *Diff. Geom. Appl.* **2005**, *22*, 19–47. [CrossRef]
3. Davies, E.T. On the curvature of Tangent Bundles. *Ann. Mat.* **1969**, *81*, 193–204. [CrossRef]
4. Dombrowski, P. On the geometry of tangent bundle. *J. Die Reine Angew. Math.* **1962**, *210*, 73–88. [CrossRef]
5. Gezer, A.; Ozkan, M. Notes on tangent bundle with deformed complete lift metric. *Turk. J. Math.* **2014**, *38*, 1038–1049. [CrossRef]
6. Ledger, A.J.; Yano, K. The tangent bundle of locally symmetric space. *J. Lond. Math. Soc.* **1965**, *40*, 487–492. [CrossRef]
7. Oproiu, V. Some new geometric structures on the tangent bundles. *Publ. Math. Debrecen* **1999**, *55*, 261–281. [CrossRef]
8. Sasaki, S. On the geometry of tangent bundles of Riemannian manifolds. *Tohoku Math. J.* **1958**, *10*, 238–354. [CrossRef]
9. Amari, S. *Information Geometry and Its Applications*; Springer: Tokyo, Japan, 2016.
10. Amari, S.; Nagaoka, H. *Method of Information Geometry*; American Mathematical Society: Providence, RI, USA, 2000.
11. Lauritzen, S. Statistical manifolds. In *Differential Geometry in Statistical Inference*; IMS Lecture Notes Monograph Series 10; Institute of Mathematical Statistics: Hyward, CA, USA, 1987; pp. 96–163.
12. Matsuzoe, H. Statistical manifolds and affine differential geometry. *Adv. Stud. Pure Math.* **2010**, *57*, 303–321.
13. Iosifidis, D. On a Torsion/Curvature Analogue of Dual Connections and Statistical Manifolds. *arXiv* **2023**, arXiv:2303.13259v1.
14. Yano, K.; Ishihara, S. *Tangent and Cotangent Bundles*; Marcel Dekker Inc.: New York, NY, USA, 1973.
15. Uohashi, K. α-connections and a symmetric cubic form on a Riemannian manifold. *Entropy* **2017**, *19*, 344. [CrossRef]
16. Călin, O.; Udrişte, C. *Geometric Modeling in Probability and Statistics*; Springer: Berlin/Heidelberg, Germany, 2014; Volume 121.
17. Dutta, S.; Genton, M. A non-Gaussian multivariate distribution with all lower-dimensional Gaussians and related families. *J. Multivar. Anal.* **2014**, *132*, 82–93. [CrossRef]
18. Yuan, M. On the geometric structure of some statistical manifolds. *Balkan J. Geom. Appl.* **2019**, *24*, 79–89.
19. Puechmorel, S. Lifting dual connections with the Riemann extension. *Mathematics* **2020**, 8, 2079. [CrossRef]

axioms MDPI

Article

Finsler Warped Product Metrics with Special Curvature Properties

Lingen Sun, Xiaoling Zhang * and Mengke Wu

College of Mathematics and Systems Science, Xinjiang University, Urumqi 830017, China; 107552100610@stu.xju.edu.cn (L.S.); 107552100637@stu.xju.edu.cn (M.W.)
* Correspondence: zhangxiaoling@xju.edu.cn

Abstract: The class of warped product metrics can often be interpreted as key space models for the general theory of relativity and theory of space–time. In this paper, we study several non-Riemannian quantities in Finsler geometry. These non-Riemannian quantities play an important role in understanding the geometric properties of Finsler metrics. In particular, we find differential equations of Finsler warped product metrics with vanishing χ-curvature or vanishing H-curvature. Furthermore, we show that, for Finsler warped product metrics, the χ-curvature vanishes if and only if the H-curvature vanishes.

Keywords: Finsler warped product metrics; χ-curvature; H-curvature

MSC: 53C30; 53C60

1. Introduction

There are several non-Riemannian quantities in Finsler geometry, such as the distortion, the (mean) Cartan torsion, the S-curvature, the (mean) Berwald curvature and the (mean) Landsberg curvature. We view the distortion and the (mean) Cartan torsion as non-Riemannian quantities of order zero, and the S-curvature, the (mean) Berwald curvature and the (mean) Landsberg curvature as non-Riemannian quantities of order one. Differentiating these quantities along geodesics, we obtain some non-Riemannian quantities of order two.

In this paper, we will consider two non-Riemannian quantities $\chi = \chi_i dx^i$ and $H = H_{ij} dx^i \otimes dx^j$ on the tangent bundle TM:

$$\chi_i := S_{.i|m} y^m - S_{|i},$$
$$H_{ij} := \frac{1}{2} S_{.i.j|m} y^m,$$

where S denotes the S-curvature of F and "|" and "." denote the horizontal and vertical covariant derivatives with respect to the Chern connection, respectively.

Shen [1] showed some relationships among the flag curvature, the S-curvature, the χ-curvature and the H-curvature. Cheng and Yuan [2] obtained a formula of χ-curvature for (α, β)-metrics. Based on this, they showed that the χ-curvature vanishes for a class of (α, β)-metrics. Shen [3] discussed several expressions for the χ-curvature of a spray. They showed that sprays, obtained by a projective deformation using the S-curvature, always have vanishing χ-curvature. They established a Beltrami theorem for sprays with vanishing χ-curvature.

The non-Riemannian quantity H was introduced by Zadeh [4] and developed by some other Finslerian geometers [5,6]. Xia [7] obtained some rigidity theorems of a compact Finsler manifold under some conditions related to H-curvature. They proved that the S-curvature for a Randers metric is almost isotropic if and only if the H-curvature almost vanishes. In particular, S-curvature is isotropic if and only if the H-curvature vanishes.

Citation: Sun, L.; Zhang, X.; Wu, M. Finsler Warped Product Metrics with Special Curvature Properties. *Axioms* **2023**, *12*, 784. https://doi.org/10.3390/axioms12080784

Academic Editor: Mića Stanković

Received: 8 July 2023
Revised: 4 August 2023
Accepted: 10 August 2023
Published: 12 August 2023

Tang [8] showed that Randers metrics have almost isotropic S-curvature if and only if they have almost vanishing H-curvature. Furthermore, Randers metrics actually have zero S-curvature if and only if they have vanishing H-curvature. Mo [9] gave a characterization of spherically symmetric Finsler metrics with almost vanishing H-curvature. Zhu [10] showed that the χ-curvature vanishes if and only if the H-curvature vanishes for general (α,β)-metrics under some conditions. Sevim and Gabrani [11] showed that, on Finsler warped product manifolds, the χ-curvature vanishes if and only if the H-curvature vanishes.

The warped product metric was introduced by Bishop and O'Neill [12] to study Riemannian manifolds with negative curvature as a generalization of Riemannian product metrics. The notion of warped products was extended to the case of Finsler manifolds by Chen-Shen-Zhao [13] and Kozma-Peter-Varga [14], respectively. These metrics are called Finsler warped product metrics.

In [15], Shen and Marcal considered a new class of Finsler metrics using the warped product notion introduced by Chen-Shen-Zhao [13], with another "warping". This metric is consistent with static spacetime. They gave partial differential equations (PDEs) characterization for the proposed metrics to be Ricci-flat. Furthermore, they explicitly constructed two types of non-Riemannian examples.

In this paper, we obtain differential equations of such metrics with vanishing χ-curvature or vanishing H-curvature. Then, we obtain that the χ-curvature vanishes if and only if the H-curvature vanishes. The main results are as follows.

Theorem 1. *Let $F = \alpha\sqrt{\phi(z,\rho)}$ be a Finsler warped product metric on an $(n+1)$-dimensional manifold $M = \mathbb{R}\times\mathbb{R}^n$ $(n \geq 2)$, where $\alpha = |\bar{y}|$, $z = \frac{y^0}{|\bar{y}|}$, $\rho = |\bar{x}|$. Then, the χ-curvature vanishes if and only if the H-curvature vanishes.*

A Finsler metric F is said to be R-quadratic if its Riemann curvature R_v is quadratic in $v \in T_uM$ [16]. Najafi-Bidabad-Tayebi [17] and Mo [6] showed that all R-quadratic Finsler metrics have vanishing H-curvature, respectively. For a R-quadratic Finsler warped product metric, we have the following result.

Corollary 1. *Let $F = \alpha\sqrt{\phi(z,\rho)}$ be a Finsler warped product metric on an $(n+1)$-dimensional manifold $M = \mathbb{R}\times\mathbb{R}^n$ $(n \geq 2)$, where $\alpha = |\bar{y}|$, $z = \frac{y^0}{|\bar{y}|}$, $\rho = |\bar{x}|$. Suppose that F is R-quadratic, then the χ-curvature vanishes.*

2. Preliminaries

Set $M = \mathbb{R}\times\mathbb{R}^n$ with the following coordinates on TM:

$$x = \left(x^0, \bar{x}\right),\ \bar{x} = \left(x^1, \ldots, x^n\right),$$
$$y = \left(y^0, \bar{y}\right),\ \bar{y} = \left(y^1, \ldots, y^n\right).$$

Furthermore, consider a Finsler metric as follows:

$$F = \alpha\sqrt{\phi(z,\rho)},$$

where $\alpha = |\bar{y}|$, $z = \frac{y^0}{|\bar{y}|}$, $\rho = |\bar{x}|$ and ϕ is a suitable function on $\mathbb{R}^2$.

Throughout this paper, our index conventions are as follows:

$$0 \leq A, B, \ldots \leq n,$$
$$1 \leq i, j, \ldots \leq n.$$

For a Finsler warped product metric $F = \alpha\sqrt{\phi(z,\rho)}$, the fundamental form $g = g_{AB}dx^A \otimes dx^B$ is given by:

$$(g_{AB}) = \left(\begin{array}{c|c} \frac{1}{2}\phi_{zz} & \frac{1}{2}\Omega_z\frac{y^j}{\alpha} \\ \hline \frac{1}{2}\Omega_z\frac{y^i}{\alpha} & \frac{1}{2}\Omega\delta_{ij} - \frac{1}{2}z\Omega_z\frac{y^iy^j}{\alpha^2} \end{array}\right),$$

where $\Omega := 2\phi - z\phi_z$. Then:

$$\det(g_{AB}) = \frac{1}{2^{n+1}}\Omega^{n-1}\Lambda,$$

where $\Lambda := 2\phi\phi_{zz} - \phi_z^2$.

Henceforth, assume F is non-degenerate. In this case, the inverse of (g_{AB}) is:

$$\left(g^{AB}\right) = \left(\begin{array}{c|c} \frac{2}{\Lambda}(\Omega - z\Omega_z) & -\frac{2}{\Lambda}\Omega_z\frac{y^j}{\alpha} \\ \hline -\frac{2}{\Lambda}\Omega_z\frac{y^i}{\alpha} & \frac{2}{\Omega}\delta^{ij} + \frac{2\phi_z\Omega_z}{\Omega\Lambda}\frac{y^iy^j}{\alpha^2} \end{array}\right).$$

Proposition 1 ([15]). *$F = \alpha\sqrt{\phi(z,\rho)}$ is strongly convex if and only if $\Omega, \Lambda > 0$.*

The spray coefficients G^A are defined by:

$$G^A := \frac{1}{4}g^{AC}\left[\left(F^2\right)_{y^Cx^B}y^B - \left(F^2\right)_{x^C}\right].$$

The Riemann curvature of F is a family of endomorphisms:

$$R_y = R^A{}_B dx^B \otimes \frac{\partial}{\partial x^A} : T_xM \to T_xM,$$

defined by:

$$R^A{}_B := 2\left(G^A\right)_{x^B} - \left(G^A\right)_{x^Cy^B}y^C + 2G^C\left(G^A\right)_{y^Cy^B} - \left(G^A\right)_{y^C}\left(G^C\right)_{x^B}.$$

For the Riemannian curvature $R^A{}_B$ of the Finsler warped product metric $F = \alpha\sqrt{\phi(z,\rho)}$, we have [15]:

$$\begin{aligned} R^0{}_0 =& [\rho^2(U+zV)W_z - (2\rho^2W+1)(U_z+V+zV_z)]\alpha^2 \\ &+ [2(V+W)(U_z+V+zV_z) - (V_z+W_z)(U+zV) \\ &+ 2U(U_{zz}+2V_z+zV_{zz}) - \frac{1}{\rho}(U_{\rho z}+V_\rho+zV_{\rho z}) \\ &- (U_z+V+zV_z)^2 - (U - zU_z - z^2V_z)V_z]\langle\bar{x},\bar{y}\rangle^2, \end{aligned}$$

$$\begin{aligned} R^0{}_j =& z[(2\rho^2W+1)(V+U_z+zV_z) - \rho^2W_z(U+zV)]\alpha y^j \\ &+ [z(U+zV)(V_z+W_z) - 2zU(U_{zz}+2V_z+zV_{zz}) \\ &+ (U - zU_z - z^2V_z)(5W - U_z) - \frac{1}{\rho}(U_\rho - zU_{\rho z} - z^2V_{\rho z})]\langle\bar{x},\bar{y}\rangle^2\frac{y^j}{\alpha} \\ &+ [(U+zV)(U_z - V + zV_z - 2W) + (V-3W)(U - zU_z - z^2V_z) \\ &+ \frac{1}{\rho}(U_\rho + zV_\rho)]\langle\bar{x},\bar{y}\rangle\alpha x^j, \end{aligned}$$

$$
\begin{aligned}
R^i{}_0 =& [\rho^2 W_z(V-W) - (2\rho^2 W+1)V_z]\alpha y^i + [(2W-V-U_z)(V_z+W_z) \\
&+ 2U(V_{zz}+W_{zz}) - \frac{1}{\rho}(V_{\rho z}+W_{\rho z})]\langle \bar{x},\bar{y}\rangle^2 \frac{y^i}{\alpha} \\
&+ [(U_z-W)W_z - 2UW_{zz} + \frac{1}{\rho}W_{\rho z}]\langle \bar{x},\bar{y}\rangle \alpha x^i,
\end{aligned}
$$

$$
\begin{aligned}
R^i{}_j =& -[2W+(2\rho^2 W+1)(V+W)]\alpha^2\delta^i_j \\
&+ [(V+W)^2 + 2U(V_z+W_z) - \frac{1}{\rho}(V_\rho+W_\rho)]\langle \bar{x},\bar{y}\rangle^2\delta^i_j \\
&+ [2W(2W-zW_z) + W_z(U-zW) - \frac{2}{\rho}W_\rho]\alpha^2 x^i x^j + [(V+W) \\
&+ z(V_z+W_z)(2\rho^2 W+1) + (\rho^2(V+W)+1)(2W-zW_z)]y^i y^j \\
&- [2zU(V_{zz}+W_{zz}) + (3U - zU_z - zV + 5zW)(V_z+W_z). \\
&- \frac{z}{\rho}(V_{\rho z}+W_{\rho z})]\langle \bar{x},\bar{y}\rangle^2 \frac{y^i y^j}{\alpha^2} + [-(2W-zW_z)^2 - 2U(W_z - zW_{zz}) \\
&+ \frac{1}{\rho}(2W_\rho - zW_{\rho z}) + W_z(U - zU_z - z^2 W_z)]\langle \bar{x},\bar{y}\rangle x^i y^j \\
&+ [-(V+W)^2 + (V_z+W_z)(U+3zW) + \frac{1}{\rho}(V_\rho+W_\rho)]\langle \bar{x},\bar{y}\rangle x^j y^i,
\end{aligned}
$$

where $\langle \bar{x},\bar{y}\rangle := \sum_{k=1}^n x^k y^k$,

$$
U := \frac{1}{2\rho\Lambda}\big(2\phi\phi_{\rho z} - \phi_\rho\phi_z\big), \quad V := \frac{1}{2\rho\Lambda}\big(\phi_\rho\phi_{zz} - \phi_z\phi_{\rho z}\big), \quad W := \frac{1}{2\rho\Omega}\phi_\rho. \tag{1}
$$

Thus, the Ricci curvature of $F = \alpha\sqrt{\phi(z,\rho)}$ is [15]:

$$
\begin{aligned}
\mathrm{Ric} :=& R^A{}_A \\
=& [-(2\rho^2 W+1)(U_z + nV + (n-3)W) - 2(nW + \rho W_\rho - \rho^2 W_z(U - zW))]\alpha^2 \\
&+ [2U(U_{zz} + nV_z + (n-2)W_z) - \frac{1}{\rho}(U_{\rho z} + nV_\rho + (n-3)W_\rho) \\
&+ nV(V+2W) + W((n-5)W + 2zW_z) + U_z(2W - U_z)]\langle \bar{x},\bar{y}\rangle^2.
\end{aligned}
$$

3. χ-Curvature

In this section, we first derive the expression for the χ-curvature of a Finsler warped product metric $F = \alpha\sqrt{\phi(z,\rho)}$. Then, we obtain differential equations of such metrics with vanishing χ-curvature.

Lemma 1 ([15]). *For a Finsler warped product metric $F = \alpha\sqrt{\phi(z,\rho)}$, the χ-curvature of F is given by:*

$$
\begin{aligned}
\chi_0 &= (\frac{1}{\rho}\Psi_{\rho z} - 2U\Psi_{zz} - 2W\Psi_z)\frac{\langle \bar{x},\bar{y}\rangle^2}{\alpha} + (2\rho^2 W+1)\Psi_z\alpha, \\
\chi_i &= [2zU\Psi_{zz} - \frac{z}{\rho}\Psi_{\rho z} + 2(U+2zW)\Psi_z]\frac{\langle \bar{x},\bar{y}\rangle^2}{\alpha^2}y^i \\
&\quad - z(2\rho^2 W+1)\Psi_z y^i - 2(U+zW)\Psi_z\langle \bar{x},\bar{y}\rangle x^i,
\end{aligned}
$$

where

$$
\Psi := U_z + (n+2)V + (n-1)W.
$$

Lemma 2 ([15]). *For $n \geq 2$, $A\alpha^2 + B\langle \bar{x}, \bar{y}\rangle^2 = 0$ if and only if $A = 0$ and $B = 0$, where A, B are functions of z and ρ.*

Lemma 3. *For $n \geq 2$:*

$$A\alpha^2 y^i + B\langle \bar{x}, \bar{y}\rangle \alpha^2 x^i + C\langle \bar{x}, \bar{y}\rangle^2 y^i = 0 \tag{2}$$

if and only if $A = 0$, $B = 0$ and $C = 0$, where A, B, C are functions of z and ρ.

Proof. "Necessity". Suppose that (2) holds. Contracting (2) with y^i, we have:

$$A\alpha^2 + (B + C)\langle \bar{x}, \bar{y}\rangle^2 = 0.$$

By Lemma 2, we obtain $A = 0, B + C = 0$.

Thus, (2) can be simplified as $B\big(\alpha^2 x^i - \langle \bar{x}, \bar{y}\rangle y^i\big) = 0$. Contracting it with x^i yields:

$$B\big(\rho^2\alpha^2 - \langle \bar{x}, \bar{y}\rangle^2\big) = 0.$$

We obtain $B = 0$. Thus, $A = 0, B = 0$ and $C = 0$.

"Sufficiency". It is obviously true. □

Theorem 2. *Let $F = \alpha\sqrt{\phi(z, \rho)}$ be a Finsler warped product metric on an $(n + 1)$-dimensional manifold $M = \mathbb{R} \times \mathbb{R}^n$ $(n \geq 2)$, where $\alpha = |\bar{y}|, z = \frac{y^0}{|\bar{y}|}, \rho = |\bar{x}|$. Then F has vanishing χ-curvature if and only if ϕ satisfies $\Psi_z = 0$.*

Proof. "Necessity". Suppose that F has vanishing χ-curvature, i.e., $\chi_0 = 0, \chi_i = 0$. For $\chi_0 = 0$, by Lemma 1 and Lemma 2, we obtain that:

$$\begin{cases} 0 = -2\rho U\Psi_{zz} + \Psi_{\rho z} - 2\rho W\Psi_z, & (3) \\ 0 = \big(2\rho^2 W + 1\big)\Psi_z. & (4) \end{cases}$$

Since $\chi_i = 0$, by Lemma 1 and Lemma 3, we obtain that:

$$\begin{cases} 0 = 2\rho z U\Psi_{zz} - z\Psi_{\rho z} + 2\rho(U + 2zW)\Psi_z, & (5) \\ 0 = (U + zW)\Psi_z. & (6) \end{cases}$$

Since $(5) - 2\rho \times (6) = -z \times (3)$, F has vanishing χ-curvature if and only if:

$$\begin{cases} 0 = \big(2\rho^2 W + 1\big)\Psi_z, \\ 0 = 2\rho z U\Psi_{zz} - z\Psi_{\rho z} + 2\rho(U + 2zW)\Psi_z, \\ 0 = (U + zW)\Psi_z. \end{cases} \tag{7}$$

We divide the problem into two cases:

Case(i) $\Psi_z = 0$. It is easy to verify that (7) holds.

Case(ii) $\Psi_z \neq 0$. We see that (7) is equivalent to:

$$\begin{cases} 0 = 2\rho^2 W + 1, & (8) \\ 0 = U + zW, & (9) \\ 0 = 2\rho U\Psi_{zz} - \Psi_{\rho z} + 2\rho W\Psi_z. & (10) \end{cases}$$

By (8), we have:

$$W = -\frac{1}{2\rho^2}.$$

Substituting it into $W := \frac{1}{2\rho\Omega}\phi_\rho$ yields:

$$2\phi - z\phi_z + \rho\phi_\rho = 0. \tag{11}$$

Plugging $W = -\frac{1}{2\rho^2}$ into (9), we obtain:

$$U = \frac{z}{2\rho^2}.$$

By (11) and $V := \frac{1}{2\rho\Lambda}\big(\phi_\rho\phi_{zz} - \phi_z\phi_{\rho z}\big)$, we have:

$$V = -\frac{1}{2\rho^2}.$$

Finally, we obtain $\Psi := U_z + (n+2)V + (n-1)W = -\frac{n}{\rho^2}$. Hence, $\Psi_z = 0$, which is a contradiction to our assumption.

"Sufficiency". It is obvious by Lemma 1.

This completes the proof of Theorem 2. □

4. *H*-Curvature

In this section, we derive the expression for the H-curvature of Finsler warped product metric $F = \alpha\sqrt{\phi(z,\rho)}$. Then, we obtain differential equations of such metrics with vanishing H-curvature.

The H-curvature can be expressed in terms of χ-curvature [8], that is:

$$H_{ij} = \frac{1}{4}\big(\chi_{i.j} + \chi_{j.i}\big). \tag{12}$$

Lemma 4. *For a Finsler warped product metric $F = \alpha\sqrt{\phi(z,\rho)}$, the H-curvature is given by:*

$$\begin{aligned}
H_{00} =& [\frac{1}{2\rho}\Psi_{\rho zz} - (U_z + W)\Psi_{zz} - U\Psi_{zzz} - W_z\Psi_z]\frac{\langle\bar{x},\bar{y}\rangle^2}{\alpha^2} + \frac{1}{2}(2\rho^2W+1)\Psi_{zz} + \rho^2W_z\Psi_z,\\
H_{0i} =& \frac{1}{2}[-\frac{z}{\rho}\Psi_{\rho zz} + 2zU\Psi_{zzz} - \frac{1}{\rho}\Psi_{\rho z} + (3U + 3zW + 2zU_z)\Psi_{zz} + (3zW_z + 3W + U_z)\Psi_z]\frac{\langle\bar{x},\bar{y}\rangle^2}{\alpha^3}y^i\\
&+ \frac{1}{2}[\frac{1}{\rho}\Psi_{\rho z} - (3U + zW)\Psi_{zz} - (3W + U_z + zW_z)\Psi_z]\frac{\langle\bar{x},\bar{y}\rangle}{\alpha}x^i\\
&- \frac{z}{2}[2\rho^2W_z\Psi_z + (2\rho^2W+1)\Psi_{zz}]\frac{y^i}{\alpha},\\
H_{ij} =& \frac{1}{2}[-2z(4U + zU_z + 2zW)\Psi_{zz} - 2z^2U\Psi_{zzz} + \frac{3z}{\rho}\Psi_{\rho z} + \frac{z^2}{\rho}\Psi_{\rho zz}\\
&- 2(6zW + zU_z + 2z^2W_z + 2U)\Psi_z]\frac{\langle\bar{x},\bar{y}\rangle^2}{\alpha^4}y^iy^j\\
&+ \frac{z}{2}[(2\rho^2W + 1 + 2z\rho^2W_z)\Psi_z + z(2\rho^2W+1)\Psi_{zz}]\frac{y^iy^j}{\alpha^2}\\
&+ \frac{1}{2}[(zU_z + 5zW + z^2W_z + 2U)\Psi_z + z(3U + zW)\Psi_{zz} - \frac{z}{\rho}\Psi_{\rho z}]\frac{\langle\bar{x},\bar{y}\rangle}{\alpha^2}(x^iy^j + x^jy^i)\\
&+ \frac{1}{2}[2zU\Psi_{zz} - \frac{z}{\rho}\Psi_{\rho z} + 2(U + 2zW)\Psi_z]\frac{\langle\bar{x},\bar{y}\rangle^2}{\alpha^2}\delta_{ij}\\
&- \frac{z}{2}(2\rho^2W+1)\Psi_z\delta_{ij} - (U + zW)\Psi_zx^ix^j.
\end{aligned}$$

Proof. For a Finsler warped product metric $F = \alpha\sqrt{\phi(z,\rho)}$:

$$\begin{aligned} H =& H_{AB}dx^A \otimes dx^B \\ =& H_{00}dx^0 \otimes dx^0 + H_{0j}dx^0 \otimes dx^j + H_{i0}dx^i \otimes dx^0 + H_{ij}dx^i \otimes dx^j, \end{aligned}$$

where $H_{00} = \frac{1}{4}(\chi_{0.0} + \chi_{0.0}), H_{0j} = \frac{1}{4}(\chi_{0.j} + \chi_{j.0}), H_{i0} = \frac{1}{4}(\chi_{i.0} + \chi_{0.i})$ and $H_{ij} = \frac{1}{4}(\chi_{i.j} + \chi_{j.i})$. Differentiating χ_A with respect to y, we obtain:

$$\begin{aligned}
\chi_{0.0} =& [\frac{1}{\rho}\Psi_{\rho zz} - 2(U_z + W)\Psi_{zz} - 2U\Psi_{zzz} - 2W_z\Psi_z]\frac{\langle\bar{x},\bar{y}\rangle^2}{\alpha^2} + (2\rho^2W + 1)\Psi_{zz} + 2\rho^2W_z\Psi_z, \\
\chi_{0.i} =& [-\frac{z}{\rho}\Psi_{\rho zz} + 2zU\Psi_{zzz} - \frac{1}{\rho}\Psi_{\rho z} + 2(zU_z + zW + U)\Psi_{zz} + 2(zW_z + W)\Psi_z]\frac{\langle\bar{x},\bar{y}\rangle^2}{\alpha^3}y^i \\
&+ 2(\frac{1}{\rho}\Psi_{\rho z} - 2U\Psi_{zz} - 2W\Psi_z)\frac{\langle\bar{x},\bar{y}\rangle}{\alpha}x^i \\
&+ [(2\rho^2W + 1 - 2z\rho^2W_z)\Psi_z - z(2\rho^2W + 1)\Psi_{zz}]\frac{y^i}{\alpha}, \\
\chi_{i.0} =& [-\frac{z}{\rho}\Psi_{\rho zz} + 2zU\Psi_{zzz} - \frac{1}{\rho}\Psi_{\rho z} + 2(zU_z + 2zW + 2U)\Psi_{zz} + 2(U_z + 2zW_z + 2W)\Psi_z]\frac{\langle\bar{x},\bar{y}\rangle^2}{\alpha^3}y^i \\
&- 2[(U + zW)\Psi_{zz} + (U_z + zW_z + W)\Psi_z]\frac{\langle\bar{x},\bar{y}\rangle}{\alpha}x^i \\
&- [z(2\rho^2W + 1)\Psi_{zz} + (2\rho^2W + 1 + 2z\rho^2W_z)\Psi_z]\frac{y^i}{\alpha}, \\
\chi_{i.j} =& [-2z^2U\Psi_{zzz} + \frac{z^2}{\rho}\Psi_{\rho zz} - 2z(4U + zU_z + 2zW)\Psi_{zz} + \frac{3z}{\rho}\Psi_{\rho z} \\
&- 2(6zW + zU_z + 2z^2W_z + 2U)\Psi_z]\frac{\langle\bar{x},\bar{y}\rangle^2}{\alpha^4}y^iy^j \\
&+ z[(2\rho^2W + 1 + 2z\rho^2W_z)\Psi_z + z(2\rho^2W + 1)\Psi_{zz}]\frac{y^iy^j}{\alpha^2} \\
&+ 2z[(U_z + W + zW_z)\Psi_z + (U + zW)\Psi_{zz}]\frac{\langle\bar{x},\bar{y}\rangle}{\alpha^2}x^iy^j \\
&+ 2[2(U + 2zW)\Psi_z + 2zU\Psi_{zz} - \frac{z}{\rho}\Psi_{\rho z}]\frac{\langle\bar{x},\bar{y}\rangle}{\alpha^2}x^jy^i \\
&+ [2zU\Psi_{zz} - \frac{z}{\rho}\Psi_{\rho z} + 2(U + 2zW)\Psi_z]\frac{\langle\bar{x},\bar{y}\rangle^2}{\alpha^2}\delta_{ij} - z(2\rho^2W + 1)\Psi_z\delta_{ij} - 2(U + zW)\Psi_z x^ix^j.
\end{aligned}$$

By simple calculations, we obtain the expression of H_{AB}. □

Lemma 5. *For* $n \geq 2$:

$$A\alpha^4x^i + B\langle\bar{x},\bar{y}\rangle^2\alpha^2x^i + C\langle\bar{x},\bar{y}\rangle\alpha^2y^i + D\langle\bar{x},\bar{y}\rangle^3y^i = 0 \tag{13}$$

if and only if $A = 0, B = 0, C = 0$ *and* $D = 0$, *where* A, B, C, D *are functions of* z *and* ρ.

Proof. "Necessity". Suppose that (13) holds. Contracting (13) with y^i yields:

$$A\langle\bar{x},\bar{y}\rangle\alpha^4 + B\langle\bar{x},\bar{y}\rangle^3\alpha^2 + C\langle\bar{x},\bar{y}\rangle\alpha^4 + D\langle\bar{x},\bar{y}\rangle^3\alpha^2 = 0,$$

that is:

$$(A + C)\alpha^2 + (B + D)\langle\bar{x},\bar{y}\rangle^2 = 0.$$

By Lemma 2, we obtain $C = -A, D = -B$.

Thus, (13) can be simplified as $A\alpha^4x^i + B\langle\bar{x},\bar{y}\rangle^2\alpha^2x^i - A\langle\bar{x},\bar{y}\rangle\alpha^2y^i - B\langle\bar{x},\bar{y}\rangle^3y^i = 0$. Contracting it with x^i yields:

$$\left(\rho^2\alpha^2 - \langle\bar{x},\bar{y}\rangle^2\right)\left[A\alpha^2 + B\langle\bar{x},\bar{y}\rangle^2\right] = 0.$$

By Lemma 2, we obtain $A = 0$ and $B = 0$. Thus, $A = 0, B = 0, C = 0$ and $D = 0$.

"Sufficiency". It is obviously true. □

Lemma 6. *For $n \geq 2$:*

$$A\langle\bar{x},\bar{y}\rangle^2y^iy^j + B\alpha^2y^iy^j + C\alpha^4x^ix^j + D\langle\bar{x},\bar{y}\rangle\alpha^2\left(x^iy^j + x^jy^i\right) + E\langle\bar{x},\bar{y}\rangle^2\alpha^2\delta_{ij} + F\alpha^4\delta_{ij} = 0 \tag{14}$$

if and only if $A = 0, B = -F = C\rho^2$ and $E = C = -D$, where A, B, C, D, E, F are functions of z and ρ. In particular, for $n > 2$, if (14) holds, then $A = B = C = D = E = F = 0$.

Proof. "Necessity". Suppose that (14) holds. Contracting (14) with y^j, we have:

$$(B + F)\alpha^2y^i + (C + D)\langle\bar{x},\bar{y}\rangle\alpha^2x^i + (A + D + E)\langle\bar{x},\bar{y}\rangle^2y^i = 0.$$

By Lemma 3, we obtain $F = -B, D = -C$ and $E = -A + C$.

Thus, (14) can be simplified as $A\langle\bar{x},\bar{y}\rangle^2y^iy^j + B\alpha^2y^iy^j + C\alpha^4x^ix^j - C\langle\bar{x},\bar{y}\rangle\alpha^2\left(x^iy^j + x^jy^i\right) - (-A + C)\langle\bar{x},\bar{y}\rangle^2\alpha^2\delta_{ij} - B\alpha^4\delta_{ij} = 0$. Contracting it with x^j yields:

$$\left(C\rho^2 - B\right)\alpha^4x^i - A\langle\bar{x},\bar{y}\rangle^2\alpha^2x^i + \left(B - C\rho^2\right)\langle\bar{x},\bar{y}\rangle\alpha^2y^i + A\langle\bar{x},\bar{y}\rangle^3y^i = 0.$$

By Lemma 5, we obtain $C\rho^2 - B = 0$ and $A = 0$. Thus, $A = 0, B = -F = C\rho^2$ and $E = C = -D$.

In this case, (14) can be rewritten as:

$$\alpha^2C\left[\rho^2y^iy^j + \alpha^2x^ix^j - \langle\bar{x},\bar{y}\rangle(x^iy^j + x^jy^i) + \langle\bar{x},\bar{y}\rangle^2\delta^{ij} - \rho^2\alpha^2\delta^{ij}\right] = 0. \tag{15}$$

Now putting $i = j$ and taking summation over i, we obtain:

$$(n - 2)C\alpha^2\left(\langle\bar{x},\bar{y}\rangle^2 - \rho^2\alpha^2\right) = 0.$$

Thus, when $n = 2$, the above equation always holds; when $n > 2$, we obtain $C = 0$. Thus, $A = B = C = D = E = F = 0$.

"Sufficiency". When $n > 2$, it is obviously true. When $n = 2$, we see that the right side of (14) is reduced to the left side of (15). Furthermore, we have that (15) holds for any i and j $(1 \leq i, j \leq 2)$. Thus, (14) holds. This completes the proof of Lemma 6. □

Theorem 3. *Let $F = \alpha\sqrt{\phi(z,\rho)}$ be a Finsler warped product metric on an $(n+1)$-dimensional manifold $M = \mathbb{R} \times \mathbb{R}^n$ $(n \geq 2)$, where $\alpha = |\bar{y}|, z = \frac{y^0}{|\bar{y}|}, \rho = |\bar{x}|$. Then, F has vanishing H-curvature if and only if ϕ satisfies $\Psi_z = 0$.*

Proof. "Necessity". Suppose that F has vanishing H-curvature, i.e, $H_{00} = 0, H_{0i} = 0, H_{ij} = 0$. Since $H_{00} = 0$, by Lemma 4 and Lemma 2, we obtain that:

$$\begin{cases} 0 = \left[2\rho U\Psi_{zz} - \Psi_{\rho z} + 2\rho W\Psi_z\right]_z, & (16) \\ 0 = \left[\left(2\rho^2W + 1\right)\Psi_z\right]_z. & (17) \end{cases}$$

For $H_{0i} = 0$, by Lemma 4 and Lemma 3, we obtain that:

$$\begin{cases} 0 = 2\rho z U\Psi_{zzz} - z\Psi_{\rho zz} + \rho(3U + 3zW + 2zU_z)\Psi_{zz} - \Psi_{\rho z} \\ \quad +\rho(3zW_z + 3W + U_z)\Psi_z, & (18) \\ 0 = \rho(3U + zW)\Psi_{zz} - \Psi_{\rho z} + \rho(3W + U_z + zW_z)\Psi_z. & (19) \end{cases}$$

Since $H_{ij} = 0$, by Lemma 4 and Lemma 6, we have that:

$$\begin{cases} 0 = 2\rho z^2 U\Psi_{zzz} - z^2\Psi_{\rho zz} + 2\rho z(4U + zU_z + 2zW)\Psi_{zz} - 3z\Psi_{\rho z} \\ \quad +2\rho\big(6zW + zU_z + 2z^2W_z + 2U\big)\Psi_z, & (20) \\ 0 = z\big(2\rho^2 W + 1\big)\Psi_{zz} + \big[(1 - z)\big(2\rho^2 W + 1\big) + 2\rho^2 zW_z\big]\Psi_z, & (21) \\ 0 = \big[z\big(2\rho^2 W + 1\big) + 2\rho^2(U + zW)\big]\Psi_z, & (22) \\ 0 = 2\rho^2 zU\Psi_{zz} - \rho z\Psi_{\rho z} + \big[2\rho^2(U + 2zW) - z\big(2\rho^2 W + 1\big)\big]\Psi_z, & (23) \\ 0 = \rho z(5U + zW)\Psi_{zz} - 2z\Psi_{\rho z} + \rho\big(zU_z + 9zW + z^2W_z + 4U\big)\Psi_z. & (24) \end{cases}$$

$(21) - z \times (17)$ yields:

$$(1 - z)\big(2\rho^2 W + 1\big)\Psi_z = 0.$$

$(22) + (23)$ yields:

$$4\rho^2(U + zW)\Psi_z + \rho z\big(2\rho U\Psi_{zz} - \Psi_{\rho z} + 2\rho W\Psi_z\big) = 0.$$

$(24) - 2z \times (19)$ yields:

$$-z(U + zW)\Psi_{zz} + \big[-z^2W_z + 4(U + zW) - z(W + U_z)\big]\Psi_z = 0.$$

Since $(18) = (19) + z \times (16)$ and $(20) = z^2 \times (16) + z \times (19) + (24)$, F has vanishing H-curvature if and only if:

$$\begin{cases} 0 = \big[2\rho U\Psi_{zz} - \Psi_{\rho z} + 2\rho W\Psi_z\big]_z, & (25) \\ 0 = \big[\big(2\rho^2 W + 1\big)\Psi_z\big]_z, & (26) \\ 0 = \rho(3U + zW)\Psi_{zz} - \Psi_{\rho z} + \rho(3W + U_z + zW_z)\Psi_z, & (27) \\ 0 = (1 - z)\big(2\rho^2 W + 1\big)\Psi_z, & (28) \\ 0 = \big[z\big(2\rho^2 W + 1\big) + 2\rho^2(U + zW)\big]\Psi_z, & (29) \\ 0 = 4\rho^2(U + zW)\Psi_z + \rho z\big(2\rho U\Psi_{zz} - \Psi_{\rho z} + 2\rho W\Psi_z\big), & (30) \\ 0 = -z(U + zW)\Psi_{zz} + \big[-z^2W_z + 4(U + zW) - z(W + U_z)\big]\Psi_z. & (31) \end{cases}$$

We divide the problem into two cases:

Case(i) $\Psi_z = 0$. It is easy to verify that (25)–(31) hold.

Case(ii) $\Psi_z \neq 0$. From (28), we can see $2\rho^2 W + 1 = 0$. Thus, $W_z = 0$. Plugging $2\rho^2 W + 1 = 0$ into (29) yields:

$$U + zW = 0.$$

Differentiating it with respect to z and combining this with $W_z = 0$, we have $U_z + W = 0$.

Substituting $2\rho^2 W + 1 = 0$ and $U + zW = 0$ into (27) yields:

$$2\rho U\Psi_{zz} - \Psi_{\rho z} + 2\rho W\Psi_z = 0.$$

It is easy to verify that (25), (26), (30) and (31) hold. Now F has vanishing H-curvature if and only if the following hold:

$$\begin{cases} 0 = 2\rho^2 W + 1, \\ 0 = U + zW, \\ 0 = 2\rho U \Psi_{zz} - \Psi_{\rho z} + 2\rho W \Psi_z. \end{cases}$$

The result is the same as in Case(ii) of Theorem 2. This is a contradiction.

"Sufficiency". It is obvious by Lemma 4.

This completes the proof of Theorem 3. □

Proof of Theorem 1. By Theorem 2 and Theorem 3, the result is obvious. □

Example 1. *(Minkowski metrics). Let $\phi = \phi(z, \rho)$ be a function defined by:*

$$\phi(z, \rho) = e^{2z},$$

where $|z| < 1$. We have the Finsler warped product metric:

$$F = \alpha\sqrt{\phi} = \alpha e^z.$$

Since $\Omega = 2(1-z)e^{2z} > 0$, $\Lambda = 4e^{4z} > 0$, by Proposition 1, we obtain that $F = \alpha\sqrt{\phi}$ gives a positive-definite metric. We know that $\Psi_z = 0$. By Theorems 2 and 3, we have that its χ-curvature and H-curvature vanish.

Example 2. *(Randers metrics). Let $\phi = \phi(z, \rho)$ be a function defined by:*

$$\phi(z, \rho) = \left(f(\rho)z + \sqrt{f^2(\rho)z^2 + g(\rho)}\right)^2,$$

where $g(\rho) > 0$ and $f(\rho)z + \sqrt{f^2(\rho)z^2 + g(\rho)} > 0$. We have the Finsler warped product metric:

$$F = \alpha\sqrt{\phi} = \alpha(f(\rho)z + \sqrt{red f^2(\rho)z^2 + g(\rho)}).$$

Since $\Omega = \frac{2g(\rho)\phi^{\frac{1}{2}}}{\sqrt{f^2(\rho)z^2+g(\rho)}} > 0$, $\Lambda = \frac{4f^2(\rho)g(\rho)\phi^{\frac{3}{2}}}{(f^2(\rho)z^2+g(\rho))^{\frac{3}{2}}} > 0$, by Proposition 1, we obtain that $F = \alpha\sqrt{\phi}$ gives a positive-definite metric. We know that $\Psi_z = 0$. By Theorems 2 and 3, we have that its χ-curvature and H-curvature vanish.

Example 3. *(Quadratic polynomial). Let $\phi = \phi(z, \rho)$ be a function defined by:*

$$\phi(z, \rho) = c_2 z^2 + c_1(\rho)z + c_0 c_1^2(\rho),$$

where c_0, c_2 are constants, $|z| < 1$, $4c_0c_2 > 1$ and $2c_0c_1(\rho) > 1$. We have the Finsler warped product metric:

$$F = \alpha\sqrt{\phi} = \alpha\sqrt{c_2 z^2 + c_1(\rho)z + c_0 c_1^2(\rho)}.$$

Since $\Omega = c_1(\rho)(z + 2c_0c_1(\rho)) > 0$, $\Lambda = c_1^2(\rho)(4c_0c_2 - 1) > 0$, by Proposition 1, we obtain that $F = \alpha\sqrt{\phi}$ gives a positive-definite metric. We know that $\Psi_z = 0$. By Theorems 2 and 3, we have that its χ-curvature and H-curvature vanish.

5. Conclusions

Non-Riemannian quantities play an important role in Finsler geometry. In this paper, we firstly obtain differential equations of Finsler warped product metrics with vanishing χ-curvature or vanishing H-curvature. Based on these, we obtain that the χ-curvature vanishes if and only if the H-curvature vanishes. Since the solution of $\Psi_z = 0$ is unknown, the classification of such metrics is to be continued.

Author Contributions: Conceptualization, L.S., X.Z. and M.W.; Methodology, X.Z. and M.W.; Writing—original draft, L.S.; Writing—review and editing, L.S. and X.Z. All authors have read and agreed to the published version of the manuscript.

Funding: This work was supported by the National Natural Science Foundation of China (No. 11961061, 11461064).

Data Availability Statement: No data were used to support this work.

Acknowledgments: This work was supported by the National Natural Science Foundation of China (No. 11961061, 11461064). The authors would like to thank Zhongmin Shen for their helpful discussion and valuable comments.

Conflicts of Interest: The authors declare no conflict of interest.

References

1. Shen, Z.M. On Some Non-Riemannian Quantities in Finsler Geometry. *Can. Math. Bull.* **2013**, *56*, 184–193. [CrossRef]
2. Cheng, X.Y.; Yuan, M.G. On conformally flat (α, β)-metrics with special curvature properties. *Acta Math. Sin. Eng. Ser.* **2015**, *31*, 879–892.
3. Shen, Z.M. On sprays with vanishing χ-curvature. *Int. J. Math.* **2022**, *32*, 2150069. [CrossRef]
4. Zadeh, H. Sur les espaces de Finsler à courbures sectionnelles constantes. *Bull. l'Acad. R. Belg.* **1988**, *74*, 281–322.
5. Najafi, B.; Shen, Z.M.; Tayebi, A. Finsler metrics of scalar flag curvature with special non-Riemannian curvature properties. *Geom. Dedicata* **2008**, *131*, 87–97. [CrossRef]
6. Mo, X.H. On the non-Riemannian quantity H of a Finsler metric. *Differ. Geom. Appl.* **2009**, *27*, 7–14.
7. Xia, Q.L. Some results on the non-riemannian quantity h of a finsler metric. *Int. J. Math.* **2011**, *22*, 925–936.
8. Tang, D.M. On the non-Riemannian quantity H in Finsler geometry. *Differ. Geom. Appl.* **2011**, *29*, 207–213.
9. Mo, X.H. A class of Finsler metrics with almost vanishing H-curvature. *Balk. J. Geom. Appl.* **2016**, *21*, 58–66.
10. Zhu, H.M. On a class of Finsler metrics with special curvature properties. *Balk. J. Geom. Appl.* **2018**, *23*, 97–108.
11. Sevim, E.S.; Gabrani, M. On H-curvature of Finsler warped product metrics. *J. Finsler Geom. Appl.* **2020**, *1*, 37–44.
12. Bishop, R.L.; O'Neill, B. Manifolds of negative curvature. *Trans. Amer. Math. Soc.* **1969**, *145*, 1–49. [CrossRef]
13. Chen, B.; Shen, Z.M.; Zhao, L.L. Constructions of Einstein Finser metrics by warped product. *Int. J. Math.* **2018**, *29*, 86–100. [CrossRef]
14. Kozma, L.; Peter, R.; Varga, C.C. Warped product of Finsler manifolds. *Ann. Univ. Sci. Budapest* **2001**, *44*, 157–170.
15. Marcal, P.; Shen, Z.M. Ricci flat Finsler metrics by warped product. *arXiv* **2020**, arXiv:2012.05699.
16. Chern, S.S.; Shen, Z.M. *Riemann-Finsler Geometry*; World Scientic Publishers: Singapore, 2004.
17. Najafi, B.; Bidabad, B.; Tayebi, A. On R-quadratic Finsler metrics. *Iran. J. Sci. Technol. Trans. A Sci.* **2007**, *31*, 439–443.

axioms

MDPI

Article

η-Ricci–Yamabe Solitons along Riemannian Submersions

Mohd Danish Siddiqi [1,*], Fatemah Mofarreh [2], Mehmet Akif Akyol [3] and Ali H. Hakami [1]

1 Department of Mathematics, College of Science, Jazan University, P.O. Box 277, Jazan 45142, Saudi Arabia; aalhakami@jazanu.edu.sa
2 Mathematical Science Department, Faculty of Science, Princess Nourah bint Abdulrahman University, Riyadh 11546, Saudi Arabia; fyalmofarrah@pnu.edu.sa
3 Department of Mathematics, Faculty of Arts and Sciences, Bingol University, Bingol 12000, Turkey; mehmetakifakyol@bingol.edu.tr
* Correspondence: msiddiqi@jazanu.edu.sa

Abstract: In this paper, we investigate the geometrical axioms of Riemannian submersions in the context of the η-Ricci–Yamabe soliton (η-RY soliton) with a potential field. We give the categorization of each fiber of Riemannian submersion as an η-RY soliton, an η-Ricci soliton, and an η-Yamabe soliton Additionally, we consider the many circumstances under which a target manifold of Riemannian submersion is an η-RY soliton, an η-Ricci soliton, an η-Yamabe soliton, or a quasi-Yamabe soliton We deduce a Poisson equation on a Riemannian submersion in a specific scenario if the potential vector field ω of the soliton is of gradient type =:grad(γ) and provide some examples of an η-RY soliton, which illustrates our finding. Finally, we explore a number theoretic approach to Riemannian submersion with totally geodesic fibers.

Keywords: η-Ricci–Yamabe soliton; Riemannian submersion; Riemannian manifold; homotopy groups

MSC: 53C25; 53C43; 11F23

Citation: Siddiqi, M.D.; Mofarreh, F.; Akyol, M.A.; Hakami, A.H. η-Ricci-Yamabe Solitons along Riemannian Submersions. *Axioms* **2023**, *12*, 796. https://doi.org/10.3390/axioms12080796

Academic Editor: Juan De Dios Pérez

Received: 19 July 2023
Revised: 9 August 2023
Accepted: 11 August 2023
Published: 17 August 2023

1. Introduction

Since Riemannian geometry's inception, the idea of Riemannian immersion has been the subject of extensive study. In fact, the Riemannian manifolds that were initially intended to be examined were surfaces embedded in $\mathbb{R}^3$ [1].

Initially, Gray and O'Neill were the first to discuss the "dual" concept of Riemannian submersion and investigated it further. Because of their applications in supergravity, the theory of relativity, and other physical theories, Riemannian submersions have received considerable attention in both mathematics and theoretical physics (see [2–7]). Studies on Riemannian submersion are reported in [8–12].

A soliton, which is related to the geometrical flow of Riemannian (semi-Riemannian) geometry, is a significant symmetry.

However, the theory of geometric flows has emerged as one of the most important geometrical theories for illuminating Riemannian geometric structures. The study of singularities of the flows involves a certain section of solutions when the metric evolves via dilations and diffeomorphisms because they appear as potential singularity models. They are frequently referred to as solitons.

In 1988, Hamilton [13] presented the ideas of Ricci flow and Yamabe flow for the first time. The limit of the solutions for the Ricci flow and the Yamabe flow, respectively, is shown to be the soliton of Ricci and the soliton of Yamabe. Geometric flow theory, including the Ricci flow and Yamabe flow, has drawn the attention of many mathematicians over the past two decades.

Under the term Ricci–Yamabe map, geometers [14] initiated research concerning a novel geometric flow that is a generalization of the Ricci and Yamabe flows. Ricci–Yamabe flow of the type (σ, ρ) is another name for this. The metrics on the Riemannian manifold defined by Guler and Crasmareanu evolve into the Ricci–Yamabe flow [14].

$$\frac{1}{2}\frac{\partial}{\partial t}g(t) = -\sigma S(t) - \frac{\rho}{2}R(t)g(t), \quad g_0 = g(0). \tag{1}$$

An interpolation of solitons between the Ricci and Yamabe soliton is considered in the Ricci–Bourguignon soliton corresponding to Ricci–Bourguignon flow but it depends on a single scalar. Ricci–Yamabe flow can either be a Riemannian flow, a semi-Riemannian flow, or a singular Riemannian flow, depending on the sign of the associated scalars σ and ρ. Such a range of options may be beneficial in various geometrical or physical models, such as the general theory of relativity.

Consequently, the Ricci–Yamabe soliton inevitably appears as the limit of the soliton of the Ricci–Yamabe flow. Ricci–Yamabe solitons are solitons to the Ricci–Yamabe flow that move only by one parameter group of diffeomorphism and scaling. Specifically, a Ricci–Yamabe soliton on the Riemannian manifold, (M, g), is a data $(g, \omega, \tau, \sigma, \rho)$ satisfying

$$\frac{1}{2}\mathcal{L}_\omega g + \sigma S + \left(\tau - \frac{\rho}{2}R\right)g = 0, \tag{2}$$

where the Ricci tensor is S, the scalar curvature is R, and the Lie-derivative along the vector field ω is $\mathcal{L}_\omega$. The manifold $(M, g, \omega, \tau, \nu)$ is referred to as a Ricci–Yamabe shrinker, expander, or stable soliton depending on the constant τ, whether $\tau < 0$, $\tau > 0$ or $\tau = 0$.

As an extension of Ricci and Yamabe solitons, Equation (2) is referred to as a Ricci–Yamabe soliton of kind (σ, ρ). We see that the Ricci–Yamabe solitons of kind $(\sigma, 0)$ and $(0, \rho)$ are, respectively, the σ-Ricci solitons and the ρ-Yamabe solitons.

The idea of an η-Ricci soliton described in [15], is an evolutionary abstraction of the Ricci soliton. As a result, we can define the new concept similarly by amending the expression (2) that explains the type of soliton by a multiple of a specific $(0, 2)$-tensor field $\eta \otimes \eta$. These findings result in a significantly more comprehensive concept, termed an η-Ricci–Yamabe soliton (briefly an η-RY soliton) of kind (σ, ρ) defined as:

$$\frac{1}{2}\mathcal{L}_\omega g + \sigma S + \left(\tau - \frac{\rho}{2}R\right)g + \nu\eta \otimes \eta = 0, \tag{3}$$

where ν is a constant. Let us reiterate that η-RY solitons of kinds $(\sigma, 0)$ or $(1, 0)$, $(0, \rho)$, or $(0, 1)$-type are an η-Ricci soliton and an η-Yamabe soliton, respectively. For more information about these specific cases, see [16–22].

According to [23], if τ in (3) is replaced with the soliton function, then we may claim that the manifold (M, g) is an almost η-RY soliton [24]. It is important to note that they originate from the Ricci–Bourguignon flow and conformal Ricci flow, which Cantino, Mazzieri and Siddiqi recently examined [25–28]. We refer to (3) as the core equation of an approximately η-RY soliton in this more extended context.

In [22], the authors proved that the total manifold of a Riemannian submersion admits a Ricci soliton. In fact, the η-Ricci–Yamabe soliton is a generalization of the η-Ricci soliton from the proceedings of the η-Yamabe soliton, Yamabe soliton, and Einstein soliton. Therefore, motivated by the previous studies, in this paper, we discuss Riemannian submersions in terms of an η-Ricci–Yamabe soliton.

Example 1. *Let us look at the instance of an Einstein soliton, which produces solutions to Einstein flow that are self-similar (for more details see [26]), so that*

$$\frac{\partial}{\partial t}g(t) = -2\left(S - \frac{R}{2}g\right).$$

As a result, an Einstein soliton appears as the limit of the Einstein flow solution, such that

$$\mathcal{L}_\omega g + 2S + (\tau - \frac{R}{2})g = 0. \tag{4}$$

When comparing Equations (3) and (4) in this situation, we find that $\sigma = 1$ and $\rho = 1$, or its type $(1,1)$, are RY solitons.

Moreover, we note a useful definition:

Definition 1. *A smooth vector field ζ on a Riemannian manifold $(\mathcal{N}, g)$ is said to be a* conformal vector field *if there exists a smooth function φ on $\mathcal{N}$ that satisfies [29]*

$$\mathcal{L}_\zeta g = 2\varphi g, \tag{5}$$

where $\mathcal{L}_\zeta g$ is the Lie derivative of ζ with respect to g. If $\varphi = 0$, then ζ is called a Killing vector field.

2. Riemannian Submersions

We present the additional context for Riemannian submersions (briefly RS) in this part.

Let $(\mathcal{N}^n, g)$ and $(\mathcal{B}^m, g_{\mathcal{B}})$ be two Riemannian manifolds (briefly RS), endowed with metrics g and $g_{\mathcal{B}}$, wherein $\dim(\mathcal{N}) > \dim(\mathcal{B})$.
A surjective mapping $\pi : (\mathcal{N}, g) \to (\mathcal{B}, g_{\mathcal{B}})$ is called a *Riemannian submersion* [30] if:
(A1)

$$\dim(\mathcal{B}) = \mathrm{Rank}(\pi).$$

In this instance, $\pi^{-1}(s) = \pi_s^{-1}$ is a submanifold $\mathcal{N}$ ($\dim(\mathcal{N}) = t$) and is referred to as a *fiber* for all $s \in \mathcal{B}$, wherein

$$\dim(\mathcal{N}) - t = \dim(\mathcal{B}).$$

If a vector field on $\mathcal{N}$ is always tangent (resp. orthogonal) to fibers, it is said to be *vertical* (resp. *horizontal*). If a vector field P on $\mathcal{N}$ is horizontal and π-related to a vector field P_* on $\mathcal{B}$, then $\pi_*(\mathrm{P}_p) = \mathrm{E}_{*\pi(p)}$ is the basis for all $s \in \mathcal{N}$ and $\mathrm{E} \in \mathcal{B}$, wherein π_* is the differential map of π.

The projections on the vertical distribution $\mathcal{K}er\pi_*$ and the horizontal distribution $\mathcal{K}er\pi_*^{\perp}$ will be indicated by the symbols V (briefly vdV) and H (briefly hdH), respectively.

The manifold $(\mathcal{N}, g)$ is regarded as the *total manifold*, and the manifold $(\mathcal{B}, g_{\mathcal{B}})$ is regarded as the *base manifold*, as is customary.
(A2) The size of the horizontal vectors are preserved by π_*.

These requirements are similar to claiming that the differential map of π_*, restricted to $\mathcal{K}er\pi_*^{\perp}$, is a linear isometry. We obtain the following information if P and Q are the fundamental vector fields, connected to $\mathrm{P}_{\mathcal{B}}$ and $\mathrm{Q}_{\mathcal{B}}$ by π:

1. $g(\mathrm{P},\mathrm{Q}) = g_{\mathcal{B}}(\mathrm{P}_{\mathcal{B}}, \mathrm{Q}_{\mathcal{B}}) \circ \pi$,
2. $h[\mathrm{P},\mathrm{Q}]$ is the basic vector field π-connected to $[\mathrm{P}_{\mathcal{B}}, \mathrm{Q}_{\mathcal{B}}]$,
3. $h(\nabla_{\mathrm{P}}\mathrm{Q})$ is the basic vector field π-connected to ${\nabla^{\mathcal{B}}}_{\mathrm{P}_{\mathcal{B}}}\mathrm{Q}_{\mathcal{B}}$.

In the case of each vertical vector field $\{V, [I,\mathrm{J}]\}$ is vertical.

O'Neill's tensors $\mathcal{T}$ and $\mathcal{A}$, which are described below:

$$\mathcal{T}_{\mathrm{I}}\mathrm{J} = V\nabla_{V\mathrm{I}}H\mathrm{J} + H\nabla_{V\mathrm{I}}V\mathrm{J}, \tag{6}$$

$$\mathcal{A}_{\mathrm{I}}\mathrm{J} = V\nabla_{H}\mathrm{I}H\mathrm{J} + H\nabla_{H\mathrm{I}}V\mathrm{J} \tag{7}$$

if any vector fields I and J exist on $\mathcal{N}$, where ∇ denotes the Levi–Civita connection of g. The skew-symmetric operators on the tangent bundle of $\mathcal{N}$ that project the vdV and the hdH are evidently $\mathcal{T}_{\mathrm{I}}$ and A_J.
If G, K are vertical vector fields on $\mathcal{N}$ and P, Q are horizontal vector fields, then we obtain

$$\mathcal{T}_{\mathrm{G}}\mathrm{K} = \mathcal{T}_{\mathrm{K}}\mathrm{G}, \tag{8}$$

$$\mathcal{A}_{\mathrm{P}}\mathrm{Q} = -\mathcal{A}_{\mathrm{Q}}\mathrm{P} = \frac{1}{2}V[\mathrm{P},\mathrm{Q}]. \tag{9}$$

Alternatively, we discover from (6) and (7)

$$\nabla_{\mathrm{G}}\mathrm{K} = \mathcal{T}_{\mathrm{G}}\mathrm{K} + \hat{\nabla}_{\mathrm{G}}\mathrm{K}, \tag{10}$$

$$\nabla_{\mathrm{G}}\mathrm{P} = \mathcal{T}_{\mathrm{G}}\mathrm{P} + H\nabla_{\mathrm{G}}\mathrm{P}, \tag{11}$$

$$\nabla_{\mathrm{P}}\mathrm{G} = \mathcal{A}_{\mathrm{P}}\mathrm{G} + V\nabla_{\mathrm{P}}\mathrm{G}, \tag{12}$$

$$\nabla_{\mathrm{P}}\mathrm{Q} = H\nabla_{\mathrm{P}}\mathrm{Q} + \mathcal{A}_{\mathrm{P}}\mathrm{Q}, \tag{13}$$

wherein $\hat{\nabla}_{\mathrm{G}}\mathrm{K} = V\nabla_{\mathrm{G}}\mathrm{K}$. Additionally, we have

$$H\nabla_{\mathrm{G}}\mathrm{P} = \mathcal{A}_{\mathrm{P}}\mathrm{G}$$

where P is basic. It is not hard to see that $\mathcal{A}$ acts on the hdH and estimates of the resistance to the integrability of this distribution while $\mathcal{T}$ operates on the fibers as the second basic form. We refer to the book [8] as well as the paper by O'Neill [30] for more information about the RS.

3. Characteristics of Curvatures on Riemannian Submersions

The following useful Riemannian submersion (RS) curvature properties are covered in this section:

Proposition 1. *For an RS π, the Riemannian curvatures of the total manifold, the base manifold, and each fiber of π denoted by R^T, $R^{\mathcal{B}}$ and $\hat{R}$, respectively, then we have*

$$R^T(\mathrm{I},\mathrm{J},\mathrm{G},\mathrm{H}) = \hat{R}(\mathrm{I},\mathrm{J},\mathrm{G},\mathrm{H}) + g(\mathcal{T}_{\mathrm{J}}\mathrm{H},\mathcal{T}_{\mathrm{I}}\mathrm{G}) - g(\mathcal{T}_{\mathrm{I}}\mathrm{H},\mathcal{T}_{\mathrm{J}}\mathrm{G}), \tag{14}$$

$$\begin{aligned} R^T(\mathrm{P},\mathrm{Q},\mathrm{R},\mathrm{L}) = {} & R^{\mathcal{B}}(\mathrm{P}_{\mathcal{B}},\mathrm{Q}_{\mathcal{B}},\mathrm{R}_{\mathcal{B}},\mathrm{L}_{\mathcal{B}})\circ\pi + 2g(\mathcal{A}_{\mathrm{P}}\mathrm{Q},\mathcal{A}_{\mathrm{R}}\mathrm{L}) \\ & -g(\mathcal{A}_{\mathrm{Q}}\mathrm{R},\mathcal{A}_{\mathrm{P}}\mathrm{L}) + g(\mathcal{A}_{\mathrm{P}}\mathrm{R},\mathcal{A}_{\mathrm{R}}\mathrm{L}). \end{aligned} \tag{15}$$

for any $\mathrm{I},\mathrm{J},\mathrm{G},\mathrm{H} \in \Gamma V(\mathcal{N})$ and $\mathrm{P},\mathrm{Q},\mathrm{R},\mathrm{L} \in \Gamma H(\mathcal{N})$.

Proposition 2. *For an RS π, Ricci curvatures of $(\mathcal{N},g)$, $(\mathcal{B},g_{\mathcal{B}})$ and any fiber of π are denoted by S, S^N and $\hat{S}$, respectively. Then, we have*

$$S(\mathrm{I},\mathrm{J}) = \hat{S}(\mathrm{I},\mathrm{J}) + g(N,\mathcal{T}_{\mathrm{I}}\mathrm{J}) - \sum_{i=1}^{n} g((\nabla_{\mathrm{P}_i}\mathcal{T})(\mathrm{I},\mathrm{J}),\mathrm{P}_i) - g(\mathcal{A}_{\mathrm{P}_1}\mathrm{I},\mathcal{A}_{\mathrm{P}_1}\mathrm{J}) \tag{16}$$

$$\begin{aligned} S(\mathrm{P},\mathrm{Q}) = {} & S^{\mathcal{B}}(\mathrm{P}^{\mathcal{B}},\mathrm{Q}^{\mathcal{B}})\circ\pi - \frac{1}{2}\{g(\nabla_{\mathrm{P}}N,\mathrm{Q}) + g(\nabla_{\mathrm{Q}}N,\mathrm{P})\}, \\ & +2\sum_{i=1}^{n} g(\mathcal{A}_{\mathrm{P}}\mathrm{P}_i,\mathcal{A}_{\mathrm{Q}}\mathrm{P}_i) + \sum_{j=1}^{r} g(\mathcal{T}_{\mathrm{P}_i}\mathrm{P},\mathcal{T}_{\mathrm{P}_i}\mathrm{Q}), \end{aligned} \tag{17}$$

$$\begin{aligned} S(\mathrm{I},\mathrm{P}) = {} & -g(\nabla_{\mathrm{I}}N,\mathrm{P}) + \sum_{j} g((\nabla_{\mathrm{I}_i}\mathcal{T})(\mathrm{I}_j,\mathrm{E}),\mathrm{P}) \\ & - \sum_{i=1}^{n}\{g((\nabla_{\mathrm{P}_i}\mathcal{A})(\mathrm{P}_i,\mathrm{P}),\mathrm{I}) + 2g(\mathcal{A}_{\mathrm{P}_i}\mathrm{P},\mathcal{T}_{\mathrm{I}}\mathrm{P}_i)\} \end{aligned} \tag{18}$$

where $\{\mathrm{I}_i\}$ and $\{\mathrm{P}_i\}$ are the orthonormal basis of vdV and hdH, respectively, and $\mathrm{I},\mathrm{E} \in \Gamma V(\mathcal{N})$, $\mathrm{P},\mathrm{Q} \in \Gamma H(\mathcal{N})$.

Using (16) and (17), we derive the following:

Proposition 3. *In an RS π , the vertical scalar curvature R_V and the horizontal scalar curvature R_H are provided as*

$$R_V = \sum_{k=1}^{s} S(\mathrm{I}_k, \mathrm{I}_k) = \sum_{k=1}^{s} \hat{S}(\mathrm{I}_k, \mathrm{I}_k) + g(N, \mathcal{T}_{\mathrm{I}_k}\mathrm{I}_k)$$

$$- \sum_{i=1}^{n} d((\nabla_{\mathrm{P}_i}\mathcal{T})(\mathrm{I}_k, \mathrm{I}_k), \mathrm{P}_i) - g(\mathcal{A}_{\mathrm{P}_i}\mathrm{I}_k, \mathcal{A}_{\mathrm{P}_i}I_k), \tag{19}$$

$$R_H = \sum_{i=1}^{r} S(\mathrm{P}_i, \mathrm{P}_i) = \sum_{i=1}^{n} \Big\{ S^{\mathcal{B}}(\mathrm{P}_i^{\mathcal{B}}, \mathrm{P}_i^{\mathcal{B}}) \circ \pi - \frac{1}{2}\{g(\nabla_{\mathrm{P}_i}N, \mathrm{P}_i) + g(\nabla_{\mathrm{P}_i}N, \mathrm{P}_i)\},$$

$$+2\sum_{i=1}^{n} g(\mathcal{A}_{\mathrm{P}}\mathrm{P}_i, \mathcal{A}_{\mathrm{Q}}\mathrm{P}_i) + \sum_{j=1}^{r} g(\mathcal{T}_{\mathrm{P}_i}\mathrm{P}, \mathcal{T}_{\mathrm{P}_i}\mathrm{Q}). \tag{20}$$

Now, Equations (19) and (20) entail that

$$R_V = \hat{R} + \|N\|^2 - \mathrm{div}(\mathrm{N}) - \|\mathcal{A}\|^2, \tag{21}$$

$$R_H = (R^{\mathcal{B}} \circ \pi) + \|\mathcal{T}\|^2 + 2\|\mathcal{A}\|^2 - \mathrm{div}(N), \tag{22}$$

Adopting (21) and (22), we turn up the scalar curvature R of the base manifold $(\mathcal{B}, g_{\mathcal{B}})$

$$R = \hat{R} + (R^{\mathcal{B}} \circ \pi) + \|N\|^2 + \|\mathcal{A}\|^2 + \|\mathcal{T}\|^2 - 2\mathrm{div}(N). \tag{23}$$

In addition, the mean curvature vector field $\mathbf{H}$ for every fiber of RS is given by $r\mathbf{H} = N$, where N is a horizontal vector field, such that

$$N = \sum_{j=1}^{r} \mathcal{T}_{\mathrm{I}_j}\mathrm{I}_j. \tag{24}$$

Additionally, any fiber π dimension is indicated by the prefix r, and the orthonormal basis for vdV is $\{\mathrm{E}_1, \mathrm{E}_2, \cdots, \mathrm{E}_r\}$. We emphasize that all fibers of RS must be minimal, if, and only if, the horizontal vector field N vanishes. From (24), we obtain

$$g(\nabla_Z N, \mathrm{P}) = \sum_{j=1}^{r} g((\nabla_Z \mathcal{T})(\mathrm{I}_j, \mathrm{I}_j), \mathrm{P}) \tag{25}$$

for any $Z \in \Gamma(T\mathcal{N})$ and $\mathrm{P} \in \Gamma H(\mathcal{N})$.

Any horizontal vector field P divergence on $\Gamma H(\mathcal{N})$, and denoted by $\mathrm{div}(\mathcal{P})$, is determined by

$$\mathrm{div}(\mathrm{P}) = \sum_{i=1}^{n} g(\nabla_{\mathrm{P}_i}\mathrm{P}, \mathrm{P}_i), \tag{26}$$

where the orthonormal basis of the horizontal space $\Gamma H(\mathcal{N})$ is $\{\mathrm{P}_1, \mathrm{P}_2, \cdots, \mathrm{P}_n\}$. Thus, taking into account (26), we have

$$\mathrm{div}(N) = \sum_{i=1}^{n}\sum_{j=1}^{r} g(\nabla_{\mathrm{P}_i}\mathcal{T})(\mathrm{I}_j, \mathrm{I}_j), \mathrm{P}_i). \tag{27}$$

4. η-Ricci–Yamabe Solitons in Riemannian Submersions

This section discusses the η-RY soliton of kind- (σ, ρ) on RS $\pi : (\mathcal{N}, g) \longrightarrow (\mathcal{B}, g_{\mathcal{B}})$ from Riemannian manifolds and the characteristics of fiber of such RS with target manifold $(\mathcal{B}, g_{\mathcal{B}})$. Throughout the study, RS stands for a Riemannian submersion between Riemannian manifolds. We discover the following conclusions as a result of Equations (10) to (13) in the case of an RS:

Theorem 1. *If $\pi : (\mathcal{N}, g) \longrightarrow (\mathcal{B}, g_{\mathcal{B}})$ is an RS. Then, the*
1. vdV is parallel with respect to the connection ∇, if the horizontal components $\mathcal{T}_{\mathrm{I}}\mathrm{J}$ and $\mathcal{A}_{\mathrm{P}}\mathrm{I}$ are eliminated, identically.
2. hdH is parallel with respect to the connection ∇, if the vertical components $\mathcal{T}_{\mathrm{I}}\mathrm{P}$ and $\mathcal{A}_{\mathrm{P}}\mathrm{Q}$ are eliminated, identically,
for any $\mathrm{P}, \mathrm{Q} \in \Gamma H(\mathcal{N})$ and $\mathrm{I}, \mathrm{J} \in \Gamma V(\mathcal{N})$.

Since $(\mathcal{N}, g)$ is an η-RY soliton, then, by (3), we find

$$2\sigma S(\mathrm{I},\mathrm{J}) + (2\tau - \rho R)g(\mathrm{I},\mathrm{J}) + 2\nu\eta(\mathrm{I})\eta(\mathrm{J}) + (\mathcal{L}_{\omega}g)(\mathrm{I},\mathrm{J}) = 0 \tag{28}$$

for each $\mathrm{I}, \mathrm{I} \in \Gamma V(\mathcal{N})$. Adopting (16), we have

$$2\sigma\hat{S}(\mathrm{I},\mathrm{J}) + g(N, \mathcal{T}_{\mathrm{I}}\mathrm{J}) + \{g(\nabla_{\mathrm{I}}\omega,\mathrm{J}) + g(\nabla_{\mathrm{J}}\omega,\mathrm{I})\} \tag{29}$$

$$-\sum_{i=1}^{n} g((\nabla_{\mathrm{P}_t}\mathcal{T})(\mathrm{I},\mathrm{F}),\mathrm{P}_i) - g(\mathcal{A}_{\mathrm{P}_t}\mathrm{I}, \mathcal{A}_{\mathrm{P}_t}\mathrm{J}) + (2\tau - \rho R)g(I,\mathrm{J}) + 2\upsilon\eta(\mathrm{I})\eta(\mathrm{J}) = 0$$

wherein ∇ is a Levi–Civita connection on $\mathcal{N}$ and $\{\mathrm{P}_i\}$ denotes an orthonormal basis of the hdH. The following equation is then obtained by using Theorem 1, the Equations (7) and (10),

$$2\sigma\hat{S}(\mathrm{I},\mathrm{J}) + [\hat{d}(\hat{\nabla}_{\mathrm{I}}\omega,\mathrm{J}) + \hat{g}(\hat{\nabla}_{\mathrm{I}}\omega,\mathrm{I})] \tag{30}$$

$$+(2\tau - \rho \hat{R\rceil}_V)\hat{g}(\mathrm{I},\mathrm{J}) + 2\nu\eta(\mathrm{I})\eta(\mathrm{J}) = 0,$$

for every $\mathrm{I}, \mathrm{J} \in \Gamma V(\mathcal{N})$. Using (21), we find

$$2\sigma\hat{S}(\mathrm{I},\mathrm{J}) + [\hat{g}(\hat{\nabla}_{\mathrm{I}}\omega,\mathrm{J}) + \hat{g}(\hat{\nabla}_{\mathrm{J}}\omega,\mathrm{I})] \tag{31}$$

$$+(2\tau - \rho\hat{R} + ||N||^2 - ||\mathcal{A}||^2 - \mathrm{div}(N))\hat{g}(\mathrm{I},\mathrm{J}) + 2\nu\eta(\mathrm{I})\eta(\mathrm{J}) = 0.$$

Defining $R = \hat{R} + ||N||^2 - ||\mathcal{A}||^2 - \mathrm{div}(N)$, then, the Equation (31) follows;

$$2\sigma\hat{S}(\mathrm{I},\mathrm{J}) + (2\tau - \rho R)\hat{g}(\mathrm{I},\mathrm{J}) + [\hat{g}(\hat{\nabla}_{\mathrm{I}}\omega,\mathrm{J}) + \hat{g}(\hat{\nabla}_{\mathrm{J}}\omega,\mathrm{I})] + 2\nu\eta(\mathrm{I})\eta(\mathrm{J}) = 0. \tag{32}$$

Let us mention here the "vertical potential vector field" (in brief $VPVF$) and the "horizontal potential vector field" ($HPVF$). Hence, we generate the following results:

Theorem 2. *Let $(\mathcal{N}, g, \omega, \tau, \nu, \sigma, \rho)$ be an η-RY soliton of kind-(σ, ρ) with a $VPVF$ ω and π be an RS from the Riemannian manifolds. If the vdV is parallel, then every fiber in an RS is an η-RY soliton.*

Remark 1. *Now, for $\sigma = 1, \rho = 0$ and $\nu \neq 0$, then, from (30), we find*

$$2\hat{S}(\mathrm{I},\mathrm{J}) + [\hat{g}(\hat{\nabla}_{\mathrm{I}}\omega,\mathrm{J}) + \hat{g}(\hat{\nabla}_{\mathrm{I}}\omega,\mathrm{J})] + 2\tau\hat{g}(\mathrm{I},\mathrm{J}) + 2\nu\eta(\mathrm{I})\eta(\mathrm{J}) = 0. \tag{33}$$

Therefore, one can obtain the following

Theorem 3. *Let $(\mathcal{N}, g, \omega, \tau, \nu, \sigma)$ be an η-Ricci soliton of kind-$(1, 0)$ with $VPVF$ ω and π be a RS. If the vdV is parallel, then every fiber in an RS is an η-Ricci soliton.*

Remark 2. *Next, setting $\sigma = 0, \rho = 1$ and $\nu \neq 0$, so (30) entails that*

$$[\hat{g}(\hat{\nabla}_{\mathrm{I}}\omega,\mathrm{J}) + \hat{g}(\hat{\nabla}_{\mathrm{I}}\omega,\mathrm{J})] + (2\tau - R)\hat{g}(\mathrm{I},\mathrm{J}) + 2\nu\eta(\mathrm{I})\eta(\mathrm{J}) = 0, \tag{34}$$

Therefore, one can obtain the following outcome:

Theorem 4. *Let $(\mathcal{N}, g, \omega, \tau, \nu, \rho)$ be an η-RY soliton of kind-$(0,1)$ with a VPVF ω and π be a RS. If the vdV is parallel, then every fiber in an RS is a η-Yamabe soliton.*

So, if the total space $(\mathcal{N}, g)$ of $RS\ \pi : (\mathcal{N}, g) \longrightarrow (\mathcal{B}, g_{\mathcal{B}})$ admits, an η-RY soliton of kind-(σ, ρ), now, in view of (3) and (16), we obtain

$$\{g(\nabla_{\mathrm{I}}\omega, \mathrm{J}) + g(\nabla_{\mathrm{J}}\omega, \mathrm{I})\} + 2\sigma\hat{S}(\mathrm{I}, \mathrm{J}) + \sum_{j=1}^{r} g(\mathcal{T}_{\mathrm{I}_j}\mathrm{I}_j, \mathcal{T}_{\mathrm{I}}\mathrm{J}) \tag{35}$$
$$- \sum_{i=1}^{n} d((\nabla_{\mathrm{P}_i}\mathcal{T})(\mathrm{I}, \mathrm{J}), \mathrm{P}_i) - g(\mathcal{A}_{\mathrm{P}_i}\mathrm{I}, \mathcal{A}_{\mathrm{P}_i}\mathrm{J}) + (2\tau - \rho\hat{R})\hat{d}(\mathrm{I}, \mathrm{J}) + 2\nu\eta(\mathrm{I})\eta(\mathrm{J}) = 0$$

where $\mathrm{I}, \mathrm{J} \in \Gamma V(\mathcal{N})$. In addition, an η-RY soliton $(\mathcal{N}, g, \omega, \tau, \nu)$ of kind- (σ, ρ) admits totally umbilical fibers and adopting (10) in (35), we obtain

$$\{g(\hat{\nabla}_{\mathrm{I}}\omega, \mathrm{G}) + g(\hat{\nabla}_{\mathrm{G}}\omega, \mathrm{I})\} + 2\sigma\hat{S}(\mathrm{I}, \mathrm{G}) + \sum_{j=1}^{r} g(\mathcal{T}_{\mathrm{I}_j}\mathrm{I}_j, \mathcal{T}_{\mathrm{I}}\mathrm{G}) \tag{36}$$
$$- \sum_{i=1}^{n} \{(\nabla_{\mathrm{P}_i}g)(\mathrm{I}, \mathrm{G})g(\mathrm{K}, \mathrm{P}_i) - g(\nabla_{\mathrm{P}_i}\mathrm{K}, \mathrm{P}_i)\hat{g}(\mathrm{I}, \mathrm{G})\}$$
$$- \sum_{i=1}^{n} g(\mathcal{A}_{\mathrm{P}_i}\mathrm{I}, \mathcal{A}_{\mathrm{P}_i}\mathrm{G}) + (2\tau - \rho R|_{\mathrm{H}})\hat{g}(\mathrm{I}, \mathrm{G}) + 2\nu\eta(\mathrm{I})\eta(\mathrm{G}) = 0.$$

Since with integrable hdH, we derive,

$$(\mathcal{L}_{\omega}\hat{g})(\mathrm{I}, \mathrm{G}) + 2\sigma\hat{S}(\mathrm{I}, \mathrm{G}) - \sum_{i=1}^{n} g(\nabla_{\mathrm{P}_i}\mathrm{K}, \mathrm{P}_i)\hat{g}(\mathrm{I}, \mathrm{G}) \tag{37}$$
$$+ r\|W\|^2\hat{g}(\mathrm{I}, \mathrm{G}) + (2\tau - \rho\hat{R})\hat{g}(\mathrm{I}, \mathrm{G}) + 2\nu\eta(\mathrm{I})\eta(\mathrm{G}) = 0$$

wherein K is the mean curvature vector of any fiber of π. By (26), we derive

$$(\mathcal{L}_{\omega}\hat{g})(\mathrm{I}, \mathrm{G}) + 2\sigma\hat{S}(\mathrm{I}, \mathrm{G}) + [2\tau - \rho(\hat{R} - \mathrm{div}(N) + r\|N\|^2]\hat{g}(\mathrm{I}, \mathrm{G}) + 2\nu\eta(\mathrm{I})\eta(\mathrm{G}) = 0. \tag{38}$$

We observe, that every fiber for π is an almost η-RY soliton. As a result, one can state the following outcome:

Theorem 5. *If $(\mathcal{N}, g, \omega, \tau, \nu, \sigma, \rho)$ be an η-RY soliton of kind-(σ, ρ) with a VPF ω and π be an RS with totally umbilical fibers and the hdH is integrable, then every fiber in an RS is an almost η-RY soliton.*

Furthermore, the following results are obtained:

Theorem 6. *If $(\mathcal{N}, g, \omega, \tau, \nu, \sigma, \rho)$ is an η-RY soliton of kind-$(1,0)$ with a VPF, ω and π are an RS with totally umbilical fibers, and the hdH is integrable, then every fiber in a RS is an almost η-Ricci soliton.*

Proof. Fix $\sigma = 1, \rho = 0$, $\nu \neq 0$ and from (38) we derive the required outcomes. □

Theorem 7. *If $(\mathcal{N}, g, \omega, \tau, \nu, \sigma, \rho)$ is an η-RY soliton of kind-$(0,1)$ with a VPF ω and π is an RS with totally umbilical fibers and the hdH is integrable, then every fiber in an RS is an almost η-quasi Yamabe soliton.*

Proof. Putting $\sigma = 0, \rho = 1$, $\nu \neq 0$ and using (38), we gain the following: □

Assuming once more the Theorem 5, we arrive at the following corollaries:

Corollary 1. *If* $(\mathcal{N}, g, \omega, \tau, v, \sigma, \rho)$ *is an* η*-RY soliton of kind-*(σ, ρ) *and* π *is an RS, and the hdH is integrable, and if every fiber of* π *is totally umbilical and admits constant mean curvature, then any fiber in RS is an almost* η*-RY soliton,*

Corollary 2. *If* $(\mathcal{N}, g, \omega, \tau, v, \sigma, \rho)$ *is an* η*-RY soliton of kind-*(σ, ρ) *and* π *is an RS, such that the hdH is integrable, and if every fiber of* π *is totally geodesic, then any fiber of an RS is an almost* η*-RY soliton,*

Remark 3. *In light of Corollaries 1 and 2, we can derive identical results for an almost* η*-Ricci soliton and an almost* η*-quasi Yamabe soliton.*

Next, we obtain the following:

Theorem 8. *If* $(\mathcal{N}, g, Z, \tau, v, \sigma, \rho)$ *is an* η*-RY soliton of kind-*(σ, ρ) *with a VPF* $Z \in \Gamma(TM)$ *and* π *is an RS and the hdH is parallel, then the following holds:*

1. $(\mathcal{B}, g_{\mathcal{B}})$ *is an* η*-Einstein if Z is a VVF,*
2. $(\mathcal{B}, g_{\mathcal{B}})$ *is an* η*-RY soliton with VPF* $Z_{\mathcal{B}}$ *if U is HVF, such that* $\pi_* Z = Z_{\mathcal{B}}$.

Proof. As far as $(\mathcal{N}, g)$, the total space of RS π admits an η-RY soliton of kind- (σ, ρ) with a VPF $Z \in \Gamma(T\mathcal{N})$; then, utilizing (3) and (17), we gain

$$[g(\nabla_{\mathrm{P}} U, \mathrm{Q}) + g(\nabla_{\mathrm{Q}} U, \mathrm{P})] + 2\sigma S_{\mathcal{B}}(\mathrm{P}_{\mathrm{B}}, \mathcal{Q}_{\mathcal{B}}) \circ \pi - (d(\nabla_{\mathcal{P}} N, \mathcal{Q}) + d(\nabla_{\mathcal{Q}} N, \mathcal{P})) \tag{39}$$

$$+2\sum_{i=1}^{n} d(\mathcal{A}_{\mathcal{P}} \mathrm{P}_i, \mathcal{A}_{\mathrm{Q}} \mathrm{P}_i) + \sum_{j=1}^{r} g(\mathcal{T}_{\mathrm{I}_j} \mathrm{P}, \mathcal{T}_{\mathrm{I}_j} \mathrm{Q}) + (2\tau - \rho R) g(\mathrm{P}, \mathrm{Q}) + 2v\eta(\mathrm{P})\eta(\mathrm{Q}) = 0$$

wherein $\mathrm{P}_{\mathcal{B}}$ and $\mathrm{Q}_{\mathcal{B}}$ are π-connected to P and Q, respectively, for any $\mathrm{P}, \mathrm{Q} \in \Gamma H(\mathcal{N})$.
Utilizing Theorems (1) to (39), we derive

$$[g(\nabla_{\mathrm{P}} Z, \mathrm{Q}) + g(\nabla_{\mathrm{Q}} Z, \mathrm{P})] + 2\sigma S_{\mathcal{B}}(\mathrm{P}_{\mathcal{B}}, \mathrm{Q}_{\mathcal{B}}) \circ \pi \tag{40}$$

$$+(2\tau - \rho R) g(\mathrm{P}, \mathrm{Q}) + 2v\eta(\mathrm{P})\eta(\mathrm{Q}) = 0.$$

1. If Z is a VVF, from (12), it follows

$$[g(\mathcal{A}_{\mathrm{P}} Z, \mathrm{Q}) + g(\mathcal{A}_{\mathrm{Q}} Z, \mathrm{P})] + 2\sigma S_{\mathcal{B}}(\mathrm{P}_{\mathcal{B}}, \mathrm{Q}_{\mathcal{B}}) \circ \pi \tag{41}$$

$$+(2\tau - \rho R|_{\mathcal{V}}) g(\mathrm{P}, \mathrm{Q}) + 2v\eta(\mathrm{P})\eta(\mathrm{Q}) = 0.$$

Since H is parallel, we obtain

$$S_{\mathcal{B}}(\mathrm{P}_{\mathcal{B}}, \mathrm{Q}_{\mathcal{B}}) \circ \pi = a g(\mathrm{P}, \mathrm{Q}) + b\eta(\mathrm{P})\eta(\mathrm{Q}) = 0. \tag{42}$$

This proves that $(\mathcal{B}, g_{\mathcal{B}})$ is an η-Einstein, wherein $a = -(\tau - \frac{R|_{\mathcal{V}}}{2})$ and $b = -v$.

2. If Z is a horizontal vector field, from (40), we obtain

$$(\mathcal{L}_Z g)(\mathrm{P}, \mathrm{Q}) + 2\sigma S_{\mathcal{B}}(\mathrm{P}_{\mathcal{B}}, \mathrm{Q}_{\mathcal{B}}) \circ \pi + (2\tau - \rho R_{\mathrm{H}}) g(\mathrm{P}, \mathrm{Q}) + 2v\eta(\mathrm{P})\eta(\mathrm{Q}) = 0. \tag{43}$$

It is observed that the total space $(\mathcal{B}, g_{\mathcal{B}})$ is an η-RY soliton with the PVF $\mathrm{E}_{\mathcal{B}}$ lying horizontally. □

Now, from (43) and assuming that the vector field Z is horizontal, we can state the following:

Theorem 9. *Let $(\mathcal{N}, g, Z, \tau, \nu, \rho)$ be an η-RY soliton of kind-$(0, \rho)$, which admits the PVF $Z \in \Gamma(T\mathcal{N})$, and π be an RS. If the hdH is parallel and the vector field Z is horizontal, then $(\mathcal{B}, g_{\mathcal{B}})$ is an η-quasi-Yamabe soliton with HPVF $\mathrm{P}_{\mathcal{B}}$, such that*

$$(\mathcal{L}_Z g)(\mathrm{P},\mathrm{Q}) + (2\tau - \rho\Big\{(R^N \circ \pi) + \|\mathcal{T}\|^2 + 2\|\mathcal{A}\|^2 - \mathrm{div}(N)\Big\})g(\mathrm{P},\mathrm{Q}) + 2\nu\eta(\mathrm{P})\eta(\mathrm{Q}) = 0. \tag{44}$$

Once more combining Theorem (1) and (17), we arrive at the following result:

Lemma 1. *If $(\mathcal{N}, g, \zeta, \tau, \nu, \sigma, \rho)$ is an η-RY soliton on RS π that admits HPVF ζ, such that H is parallel, then the vector field N on hdH is Killing.*

Since $(\mathcal{N}, g, \zeta, \tau, \nu)$ is an η-RY soliton of kind-(σ, ρ), and again using (17) in (3), we find that

$$(\mathcal{L}_\zeta g)(\mathrm{P},\mathrm{Q}) + 2\sigma S^{\mathcal{B}}(\mathrm{P}_{\mathcal{B}}, \mathrm{Q}_{\mathcal{B}}) \circ \pi - \{g(\nabla_{\mathrm{P}} N, \mathrm{Q}) + g(\nabla_{\mathcal{Q}} N, \mathcal{P})\} \tag{45}$$

$$+2\sum_i g(\mathcal{A}_{\mathrm{P}}\mathrm{P}_i, \mathcal{A}_{\mathrm{Q}}\mathrm{P}_i) + \sum_j g(\mathcal{T}_{Z_j}\mathrm{P}, \mathcal{T}_{Z_j}\mathrm{Q}) + (2\tau - \rho R|_H)g(\mathrm{P},\mathrm{Q}) + 2\nu\eta(\mathrm{P})\eta(\mathrm{Q}) = 0.$$

For any $\mathrm{P},\mathrm{Q} \in \Gamma H(\mathcal{N})$, where $\{\mathcal{P}_i\}$ denotes an orthonormal basis of H. Equation (45) is derived from Theorem 1 as follows:

$$(\mathcal{L}_\zeta d)(\mathcal{P}, \mathcal{Q}) + 2\sigma S^{\mathcal{B}}(\mathcal{P}_{\mathcal{B}}, \mathcal{Q}_{\mathcal{B}}) \circ \pi + (2\tau - \rho R|_H)d(\mathcal{P}, \mathcal{Q}) + 2\nu\eta(\mathcal{P})\eta(\mathcal{Q}) = 0. \tag{46}$$

We may determine that ζ is a conformal Killing vector field $(CKVF)$ because the Riemannian manifold $(\mathcal{B}, d_{\mathcal{N}})$ is an η-Einstein. As a result, we can state the following outcome:

Theorem 10. *Let $(\mathcal{B}, d, \zeta, \tau, \nu, \sigma, \rho)$ be an η-RY soliton of kind (σ, ρ) on RS to an η-Einstein which admits HPVF ζ, such that hdH is parallel. Then, the vector field ζ on hdH is CKVF.*

5. Examples

Example 2. *Let $\mathcal{N}^6 = \{(\theta_1, \theta_2, \theta_3, \theta_4, \theta_5, \theta_6) | \theta_6 \neq 0\}$ be a 6-dimensional differentiable manifold where (θ_i) signifies the standard coordinates of a point in $\mathbb{R}^6$, and $i = 1, 2, 3, 4, 5, 6$.*

Let

$$\delta_1 = \partial\theta_1, \quad \delta_2 = \partial\theta_2, \quad \delta_3 = \partial\theta_3,$$

$$\delta_4 = \partial\theta_4, \quad \delta_5 = \partial\theta_5, \quad \delta_6 = \partial\theta_6$$

be the basis for the tangent space $T(\mathcal{N}^6)$ since it consists of a set of linearly independent vector fields at each point of the manifold $\mathcal{N}^6$. A definite positive metric d on $\mathcal{N}^6$ is defined as follows: with $i, j = 1, 2, 3, 4, 5, 6$, and it is defined as

$$d = \sum_{i.j=1}^{6} dx\theta_i \otimes d\theta_j.$$

Let γ be a 1-form such that $\gamma(U) = d(U, P)$ where $\delta_6^\sharp = P$. Thus, $(\mathcal{N}^6, d)$ is a Riemannian manifold. In addition, $\bar{\nabla}$ is the Levi–Civita connection with respect to d. Then, we have

$$[\delta_1, \delta_2] = 0, \quad [\delta_1, \delta_6] = \delta_1, \quad [\delta_2, \delta_6] = \delta_2, \quad [\delta_3, \delta_6] = \delta_3,$$

$$[\delta_4, \delta_6] = \delta_4, \quad [\delta_5, \delta_6] = \delta_6, \quad [\delta_i, \delta_j] = 0,$$

where $1 \leq i \neq j \leq 5$.
The induced connection $\hat{\nabla}$ for the metric $\hat{g}$ is described as

$$2g(\hat{\nabla}_{UV,W}) = Ug(V,W) + Vg(W,U) - Wg(U;V) \\ - g(U,[V,W]) - g(V,[U,W]) + g(W,[U,V]),$$

where the metric g corresponds to the Levi–Civita connection denoted by the symbol ∇.
The following equations are obtained by combining Koszul's formula with (10).

$$\hat{\nabla}_{\delta_1}\delta_1 = \delta_6,\ \hat{\nabla}_{\delta_2}\delta_2 = \delta_6,\ \hat{\nabla}_{\delta_3}\delta_3 = \delta_6,\ \hat{\nabla}_{\delta_4}\delta_4 = \delta_6, \hat{\nabla}_{\delta_5}\delta_5 = \delta_6 \tag{47}$$

$$\hat{\nabla}_{\delta_6}\delta_6 = 0,\ \hat{\delta}_{\delta_6}\delta_i = 0,\ \hat{\nabla}_{\delta_i}\delta_6 = \delta_i,\ 1 \leq i \leq 5$$

wherein $1 \leq i, j \leq 5$, we have $\hat{\nabla}_{\delta_i}\delta_i = 0$.

The non-vanishing components of $\hat{R}$, $\hat{S}$, and $\hat{R}$ of the fiber may now be computed from Equations (14) and (47).

$$\hat{R}(\delta_1,\delta_2)\delta_1 = \delta_2,\quad \hat{R}(\delta_1,\delta_2)\delta_2 = -\delta_1,\quad \hat{R}(\delta_1,\delta_3)\delta_1 = -\delta_3, \hat{R}(\delta_1,\delta_3)\delta_3 = \delta_1 \tag{48}$$

$$\hat{R}(\delta_1,\delta_4)\delta_1 = -\delta_4,\quad \hat{R}(\delta_1,\delta_4)\delta_4 = \delta_1,\quad \hat{R}(\delta_1,\delta\delta_5)\delta_1 = -\delta_5, \hat{R}(\delta_1,\delta_5)\delta_5 = \delta_1$$

$$\hat{R}(\delta_1,\delta_6)\delta_1 = -\delta_6,\quad \hat{R}(\delta_1,\delta_6)\delta_6 = -\delta_1,\quad \hat{R}(\delta_2,\delta_3)\delta_2 = -\delta_3, \hat{R}(\delta_2,\delta_3)\delta_3 = \delta_2$$

$$\hat{R}(\delta_2,\delta_4)\delta_2 = \delta_4,\quad \hat{R}(\delta_2,\delta_4)\delta_4 = -\delta_2,\quad \hat{R}(\delta_2,\delta_5)\delta_2 = \delta_5, \hat{R}(\delta_2,\delta_5)\delta_5 = -\delta\delta_2$$

$$\hat{R}(\delta_2,\delta_6)\delta_2 = \delta_6,\quad \hat{R}(\delta_2,\delta_6)\delta_6 = -\delta_2,\quad \hat{R}(\delta_3,\delta_4)\delta_3 = \delta_4, \hat{R}(\delta_3,\delta_4)\delta_4 = \delta_5$$

$$\hat{R}(\delta_3,\delta_5)\delta_5 = -\delta_3,\quad \hat{R}(\delta_3,\delta_6)\delta_3 = -\delta_6,\quad \hat{R}(\delta_3,\delta_6)\delta_3 = -\delta_6, \hat{R}(\delta_3,\delta_6)\delta_6 = -\delta_3$$

$$\hat{R}(\delta_4,\delta_5)\delta_4 = \delta_5,\quad \hat{R}(\delta_4,\delta_5)\delta_5 = -\delta_4,\quad \hat{R}(\delta_4,\delta_6)\delta_4 = -\delta_6,$$

$$\hat{R}(\delta_4,\delta_6)\delta_6 = -\delta_4,\quad \hat{R}(\delta_5,\delta_6)\delta_5 = -\delta_6,\quad \hat{R}(\delta_5,\delta_6)\delta_6 = -\delta_5.$$

$$\hat{S}(\delta_i,\delta_j) = \begin{bmatrix} -3 & 0 & 0 & 0 & 0 & 0 \\ 0 & -3 & 0 & 0 & 0 & 0 \\ 0 & 0 & -3 & 0 & 0 & 0 \\ 0 & 0 & 0 & -3 & 0 & 0 \\ 0 & 0 & 0 & 0 & -3 & 0 \\ 0 & 0 & 0 & 0 & 0 & -5 \end{bmatrix}.$$

$$\hat{R} = Trace(\hat{S}) = -20. \tag{49}$$

From Equation (16), we have

$$\frac{1}{2}[\hat{g}(\hat{\nabla}_{\delta_i}\delta_6,\delta_i) + \hat{g}(\hat{\nabla}_{\delta_i}\delta_6,\delta_i)] + \sigma\hat{S}(\delta_i,\delta_i) + (\tau - \frac{1}{2}\rho\hat{R})\hat{g}(\delta_i,\delta_i) + 2\nu\Omega^i_j = 0 \tag{50}$$

wherein, for all $i \in \{1,2,3,4,5,6\}$. Thus, $\tau = 10\rho - 3\sigma - 1$ and $\nu = 23\sigma - 30\rho - 1$, and the data $(\hat{g},\delta_6,\tau,\nu,\sigma,\rho)$ is an η-RY soliton, verified by Equation (16). Therefore, the data $(\omega,\hat{d},\tau,\nu,\sigma,\rho)$ admits increasing, decreasing and stable η-RY solitons referring to $(3\sigma+1) > 10\rho$, $(3\alpha+1) < 10\rho$ or $(3\sigma+1) = 10\rho$, respectively

The two basic instances for a specific value of σ and ρ are as follows:

Case 1. For an η-Ricci–Yamabe soliton of type (σ, ρ), if $\sigma = 1, \rho = 0$, we gain $\tau = -4$ and $\nu = 22$. Then, we say $(\hat{d}, \delta_6, \tau, \nu, 1, 0)$ is an η-Ricci soliton which is shrinking. This case illustrates Theorem 3.

Case 2. For an η-RY soliton of kind (σ, ρ) if $\sigma = 0, \rho = 0$, we derive $\tau = 9$ and $\nu = -31$; then, we have the data $(\hat{g}, \delta_6, \tau, \nu, 0, 1)$ is an η-Yamabe soliton is expanding. This illustrates Theorem 4.

Example 3. *Let* $\pi : \mathbb{R}^6 \to \mathbb{R}^3$ *be a submersion defined by*

$$\pi(x_1, x_2, ...x_6) = (y_1, y_2, y_3),$$

where

$$y_1 = \frac{x_1 + x_2}{\sqrt{2}},\ y_2 = \frac{x_3 + x_4}{\sqrt{2}} \text{ and } y_3 = \frac{x_5 + x_6}{\sqrt{2}}.$$

The Jacobi matrix of π has rank 3 at that point. This indicates that π is a submersion. Simple calculations produce

$$(\mathcal{K}er\pi_*) = Span\{V_1 = \frac{1}{\sqrt{2}}(-\partial x_1 + \partial x_2), V_2 = \frac{1}{\sqrt{2}}(-\partial x_3 + \partial x_4),$$
$$V_3 = \frac{1}{\sqrt{2}}(-\partial x_5 + \partial x_6)\},$$

and

$$(\mathcal{K}er\pi_*)^{\perp} = Span\{H_1 = \frac{1}{\sqrt{2}}(\partial x_1 + \partial x_2), H_2 = \frac{1}{\sqrt{2}}(\partial x_3 + \partial x_4),$$
$$H_3 = \frac{1}{\sqrt{2}}(\partial x_5 + \partial x_6)\},$$

Also, direct computation yields

$$\pi_*(H_1) = \partial y_1, \pi_*(H_2) = \partial y_2 \text{ and } \pi_*(H_3) = \partial y_3.$$

It is easy to observe that

$$g_{\mathbb{R}^6}(H_i, H_i) = g_{\mathbb{R}^3}(\pi_*(H_i), \pi_*(H_i)),\ i = 1, 2, 3$$

Hence, ψ is a RS.

Next, we estimate the components of $\hat{R}$, $\hat{S}$ and $\hat{R}$ for $\mathcal{K}er\pi_*$ and $\mathcal{K}er\pi_*^{\perp}$, respectively. For the vertical space, we gain

$$\hat{R}(V_1, V_2)V_1 = -2V_2, \quad \hat{R}(V_1, V_2)V_2 = 2V_1, \quad \hat{R}(V_1, V_3)V_1 = -2V_3 \tag{51}$$

$$\hat{R}(V_1, V_2)V_3 = V_1, \quad \hat{R}(V_2, V_3)V_3 = V_2, \quad \hat{R}(V_2, V_3)V_2 = V_2.$$

$$\hat{S}(V_i^{\sharp}, V_j^{\sharp}) = \begin{bmatrix} 2 & 0 & 0 \\ 0 & 2 & 0 \\ 0 & 0 & 1 \end{bmatrix}.$$

$$\hat{R} = Trace(\hat{S}) = 5. \tag{52}$$

Using (3), we find $\tau = \frac{5\rho}{2} - \sigma$ and $\nu = \alpha$. Therefore, $(\mathcal{K}er\psi_*, g)$ admits the increasing, decreasing and stable η-RY solitons referring to $\frac{5\rho}{2} < \sigma$, $\frac{5\rho}{2} > \sigma$ or $\frac{5\rho}{2} = \sigma$, respectively.

Moreover, we also have the following cases for particular values of α and β, such as:
Case 1. In an η-RY soliton of type (σ,ρ) for $\sigma=1,\sigma=0$, we find $\tau=-2$ and $\nu=1$, then $(\mathcal{K}er\psi_*,d)$ admitting a shrinking η-Ricci soliton.
Case 2. In an η-RY soliton of type (σ,τ) for $\sigma=0,\rho=1$, we find $\tau=\frac{5}{2}$ and $\nu=0$; then, we have $(\mathcal{K}er\pi_*,g)$ admitting an expanding Yamabe soliton.

In a similar way, for the horizontal space, we derive

$$R^{\mathcal{B}}(\pi_*(H_1),\pi_*(H_2))\pi_*(H_1)=\frac{1}{2}(\partial x_3+\partial x_4),\qquad R^{\mathcal{B}}(\pi_*(H_1),\pi_*(H_3))\pi_*(H_3)=\frac{1}{\sqrt{2}}(\partial x_6-\partial x_5),$$

$$R^{\mathcal{B}}(\pi_*(H_1),\pi_*(H_3))\pi_*(H_1)=\frac{1}{2}\partial x_6,\qquad R^{\mathcal{B}}(\pi_*(H_2),\pi_*(H_3))\pi_*(H_2)=(\frac{1}{\sqrt{2}}-1)\partial x_6,$$

$$R^{\mathcal{B}}(\pi_*(H_2),\pi_*(H_3))\pi_*(H_3)=-\frac{1}{2}(\partial x_3+\partial x_4),$$
$$R^{\mathcal{B}}(\pi_*(H_1),\pi_*(H_2))\pi_*(H_2)=\frac{1}{2\sqrt{2}}(\partial x_1+\partial x_2).$$

and

$$S^{\mathcal{B}}(\pi_*H_i,\pi_*H_j)=\begin{bmatrix}-\frac{3}{2\sqrt{2}} & 0 & 0\\ 0 & -\frac{3}{2\sqrt{2}} & 0\\ 0 & 0 & -\frac{1}{\sqrt{2}}\end{bmatrix}.$$

$$R^{\mathcal{B}}=Trace(S^{\mathcal{B}})=-2\sqrt{2}. \tag{53}$$

Again using (3), we derive $\tau=\frac{3\sigma}{2\sqrt{2}}-\sqrt(2)\rho$ and $\nu=-\frac{\sigma}{2\sqrt{2}}$. Therefore, $((\mathcal{K}er\pi_*^{\perp}),g)$ admits the expanding, shrinking and steady η-RY solitons referring to $\frac{53\sigma}{2\sqrt{2}}>\sqrt(2)\rho$, $\frac{3\sigma}{2\sqrt{2}}<\sqrt(2)\rho$ or $\frac{3\sigma}{2\sqrt{2}}=\sqrt(2)\rho$, respectively.

Also, we have obtained the following cases for particular values of σ and ρ, such as:

Case 1. In an η-RY soliton of type (σ,ρ) for $\sigma=1,\rho=0$, we find $\tau=\frac{3}{2\sqrt{2}}$ and $\nu=-\frac{1}{2\sqrt{2}}$; then, $((\mathcal{K}er\pi_*^{\perp}),g)$ is admitting an expanding η-Ricci soliton.

Case 2. In an η-RY soliton of type (σ,ρ) for $\sigma=0,\rho=1$, we find $\tau=-\sqrt{2}$ and $\nu=0$; then, we have $((\mathcal{K}er\pi_*^{\perp}),g)$ is admitting a shrinking Yamabe soliton.

6. η-Ricci–Yamabe Soliton with a Potential Vector Field $\omega=grad(\gamma)$

Let the potential vector field $\omega=grad(\gamma)$ on $\mathcal{N}$; then, $(\mathcal{N},g,\omega,\tau,\nu,\sigma,\rho)$ is said to be a gradient η-RY soliton, which is indicated by $(\mathcal{N},g,\omega,\tau,\nu,\sigma,\rho)$.

Now, consider the equation η-Ricci–Yamabe soliton for an r-dimensional fiber in RS.

$$2\sigma\hat{S}(\mathrm{I},\mathrm{J})=-[\hat{g}(\hat{\nabla}_{\mathrm{I}}\omega,\mathrm{J})+\hat{g}(\hat{\nabla}_{\mathrm{J}}\omega,\mathrm{I})]-(2\tau-\rho R)\hat{d}(\mathrm{I},\mathrm{J})-2\nu\eta(\mathrm{I})\eta(\mathrm{J}). \tag{54}$$

Contracting the Equation (54), we obtain

$$\mathrm{div}(\omega)=-r\tau+R\Big(\frac{\rho}{2}-\sigma\Big)-\nu. \tag{55}$$

As a result, the following theorems exist:

Theorem 11. *If $(\mathcal{N},g,\omega,\tau,\nu,\sigma,\rho)$ is an η-RY soliton of kind (σ,ρ) with gradient PVF $\omega=grad(\gamma)$, and the vd$\mathcal{V}$ is parallel, then every fiber in RS is an η-RY soliton, and the Poisson equation satisfied by γ becomes*

$$\Delta(\gamma)=-r\tau+R\Big(\frac{\rho}{2}-\sigma\Big)-\nu. \tag{56}$$

Theorem 12. *Let* $(\mathcal{N}, g, \omega, \tau, \nu, \rho)$ *be an* η*-RY soliton of kind* $(0, \rho)$ *with gradient PVF* $\omega = grad(\gamma)$ *and the* $vd\mathcal{V}$ *is parallel, then every fiber in RS is an* η*-Yamabe soliton, and the Poisson equation satisfied by* γ *becomes*

$$\Delta(\gamma) = -r\tau + R\left(\frac{\rho}{2}\right) - \nu. \tag{57}$$

Remark 4. *If* $\nu = 0$ *in (56) and (57), we can easily obtain similar types of results for the RY soliton and Yamabe soliton from Theorems (11) and (12), respectively.*

7. Physical Applications of Solitons

As far as a physically relevant model having a solitonic solution is concerned, the theory of collapse condensates with the inter-atomic attraction and spin-orbit coupling (SOC) [31], which is a fundamentally important effect in physical models, chiefly, Bose–Einstein condensates (BEC) [32]. The SOC emulation proceeds by mapping the spinor wave function of electrons into a pseudo-spinor mean-field wave function in BEC, whose components represent two atomic states in the condensate. While SOC in bosonic gases is a linear effect, there is interplay with the intrinsic BEC non-linearity, including several types of one dimensional $(1D)$ solitons [33]. An experimental realization of SOC in two-dimensional $(2D)$ geometry has been reported too [34], which suggests, in particular, the possibility of creation of a $2D$ gap soliton [35], supported by a combination of SOC and a spatially periodic field.

A fundamental problem that impedes the creation of $2D$ and $3D$ solitons in BES, nonlinear optics, and other nonlinear settings, is that the ubiquitous cubic self-attraction, which usually rise to solitons, simultaneously derives the critical and supercritical collapse in the $2D$ and $3D$ cases, respectively [36]. Although SOC modifies the conditions of the existence of the solutions and of the blow-up, it does not arrest the collapse completely [33]. The collapse destabilizes formally existing solitons, which results in stabilization of $2D$ and $3D$ solitons [32].

In the presence of SOC, the evolution of the wave function is described by a system-coupled nonlinear PDE in the Schrödinger form [37]

$$i\hbar\frac{\partial\Psi}{\partial} = \left[-\frac{\hbar^2}{2M}\Delta + \hat{H}_{so} + \frac{1}{2}(B.\hat{\sigma}) - g_2|\Psi|^2\right]\Psi, \tag{58}$$

where M is the mass of the particle, $\hat{H}_{so}$ is the SOC Hamiltonian, B is the effective magnetic field, $\hat{\sigma}$ is the spin operator and g_2 is the coupling constant.

The key point in understanding the role of the SOC in the collapse process is the modified velocity

$$v = k + \nabla_k \hat{H}_{so}, \tag{59}$$

where $k = -i\frac{\partial}{\partial r}$, including the velocity and $\nabla_k \hat{H}_{so}$ $(\nabla_k \equiv \frac{\partial}{\partial k})$, are directly related to the particle spin.

Let the first form Rashaba spin-orbit coupling

$$\hat{H}_{so} \equiv \hat{H}_R = \alpha(k_x\hat{\sigma}_y - k_y\hat{\sigma}_x), \tag{60}$$

with coupling constant α and $k = (k_x, k_y)$. The corresponding spin-dependent term in the velocity operators in Equation (59) becomes (for more details see [33])

$$\frac{\partial \hat{H}_R}{\partial k_x} = \alpha\hat{\sigma}_y, \qquad \frac{\partial \hat{H}_R}{\partial k_y} = -\alpha\hat{\sigma}_x. \tag{61}$$

In particular, in the $2D$ case, the nonlinear Schrödinger equation with cubic self-attraction term gives rise to degenerate families of the fundamental *Townes solitons* [38] with vorticity $S = 0$, which means decaying solutions. Hence, Townes solitons, that play the role of separation between the type of dynamical behavior, are the completable unstable and total norm of the spinor wave function that does not exceed a critical value. Further, it also produces stable dipole and quadrupole bound states of fundamental solitons with opposite signs.

8. Application of Riemannian Submersions to Number Theory

The Hopf fibration [39] is a Riemannian submersion $\pi : (\mathcal{N}^n, g) \to (\mathcal{B}^b, g_{\mathcal{B}})$ with totally geodesic fibers. In addition, a large class of Riemannian submersions are Riemannian submersions between spheres of higher dimensions, such as

$$\pi : \mathbb{S}^{r+m} \longrightarrow \mathbb{S}^m$$

whose fibers have dimension m. The Hopf fibration asserts that the fibration generalizes the idea of a fiber bundle and plays a significant role in algebraic topology, number theory and groups theory [40].

Every fiber in a fibration is closely connected to the homotopy group and satisfies the homotopy property [41]. The homotopy group of spheres $\mathbb{S}^n$ essentially describes how several spheres of different dimensions may twist around one another. For the j-th homotopy group $\Phi_j(\mathbb{S}^r)$, the j-dimensional sphere $\mathbb{S}^j$ can be mapped continuously to the r-dimensional sphere $\mathbb{S}^r$.

Now, we can make the following remark :

Remark 5. *To determine the homotopy groups for positive k using the formula $\pi_{r+k}(\mathbb{S}^r)$. The homotopy groups $\pi_{r+k}(\mathbb{S}r)$ with $r > k+1$ are known as stable homotopy groups of spheres and are denoted by $\pi_k^{\mathbb{S}}$; they are finite abelian groups for $k \neq 0$. In view of Freudenthal's suspension theorem [42], the groups are known as unstable homotopy groups of spheres for $r \leq k+1$.*

Now, in the light of Corollary 2 and using the above facts (5), we gain the following outcomes.

Theorem 13. *If $(\mathcal{N}, g, \omega, \tau, \nu, \sigma, \rho)$ is an η-RY soliton of kind (σ, ρ) and π is an RS , such that the hdH is integrable, if every fiber of π is totally geodesic and any fiber of RS is an almost η-RY soliton, then the homotopy group of RS is $\pi_n(\mathcal{B}^b)$.*

Example 4. *Let us adopt the example (3); we have Riemannian submersion ,*

$$\pi : \mathbb{R}^6 \cong \mathbb{S}^6 \to \mathbb{R}^3 \cong \mathbb{S}^3$$

defined in (3).

Then, according to Hopf-fibration of the fiber bundle, we have homotopy groups

$$\pi_6(\mathbb{S}^3) = \pi_{3+3}\mathbb{S}^3. \tag{62}$$

Therefore, the above remark entails that $r \leq k+1$ i.e., $3 \leq 3+1$. Thus, the homotopy groups $\pi_6(\mathbb{S}^3)$ are unstable homotopy groups.

Remark 6. *For a prime number p, the homotopy p-exponent of a topological space $\mathcal{T}$, denoted by $Exp_p(U)$, is defined to be a largest $e \in \mathbb{N} = \{0, 1, 2, \cdots\}$ such that some homotopy group $\Phi_j(\mathcal{T})$ has an element of order p^e. Cohen et al. [43] proved that the*

$$Exp_p(\mathcal{S}^{2n+1}) = n \quad if \quad p \neq 2.$$

For a prime number p and an integer z, the p-adic order of z is given by $Ord_p(z) = sup\{z \in \mathbb{N} : p^z | z\}$.

Through the above observation, in 2007, Davis and Sun proved an interesting inequality in terms of homotopy groups. For more details see ([44] Theorem 1.1 Page 2). According to these authors, for any prime p and $z = 2, 3, \cdots$ some homotopy group $\pi_i(SU(n))$ contains an element of order $p^{n-1+Ord_p(\lfloor n/p \rfloor !)}$, i.e., then the strong and elegant lower bound for the homotopy p-exponent of a homotopy group is

$$Exp_p(SU(n)) \geq n - 1 + Ord_p\left(\left\lfloor \frac{n}{p} \right\rfloor !\right), \tag{63}$$

where $S(U)(n)$ is a special unitary group of degree n.

Therefore, using Davis and Sun's result (Theorem 1.1 [44]) with Theorem 13, we gain an interesting inequality

Theorem 14. *For any prime number p and $s = 2, 3, \cdots$, some homotopy group $\pi_n(\mathcal{B}^b)$ of Riemannian submersion π with totally geodesic fiber where the fiber is an almost η-RY-soliton of π, contains an element of order $p^{s-1+Ord_p(\lfloor s/p \rfloor !)}$, we derive the inequality*

$$Exp_p(\pi_n(\mathcal{B}^b)) \geq n - 1 + Ord_p\left(\left\lfloor \frac{b}{p} \right\rfloor !\right). \tag{64}$$

Example 5. *Again considering the case of example (4), we have that a homotopy group of Riemannian submersion π with totally geodesic fiber is $\pi_6(\mathbb{R}^3)$. Equation (14) also holds for homotopy group $\pi_6(\mathbb{R}^3)$ of Riemannian submersion π such that*

$$Exp_p(\pi_6(\mathbb{R}^3)) \geq 2 + Ord_p\left(\left\lfloor \frac{3}{p} \right\rfloor !\right). \tag{65}$$

The geometric interpretation of the Hopf fibration can be obtained considering rotations of the 2-sphere in 3-dimensional space. Therefore, the rotation group $SO(3)$, spin group $Spin(3)$, diffeomorphic to the 3-sphere and $Spin(3)$, can be identified with the special unitary group $SU(2)$. Indeed, there are p-local equivalences

$$SO(3) \cong Spin(3) \cong SU(2).$$

Thus, in view (65), we obtain

$$Exp_p(SU(2)) \geq 2 + Ord_p\left(\left\lfloor \frac{2}{p} \right\rfloor !\right). \tag{66}$$

$$Exp_p(Spin(3)) \geq 2 + Ord_p\left(\left\lfloor \frac{3}{p} \right\rfloor !\right). \tag{67}$$

Remark 7. *Each homotopy group is the product of cyclic groups of order p. In [45] Hirsi, a useful classification of homotopy groups of spheres is provided. Again, in light of example (4) $\pi_6(\mathbb{R}^3) = \pi_{3+3}(\mathbb{R}^3) = 12 = 2^2.3 = \mathbb{Z}_{12} = \mathbb{Z}_4 \times \mathbb{Z}_2 \times \mathbb{Z}_3$ or $\mathbb{Z}_4 \times \mathbb{Z}_3$.*

Remark 8. *In [46], Herstien noted the following facts about any group of order type p^2q:*
1. If G is a group of order p^2q, p, q are primes, then group G has a non-trivial normal subgroup.
2. If G is a group of order p^2q, p, q are primes, then either a p-Sylow subgroup or a q-Sylow subgroup of G must be normal.

Therefore, in light of the above remarks, we can make the following remark:

Remark 9. *The order of a homotopy group $\pi_6(\mathbb{R}^3)$ of Riemanian submersion ψ can be expressed as $2^2.3$. Therefore, The homotopy group $\pi_6(\mathbb{R}^3)$ of Riemanian submersion π has a non-trivial normal* *subgroup. In addition, the homotopy group* $\pi_6(\mathbb{R}^3)$ *of Riemanian submersion* π *with a 2-Sylow subgroup or a 3-Sylow subgroup of* $\pi_6(\mathbb{R}^3)$ *must be normal.*

Remark 10. *In light of Remark 9, we can also find some results for the p-Sylow subgroup of the group of spin of Riemannian submersion and the unitary group of Riemannian submersion. These facts distinguish this manuscript from previously published works based on submersion.*

Author Contributions: Conceptualization M.D.S. and M.A.A.; formal analysis, M.D.S., M.A.A. and A.H.H.; investigation, M.D.S., M.A.A. and F.M.; methodology, M.D.S.; project administration and funding F.M.; validation, M.D.S., F.M. and A.H.H.; writing—original draft, M.D.S. All authors have read and agreed to the published version of the manuscript.

Funding: The author, Fatemah Mofarreh, expresses her gratitude to Princess Nourah bint Abdulrahman University Researchers Supporting Project number (PNURSP2023R27), Princess Nourah bint Abdulrahman University, Riyadh, Saudi Arabia.

Acknowledgments: We thank the anonymous reviewers for their careful reading of our manuscript and their many insightful comments and suggestions. The author, Fatemah Mofarreh, expresses her gratitude to Princess Nourah bint Abdulrahman University Researchers Supporting Project number (PNURSP2023R27), Princess Nourah bint Abdulrahman University, Riyadh, Saudi Arabia.

Conflicts of Interest: The authors declare no conflict of interest.

Abbreviations

η-RY soliton	η-Ricci–Yamabe soliton
RS	Riemannian submersion
vdV	vertical distribution vector field
hdH	horizontal distribution vector field
PVF	potential vector field
HPVF	horizontal potential vector field
CKVF	conformal Killing vector field

References

1. Nash. J.N. The embedding problem for Riemannian manifolds. *Ann. Math.* **1956**, *63*, 20–63. [CrossRef]
2. Bourguignon, J.P.; Lawson, H.B. Stability and isolation phenomena for Yang-mills fields. *Commun. Math. Phys.* **1981**, *79*, 189–230. [CrossRef]
3. Bourguignon, J.P.; Lawson, H.B. A mathematician's visit to Kaluza-Klein theory. *Rend. Semin. Mat. Torino Fasc. Spec* **1989**, 143–163.
4. Ianus, S.; Visinescu, M. Kaluza-Klein theory with scalar fields and generalized Hopf manifolds. *Class. Quantum Gravity* **1987**, *4*, 1317–1325. [CrossRef]
5. Ianus, S.; Visinescu, M. Space-time compaction and Riemannian submersions. In *The Mathematical Heritage of C. F. Gauss*; Rassias, G., Ed.; World Scientific: River Edge, NJ, USA, 1991; pp. 358–371.
6. Mustafa, M.T. Applications of harmonic morphisms to gravity. *J. Math. Phys.* **2000**, *41*, 6918–6929. [CrossRef]
7. Watson, B. G, G′-Riemannian submersions and nonlinear gauge field equations of general relativity. In *Global Analysis-Analysis on manifolds, Dedicated M. Morse. Teubner-Texte Math*; Rassias, T., Ed.; Teubner: Leipzig, Germany 1983; Volume 57, pp. 324–349.
8. Falcitelli, M.; Ianus, S.; Pastore, A.M. *Riemannian Submersions and Related Topics*; World Scientific: River Edge, NJ, USA, 2004.
9. Şahin, B. *Riemannian Submersions, Riemannian Maps in Hermitian Geometry and Their Applications*; Elsevier: Amsterdam, The Netherlands; Academic Press: Cambridge, MA, USA, 2017.
10. Şahin, B. Anti-invariant Riemannian submersions from almost Hermitian manifolds. *Cent. Eur. J. Math.* **2010**, *8*, 437–447. [CrossRef]
11. Taştan, H.M.; Özdemir, F.; Sayar, C. On anti-invariant Riemannian submersions whose total manifolds are locally product Riemannian. *J. Geom.* **2017**, *108*, 411–422. [CrossRef]
12. Shahid, A.; Tanveer, F. Anti-invariant Riemannian submersions from nearly Kählerian manifolds. *Filomat* **2013**, *27*, 1219–1235.
13. Hamilton, R.S. The Ricci flow on surfaces, Mathematics and general relativity. *Contemp. Math. Am. Math. Soc.* **1988**, *71*, 237–262.
14. Güler, S.; Crasmareanu, M. Ricci-Yamabe maps for Riemannian flow and their volume variation and volume entropy. *Turk. J. Math.* **2019**, *43*, 2631–2641. [CrossRef]
15. Cho, J.T.; Kimura, M. Ricci solitons and Real hypersurfaces in a complex space form. *Tohoku Math. J.* **2009**, *61*, 205–212. [CrossRef]
16. Blaga, A.M. η-Ricci solitons on para-Kenmotsu manifolds. *Balk. J. Geom. Its Appl.* **2015**, *20*, 1–13.

17. Chen, B.Y.; Desahmukh, S. Yamabe and quasi-Yamabe soliton on euclidean submanifolds. *Mediterr. J. Math.* **2018**, *15*, 194 [CrossRef]
18. Güler, S. On a Class of Gradient Almost Ricci Solitons. *Bull. Malays. Math. Sci. Soc.* **2020**, *43*, 3635–3650. [CrossRef]
19. Siddiqi, M.D. η-Ricci solitons in 3-dimensional normal almost contact metric manifolds. *Bull. Transilv. Univ. Bras. Ser. Iii Math. Inform. Phys.* **2018**, *11*, 215–234.
20. Siddiqi, M.D.; Alkhaldi, A.H.; Khan, M.A.; Siddiqui, A.N. Conformal η-Ricci solitons on Riemannian submersions under canonical variation. *Axioms* **2022**, *11*, 594. [CrossRef]
21. Chaubey, S.K.; Siddiqi, M.D.; Yadav, S. Almost η-Ricci-Bourguignon solitons on submersions from Riemannian submersions *Balk. J. Geom. Its Appl.* **2022**, *27*, 24–38.
22. Meriç, Ş.E.; Kılıç, E. Riemannian submersions whose total manifolds admit a Ricci soliton. *Int. J. Geom. Meth. Mod. Phys.* **2019** *16*, 1950196. [CrossRef]
23. Pigola, S.; Rigoli, M.; Rimoldi, M.; Setti, A. Ricci almost solitons. *Ann. Della Sc. Norm. Super.-Pisa-Cl. Sci.* **2011**, *10*, 757–799 [CrossRef]
24. Siddiqi, M.D.; De, U.C.; Deshmukh, S. Estimation of almost Ricci-Yamabe solitons on static spacetime. *Filomat* **2022**, *32*, 397–407 [CrossRef]
25. Catino, G.; Cremaschi, L.; Djadli, Z.; Mantegazza, C.; Mazzieri, L. The Ricci-Bourguignon flow. *Pacific J. Math.* **2017**, *287*, 337–370 [CrossRef]
26. Catino, G.; Mazzieri, L. Gradient Einstein solitons. *Nonlinear Anal.* **2016**, *132*, 66–94. [CrossRef]
27. Siddiqi, M.D. Ricci ρ-soliton and geometrical structure in a dust fluid and viscous fluid sapcetime. *Bulg. J. Phys.* **2019**, *46*, 163–173
28. Siddiqi, M.D.; Siddqui, S.A. Conformal Ricci soliton and Geometrical structure in a perfect fluid spacetime. *Int. J. Geom. Methods Mod. Phys.* **2020**, *17*, 2050083. [CrossRef]
29. Deshmukh, S. Conformal vector fields and eigenvectors of Laplacian operator. *Math. Phy. Anal. Geom.* **2012**, *15*, 163–172 [CrossRef]
30. O'Neill, B. The fundamental equations of a submersion. *Mich. Math. J.* **1966**, *13* 458–469. [CrossRef]
31. Galitski, V.; Spielman, I.B. Spin-orbit coupling in quantum gases. *Nature* **2013**, *494*, 49–54. [CrossRef] [PubMed]
32. Sakaguchi, H.; Sherman, Y.E.; Malomed, B.A. Vortex solitons in two-dimensional spin-orbit coupled Bose-Einstein condensates: Effect of the Rashaba-Dresselhaus coupling and Zeman splitting. *Phys. Rev. E* **2016**, *94*, 032202. [CrossRef]
33. Mardonov, S.; Sherman, E.Y.; Muga, J.G.; Wang, H.W.; Ban, Y.; Chen, X. Collapse of spin-orbit-coupled Bose-Einstein condensates. *Phys. Rev. A* **2015**, 91, 043604. [CrossRef]
34. Malomed, B.A. Multidimensional solitons: Well-established results and novel findings. *Eur. Phys. J. Spec. Top.* **2016**, 225, 2507 [CrossRef]
35. Kartashov, Y.V.; Konotop, V.V.; Abdullaev, F.K. Gap solitons in a spin-orbit-coupled Bose-Einstein condensate. *Phys. Rev. Lett.* **2013**, 11, 060402. [CrossRef]
36. Baizakov, B.B.; Malomed, B.A.; Salerno, M. Multidimensional solitons in a low-dimensional periodic potential. *Phys. Rev. A* **2004**, 70, 053613. [CrossRef]
37. Fibich, G. *The Nonlinear Schrödinger Equation: Singular Solution and Optical Collapse*; Springer: Berlin/Heidelberg, Germany, 2015
38. Chiao, R.Y.; Garmire, E.; Townes, C.H. Self-trapping of optical beam. *Phys. Rev. Lett.* **1964**, 13, 479. [CrossRef]
39. Hopf, H. Über die Abbildungen der dreidimensional Sphäre auf die Kugelfäche. *Math. Ann.* **1931**, 104, 637–665. [CrossRef]
40. Pontryagin, L. Smooth manifold and their application in homotopy theory. *Am. Math. Soc. Ser.* **1959**, *11*, 1–114. .
41. Scorpan, A. *The Wild World of 4-Manifold*; American Mathematical Society: Washington, DC, USA, 2005.
42. Cohen, F.R.; Joel, M. The decoposition of stable homotopy. *Ann. Math.* **1968**, *87*, 305–320. [CrossRef]
43. Cohen, F.R.; Moore, J.C.; Neisendorfer, J.A. The double suspension and exponents of the homotopy group of spheres. *Ann. Math.* **1979**, *110*, 549–565. [CrossRef]
44. Davis, D.M.; Sun, Z.W. A number theoretic approach to homotopy exponents of $S(U)$. *J. Pure Appl. Algebra* **2007**, *209*, 57–69 [CrossRef]
45. Hirosi, T. *Composition Methods in Homotopy Groups of Spheres*; Annals of Mathematics Studies—Princeton University Press: Princeton, NJ, USA, 1962.
46. Herstein, I.N. *Topics in Algebra*, 2nd ed.; Wiley & Sons: New York, NY, USA, 1975.

axioms

MDPI

Article

A Solitonic Study of Riemannian Manifolds Equipped with a Semi-Symmetric Metric ξ-Connection

Abdul Haseeb [1,*], Sudhakar Kumar Chaubey [2], Fatemah Mofarreh [3] and Abdullah Ali H. Ahmadini [1]

1 Department of Mathematics, College of Science, Jazan University, Jazan 45142, Saudi Arabia
2 Section of Mathematics, Department of IT, University of Technology and Applied Sciences, Shinas 324, Oman
3 Mathematical Science Department, Faculty of Science, Princess Nourah bint Abdulrahman University, Riyadh 11546, Saudi Arabia
* Correspondence: malikhaseeb80@gmail.com or haseeb@jazanu.edu.sa

Abstract: The aim of this paper is to characterize a Riemannian 3-manifold M^3 equipped with a semi-symmetric metric ξ-connection $\tilde{\nabla}$ with ρ-Einstein and gradient ρ-Einstein solitons. The existence of a gradient ρ-Einstein soliton in an M^3 admitting $\tilde{\nabla}$ is ensured by constructing a non-trivial example, and hence some of our results are verified. By using standard tensorial technique, we prove that the scalar curvature of $(M^3, \tilde{\nabla})$ satisfies the Poisson equation $\Delta R = \frac{4(2-\sigma-6\rho)}{\rho}$.

Keywords: Riemannian manifolds; ρ-Einstein solitons; Einstein manifolds; Poisson equation

MSC: 53E20; 53C25; 53C21

Citation: Haseeb, A.; Chaubey, S.K.; Mofarreh, F.; Ahmadini, A.A.H. A Solitonic Study of Riemannian Manifolds Equipped with a Semi-Symmetric Metric ξ-Connection. *Axioms* **2023**, *12*, 809. https://doi.org/10.3390/axioms12090809

Academic Editor: Mića Stanković

Received: 24 July 2023
Revised: 14 August 2023
Accepted: 16 August 2023
Published: 23 August 2023

1. Introduction

The Ricci and other geometric flows are active topics of current research in mathematics, physics and engineering. The Ricci flow [1] is defined on a Riemannian n-manifold (M^n, g) by an evolution equation for metric $g(t)$ of the form $\frac{\partial g}{\partial t} = -2S$, where S is the Ricci tensor of M^n and t indicates the time. The metric g on M^n satisfies the Ricci soliton (in short, RS) equation $\pounds_E g + 2\sigma g + 2S = 0$, where E is a vector field on M^n, $\sigma \in \mathcal{R}$ (the set of real numbers), and $\pounds_E$ represents the Lie derivative operator in the direction of E on M^n. A RS is called expanding (steady or shrinking) if $\sigma > 0$ ($\sigma = 0$ or $\sigma < 0$). If $E = 0$ or Killing, then the RS is called a trivial RS, and M^n becomes an Einstein manifold. Thus the RS is a basic generalization of an Einstein manifold [2]. If $\mathcal{F}$ is a smooth function such that $E = \mathcal{DF}$ for the gradient operator $\mathcal{D}$ of g, then the RS is described as a gradient Ricci soliton (GRS), E is referred to as the potential vector field, and $\mathcal{F}$ is called the potential function. Thus, the RS equation becomes $Hess\mathcal{F} + \sigma g + S = 0$, where $Hess\mathcal{F}$ is the Hessian of $\mathcal{F}$ and $(Hess\mathcal{F})(\zeta_1, \zeta_2) = g(\nabla_{\zeta_1} D\mathcal{F}, \zeta_2)$ for all vector fields ζ_1 and ζ_2 on M^n. Here, ∇ stands for the Levi–Civita connection.

The notion of Ricci–Bourguignon flow, a natural generalization of Ricci flow, has been proposed in [3] and is described on an M^n as:

$$\frac{\partial g}{\partial t} = -2(S - \rho R g), \quad g(0) = g_0, \tag{1}$$

where R is the scalar curvature and $\rho \in \mathcal{R}$. It is to be noticed that for the specific values of ρ, the following cases for the tensor $S - \rho R g$ appeared in (1) [4] are obtained:

(i) $\rho = \frac{1}{2}$, the Einstein tensor $S - \frac{R}{2}g$, (for Einstein soliton),
(ii) $\rho = \frac{1}{n}$, the trace-less Ricci tensor $S - \frac{R}{n}g$,
(iii) $\rho = \frac{1}{2(n-1)}$, the Schouten tensor $S - \frac{R}{2(n-1)}g$, (for Schouten soliton),
(iv) $\rho = 0$, the Ricci tensor S (for RS).

An (M^n, g), $n \geq 3$ is said to be a ρ-Einstein soliton (or ρ-ES) (g, E, σ, ρ) if

$$\pounds_E g + 2S + 2(\sigma - \rho R)g = 0. \tag{2}$$

Similar to the RS, a ρ-ES is called expanding (steady or shrinking) if $\sigma > 0$ ($\sigma = 0$ or $\sigma < 0$). If $E = \mathcal{D}\mathcal{F}$, then (M^n, g) is called a gradient ρ-Einstein soliton (or gradient ρ-ES). Hence, (2) takes the form

$$Hess\mathcal{F} + S + (\sigma - \rho R)g = 0, \tag{3}$$

where $Hess\mathcal{F}$ denotes the Hessian of $\mathcal{F} \in C^\infty(M^n)$ and defined by $Hess\mathcal{F} = \nabla\nabla\mathcal{F}$. Recently, ρ-Einstein solitons have been studied by several authors, such as [5–12]. On the other hand, we recommend the papers [13–19] for the studies of Ricci, Yamabe, Ricci-Yamabe, η-Ricci-Yamabe and quasi-Yamabe solitons on different geometric structures.

In this paper, we have made an effort to the solitonic study of a 3-dimensional Riemannian manifold M^3 equipped with a semi-symmetric metric ξ-connection $\tilde{\nabla}$. To achieve the goal, we present our work as follows: In Section 2, we gather the basic information of a Riemannian 3-manifold equipped with a semi-symmetric metric ξ-connection $(M^3, \tilde{\nabla}, g)$, definitions and Lemmas. The properties of ρ-ES in $(M^3, \tilde{\nabla}, g)$ are studied in Section 3. We address the properties of gradient ρ-ES in $(M^3, \tilde{\nabla}, g)$ in Section 4. In the last section, we model a non-trivial example of $(M^3, \tilde{\nabla}, g)$ admitting a gradient ρ-ES, and prove our results.

2. Riemannian Manifolds with a Semi-Symmetric Metric ξ-Connection

In 1970, Yano [20] investigated the properties of a semi-symmetric metric connection $\tilde{\nabla}$ on Riemannian n-manifolds M^n and defined by $\tilde{\nabla}_{\zeta_1}\zeta_2 = \nabla_{\zeta_1}\zeta_2 + \eta(\zeta_2)\zeta_1 - g(\zeta_1, \zeta_2)\xi$ for all ζ_1 and ζ_2 on M^n, where η is a 1-form associated with the unit vector field ξ such that $g(\xi, \xi) = \eta(\xi) = 1$ and $g(\zeta_1, \xi) = \eta(\zeta_1)$. Later, the properties of the semi-symmetric metric connection $\tilde{\nabla}$ have been explored by several researchers. One of these properties is the curvature invariant respecting to the semi-symmetric metric connection $\tilde{\nabla}$ and the Levi–Civita connection ∇. For example, the conformal curvature tensors corresponding to the semi-symmetric connection (Yano's sense) and the Levi–Civita connection coincide. Similar results for different curvature tensors have been established by many geometers. A connection $\tilde{\nabla}$ is said to be semi-symmetric metric ξ-connection if and only if $\tilde{\nabla}\xi = 0$. Afterwards, the properties of semi-symmetric metric ξ-connection have been studied in [21–24].

In an $(M^n, \tilde{\nabla}, g)$, we have [21]

$$\nabla_{\zeta_1}\xi = -\zeta_1 + \eta(\zeta_1)\xi, \;\; g(\xi, \xi) = 1, \text{ and } \eta(\zeta_1) = g(\zeta_1, \xi) \tag{4}$$

for any ζ_1 on M^n. Next, we have [21]

$$(\nabla_{\zeta_1}\eta)\zeta_2 = -g(\zeta_1, \zeta_2) + \eta(\zeta_1)\eta(\zeta_2), \tag{5}$$

$$K(\zeta_1, \zeta_2)\xi = \eta(\zeta_1)\zeta_2 - \eta(\zeta_2)\zeta_1, \tag{6}$$

$$K(\zeta_1, \xi)\zeta_2 = g(\zeta_1, \zeta_2)\xi - \eta(\zeta_2)\zeta_1, \tag{7}$$

$$S(\zeta_1, \xi) = -(n-1)\eta(\zeta_1) \iff Q\xi = -(n-1)\xi, \tag{8}$$

$$(\pounds_\xi g)(\zeta_1, \zeta_2) = 2\{-g(\zeta_1, \zeta_2) + \eta(\zeta_1)\eta(\zeta_2)\}, \tag{9}$$

for all ζ_1, ζ_2 on M^n. Here, K and Q represent the curvature tensor and the Ricci operator of M^n, respectively.

Definition 1. *An M^n is said to be quasi-Einstein if its $S(\neq 0)$ satisfies*

$$S(\zeta_1, \zeta_2) = \mathfrak{l}\, g(\zeta_1, \zeta_2) + \mathfrak{m}\, \eta(\zeta_1)\eta(\zeta_2),$$

where $\mathfrak{m}$ and $\mathfrak{l}$ are smooth functions on M^n. If $\mathfrak{m} = 0$, then the manifold is called an Einstein manifold.

Definition 2. *A partial differential equation $\Delta u = v$ on a complete M^n is called a Poisson equation for some smooth functions u and v.*

Remark 1 ([21,22]). *An $(M^3, \tilde{\nabla}, g)$ is a quasi-Einstein manifold of the form*

$$S(\zeta_1, \zeta_2) = \left(1 + \frac{R}{2}\right) g(\zeta_1, \zeta_2) - \left(3 + \frac{R}{2}\right) \eta(\zeta_1)\eta(\zeta_2). \tag{10}$$

Remark 2 ([21,22]). *In an $(M^3, \tilde{\nabla}, g)$, we have*

$$\xi(R) = 2(R+6), \tag{11}$$

$$\eta(\nabla_\xi \mathcal{D}R) = 4(R+6), \tag{12}$$

where $\mathcal{D}$ is the gradient operator of g. From (11), it is noticed that R of M^3 is constant if and only if $R = -6$.

3. ρ-ES on $(M^3, \tilde{\nabla}, g)$

First, we prove the following theorem.

Theorem 1. *If $(M^3, \tilde{\nabla}, g)$ admits a ρ-ES (g, E, σ, ρ), then its scalar curvature R satisfies the Poisson equation $\Delta R = \frac{4(2-\sigma-6\rho)}{\rho}$, provided $\rho \neq 0$.*

Proof. Let the metric of an $(M^3, \tilde{\nabla}, g)$ be a ρ-ES (g, E, σ, ρ), then in view of (10), (2) leads to

$$\begin{aligned} (\pounds_E g)(\zeta_1, \zeta_2) &= -2\{1 + \sigma + (\frac{1}{2} - \rho)R\} g(\zeta_1, \zeta_2) \\ &\quad + (R+6)\eta(\zeta_1)\eta(\zeta_2), \end{aligned} \tag{13}$$

for any vector fields ζ_1, ζ_2 on M^3.

Taking covariant derivative of (13) respecting to ζ_3, we find

$$\begin{aligned} (\nabla_{\zeta_3} \pounds_E g)(\zeta_1, \zeta_2) &= (\zeta_3 R)\{(2\rho - 1) g(\zeta_1, \zeta_2) + \eta(\zeta_1)\eta(\zeta_2)\} \\ &\quad - (R+6)\{g(\zeta_1, \zeta_3)\eta(\zeta_2) + g(\zeta_2, \zeta_3)\eta(\zeta_1) - 2\eta(\zeta_1)\eta(\zeta_2)\eta(\zeta_3)\}. \end{aligned} \tag{14}$$

As g is parallel with respect to ∇, then the formula [25]

$$(\pounds_E \nabla_{\zeta_1} g - \nabla_{\zeta_1} \pounds_E g - \nabla_{[E, \zeta_1]} g)(\zeta_2, \zeta_3) = -g((\pounds_E \nabla)(\zeta_1, \zeta_2), \zeta_3) - g((\pounds_E \nabla)(\zeta_1, \zeta_3), \zeta_2)$$

turns to

$$(\nabla_{\zeta_1} \pounds_E g)(\zeta_2, \zeta_3) = g((\pounds_E \nabla)(\zeta_1, \zeta_2), \zeta_3) + g((\pounds_E \nabla)(\zeta_1, \zeta_3), \zeta_2).$$

Since $\pounds_E \nabla$ is symmetric, therefore we have

$$2g((\pounds_E \nabla)(\zeta_1, \zeta_2), \zeta_3) = (\nabla_{\zeta_1} \pounds_E g)(\zeta_2, \zeta_3) + (\nabla_{\zeta_2} \pounds_E g)(\zeta_1, \zeta_3) - (\nabla_{\zeta_3} \pounds_E g)(\zeta_1, \zeta_2),$$

which in view of (14) gives

$$2g((\pounds_E \nabla)(\zeta_1, \zeta_2), \zeta_3) = (\zeta_1 R)\{(2\rho - 1)g(\zeta_2, \zeta_3) + \eta(\zeta_2)\eta(\zeta_3)\} + (\zeta_2 R)\{(2\rho - 1)g(\zeta_1, \zeta_3) + \eta(\zeta_1)\eta(\zeta_3)\} - (\zeta_3 R)\{(2\rho - 1)g(\zeta_1, \zeta_2) + \eta(\zeta_1)\eta(\zeta_2)\} - 2(R + 6)\{g(\zeta_1, \zeta_2)\eta(\zeta_3) - \eta(\zeta_1)\eta(\zeta_2)\eta(\zeta_3)\},$$

from which it follows that

$$2(\pounds_E \nabla)(\zeta_1, \zeta_2) = (\zeta_1 R)\{(2\rho - 1)\zeta_2 + \eta(\zeta_2)\xi\} + (\zeta_2 R)\{(2\rho - 1)\zeta_1 + \eta(\zeta_1)\xi\} - (\mathcal{D}R)\{(2\rho - 1)g(\zeta_1, \zeta_2) + \eta(\zeta_1)\eta(\zeta_2)\} - 2(R + 6)\{g(\zeta_1, \zeta_2)\xi - \eta(\zeta_1)\eta(\zeta_2)\xi\}. \tag{15}$$

Replacing ζ_2 by ξ and ζ_1 by ζ_2 in (15), we have

$$(\pounds_E \nabla)(\zeta_2, \xi) = \rho g(\mathcal{D}R, \zeta_2)\xi - \rho(\mathcal{D}R)\eta(\zeta_2) + (R + 6)\{(2\rho - 1)\zeta_2 + \eta(\zeta_2)\xi\}. \tag{16}$$

The covariant differentiation of (16) respecting to ζ_1 yields

$$(\nabla_{\zeta_1} \pounds_E \nabla)(\zeta_2, \xi) = 2(\zeta_1 R)\{(2\rho - 1)\zeta_2 + \eta(\zeta_2)\xi\} + (\zeta_2 R)\{(\rho - 1)\zeta_1 + \eta(\zeta_1)\xi\} - (\mathcal{D}R)\{(\rho - 1)g(\zeta_1, \zeta_2) + \eta(\zeta_1)\eta(\zeta_2)\} - 3(R + 6)\{g(\zeta_1, \zeta_2)\xi - \eta(\zeta_1)\eta(\zeta_2)\xi\} - (R + 6)\{(2\rho - 1)\eta(\zeta_1)\zeta_2 + \eta(\zeta_2)\zeta_1\} + \rho g(\nabla_{\zeta_1} \mathcal{D}R, \zeta_2)\xi - \rho(\nabla_{\zeta_1} \mathcal{D}R)\eta(\zeta_2), \tag{17}$$

where (4), (5) and (16) being used.

Again from [25], we have

$$(\pounds_E K)(\zeta_1, \zeta_2)\zeta_3 = (\nabla_{\zeta_1} \pounds_E \nabla)(\zeta_2, \zeta_3) - (\nabla_{\zeta_2} \pounds_E \nabla)(\zeta_1, \zeta_3), \tag{18}$$

which by putting $\zeta_3 = \xi$ and using (17) becomes

$$(\pounds_E K)(\zeta_1, \zeta_2)\xi = g(\mathcal{D}R, \zeta_1)\{(3\rho - 1)\zeta_2 + \eta(\zeta_2)\xi\} - g(\mathcal{D}R, \zeta_2)\{(3\rho - 1)\zeta_1 + \eta(\zeta_1)\xi\} + 2(R + 6)(\rho - 1)\{\eta(\zeta_2)\zeta_1 - \eta(\zeta_1)\zeta_2\} + \rho g(\nabla_{\zeta_1} \mathcal{D}R, \zeta_2)\xi - \rho g(\nabla_{\zeta_2} \mathcal{D}R, \zeta_1)\xi - \rho(\nabla_{\zeta_1} \mathcal{D}R)\eta(\zeta_2) + \rho(\nabla_{\zeta_2} \mathcal{D}R)\eta(\zeta_1). \tag{19}$$

Contracting (19) respecting to ζ_1 then using (4) and (11) we lead to

$$(\pounds_E S)(\zeta_2, \xi) = (1 - 6\rho)\zeta_2(R) + 2(R + 6)(2\rho - 1)\eta(\zeta_2) + \rho g(\nabla_\xi \mathcal{D}R, \zeta_2)\xi - \rho(\Delta R)\eta(\zeta_2). \tag{20}$$

By putting $\zeta_2 = \xi$ in (20) then using (4), (11) and (12), we find

$$(\pounds_E S)(\xi, \xi) = -4\rho(R + 6) - \rho(\Delta R). \tag{21}$$

The Lie derivative of (8) respecting to E leads to

$$(\pounds_E S)(\xi, \xi) = 4\eta(\pounds_E \xi). \tag{22}$$

Putting $\zeta_1 = \zeta_2 = \xi$ in (13) infers

$$(\pounds_E g)(\xi,\xi) = -2\sigma + 2\rho R + 4. \tag{23}$$

The Lie derivative of $g(\xi,\xi) = 1$ gives

$$(\pounds_E g)(\xi,\xi) = -2\eta(\pounds_E \xi). \tag{24}$$

Now combining (21)–(25) we deduce

$$\Delta R = \frac{4(2-\sigma-6\rho)}{\rho}, \text{ provided } \rho \neq 0. \tag{25}$$

This completes the proof. □

It is well-known that the ρ-ES Equation (2) on M^n with the soliton constant $\rho = \frac{1}{2}, \frac{1}{n}, \frac{1}{2(n-1)}$ reduces to the Einstein soliton, traceless Ricci soliton, Schouten soliton, respectively. It is also known that a smooth function $\mathfrak{f}$ on an M^n is called harmonic, subharmonic or superharmonic if $\Delta\mathfrak{f} = 0$, ≥ 0 or ≤ 0, respectively. These facts together with Theorem 1 state the following:

Corollary 1. *Let $(M^3, \tilde{\nabla}, g)$ admit a ρ-ES, then we have*

Value of ρ	*Solitons*	*Poisson equation*	*Condition for R to be subharmonic and superharmonic*
$\rho = \frac{1}{2}$	*Einstein soliton*	$\Delta R = -8(\sigma+1)$	*(i) R is subharmonic if $\sigma \leq -1$,* *(ii) R is superharmonic if $\sigma \geq -1$,*
$\rho = \frac{1}{3}$	*traceless Ricci soliton*	$\Delta R = -12\sigma$	*(i) R is subharmonic if $\sigma \leq 0$,* *(ii) R is superharmonic if $\sigma \geq 0$,*
$\rho = \frac{1}{4}$	*Schouten soliton*	$\Delta R = 16(\frac{1}{2}-\sigma)$	*(i) R is subharmonic if $\sigma \leq \frac{1}{2}$,* *(ii) R is superharmonic if $\sigma \geq \frac{1}{2}$.*

Remark 3. *The ρ-ES on an M^n with $\rho = 0$ reduces to the RS. The properties of RS on $(M^3, \tilde{\nabla}, g)$ have been explored by Chaubey and De [22]. Thus, we can say that the Theorem 1 generalizes the study of Einstein soliton, traceless RS and the Schouten soliton on $(M^3, \tilde{\nabla}, g)$.*

It is well-known that the Poisson equation $\Delta u = v$ with $v = 0$ becomes a Laplace equation. Suppose that an $(M^3, \tilde{\nabla}, g)$ does not admit RS. Then, Theorem 1 and above discussion state:

Corollary 2. *If $(M^3, \tilde{\nabla}, g)$ admits a ρ-ES, which is not a RS ($\rho \neq 0$), then R of M^3 satisfies Laplace equation if and only if $\sigma = 2(1-3\rho)$.*

Let $(M^3, \tilde{\nabla}, g)$ admit a ρ-ES. If R of M^3 satisfies the Laplace equation, then $\sigma = 2(1-3\rho)$. The ρ-ES under consideration to be steady, shrinking or expanding if ρ is equal to, less than or greater than $\frac{1}{3}$. Thus, we write our corollary as

Corollary 3. *Let the metric of an $(M^3, \tilde{\nabla}, g)$ be ρ-ES, which is not a RS ($\rho \neq 0$). If R of M^3 satisfies the Laplace equation, then the ρ-ES is steady, shrinking or expanding if $\rho = \frac{1}{3}$, $\rho < \frac{1}{3}$ or $\rho > \frac{1}{3}$, respectively.*

4. Gradient ρ-ES on $(M^3, \tilde{\nabla}, g)$

Theorem 2. *Let $(M^3, \tilde{\nabla}, g)$ admit a gradient ρ-ES. Then, either M^3 is Einstein or the gradient ρ-ES is steady type gradient traceless RS.*

Proof. Let the metric of an $(M^3, \tilde{\nabla}, g)$ be a gradient ρ-ES. Then, (3) can be written as

$$\nabla_{\zeta_1}\mathcal{D}\mathcal{F} + Q\zeta_1 + (\sigma - \rho R)\zeta_1 = 0, \tag{26}$$

for all ζ_1 on M^3.

The covariant differentiation of (26) with respect to ζ_2 leads to

$$\nabla_{\zeta_2}\nabla_{\zeta_1}\mathcal{DF} = -(\nabla_{\zeta_2}Q)\zeta_1 - Q(\nabla_{\zeta_2}\zeta_1) - (\sigma - \rho R)\nabla_{\zeta_2}\zeta_1 + \rho\zeta_2(R)\zeta_1. \tag{27}$$

Interchanging ζ_1 and ζ_2 in (27) leads to

$$\nabla_{\zeta_1}\nabla_{\zeta_2}\mathcal{DF} = -(\nabla_{\zeta_1}Q)\zeta_2 - Q(\nabla_{\zeta_1}\zeta_2) - (\sigma - \rho R)\nabla_{\zeta_1}\zeta_2 + \rho\zeta_1(R)\zeta_2. \tag{28}$$

By plugging of (26)–(28), we find

$$K(\zeta_1,\zeta_2)\mathcal{DF} = -(\nabla_{\zeta_1}Q)\zeta_2 + (\nabla_{\zeta_2}Q)\zeta_1 + \rho\{\zeta_1(R)\zeta_2 - \zeta_2(R)\zeta_1\}.$$

Contracting the forgoing equation along ζ_1, we obtain

$$S(\zeta_2,\mathcal{DF}) = \frac{(1-4\rho)}{2}\zeta_2(R). \tag{29}$$

In account of (10), we have

$$S(\zeta_2,\mathcal{DF}) = (1+\frac{R}{2})\zeta_2(\mathcal{F}) - (3+\frac{R}{2})\eta(\zeta_2)\xi(\mathcal{F}). \tag{30}$$

Thus, from (29) and (30), it follows that

$$(1-4\rho)\zeta_2(R) = (R+2)\zeta_2(\mathcal{F}) - (R+6)\eta(\zeta_2)\xi(\mathcal{F}). \tag{31}$$

By putting $\zeta_2 = \xi$ in (31), then using (4) and (11), we find

$$\xi(\mathcal{F}) = -\frac{1}{2}(1-4\rho)(R+6). \tag{32}$$

By using (32) and (31) turns to

$$(1-4\rho)\zeta_2(R) = (R+2)\zeta_2(\mathcal{F}) + \frac{1}{2}(R+6)^2(1-4\rho)\eta(\zeta_2). \tag{33}$$

The covariant differentiation of (33) along ζ_1 leads to

$$\begin{aligned}(1-4\rho)g(\nabla_{\zeta_1}\mathcal{D}R,\zeta_2) &= \zeta_1(R)\zeta_2(\mathcal{F}) + (R+2)g(\nabla_{\zeta_1}\mathcal{DF},\zeta_2)\\ &\quad+(R+6)(1-4\rho)\zeta_1(R)\eta(\zeta_2)\\ &\quad+\frac{1}{2}(R+6)^2(1-4\rho)\{\eta(\zeta_1)\eta(\zeta_2) - g(\zeta_1,\zeta_2)\}.\end{aligned} \tag{34}$$

Interchanging ζ_1 and ζ_2 in (34), we have

$$\begin{aligned}(1-4\rho)g(\nabla_{\zeta_2}\mathcal{D}R,\zeta_1) &= \zeta_2(R)\zeta_1(\mathcal{F}) + (R+2)g(\nabla_{\zeta_2}\mathcal{DF},\zeta_1)\\ &\quad+(R+6)(1-4\rho)\zeta_2(R)\eta(\zeta_1)\\ &\quad+\frac{1}{2}(R+6)^2(1-4\rho)\{\eta(\zeta_1)\eta(\zeta_2) - g(\zeta_1,\zeta_2)\}.\end{aligned} \tag{35}$$

Equating the left hand sides of last two equations gives

$$\begin{aligned}&\zeta_1(R)\zeta_2(\mathcal{F}) + (R+6)(1-4\rho)\zeta_1(R)\eta(\zeta_2)\\ &-\zeta_2(R)\zeta_1(\mathcal{F}) - (R+6)(1-4\rho)\zeta_2(R)\eta(\zeta_1) = 0,\end{aligned}$$

which by replacing $\zeta_2 = \xi$ then using (4), (11) and (32) takes the form

$$(R+6)\{(1-4\rho)\zeta_1(R) - 4\zeta_1(\mathcal{F}) - 4(R+6)(1-4\rho)\eta(\zeta_1)\} = 0.$$

Thus, we have either $R = -6$, or $(1-4\rho)\zeta_1(R) = 4\zeta_1(\mathcal{F}) + 4(R+6)(1-4\rho)\eta(\zeta_1)$. If we firstly suppose that $R \neq -6$ and $(1-4\rho)\zeta_1(R) = 4\zeta_1(\mathcal{F}) + 4(R+6)(1-4\rho)\eta(\zeta_1)$, which by virtue of (33) turns to

$$(R-2)\{2\zeta_1(\mathcal{F}) + (R+6)(1-4\rho)\eta(\zeta_1)\} = 0, \tag{36}$$

which refers that either $R = 2$ or $\zeta_1(\mathcal{F}) = -\frac{1}{2}(R+6)(1-4\rho)\eta(\zeta_1)$. From (11), it is obvious that if R is constant, then its value must be -6, which shows that $R = 2$ is inadmissible. Thus, we have $\zeta_1(\mathcal{F}) = -\frac{1}{2}(R+6)(1-4\rho)\eta(\zeta_1)$, which is equivalent to

$$\mathcal{DF} = -\frac{1}{2}(R+6)(1-4\rho)\xi = \xi(\mathcal{F})\xi. \tag{37}$$

Thus, the gradient of $\mathcal{F}$ is pointwise collinear with ξ. Now, taking the covariant derivative of (37) with respect to ζ_1 and using (4), we have

$$\nabla_{\zeta_1}\mathcal{DF} = \zeta_1(\xi(\mathcal{F}))\xi - \xi(\mathcal{F})(\zeta_1 - \eta(\zeta_1)\xi). \tag{38}$$

Therefore, from (26) and (38), we obtain

$$Q\zeta_1 + (\sigma - \rho R)\zeta_1 = -\zeta_1(\xi(\mathcal{F}))\xi + \xi(\mathcal{F})(\zeta_1 - \eta(\zeta_1)\xi). \tag{39}$$

Now, by replacing ζ_1 by ξ in (39) then using (8), (11) and (32) we lead to

$$\sigma = (1-3\rho)(R+8). \tag{40}$$

Let us suppose that $\rho = \frac{1}{3}$, that is, the gradient ρ-ES on an M^3 is gradient traceless RS. This fact together with Equation (40) leads to $\sigma = 0$. Thus, the gradient traceless RS is steady. This completes the proof. □

Theorem 3. *Let an $(M^3, \tilde{\nabla}, g)$ be a non-gradient traceless RS. Then, the gradient ρ-ES is trivial soliton with constant $\sigma = 2(1-3\rho)$. Also, the ρ-ES is shrinking and expanding according to $\rho > \frac{1}{3}$ and $\rho < \frac{1}{3}$.*

Proof. Now, we suppose that $\rho \neq \frac{1}{3}$. Thus, (40) leads to

$$R = \frac{\sigma}{1-3\rho} - 8, \tag{41}$$

which informs that R is constant and hence (11) infers that $R = -6$. This contradicts our hypothesis $R \neq -6$.

Secondly, we consider that $R = -6$ and $(1-4\rho)\zeta_1(R) \neq 4\zeta_1(\mathcal{F}) + 4(R+6)(1-4\rho)\eta(\zeta_1)$. For $R = -6$, (33) informs that $\mathcal{F} \in \mathcal{R}$ and hence the GRBS on the manifold is trivial. Moreover, the Riemannian 3-manifold under assumption is an Einstein manifold with $\sigma = 2(1-3\rho)$. This completes the proof. □

Let us suppose that an $(M^3, \tilde{\nabla}, g)$ admits a proper gradient ρ-ES. Then, the ρ-ES reduces to the gradient traceless RS and $\rho = \frac{1}{3}, \sigma = 0$. Using these facts in (26) and then contracting the foregoing equation over ζ_1 gives $\Delta\mathcal{F} = 0$.

A smooth function $\mathfrak{h}$ on an M^n is called harmonic if $\Delta\mathfrak{h} = 0$.

The above discussions state the following:

Corollary 4. *Let a complete $(M^3, \tilde{\nabla}, g)$ admit a proper gradient ρ-ES. Then the gradient function of the gradient ρ-ES is harmonic.*

Contracting (38) over ζ_1, we find

$$\Delta\mathcal{F} = \xi(\xi(\mathcal{F})) - 2\xi(\mathcal{F}).$$

Again, considering $\sigma = 0,\ \rho = \frac{1}{3}$ and then contracting (26) over ζ_1, we conclude that

$$\Delta\mathcal{F} = 0.$$

The last two equations show that $\xi(\xi(\mathcal{F})) - 2\xi(\mathcal{F}) = 0$. Let $\xi = \frac{\partial}{\partial t}$. Thus, we notice that the potential function $\mathcal{F}$ satisfies the PDE

$$\frac{\partial^2 \mathcal{F}}{\partial t^2} - 2\frac{\partial \mathcal{F}}{\partial t} = 0.$$

It is obvious that $\mathcal{F} = Ae^{2t} + B$ for smooth functions A and B, which are independent of t is the solution of the above PDE. Now, we list our results in the following:

Corollary 5. *Let the metric of a complete $(M^3, \tilde{\nabla}, g)$ admit a proper gradient ρ-ES. Then, the potential function $\mathcal{F}$ of such soliton satisfies the PDE $\frac{\partial^2 \mathcal{F}}{\partial t^2} - 2\frac{\partial \mathcal{F}}{\partial t} = 0$, and it can be evaluated by $\mathcal{F} = Ae^{2t} + B$.*

5. Example

We consider the manifold $M^3 = \{(w_1, w_2, w_3) \in \mathcal{R}^3\}$, where (w_1, w_2, w_3) are the usual coordinates in $\mathbb{R}^3$. Let u_1, u_2 and u_3 be the vector fields on M^3 given by

$$u_1 = e^{\mathfrak{b}w_3 + w_1}\frac{\partial}{\partial w_1}, \quad u_2 = e^{\mathfrak{b}w_3 + w_2}\frac{\partial}{\partial w_2}, \quad u_3 = \frac{1}{\mathfrak{b}}\frac{\partial}{\partial w_3} = \xi,$$

where $\mathfrak{b}(\neq 0) \in \mathcal{R}$. Then, $\{u_1, u_2, u_3\}$ forms a basis in the module of the vector fields of M^3.
Let the Riemannian metric g be defined by

$$g(u_p, u_q) = \begin{cases} 1, & 1 \leq p = q \leq 3, \\ 0, & otherwise. \end{cases}$$

Hence, M^3 is a Riemannian manifold of dimension 3. Let the 1-form η on M^3 be defined by $\eta(\zeta_1) = g(\zeta_1, u_3) = g(\zeta_1, \xi)$ for all ζ_1 on M^3. Now, by direct computations, we obtain

$$[u_1, u_2] = 0, \ [u_1, u_3] = -u_1, \ [u_2, u_3] = -u_2.$$

By using Koszul's formula, we obtain

$$\nabla_{u_p} u_q = \begin{cases} -u_p, & p = 1,2, q = 3, \\ u_3, & 1 \leq p = q \leq 2, \\ 0, & otherwise. \end{cases}$$

Now we suppose that $\zeta_1 = \zeta_1^1 u_1 + \zeta_1^2 u_2 + \zeta_1^3 u_3$, then for $\xi = u_3$ it follows that $\nabla_{\zeta_1}\xi = -\zeta_1 + \eta(\zeta_1)\xi$. It can be easily seen that $\tilde{\nabla}$ defined on M^3 satisfies the conditions

$$\tilde{T}(\zeta_1, \zeta_2) = -\eta(\zeta_1)\zeta_2 + \eta(\zeta_2)\zeta_1, \ \tilde{\nabla}g = 0, \text{ and } \tilde{\nabla}\xi = 0,$$

for arbitrary vector fields ζ_1 and ζ_2 on M^3, where $\tilde{T}$ indicates the torsion tensor of $\tilde{\nabla}$. Thus, we can say that $\tilde{\nabla}$ is a semi-symmetric metric ξ-connection on M^3.

The non-zero constituents of K are obtained as follows:

$$K(u_1, u_3)u_1 = u_3, \ K(u_1, u_2)u_1 = u_2, \ K(u_2, u_3)u_2 = u_3,$$

$$K(u_1, u_2)u_2 = K(u_1, u_3)u_3 = -u_1, \ K(u_2, u_3)u_3 = -u_2.$$

By using above components of the curvature tensor K we obtain

$$S(u_p, u_q) = -2, \ 1 \leq p = q \leq 3,$$

from which we obtain $R = -6$.

Now, by taking $\mathcal{DF} = (u_1\mathcal{F})u_1 + (u_2\mathcal{F})u_2 + (u_3\mathcal{F})u_3$, we have

$$\nabla_{u_1}\mathcal{DF} = (u_1(u_1\mathcal{F}) - u_3\mathcal{F})u_1 + (u_1(u_2\mathcal{F}))u_2 + (u_1(u_3\mathcal{F}) + u_1\mathcal{F})u_3,$$

$$\nabla_{\mathcal{E}_2}\mathcal{DF} = (u_2(u_1\mathcal{F}))u_1 + (u_2(u_2\mathcal{F}) - u_3\mathcal{F})u_2 + (u_2(u_3\mathcal{F}) + \mathcal{F}_2\mathcal{F})\mathcal{F}_3,$$

$$\nabla_{\mathcal{E}_3}\mathcal{DF} = (u_3(u_1\mathcal{F}))u_1 + (u_3(u_2\mathcal{F}))u_2 + (u_3(u_3\mathcal{F}))u_3.$$

Thus, by virtue of (26), we obtain

$$\begin{cases} u_1(u_1\mathcal{F}) - u_3\mathcal{F} = 2 - 6\rho - \sigma, \\ u_2(u_2\mathcal{F}) - u_3\mathcal{F} = 2 - 6\rho - \sigma, \\ u_3(u_3\mathcal{F}) = 2 - 6\rho - \sigma, \\ u_1(u_2\mathcal{F}) = 0, \\ u_2(u_1\mathcal{F}) = 0, \\ u_2(u_3\mathcal{F}) + u_2\mathcal{F} = 0. \end{cases} \tag{42}$$

Thus, the relations in (42) are, respectively, amounting to

$$e^{2(\mathfrak{b}w_3+w_1)}[\frac{\partial^2\mathcal{F}}{\partial w_1^2} + \frac{\partial\mathcal{F}}{\partial w_1}] - \frac{1}{\mathfrak{b}}\frac{\partial\mathcal{F}}{\partial w_3} = 2 - 6\rho - \sigma,$$

$$e^{2(\mathfrak{b}w_3+w_1)}[\frac{\partial^2\mathcal{F}}{\partial w_2^2} + \frac{\partial\mathcal{F}}{\partial w_2}] - \frac{1}{\mathfrak{b}}\frac{\partial\mathcal{F}}{\partial w_3} = 2 - 6\rho - \sigma,$$

$$\frac{1}{\mathfrak{b}^2}\frac{\partial^2\mathcal{F}}{\partial w_3^2} = 2 - 6\rho - \sigma,$$

$$\frac{\partial^2\mathcal{F}}{\partial w_1\partial w_2} = 0,$$

$$\frac{\partial^2\mathcal{F}}{\partial w_2\partial w_1} = 0,$$

$$\frac{1}{\mathfrak{b}}[\frac{\partial^2\mathcal{F}}{\partial w_2\partial w_3} + \frac{\partial\mathcal{F}}{\partial w_2}] = 0.$$

From the above relations, it is noticed that $\mathcal{F} \in \mathcal{R}$ for $\sigma = 2 - 6\rho$. Hence, the Equation (26) is satisfied. Thus, g is a gradient ρ-ES with the soliton vector field $E = \mathcal{DF}$, where $\mathcal{F} \in \mathcal{R}$ and $\sigma = 2 - 6\rho$. For $\rho = \frac{1}{3}$, we obtain $\sigma = 0$, i.e., the gradient ρ-ES is trivial with constant $\sigma = 2 - 6\rho$. Thus, Theorem 2 is verified.

6. Results and Discussion

It is well known that the ρ-Einstein soliton Equation (2) with $\rho = 0$ becomes the Ricci soliton equation, which has been studied in [22]. Thus, we can say that the ρ-Einstein soliton is a natural generalization of Ricci soliton. In this manuscript, we have explored the properties of ρ-Einstein solitons in Riemannian geometry, which generalizes the results of [22].

7. Conclusions

To prove the curvatures invariant, Chauey et al. [23] defined the notion of semi-symmetric metric P-connection in Riemannian setting, which is a particular case of Riemannian concircular structure manifold [26]. This topic has great applications in differential equations. We proved that the scalar curvature of Riemannian 3-manifolds endowed with a semi-symmetric metric ξ-connection and Ricci–Bourguignon soliton satisfies the Poisson and Laplace equations. It is well known that the Poisson and Laplace equations play a crucial role in the development of engineering, physics, mathematics, etc. We have also established the conditions for which the scalar curvature is harmonic, sub-harmonic and super-harmonic. We also established the existence condition of a gradient ρ-Einstein soliton in the Riemannian 3-manifolds, and consequently we proved some results. To verify our results, we constructed a non-trivial example of a three-dimensional Riemannian manifold equipped with a semi-symmetric metric ξ-connection. These topics are modern and have a lot of scope for researchers.

Author Contributions: Conceptualization, A.H., S.K.C., F.M. and A.A.H.A.; methodology, A.H., S.K.C., F.M. and A.A.H.A.; investigation, A.H., S.K.C., F.M. and A.A.H.A.; writing—original draft preparation, A.H., S.K.C. and F.M.; writing—review and editing, A.H., S.K.C. and A.A.H.A. All authors have read and agreed to the published version of the manuscript.

Funding: The third author, Fatemah Mofarreh, expresses her gratitude to Princess Nourah bint Abdulrahman University researchers Supporting Project number (PNURSP2023R27), Princess Nourah bint Abdulrahman University, Riyadh, Saudi Arabia.

Institutional Review Board Statement: Not applicable.

Informed Consent Statement: Not applicable.

Data Availability Statement: Not applicable.

Acknowledgments: The authors are thankful to the editor and anonymous referees for the constructive comments given to improve the quality of the paper. The third author, Fatemah Mofarreh, expresses her gratitude to Princess Nourah bint Abdulrahman University Researchers Supporting Project number (PNURSP2023R27), Princess Nourah bint Abdulrahman University, Riyadh, Saudi Arabia.

Conflicts of Interest: The authors declare no conflict of interest.

References

1. Hamilton, R.S. The Ricci Flow on Surfaces, Mathematics and General Relativity. *Contemp. Math.* **1988**, *71*, 237–262.
2. Besse, A.L. *Einstein Manifolds, Classics in Mathematics*; Springer: Berlin/Heidelberg, Germany, 2008.
3. Bourguignon, J.P. Ricci curvature and Einstein metrics. *Glob. Differ. Geom. Glob. Anal. Lect. Notes Math.* **1981**, *838*, 42–63.
4. Bourguignon, J.P.; Lawson, H.B. Stability and isolation phenomena for Yang-mills fields. *Commun. Math. Phys.* **1981**, *79*, 189–230. [CrossRef]
5. Dwivedi, S. Some results on Ricci-Bourguignon solitons and almost solitons. *Can. Math. Bull.* **2020**, *64*, 1–15. [CrossRef]
6. Huang, G. Integral pinched gradient shrinking ρ-Einstein solitons. *J. Math. Anal. Appl.* **2017**, *451*, 1045–1055. [CrossRef]
7. Mondal, C.K.; Shaikh, A.A. Some results on η-Ricci Soliton and gradient ρ-Einstein soliton in a complete Riemannian manifold. *Commun. Korean Math. Soc.* **2019**, *34*, 1279–1287.
8. Patra, D.S. Some characterizations of ρ-Einstein solitons on Sasakian manifolds. *Can. Math. Bull.* **2022**, *65*, 1036–1049. [CrossRef]
9. Shaikh, A.A.; Cunha, A.W.; Mandal, P. Some characterizations of ρ-Einstein solitons. *J. Geom. Phys.* **2021**, *166*, 104270. [CrossRef]
10. Shaikh, A.A.; Mandal, P.; Mondal, C.K. Diameter estimation of gradient ρ-Einstein solitons. *J. Geom. Phys.* **2022**, *177*, 104518. [CrossRef]
11. Suh, Y.J. Ricci-Bourguignon solitons on real hypersurfaces in the complex hyperbolic quadric. *Rev. Real Acad. Cienc. Exactas Fis. Nat. Ser. A-Mat.* **2022**, *116*, 110. [CrossRef]
12. Venkatesha, V.; Kumara, H.A. Gradient ρ-Einstein soliton on almost Kenmotsu manifolds. *Ann. Dell' Univ. Ferrara* **2019**, *65*, 375–388. [CrossRef]
13. Blaga, A.M. Some geometrical aspects of Einstein, Ricci and Yamabe solitons. *J. Geom. Symmetry Phys.* **2019**, *52*, 17–26. [CrossRef]
14. Chen, Z.; Li, Y.; Sarkar, S.; Dey, S.; Bhattacharyya, A. Ricci soliton and certain related metrics on a three-dimensional trans-Sasakian manifold. *Universe* **2022**, *8*, 595. [CrossRef]

15. Li, Y.; Haseeb, A.; Ali, M. LP-Kenmotsu manifolds admitting η-Ricci-Yamabe Solitons and spacetime. *J. Math.* **2022**, *2022*, 6605127. [CrossRef]
16. Suh, Y.J. Yamabe and gradient Yamabe solitons in the complex hyperbolic two-plane Grassmannians. *Rev. Math. Phys.* **2022**, *34*, 2250024. [CrossRef]
17. Suh, Y.J. Yamabe and quasi-Yamabe solitons on hypersurfaces in the complex hyperbolic space. *Mediterr. J. Math.* **2023**, *20*, 69. [CrossRef]
18. Turki, N.B.; Blaga, A.M.; Deshmukh, S. Soliton-type equations on a Riemannian manifold. *Mathematics* **2022**, *10*, 633. [CrossRef]
19. Yoldas, H.I. On Kenmotsu manifolds admitting η-Ricci-Yamabe solitons. *Int. J. Geom. Methods Mod. Phys.* **2021**, *18*, 2150189. [CrossRef]
20. Yano, K. On semi-symmetric metric connections. *Rev. Roumaine Math. Pures Appl.* **1970**, *15*, 1579–1586.
21. Chaubey, S.K.; De, U.C. Characterization of three-dimensional Riemannian manifolds with a type of semi-symmetric metric connection admitting Yamabe soliton. *J. Geom. Phys.* **2020**, *157*, 103846. [CrossRef]
22. Chaubey, S.K.; De, U.C. Three dimensional Riemannian manifolds and Ricci solitons. *Quaest. Math.* **2022**, *45*, 765–778. [CrossRef]
23. Chaubey, S.K.; Lee, J.W.; Yadav, S. Riemannian manifolds with a semisymmetric metric P-connection. *J. Korean Math. Soc.* **2019**, *56*, 1113–1129.
24. Haseeb, A.; Chaubey, S.K.; Khan, M.A. Riemannian 3-manifolds and Ricci-Yamabe solitons. *Int. J. Geom. Methods Mod. Phys.* **2023**, *20*, 2350015. [CrossRef]
25. Yano, K. *Integral Formulas in Riemannian Geometry, Pure and Applied Mathematics*; Marcel Dekker: New York, NY, USA, 1970; Volume I.
26. Chaubey, S.K.; Suh, Y.J. Riemannian concircular structure manifolds. *Filomat* **2022**, *36*, 6699–6711. [CrossRef]

axioms

Article

Kinematic Geometry of a Timelike Line Trajectory in Hyperbolic Locomotions

Areej A. Almoneef [1] and Rashad A. Abdel-Baky [2,*]

[1] Department of Mathematical Sciences, College of Science, Princess Nourah bint Abdulrahman University, P.O. Box 84428, Riyadh 11671, Saudi Arabia
[2] Department of Mathematics, Faculty of Science, University of Assiut, Assiut 71516, Egypt
* Correspondence: rbaky@live.com

Abstract: This study utilizes the axodes invariants to derive novel hyperbolic proofs of the Euler-Savary and Disteli formulae. The inflection circle, which is widely recognized, is situated on the hyperbolic dual unit sphere, in accordance with the principles of the kinematic theory of spherical locomotions. Subsequently, a timelike line congruence is defined and its spatial equivalence is thoroughly studied. The formulated assertions degenerate into a quadratic form, which facilitates a comprehensive understanding of the geometric features of the inflection line congruence.

Keywords: axodes; Disteli's formulae; inflection circle

MSC: 53A15; 53A17; 53A25; 53A35

1. Introduction

Line geometry has an alliance with spatial locomotions and has thus found implementations in mechanism layout and robot kinematics. In locomotion, it is interested in inspecting the essential characteristics of the line path from the connotations of ruled surface. It is well known in spatial locomotions that the instantaneous screw axis ($\mathbb{ISA}$) of a movable body traces a couple of ruled surfaces, named the mobile and immobile axodes, with $\mathbb{ISA}$ as its tracing line in the movable space and in the steady space, respectively. Through locomotion, the axodes slide and roll relative to each other in a specific path such that the contact amidst the axodes is permanently maintained on the length of the two matting rulings (one being in all axodes), which define the $\mathbb{ISA}$ at any instant. It is essential that not only does an assured locomotion confer a rise to a unique set of axodes but the converse furthermore stratifies. This shows that, should the axodes of any locomotion be renowned, the evident locomotion can be reconstructed without knowledge of the physical features of the mechanism, their explications, given dimensions, or the manners by which they are united. There exists major literature on the topic including sundry monographs [1–5].

On the other hand, dual numbers have been employed to study the locomotion of a line space; and they may even serve as more effective tools for this purpose. According to the E Study map in the theory of dual numbers, it may be concluded that there exists a bijection between the set of the dual points on dual unit sphere ($\mathcal{DUS}$) in the dual 3-space $\mathbb{D}^3$ and the set of all directed lines in Euclidean 3-space $\mathbb{E}^3$. By use of this map, a one-parameter set of points (a dual curve) on $\mathcal{DUS}$ can be associated with a one-parameter set of directed lines (ruled surface) in $\mathbb{E}^3$ [6–12]. In the Minkowski three-space $\mathbb{E}_1^3$, since the Lorentzian metric can be positive, negative or zero. Conversely, in the Euclidean three-space $\mathbb{E}^3$, the metric is exclusively positive definite. Therefore, the kinematic and geometrical clarifications hold significant importance in $\mathbb{E}_1^3$ [13–19].

In this paper, we utilized the E. Study map for investigating the kinematic-geometry of a timelike ($\mathbb{T}-like$) line trajectory in one-parameter hyperbolic spatial locomotions. Then

Citation: Almoneef, A.A.; Abdel-Baky, R.A. Kinematic Geometry of a Timelike Line Trajectory in Hyperbolic Locomotions. *Axioms* **2023**, *12*, 915. https://doi.org/10.3390/axioms12100915

Academic Editor: Mića Stanković

Received: 5 August 2023
Revised: 7 September 2023
Accepted: 19 September 2023
Published: 26 September 2023

we gained new dual versions of Euler–Savary formula ($\mathbb{ES}$), resulting in distinct statements that are based on the axode invariants. Lastly, we explored a theoretical narration of the infliction circle of planar locomotions.

2. Preliminaries

In this section, we list notations of dual Lorentzian vectors and E. Study map (See [13–19]): A non-null oriented line L in Minkowski 3-space $\mathbb{E}_1^3$ can be appointed by a point $\mathbf{q} \in L$ and a normalized vector $\mathbf{u}$ of L, that is, $\|\mathbf{u}\|^2 = \pm 1$. To have coordinates for L, one must have the moment vector $\mathbf{u}^* = \mathbf{q} \times \mathbf{u}$ in $\mathbb{E}_1^3$. If $\mathbf{q}$ is reciprocal by any point $\mathbf{p} = \mathbf{q}+t\mathbf{u}$, $t \in \mathbb{R}$ on L, this offers that $\mathbf{u}^*$ is independent of $\mathbf{q}$ on L. The two non-null vectors $\mathbf{u}$ and $\mathbf{u}^*$ satisfy that

$$\langle \mathbf{u}, \mathbf{u} \rangle = \pm 1, \ \ \langle \mathbf{u}^*, \mathbf{u} \rangle = 0. \tag{1}$$

The 6-component u_i, $u_i^*(i = 1, 2, 3)$ of $\mathbf{u}$ and $\mathbf{u}^*$ are the normalized Plŭcker coordinates of L [1–4].

A dual number $\widehat{u}$ is a number $u + \varepsilon u^*$, where $(u, u^*) \in \mathbb{R} \times \mathbb{R}$, ε is a dual unit with $\varepsilon \neq 0$, and $\varepsilon^2 = 0$. Thus, the set

$$\mathbb{D}^3 = \{\widehat{\mathbf{u}} := \mathbf{u} + \varepsilon \mathbf{u}^* = (\widehat{u}_1, \widehat{u}_2, \widehat{u}_3)\}, \tag{2}$$

with the Lorentzian scalar product

$$\langle \widehat{\mathbf{u}}, \widehat{\mathbf{u}} \rangle = -\widehat{u}_1^2 + \widehat{u}_2^2 + \widehat{u}_3^2, \tag{3}$$

explain dual Lorentzian three-space $\mathbb{D}_1^3$. Then, a point $\widehat{u} = (\widehat{u}_1, \widehat{u}_2, \widehat{u}_3)^t$ has dual coordinates $\widehat{u}_i = (u_i + \varepsilon u_i^*) \in \mathbb{D}$. If $\mathbf{u} \neq \mathbf{0}$, the norm $\|\widehat{\mathbf{u}}\|$ of $\widehat{\mathbf{u}} = \mathbf{u} + \varepsilon \mathbf{u}^*$ is

$$\|\widehat{\mathbf{u}}\| = \sqrt{|\langle \widehat{\mathbf{u}}, \widehat{\mathbf{u}} \rangle|} = \|\mathbf{u}\|(1+\varepsilon \frac{\langle \mathbf{u}, \mathbf{u}^* \rangle}{\|\mathbf{u}\|^2}). \tag{4}$$

So, if $\|\widehat{\mathbf{u}}\|^2 = -1$ $(\|\widehat{\mathbf{u}}\|^2 = 1)$, the vector $\widehat{\mathbf{u}}$ is a $\mathbb{T}-$*like* (spacelike ($\mathbb{S}-$*like*)) dual unit vector. Then,

$$\|\widehat{\mathbf{u}}\|^2 = \pm 1 \Longleftrightarrow \|\mathbf{u}\|^2 = \pm 1, \ \langle \mathbf{u}, \mathbf{u}^* \rangle = 0. \tag{5}$$

The dual hyperbolic, and Lorentzian (de Sitter space) unit spheres with the center $\widehat{\mathbf{0}}$, respectively, are [13–19]:

$$\mathbb{H}_+^2 = \left\{ \widehat{\mathbf{u}} \in \mathbb{D}_1^3 \mid -\widehat{u}_1^2 + \widehat{u}_2^2 + \widehat{u}_3^2 = -1 \right\}, \tag{6}$$

and

$$\mathbb{S}_1^2 = \left\{ \widehat{\mathbf{u}} \in \mathbb{D}_1^3 \mid -\widehat{u}_1^2 + \widehat{u}_2^2 + \widehat{u}_3^2 = 1 \right\}. \tag{7}$$

Therefore, presented here is the map provided by E. Study: the ring-shaped hyperboloid may be bijectively mapped to the set of $\mathbb{S}-$*like* lines. Similarly, the common asymptotic cone can be bijectively mapped to the set of null-lines. Lastly, the oval-shaped hyperboloid can be bijectively mapped to the set of $\mathbb{T}-$*like* lines (see Figure 1). Then, a regular curve on $\mathbb{H}_+^2$ matches a $\mathbb{T}-$*like* ruled surface in $\mathbb{E}_1^3$. Also, a regular curve on $\mathbb{S}_1^2$ matches a $\mathbb{S}-$*like* or $\mathbb{T}-$*like* ruled surface in $\mathbb{E}_1^3$ [13–19].

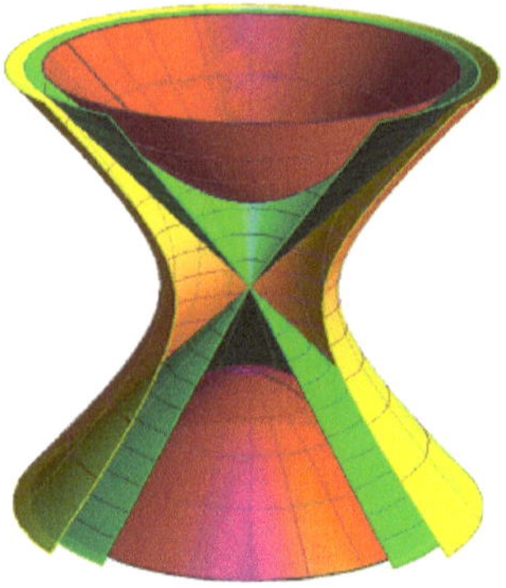

Figure 1. The dual hyperbolic and dual Lorentzian unit spheres.

Hyperbolic Dual Spherical Locomotions

Let us address that $\mathbb{H}^2_{+m}$ and $\mathbb{H}^2_{+f}$ are two hyperbolic $\mathcal{DUS}$ centered at the origin $\widehat{\mathbf{0}}$ in $\mathbb{D}^3_1$. Let $\{\widehat{\mathbf{u}}\} = \{\widehat{\mathbf{0}};\ \widehat{\mathbf{u}}_1(\mathbb{T}-like),\ \widehat{\mathbf{u}}_2,\ \widehat{\mathbf{u}}_3\}$, and $\{\widehat{\zeta}\} = \{\widehat{\mathbf{0}};\ \widehat{\zeta}_1(\mathbb{T}-like),\ \widehat{\zeta}_2,\ \widehat{\zeta}_3\}$ be two orthonormal dual frames of $\mathbb{H}^2_{+m}$ and $\mathbb{H}^2_{+f}$, respectively. If we say $\{\widehat{\zeta}\}$ is stationary, whereas the elements of $\{\widehat{\mathbf{u}}\}$ are functions of a real parameter $t \in \mathbb{R}$ (say the time). Then, we say that $\mathbb{H}^2_{+m}$ moves on $\mathbb{H}^2_{+f}$. This is a one-parameter Lorentzian dual spherical ($\mathcal{DS}$) locomotion and will signalize by $\mathbb{H}^2_{+m}/\mathbb{H}^2_{+f}$. Via the E. Study map, the hyperbolic $\mathcal{DUS}$ $\mathbb{H}^2_{+m}$ and $\mathbb{H}^2_{+f}$ matches the hyperbolic line spaces $\mathbb{L}_m$(mobile) and $\mathbb{L}_f$ (immobile), respectively. Therefore, $\mathbb{L}_m/\mathbb{L}_f$ is the mobile hyperbolic line space against the hyperbolic immobile space $\mathbb{L}_f$, because at any instant the instantaneous screw axis ($\mathbb{ISA}$) of $\mathbb{L}_m/\mathbb{L}_f$ creates a mobile $\mathbb{T}-like$ axode π_m in $\mathbb{L}_m$, and immobile $\mathbb{T}-like$ axode π_f in $\mathbb{L}_f$. Therefore, we insert an orthonormal dual frame $\{\widehat{\mathbf{r}}\} = \{\widehat{\mathbf{0}};\ \widehat{\mathbf{r}}_1(\mathbb{T}-like),\ \widehat{\mathbf{r}}_2,\ \widehat{\mathbf{r}}_3\}$, which is specified as follows: we set $\widehat{\mathbf{r}}_1(t) = \mathbf{r}_1(t) + \varepsilon\mathbf{r}_1^*(t)$, as the $\mathbb{ISA}$ and $\widehat{\mathbf{r}}_2(t) := \mathbf{r}_2(t) + \varepsilon\mathbf{r}_2^*(t) = \frac{d\widehat{\mathbf{r}}_1}{dt}\left\|\frac{d\widehat{\mathbf{r}}_1}{dt}\right\|^{-1}$ as the joint central normal of two disjoint screw axes. A third dual unit vector is designated as $\widehat{\mathbf{r}}_3(t) = \widehat{\mathbf{r}}_1 \times \widehat{\mathbf{r}}_2$. Then,

$$\left.\begin{array}{l}\widehat{\mathbf{r}}_1\times\widehat{\mathbf{r}}_2=\widehat{\mathbf{r}}_3,\ \widehat{\mathbf{r}}_1\times\widehat{\mathbf{r}}_3=-\widehat{\mathbf{r}}_2,\ \widehat{\mathbf{r}}_2\times\widehat{\mathbf{r}}_3=-\widehat{\mathbf{r}}_1,\\ -\langle\widehat{\mathbf{r}}_1,\widehat{\mathbf{r}}_1\rangle=\langle\widehat{\mathbf{r}}_2,\widehat{\mathbf{r}}_2\rangle=\langle\widehat{\mathbf{r}}_3,\widehat{\mathbf{r}}_3\rangle=1.\end{array}\right\} \tag{8}$$

The set $\{\widehat{\mathbf{r}}\}$ is the relative Blaschke frame, and $\widehat{\mathbf{r}}_1$, $\widehat{\mathbf{r}}_2$, and $\widehat{\mathbf{r}}_3$ are intersected at the joint striction (central) point $\mathbf{s}(t)$ of the $\mathbb{T}-like$ axodes π_i $(i = m,\ f)$. The dual arc length $d\widehat{s}_i = ds_i + \varepsilon ds_i^*$ of $\widehat{\mathbf{r}}_1(t)$ is

$$d\widehat{s}_i = \left\|\frac{d\widehat{\mathbf{r}}_1}{dt}\right\|dt = \widehat{\sigma}(t)dt. \tag{9}$$

$\widehat{\sigma}(t) = \sigma(t) + \varepsilon\sigma^*(t)$ is the first order asset of the locomotions $\mathbb{H}^2_{+m}/\mathbb{H}^2_{+f}$. We set $d\widehat{s} = ds + \varepsilon ds^*$ to represent $d\widehat{s}_i$, since they are equal to each other. Then,

$$\mu(s) := \frac{\sigma^*}{\sigma} = \frac{ds^*}{ds}, \tag{10}$$

is the distribution parameter ($\mathcal{D}-par$) of the $\mathbb{T}-like$ axodes. Via the E. Study map: for the locomotion the $\mathbb{T}-like$ axodes have the $\mathbb{ISA}$ in mutual; that is the mobile axode osculating with the immobile axode along the $\mathbb{ISA}$ in the first order (compared with [1–3]). For the Blaschke formulae with respect to $\mathbb{H}^2_{+i}(i=m,f)$, we find

$$\mathbb{H}^2_{+r}/\mathbb{H}^2_{+i}:\begin{pmatrix}\widehat{\mathbf{r}}_1'\\ \widehat{\mathbf{r}}_2'\\ \widehat{\mathbf{r}}_3'\end{pmatrix}=\widehat{\varrho}_i\times\begin{pmatrix}\widehat{\mathbf{r}}_1\\ \widehat{\mathbf{r}}_2\\ \widehat{\mathbf{r}}_3\end{pmatrix},\ ('=\frac{d}{d\widehat{s}}).\tag{11}$$

where $\widehat{\varrho}_i:=\varrho_i+\varepsilon\varrho_i^*=\widehat{\beta}_i\widehat{\mathbf{r}}_1-\widehat{\mathbf{r}}_3$ is the Darboux vector, and

$$\widehat{\beta}_i=\beta_i+\varepsilon(\Omega_i+\lambda\beta_i)=\det(\widehat{\mathbf{r}}_1,\widehat{\mathbf{r}}_1',\widehat{\mathbf{r}}_1''),\tag{12}$$

is the radii of curvature of the $\mathbb{T}-like$ axodes π_i. $\beta_i(s)$, $\Omega_i(s)$ and $\lambda(s)$ are the curvature (construction) functions of the $\mathbb{T}-like$ axodes. By setting $\left|\widehat{\beta}_i\right|<1$, the $\mathbb{S}-like$ Disteli-axes ($\mathbb{DA}$) of the $\mathbb{T}-like$ axodes is

$$\widehat{\mathbf{b}}_i(\widehat{s})=\frac{\widehat{\varrho}_i}{\|\widehat{\varrho}_i\|}=\frac{\widehat{\beta}_i\widehat{\mathbf{r}}_1-\widehat{\mathbf{r}}_3}{\sqrt{1-\widehat{\beta}_i^2}}=\sinh\widehat{\varphi}_i\widehat{\mathbf{r}}_1-\cosh\widehat{\varphi}\widehat{\mathbf{r}}_3,\tag{13}$$

where $\widehat{\varphi}_i(\widehat{s})=\varphi_i+\varepsilon\varphi_i^*$ is a Lorentzian $\mathbb{T}-like$ dual angle (radius of curvature) among $\widehat{\mathbf{r}}_1$ and $\widehat{\mathbf{b}}_i$. Then,

$$\widehat{\beta}_f-\widehat{\beta}_m=\tanh\widehat{\varphi}_f-\tanh\widehat{\varphi}_m.\tag{14}$$

Equation (14) is a novel dual hyperbolic version of the $\mathbb{ES}$ formula (Compared with [1–3]). Via the real and the dual parts, respectively, we attain

$$\tanh\varphi_f-\tanh\varphi_m=\beta_f-\beta_m,\tag{15}$$

and

$$\frac{\varphi_f^*}{\cosh^2\vartheta_f}-\frac{\varphi_m^*}{\cosh^2\vartheta_m}+\lambda\left(\beta_f-\beta_m\right)=\Omega_m-\Omega_f.\tag{16}$$

Equation (15) in conjuction with (16) are new Disteli formulae ($\mathbb{DF}$) for the $\mathbb{T}-like$ axodes of the locomotion $\mathbb{L}_m/\mathbb{L}_f$.

Now let us assume that $\{\widehat{\mathbf{r}}\}$ is stabilized in $\mathbb{H}^2_{+m}$. Then,

$$\mathbb{H}^2_{+m}/\mathbb{H}^2_{+f}:\begin{pmatrix}\widehat{\mathbf{r}}_1'\\ \widehat{\mathbf{r}}_2'\\ \widehat{\mathbf{r}}_3'\end{pmatrix}=\widehat{\varrho}\times\begin{pmatrix}\widehat{\mathbf{r}}_1\\ \widehat{\mathbf{r}}_2\\ \widehat{\mathbf{r}}_3\end{pmatrix},\tag{17}$$

where

$$\widehat{\varrho}:=\widehat{\varrho}_f-\widehat{\varrho}_m=\widehat{\varrho}\widehat{\mathbf{r}}_1,\tag{18}$$

is the relative Darboux vector. $\|\widehat{\varrho}\|=\widehat{\varrho}=\widehat{\varrho}+\varepsilon\widehat{\varrho}^*=\beta_r+\varepsilon(\Omega_r+\lambda\beta_r)$ is the relative radii of curvature; $\varrho=\beta_f-\beta_m$, and $\varrho^*=\Omega_f-\Omega_m-\lambda\left(\beta_f-\beta_m\right)$ are the rotational angular speed and translational angular speed of the locomotion $\mathbb{L}_m/\mathbb{L}_f$, as well they are both invariants in kinematics, respectively. As a result, the following corollary can be stated:

Corollary 1. *For the locomotion $\mathbb{L}_m/\mathbb{L}_f$, at any instant $t\in\mathbb{R}$, the pitch is*

$$h(s):=\frac{\varrho^*}{\varrho}=\frac{\Omega_f-\Omega_m}{\beta_f-\beta_m}-\lambda.\tag{19}$$

In this study, we deviate from the exclusive use of translational locomotion, namely when $\varrho^*\neq 0$. Moreover, we impose the condition of excluding zero divisors, denoted by $\varrho=0$. Consequently, our investigation will solely focus on non-torsional locomotions, ensuring that the axodes associated with these motions are non-developable $\mathbb{T}-like$ ruled surfaces, characterized by $\lambda \neq 0$.

3. Timelike Line with Particular Trajectories

Through the locomotion $\mathbb{L}_m/\mathbb{L}_f$, any fixed $\mathbb{T}-like$ line $\widehat{\mathbf{x}}$ connected with the mobile space $\mathbb{L}_m$-space, normally, creates a $\mathbb{T}-like$ ruled surface $(\widehat{x})$ in the immobile $\mathbb{L}_f$-space. Then,

$$\widehat{\mathbf{x}}(\widehat{s}){=}\widehat{x}^t\widehat{\mathbf{r}},\ \widehat{x}=\begin{pmatrix}\widehat{x}_1\\ \widehat{x}_2\\ \widehat{x}_3\end{pmatrix}=\begin{pmatrix}x_1+\varepsilon x_1^*\\ x_2+\varepsilon x_2^*\\ x_3+\varepsilon x_3^*\end{pmatrix},\ \widehat{\mathbf{r}}=\begin{pmatrix}\widehat{\mathbf{r}}_1\\ \widehat{\mathbf{r}}_2\\ \widehat{\mathbf{r}}_3\end{pmatrix}, \tag{20}$$

where

$$\begin{aligned}-x_1^2+x_2^2+x_3^2&=-1,\\ -x_1x_1^*+x_2x_2^*+x_3x_3^*&=0.\end{aligned} \tag{21}$$

The velocity and the acceleration vectors of $\widehat{\mathbf{x}} \in \mathbb{H}^2_{+m}$, respectively, are

$$\widehat{\mathbf{x}}' = \widehat{\boldsymbol{\varrho}}\times\widehat{\mathbf{x}} = \widehat{\varrho}(-\widehat{x}_3\widehat{\mathbf{r}}_2+\widehat{x}_2\widehat{\mathbf{r}}_3), \tag{22}$$

and

$$\widehat{\mathbf{x}}'' = -\widehat{x}_3\widehat{\varrho}\widehat{\mathbf{r}}_1-(\widehat{x}_2\widehat{\varrho}^2+\widehat{x}_3\widehat{\varrho}')\widehat{\mathbf{r}}_2+(\widehat{x}_2\widehat{\varrho}'-\widehat{x}_1\widehat{\varrho}-\widehat{x}_3\widehat{\varrho}^2)\widehat{\mathbf{r}}_3. \tag{23}$$

So, we have

$$\widehat{\mathbf{x}}'\times\widehat{\mathbf{x}}'' = -\widehat{\varrho}^2\left[(\widehat{x}_1^2-1)\widehat{\varrho}\widehat{\mathbf{r}}_1+\widehat{x}_3\widehat{\mathbf{x}}\right]. \tag{24}$$

The dual arc-length of the dual curve $\widehat{\mathbf{x}}(\widehat{s})$ is

$$d\widehat{w} := dw+\varepsilon dw^* = \left\|\widehat{\mathbf{x}}'\right\|d\widehat{s} = \widehat{\varrho}\sqrt{\widehat{x}_1^2-1}d\widehat{s},\ \text{with } |\widehat{x}_1|>1. \tag{25}$$

Then, the $\mathcal{D}-par$ of $(\widehat{x})$ is

$$\mu(w) := \frac{dw^*}{dw} = h-\frac{x_1x_1^*}{x_1^2-1}. \tag{26}$$

Moreover, the Balschke frame is

$$\widehat{\mathbf{x}}{=}\widehat{\mathbf{x}}(\widehat{s}),\ \widehat{\mathbf{t}}(\widehat{s}) = \widehat{\mathbf{x}}'\left\|\widehat{\mathbf{x}}'\right\|^{-1},\ \widehat{\mathbf{g}}(\widehat{s}) = \widehat{\mathbf{x}}\times\widehat{\mathbf{t}}, \tag{27}$$

where

$$\left.\begin{aligned}&\widehat{\mathbf{x}}\times\widehat{\mathbf{t}}=\widehat{\mathbf{g}},\ \widehat{\mathbf{x}}\times\widehat{\mathbf{g}}=-\widehat{\mathbf{t}},\ \widehat{\mathbf{t}}\times\widehat{\mathbf{g}}=\widehat{\mathbf{x}},\\ &-\langle\widehat{\mathbf{x}},\widehat{\mathbf{x}}\rangle=\langle\widehat{\mathbf{t}},\widehat{\mathbf{t}}\rangle=\langle\widehat{\mathbf{g}},\widehat{\mathbf{g}}\rangle=1.\end{aligned}\right\} \tag{28}$$

The dual unit vectors $\widehat{\mathbf{x}}$, $\widehat{\mathbf{t}}$, and $\widehat{\mathbf{g}}$ are three simultaneous alternately orthogonal lines in Minkowski three-space $\mathbb{E}^3_1$. Their joint point is the central point $\mathbf{c}$ on the ruling $\widehat{\mathbf{x}}$. $\widehat{\mathbf{g}}(\widehat{s})$ is the mutual orthogonal to $\widehat{\mathbf{x}}(\widehat{w})$ and $\widehat{\mathbf{x}}(\widehat{w}+d\widehat{w})$, and it is named the central tangent of $(\widehat{\mathbf{x}})$ at the central point. The trace of $\mathbf{c}$ is the striction curve. The line $\widehat{\mathbf{t}}$ is the central normal of $(\widehat{\mathbf{x}})$ at $\mathbf{c}$. So, the Blaschke formulae are

$$\frac{d}{d\widehat{w}}\begin{pmatrix}\widehat{\mathbf{x}}\\ \widehat{\mathbf{t}}\\ \widehat{\mathbf{g}}\end{pmatrix}=\begin{pmatrix}0&1&0\\ 1&0&\widehat{\beta}\\ 0&-\widehat{\beta}&0\end{pmatrix}\begin{pmatrix}\widehat{\mathbf{x}}\\ \widehat{\mathbf{t}}\\ \widehat{\mathbf{g}}\end{pmatrix}=\widehat{\boldsymbol{\eta}}(\widehat{w})\times\begin{pmatrix}\widehat{\mathbf{x}}\\ \widehat{\mathbf{t}}\\ \widehat{\mathbf{g}}\end{pmatrix}, \tag{29}$$

where $\widehat{\boldsymbol{\eta}} = \boldsymbol{\eta}+\varepsilon\boldsymbol{\eta}^* = \widehat{\beta}\widehat{\mathbf{x}}-\widehat{\mathbf{g}}$ is the Darboux vector, and

$$\widehat{\beta}(\widehat{w}) = \beta+\varepsilon(\Omega+\mu\beta) = \det(\widehat{\mathbf{x}},\frac{d\widehat{\mathbf{x}}}{d\widehat{w}},\frac{d^2\widehat{\mathbf{x}}}{d\widehat{w}^2}) = \frac{\widehat{x}_1\widehat{\varrho}(\widehat{x}_1^2-1)+\widehat{x}_3}{\widehat{\varrho}(\widehat{x}_1^2-1)^{\frac{3}{2}}}, \tag{30}$$

is the radii of curvature of $\widehat{\mathbf{x}}(\widehat{w})$. The tangent vector of $\mathbf{c}(w)$ is

$$\frac{d\mathbf{c}}{dw} = -\Omega\mathbf{x} + \mu\mathbf{g}, \tag{31}$$

which is a **S***like* (a **T***like*) curve if $|\mu| > |\Omega|$ ($|\mu| < |\Omega|$). $\beta(w)$, $\Omega(w)$ and $\mu(w)$ are construction parameters of the $\mathbb{T}-like$ ruled surface $(\widehat{x})$. Under the hypothesis that $\left|\widehat{\beta}\right| < 1$, we specify the $\mathbb{S}-like$ $\mathbb{DA}$ as follows:

$$\widehat{\mathbf{b}}(\widehat{w}) = \frac{\widehat{\overline{\varrho}}}{\left\|\widehat{\overline{\varrho}}\right\|} = \frac{\widehat{\beta}\widehat{\mathbf{x}} - \widehat{\mathbf{g}}}{\sqrt{1-\widehat{\beta}^2}} = \sinh\widehat{\vartheta}\widehat{\mathbf{x}} - \cosh\widehat{\vartheta}\widehat{\mathbf{g}}, \tag{32}$$

where $\widehat{\vartheta}(\widehat{w}) = \vartheta + \varepsilon\vartheta^*$ is radii of curvature through $\widehat{\mathbf{x}}$ and $\widehat{\mathbf{b}}$. Then,

$$\tanh\widehat{\vartheta} = \frac{\widehat{x}_1\widehat{\varrho}(\widehat{x}_1^2 - 1) + \widehat{x}_3}{\widehat{\varrho}(\widehat{x}_1^2 - 1)^{\frac{3}{2}}} = \widehat{\beta}(\widehat{w}). \tag{33}$$

Further, we may have

$$\left.\begin{aligned} \widehat{\kappa}(\widehat{w}) &= \kappa + \varepsilon\kappa^* = \sqrt{1-\widehat{\beta}^2} = \tfrac{1}{\cosh\widehat{\vartheta}}, \\ \widehat{\tau}(\widehat{w}) &= \tau + \varepsilon\tau^* = \pm\tfrac{d\widehat{\vartheta}}{d\widehat{w}} = \pm\tfrac{1}{1-\widehat{\beta}^2}\tfrac{d\widehat{\beta}}{d\widehat{w}}, \end{aligned}\right\} \tag{34}$$

where $\widehat{\kappa}(\widehat{w})$ is the dual curvature, and $\widehat{\tau}(\widehat{w})$ is the dual torsion of the dual curve $\widehat{\mathbf{x}}(\widehat{w})$. Via Equations (26) and (31), $(\widehat{x})$ is a $\mathbb{T}-like$ tangential developable surface if and only if $\frac{d\mathbf{c}}{dw} \parallel \mathbf{x}$, that is,

$$\mu = 0 \Leftrightarrow h(x_1^2 - 1) - x_1x_1^* = 0, \tag{35}$$

which represents that the developable conditions of a $\mathbb{T}-like$ line trajectory are only founded on x_1, x_1^* and h.

Theorem 1. *For the locomotion $\mathbb{L}_m/\mathbb{L}_f$, the $\mathbb{T}-like$ line trajectory has torsional rulings at those instants at which it belongs to the quadratic $\mathbb{T}-like$ line complex pointed out by Equation (35).*

In any quadratic $\mathbb{T}-like$ line complex the lines of this complex passing through a point mostly form a quadratic $\mathbb{T}-like$ cone. Primarily, for some points, this $\mathbb{T}-like$ cone reduces to a couple of $\mathbb{T}-like$ planes. Such points are the singular points of the $\mathbb{T}-like$ line complex. Thus, when $(\widehat{x})$ is a $\mathbb{T}-like$ cone, the conditions are $\mu = 0$, and $\Omega = 0$ define a quadratic $\mathbb{T}-like$ line congruence given by the mutual lines of the two quadratic $\mathbb{T}-like$ line complexes ($\mu = 0$, and $\Omega = 0$).

Theorem 2. *For the locomotion $\mathbb{L}_m/\mathbb{L}_f$, the set of $\mathbb{T}-like$ lines correlated with the mobile $\mathbb{T}-like$ axode are rulings of a quadratic $\mathbb{T}-like$ cone in $\mathbb{L}_f$. Moreover, this family of $\mathbb{T}-like$ lines belong to a quadratic $\mathbb{T}-like$ line congruence.*

3.1. The Euler–Savary and Disteli Formulae

In the context of planar locomotions, the $\mathbb{ES}$ formula associates the locus of a point to its curvature center and is the main ingredient for a graphical structure producing one assigned the other [1–3]. In 1914, Disteli [20] assigned a curvature axis for the ruling of a ruled surface and extended the planar $\mathbb{ES}$ formula to spatial locomotions. However, the $\mathbb{DF}$ of a line trajectory had been acquired in [4–8,21], around inscription should be refind as follows: we shall define a new manner to have $\mathbb{DF}$ by dual function approximations. Thus, we request the $\mathbb{T}-like$ line $\widehat{\mathbf{x}} \in \mathbb{L}_m$, which at a steady dual angle from a steady $\mathbb{S}-like$ line $\widehat{\mathbf{y}} \in \mathbb{L}_f$. So, if $\widehat{\psi} = \psi + \varepsilon\psi^*$ is the dual angle of $\widehat{\mathbf{x}}$($\mathbb{T}-like$), and $\widehat{\mathbf{y}}$ ($\mathbb{S}-like$), then

$$\widehat{\psi} = \sinh^{-1}(\langle \widehat{\mathbf{x}}, \widehat{\mathbf{y}} \rangle). \tag{36}$$

For $\widehat{\psi}$ is steady up to the 2nd order at $\widehat{w} = \widehat{w}_0$, we have

$$\widehat{\psi}'_{|\widehat{w}=\widehat{w}_0} = 0, \ \widehat{\mathbf{x}}'_{|\widehat{w}=\widehat{w}_0} = \mathbf{0}, \tag{37}$$

and

$$\widehat{\psi}''_{|\widehat{w}=\widehat{w}_0} = 0, \ \widehat{\mathbf{x}}''_{|\widehat{w}=\widehat{w}_0} = \mathbf{0}. \tag{38}$$

Hence, for the 1st order $\langle \widehat{\mathbf{x}}', \widehat{\mathbf{y}} \rangle = 0$, and for the 2nd order $\langle \widehat{\mathbf{x}}'', \widehat{\mathbf{y}} \rangle = 0$. Therefore, $\widehat{\psi}$ will be steady in the 2nd approximation if and only if $\widehat{\mathbf{y}}$ is the $\mathbb{S}-like$ $\mathbb{DA}$ $\widehat{\mathbf{b}}$ of $(\widehat{x})$, that is,

$$\widehat{\psi}' = \widehat{\psi}'' = 0 \Leftrightarrow \pm\widehat{\mathbf{y}} = \frac{\widehat{\mathbf{x}}' \times \widehat{\mathbf{x}}''}{\left\| \widehat{\mathbf{x}}' \times \widehat{\mathbf{x}}'' \right\|} = \widehat{\mathbf{b}}. \tag{39}$$

Hence, from Equations (32) and (39), the following corollary can be given.

Corollary 2. *$(\widehat{x})$ is a steady-$\mathbb{DA}$ $\mathbb{T}-like$ ruled surface if and only if* $\widehat{\vartheta}' = 0$.

Via this corollary, and based on Equation (34), it can be concluded that the rulings of $(\widehat{x})$ are the constant dual angle $\widehat{\vartheta}$ with respect to the $\mathbb{S}-like$ $\mathbb{DA}$ if and only if $\widehat{\beta}' = 0$. Therefore, the $\mathbb{T}-like$ ruled surface $(\widehat{x})$ is generated locally by a one-parameter hyperbolic spatial locomotion with pitch $h(s)$ along the steady $\mathbb{DA}$ $\widehat{\mathbf{b}}$, This locomotion is performed by the $\mathbb{T}-like$ line $\widehat{\mathbf{x}}$, which is positioned at a constant hyperbolic distance ϑ^* and a constant angle ϑ relative to $\widehat{\mathbf{b}}$. This indicates that the striction curve of $(\widehat{x})$ can be classified as either a $\mathbb{S}-like$ or $\mathbb{T}-like$ cylindrical helix. The corollary shown below can be used to identify the circumstances of steady $\mathbb{DA}$.

Corollary 3. *$(\widehat{x})$ is a steady $\mathbb{DA}$ $\mathbb{T}-like$ ruled surface if and only if*

$$\widehat{\beta}' = 0 \Leftrightarrow \frac{d\beta}{dw} = 0, \textit{and } \frac{d\Omega}{dw} + \beta\frac{d\mu}{dw} = 0. \tag{40}$$

Furthermore, from Equations (32) and (39), we find that

$$\widehat{\vartheta} = \sinh^{-1}(\langle \widehat{\mathbf{x}}, \widehat{\mathbf{b}} \rangle), \langle \widehat{\mathbf{x}}', \widehat{\mathbf{b}} \rangle = 0, \langle \widehat{\mathbf{x}}'', \widehat{\mathbf{b}} \rangle = 0. \tag{41}$$

So, $\widehat{\mathbf{b}}$ is the osculating circle of $\widehat{\mathbf{x}}(\widehat{u}) \in \mathbb{H}^2_{+f}$. Further, it can be seen from Equations (22), (27) and (32) that

$$\langle \widehat{\mathbf{t}}, \widehat{\mathbf{r}}_1 \rangle = \langle \widehat{\mathbf{t}}, \widehat{\mathbf{x}} \rangle = \langle \widehat{\mathbf{t}}, \widehat{\mathbf{b}} \rangle = 0, \tag{42}$$

Then, all $\widehat{\mathbf{r}}_1$, $\widehat{\mathbf{x}}$ and $\widehat{\mathbf{b}}$ belong to a $\mathbb{T}-like$ line congruence whose focus line is the $\mathbb{S}-like$ line $\widehat{\mathbf{t}}$. This can be realized as follows: we set $\widehat{\mathbf{t}}$ with respect to the set $\{\widehat{\mathbf{r}}\}$ by its intercept distance φ^*, control on the $\mathbb{ISA}$ and the angle φ, control with respect to $\widehat{\mathbf{r}}_2$. We set the dual angle $\widehat{\alpha} = \alpha + \varepsilon\alpha^*$, which realizes the attitude of $\widehat{\mathbf{b}}$ over $\widehat{\mathbf{t}}$. These dual angles are all estimated relative to the $\mathbb{ISA}$ (see Figure 2). The following governs the signals: (ϑ, ϑ^*) and (α, α^*) are via the right-hand screw rule with the thumb pointing on $\widehat{\mathbf{t}}$; the sense of $\widehat{\mathbf{t}}$ is such that $\widehat{\vartheta} = \vartheta + \varepsilon\vartheta^* \geq 0$, and $0 \leq \varphi \leq 2\pi$, $\varphi^* \in \mathbb{R}$ are explained with the thumb in the direction of the $\mathbb{ISA}$. Since $\widehat{\mathbf{x}}$ is a $\mathbb{T}-like$ dual unit vector, we can write out the components of $\widehat{\mathbf{x}}$ in the following form:

$$\widehat{\mathbf{x}} = \cosh\widehat{\vartheta}\widehat{\mathbf{r}}_1 + \sinh\widehat{\vartheta}\widehat{\mathbf{m}}, \text{ with } \widehat{\mathbf{m}} = \cos\widehat{\varphi}\widehat{\mathbf{r}}_2 + \sin\widehat{\varphi}\widehat{\mathbf{r}}_3. \tag{43}$$

Therefore, the Blaschke frame of $\widehat{\mathbf{x}} = \widehat{\mathbf{x}}(\widehat{w})$ can be written as

$$\begin{pmatrix} \widehat{\mathbf{x}} \\ \widehat{\mathbf{t}} \\ \widehat{\mathbf{g}} \end{pmatrix} = \begin{pmatrix} \cosh\widehat{\vartheta} & \sinh\widehat{\vartheta}\cos\widehat{\varphi} & \sinh\widehat{\vartheta}\sin\widehat{\varphi} \\ 0 & -\sin\widehat{\varphi} & \cos\widehat{\varphi} \\ -\sinh\widehat{\vartheta} & -\cosh\widehat{\vartheta}\cos\widehat{\varphi} & -\cosh\widehat{\vartheta}\sin\widehat{\varphi} \end{pmatrix} \begin{pmatrix} \widehat{\mathbf{r}}_1 \\ \widehat{\mathbf{r}}_2 \\ \widehat{\mathbf{r}}_3 \end{pmatrix}. \tag{44}$$

Comparably, the $\mathbb{S}-like$ $\mathbb{DA}$ is

$$\widehat{\mathbf{b}} = \sinh\widehat{\alpha}\widehat{\mathbf{r}}_1 + \cosh\widehat{\alpha}\widehat{\mathbf{m}}, \text{ with } \widehat{\alpha} = \alpha + \varepsilon\alpha^* \geq 0. \tag{45}$$

Substituting from Equations (23) and (45) into the third term of Equation (38) yields

$$\widehat{\varrho}\widehat{x}_3\tanh\widehat{\alpha} - (\widehat{x}_2^2\widehat{\varrho} + \widehat{x}_3\widehat{\varrho}^{'})\cos\widehat{\varphi} + (-\widehat{x}_1\widehat{\varrho} + \widehat{x}_2\widehat{\varrho}^{'} - \widehat{x}_3^2\widehat{\varrho})\sin\widehat{\varphi} = 0. \tag{46}$$

Into Equation (46) we substitute from Equation (43) to obtain

$$\coth\widehat{\alpha} - \coth\widehat{\vartheta} = \frac{\widehat{\varrho}}{\sin\widehat{\varphi}}. \tag{47}$$

Equation (47) is a new hyperbolic $\mathbb{ES}$ formula that fastens a $\mathbb{T}-like$ ruled surface and its osculating circle in terms of the dual angle $\widehat{\varphi}$ as well as the second order invariant $\widehat{\varrho}$. Via the real and the dual parts, respectively, we obtain

$$\coth\alpha - \coth\vartheta = \frac{\varrho}{\sin\varphi}, \tag{48}$$

and

$$\varphi^* = \frac{1}{\varrho}\left[\left(\frac{\alpha^*}{\sinh^2\alpha} - \frac{\vartheta^*}{\sinh^2\vartheta}\right)\sin\varphi + \frac{\varrho}{\sin\varphi}(h-\mu)\right]\tan\varphi. \tag{49}$$

Equation (48) with (49) are novel $\mathbb{DF}$ in the context of one-parameter hyperbolic spatial locomotions. The former equation establishes a relationship between the positions of the $\mathbb{T}-like$ line in the space $\mathbb{L}_m$ and the $\mathbb{S}-like$ $\mathbb{DA}$ denoted as $\widehat{\mathbf{b}}$. Based on the information provided in Figure 2, the presence of the signal α^* ($+$ or $-$) in Equation (49) indicates whether the positions of the $\mathbb{DA}$ $\widehat{\mathbf{b}}$ are located on the positive or negative direction of the mutual central normal $\widehat{\mathbf{t}}$.

However, we can derive the Equation (47) as follows: the hyperbolic radii of curvature $\widehat{\psi}$ can be written as (see Figure 2):

$$\widehat{\psi} = \widehat{\vartheta} - \widehat{\alpha} \Leftrightarrow \psi = \vartheta - \alpha,\ \psi^* = \vartheta^* - \alpha^*. \tag{50}$$

Then, we have

$$\widehat{\beta}(\widehat{u}) := \tanh\widehat{\psi} = \tanh\left(\widehat{\vartheta} - \widehat{\alpha}\right). \tag{51}$$

Substituting Equation (51), into Equation (33), with awareness of (43), we obtain

$$\tanh(\widehat{\vartheta} - \widehat{\alpha}) = \frac{\sin\widehat{\varphi}}{\widehat{\varrho}\sinh^2\widehat{\vartheta}}, \tag{52}$$

After some algebraic manipulations, we find

$$\coth\widehat{\alpha} - \coth\widehat{\vartheta} = \frac{\widehat{\varrho}}{\sin\widehat{\varphi}}. \tag{53}$$

as asserted. Moreover, in the case of axodes, it is possible to derive a second dual formulation of the $\mathbb{ES}$ formulae in the following manner: from Equations (22) and (43), one finds facilely

$$d\widehat{s} = \widehat{\varrho}\sinh\widehat{\vartheta}dt. \tag{54}$$

Moreover, from Equation (44) we have

$$\widehat{\mathbf{r}}_1 = \cosh\widehat{\vartheta}\widehat{\mathbf{x}} + \sinh\widehat{\vartheta}\widehat{\mathbf{g}}. \tag{55}$$

A simple computation offers that

$$\widehat{\mathbf{r}}_1^{'} := (\frac{dt}{d\widehat{s}})\frac{d\widehat{\mathbf{r}}_1}{dt} = (\sinh\widehat{\vartheta}\widehat{\mathbf{x}} + \cosh\widehat{\vartheta}\widehat{\mathbf{g}})\widehat{\vartheta}^{'} + (\cosh\widehat{\vartheta} - \widehat{\gamma}\sinh\widehat{\vartheta})\widehat{\mathbf{t}}, \tag{56}$$

and

$$\widehat{\mathbf{r}}_1^{'} = \left(\frac{1}{\widehat{\varrho}\sinh\widehat{\vartheta}}\right)\widehat{\mathbf{r}}_2. \tag{57}$$

The amalgamation of Equations (44) and (57) leads to

$$\widehat{\mathbf{r}}_1^{'} = \frac{1}{\widehat{\varrho}\sinh\widehat{\vartheta}}(-\sinh\widehat{\vartheta}\cos\widehat{\varphi}\widehat{\mathbf{x}} - \sin\widehat{\varphi}\widehat{\mathbf{t}} - \cosh\widehat{\vartheta}\cos\widehat{\varphi}\widehat{\mathbf{g}}). \tag{58}$$

Then, by equating the coefficients of $\widehat{\mathbf{x}}$, $\widehat{\mathbf{t}}$, and $\widehat{\mathbf{g}}$ in Equations (56) and (58), we have

$$\widehat{\vartheta}^{'}\sinh\widehat{\vartheta} + \frac{1}{\widehat{\varrho}}\cos\widehat{\varphi} = 0, \tag{59}$$

and

$$\cosh\widehat{\vartheta} - \widehat{\beta}\sinh\widehat{\vartheta} = -\frac{\sin\widehat{\varphi}}{\widehat{\varrho}\sinh\widehat{\vartheta}}. \tag{60}$$

Substituting this into the left hand side of Equation (53), one finds

$$\cosh\widehat{\vartheta} - \widehat{\beta}\sinh\widehat{\vartheta} = -\frac{1}{\sinh\widehat{\vartheta}}(\frac{1}{\coth\widehat{\alpha} - \coth\widehat{\vartheta}}). \tag{61}$$

Finally, by substituting $\widehat{\varrho} := \widehat{\beta}_f - \widehat{\beta}_m = \tanh\widehat{\varphi}_f - \tanh\widehat{\varphi}_m$ into Equation (59), one obtains

$$\tanh\widehat{\varphi}_m - \tanh\widehat{\varphi}_f = \frac{\cos\widehat{\varphi}}{\widehat{\vartheta}^{'}\sinh\widehat{\vartheta}}. \tag{62}$$

Equation (62) presents a novel hyperbolic dual variant of the widely recognized $\mathbb{ES}$ formula in the context of conventional spherical kinematics, as discussed in References [1–9,21]. This narrative provides a link between the two $\mathbb{T}-$*like* axodes in the locomotion of $\mathbb{L}_m/\mathbb{L}_f$. It should be noted that the striction point is the origin of the relative Blaschke frame, denoted as $\mathbf{s} = \mathbf{0}$, see Figure 2.

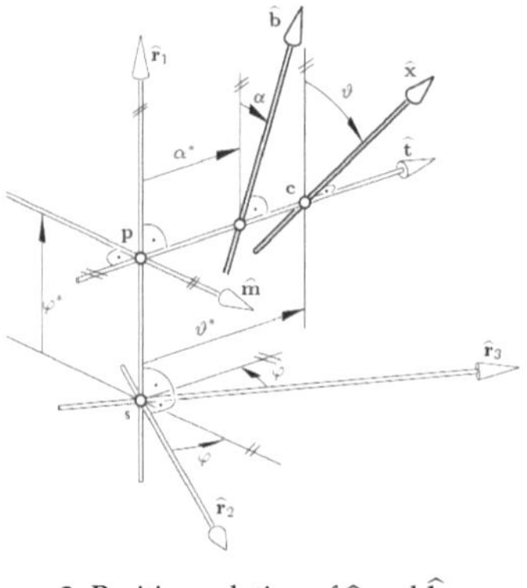

Figure 2. Position relation of $\widehat{\mathbf{x}}$ and $\widehat{\mathbf{b}}$.

3.2. A Timelike Line Congruence

We present a method for locating a $\mathbb{T}-like$ line congruence. Therefore, from the real and the dual parts of $\widehat{\mathbf{x}}$ in Equation (43), respectively, we obtain

$$\mathbf{x}(\vartheta,\varphi)=(\cosh\vartheta,\sinh\vartheta\cos\varphi,\sinh\vartheta\sin\varphi), \tag{63}$$

and

$$\mathbf{x}^*(\vartheta,\varphi,\vartheta^*,\varphi^*)=\begin{pmatrix}\vartheta^*\sinh\vartheta\\ \vartheta^*\cosh\vartheta\cos\varphi-\varphi^*\sinh\vartheta\sin\varphi\\ \vartheta^*\cosh\vartheta\sin\varphi+\varphi^*\sinh\vartheta\cos\varphi\end{pmatrix}. \tag{64}$$

Since $\mathbf{x}^*=\epsilon\times\mathbf{x}$, we possess the system of linear equations in $\epsilon_i(\mathrm{i}=1,2,3)$:

$$\left.\begin{array}{r}-\epsilon_2\sinh\vartheta\sin\varphi+\epsilon_3\sinh\vartheta\cos\varphi=x_1^*,\\ -\epsilon_1\sinh\vartheta\sin\varphi+\epsilon_3\cosh\vartheta=x_2^*,\\ \epsilon_1\sinh\vartheta\cos\varphi-\epsilon_2\cosh\vartheta=x_3^*.\end{array}\right\} \tag{65}$$

The coefficient matrix of unknowns $\epsilon_i(\mathrm{i}=1,2,3)$ is the skew-adjoint matrix

$$\begin{pmatrix}0 & -\sinh\vartheta\sin\varphi & \sinh\vartheta\cos\varphi\\ -\sinh\vartheta\sin\varphi & 0 & \cosh\vartheta\\ \sinh\vartheta\cos\varphi & -\cosh\vartheta & 0\end{pmatrix}, \tag{66}$$

and thus its rank is 2 with $\vartheta\neq0$, and $\varphi\neq2\pi k$ (k is an integer). The rank of the augmented matrix

$$\begin{pmatrix}0 & -\sinh\vartheta\sin\varphi & \sinh\vartheta\cos\varphi & x_1^*\\ -\sinh\vartheta\sin\varphi & 0 & \cosh\vartheta & x_2^*\\ \sinh\vartheta\cos\varphi & -\cosh\vartheta & 0 & x_3^*\end{pmatrix}, \tag{67}$$

is also 2. Hence, this system possesses an infinite number of solutions that are specified by

$$\begin{array}{c}\epsilon_2=(\epsilon_1-\varphi^*)\tanh\vartheta\cos\varphi-\vartheta^*\sin\varphi,\\ \epsilon_3=(\epsilon_1-\varphi^*)\tanh\vartheta\sin\varphi+\vartheta^*\cos\varphi,\\ \epsilon_1=\epsilon_1(\vartheta,\varphi).\end{array} \tag{68}$$

Since ϵ_1 can be arbitrary, we may then put $\epsilon_1=\varphi^*$. In this affair, we have

$$\epsilon(\varphi,\varphi^*)=(\varphi^*,-\vartheta^*\sin\varphi,\vartheta^*\cos\varphi), \tag{69}$$

which is the base (director) surface of the $\mathbb{T}-like$ line congruence. Let $\boldsymbol{\xi}(\xi_1,\xi_2,\xi_3)$ be a point on the directed $\mathbb{T}-like$ line $\widehat{\mathbf{x}}$. We can write that

$$(\widehat{x}):\left.\begin{array}{l}\xi_1(\varphi,\varphi^*,\rho)=\varphi^*+\rho\cosh\vartheta,\\ \xi_2(\varphi,\varphi^*,\rho)=-\vartheta^*\sin\varphi+\rho\sinh\vartheta\cos\varphi,\\ \xi_3(\varphi,\varphi^*,\rho)=\vartheta^*\cos\varphi+\sinh\vartheta\sin\varphi,\end{array}\right\} \tag{70}$$

where $\rho\in\mathbb{R}$. Given that φ and φ^* are two independent variables, it may be said that $\widehat{x}$ is a $\mathbb{T}-like$ line congruence in $\mathbb{L}_f$-space in general. If we define $\varphi^*=h\varphi$ and φ as the parameter for locomotion, then $(\widehat{x})$ can be considered as a $\mathbb{T}-like$ ruled in $\mathbb{L}_f$-space. As a result, the director surface represented by Equation (69) is constrained by the striction curve on $(\widehat{x})$, which implies that

$$\mathbf{c}(\varphi)=(h\varphi,-\vartheta^*\sin\varphi,\vartheta^*\cos\varphi). \tag{71}$$

The curvature $\kappa_c(\varphi)$ and torsion $\tau_c(\varphi)$ can be given by

$$\kappa_c(\varphi)=\frac{\vartheta^*}{\vartheta^{*2}-h^2},\ \tau_c(\varphi)=\frac{h}{\vartheta^{*2}-h^2}. \tag{72}$$

Then, $\mathbf{c}(\varphi)$ is a $\mathbb{S}-like$ ($|\vartheta^*| > |h|$) or $\mathbb{T}-like$ ($|\vartheta^*| < |h|$) cylindrical helix with the $\mathbb{ISA}$ as its axis. Further, the $\mathbb{T}-like$ ruled surface is

$$(\widehat{x}): \left.\begin{array}{l} \xi_1(\varphi,\rho) = h\varphi + \rho\cosh\vartheta, \\ \xi_2(\varphi,\rho) = -\vartheta^*\sin\varphi + \rho\sinh\vartheta\cos\varphi, \\ \xi_3(\varphi,\rho) = \vartheta^*\cos\varphi + \sinh\vartheta\sin\varphi. \end{array}\right\} \tag{73}$$

The constants h, ϑ and ϑ^* can control the shape of $(\widehat{x})$. In the case of $0 \leq \varphi \leq 2\pi$, and $\vartheta^* \neq 0$, we attain

$$(\widehat{x}): -\frac{\Psi_1^2}{\varpi^2} + \frac{\xi_2^2}{\vartheta^{*2}} + \frac{\xi_3^2}{\vartheta^{*2}} = 1, \tag{74}$$

where $\varpi = \vartheta^*\tanh\vartheta$, and $\Psi_1 = \xi_1 - h\varphi$. So, $(\widehat{x})$ is a two-parameter family of one-sheeted hyperboloids. The intersection of each hyperboloid and the $\mathbb{S}-like$ plane $\xi_1 = h\varphi$ is a one-parameter family of Lorentzian cylinder (c): $\xi_2^2 + \xi_3^2 = \vartheta^{*2}$ which is the envelope of $(\widehat{x})$. The $\mathbb{T}-like$ ruled surface $(\widehat{x})$ can be classified into 4-kinds via their striction curves:

(a) $\mathbb{T}-like$ Archimedes with its striction curve is a $\mathbb{T}-like$ cylindrical helix for $h = \vartheta^* = 1$, $\vartheta = 1.1$, $-4 \leq v \leq 4$, and $0 \leq \varphi \leq 2\pi$ (Figure 3).
(b) Lorentzian sphere with its striction curve is a $\mathbb{S}-like$ circle for $h = 0$, $\vartheta^* = 1$, $\vartheta = 1.1$, $-4 \leq v \leq 4$, and $0 \leq \varphi \leq 2\pi$ (Figure 4).
(c) $\mathbb{T}-like$ helicoid with its striction curve is a $\mathbb{T}-like$ line for $h = 1$, $\vartheta^* = 0$, $\vartheta = 1.1$, $-4 \leq v \leq 4$, and $0 \leq \varphi \leq 2\pi$ (Figure 5).
(d) $\mathbb{T}-like$ cone with its striction curve is a stationary point for $h = \vartheta^* = 0$, $\vartheta = 1.1$, $-4 \leq v \leq 4$, and $0 \leq \varphi \leq 2\pi$ (Figure 6).

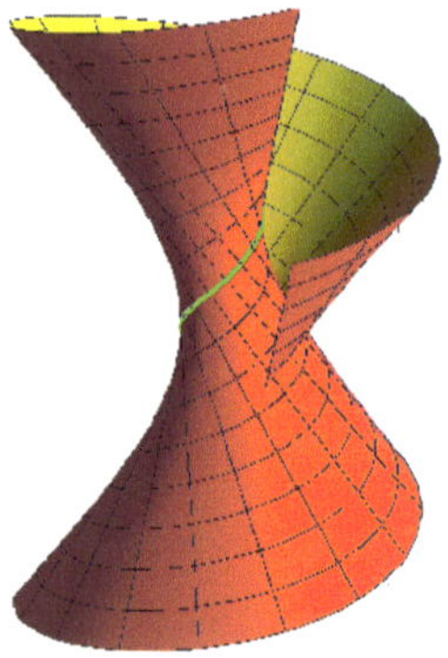

Figure 3. $\mathbb{T}-like$ Archimedes.

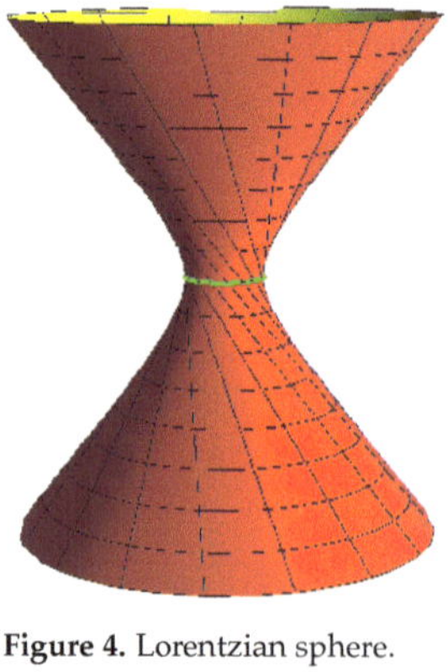

Figure 4. Lorentzian sphere.

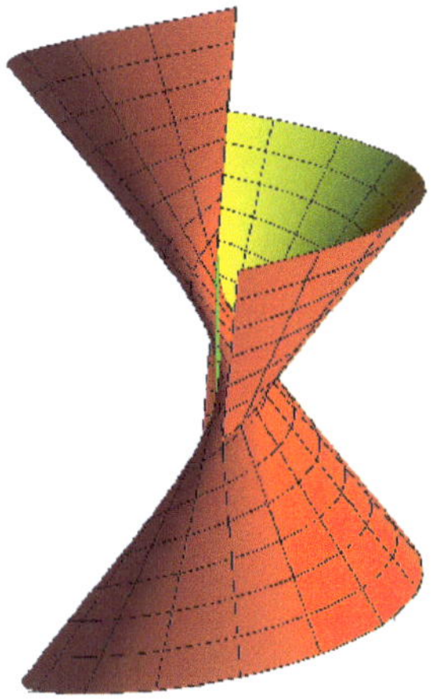

Figure 5. $\mathbb{T}-like$ helicoid.

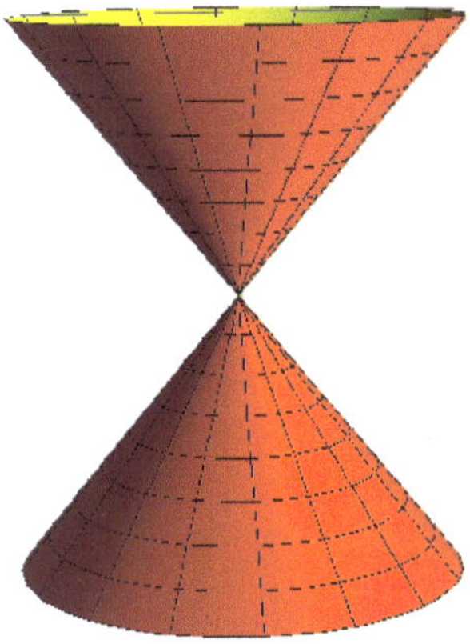

Figure 6. $\mathbb{T}-like$ cone.

4. Inflection Timelike Line Congruence

This section demonstrates how a line congruence, which we refer to as an inflection $\mathbb{T}-like$ line congruence, is the spatial equivalent of the inflection circle of planar kinematics. Hence, we establish that the locus comprising the entire set of lines exhibiting a dual geodesic curvature of zero corresponds to the spatial equivalent of the circle of inflection for planar locomotios. Then, from Equation (34), we have

$$\widehat{\beta}(\widehat{w}) = 0 \Leftrightarrow \widehat{\kappa}(\widehat{w}) = 1. \tag{75}$$

Furthermore, from Equations (30) and (33), we can see that

$$\widehat{\beta}(\widehat{w}) = 0 \Leftrightarrow \tanh \widehat{\psi} = 0 \Leftrightarrow \psi = \psi^* = 0 \Leftrightarrow \beta = 0, \text{and } \Omega = 0. \tag{76}$$

In this particular case, the lines denoted as $\widehat{\mathbf{x}}$, $\widehat{\mathbf{t}}$, and $\widehat{\mathbf{b}}$ represent the Blaschke frame. These lines intersect at the striction point of the $\mathbb{T}-like$ ruled surface denoted as $(\widehat{x})$. Based on the Equations (31) and (76), it may be inferred that the striction curve is a $\mathbb{S}-like$ curve, that is, $\frac{d\mathbf{c}}{dw} \parallel \mathbf{g}$. Given that $\widehat{\beta}(\widehat{w}) = 0$, we can derive the ODE $\frac{d^2\widehat{\mathbf{t}}}{d\widehat{w}^2} - \widehat{\mathbf{t}} = \mathbf{0}$ from Equation (29). Furthermore, by setting $\widehat{\mathbf{t}}(0) = (0, 1, 0)$, the solution of the ODE is obtained as follows:

Since $\widehat{\beta}(\widehat{w}) = 0$, from Equation (29) we have the ODE, $\frac{d^2\widehat{\mathbf{t}}}{d\widehat{w}^2} - \widehat{\mathbf{t}} = \mathbf{0}$. Moreover, we may write $\widehat{\mathbf{t}}(0) = (0, 1, 0)$, and the solution of the ODE becomes

$$\widehat{\mathbf{t}}(\widehat{w}) = \left(\widehat{b}_1 \sinh \widehat{w}, \cosh \widehat{w} + \widehat{b}_2 \sinh \widehat{w}, \widehat{b}_3 \sinh \widehat{w}\right), \tag{77}$$

for dual constants $\widehat{b}_1$, $\widehat{b}_2$, and $\widehat{b}_3$. Since $\left\|\widehat{\mathbf{t}}\right\|^2 = 1$, we obtain $\widehat{b}_2 = 0$, and $\widehat{b}_1^2 - \widehat{b}_3^2 = 1$, it shows that $\widehat{\mathbf{x}}(\widehat{w})$ can be specified by

$$x(\widehat{w}) = \left(\widehat{b}_1 \cosh \widehat{w} + \widehat{d}_1, \sinh \widehat{w}, \widehat{b}_3 \cosh \widehat{w} + \widehat{d}_3\right), \tag{78}$$

for dual constants $\widehat{d}_2$, and $\widehat{d}_3$ satisfying $\widehat{b}_1\widehat{d}_1 - \widehat{b}_3\widehat{d}_3 = 0$. We make change the coordinates by

$$\begin{pmatrix} \widetilde{x}_1 \\ \widetilde{x}_2 \\ \widetilde{x}_3 \end{pmatrix} = \begin{pmatrix} \widehat{b}_1 & 0 & -\widehat{b}_3 \\ 0 & 1 & 0 \\ -\widehat{b}_3 & 0 & \widehat{b}_1 \end{pmatrix} \begin{pmatrix} \widehat{x}_1 \\ \widehat{x}_2 \\ \widehat{x}_3 \end{pmatrix}. \tag{79}$$

Then, $\widehat{\mathbf{x}}(\widehat{w})$ turns into

$$\widehat{\mathbf{x}}(\widehat{w}) = \cosh \widehat{w}\widehat{\mathbf{r}}_1 + \sinh \widehat{w}\widehat{\mathbf{r}}_2, \tag{80}$$

for $\widehat{b}_1\widehat{d}_3 - \widehat{b}_3\widehat{d}_1 = 0$. Let $\boldsymbol{\chi}(\chi_1, \chi_2, \chi_3)$ be a point on $\widehat{\mathbf{x}}(\widehat{w})$, then

$$(\widehat{x}) : \boldsymbol{\chi}(w, w^*, \rho) = (0, 0, w^*) + \rho(\cosh w, \sinh w, 0), \quad \rho \in \mathbb{R}, \tag{81}$$

which yields that

$$\chi_1 = \rho \cosh w, \ \chi_2 = \rho \sinh w, \ \ \chi_3 = w^*. \tag{82}$$

So, if we take $w^* = hw$, h signaling the pitch of the locomotion $\mathbb{L}_m/\mathbb{L}$. Then,

$$\chi_3 = \frac{1}{h} \coth^{-1} \frac{\chi_1}{\chi_2}, \tag{83}$$

which is a one-parameter family of $\mathbb{T}-$*like* helicoid of the second kind; where for $h = 1$, $-3 \leq w \leq 3$, $-1 \leq \rho \leq 1$, a member is shown in (Figure 7).

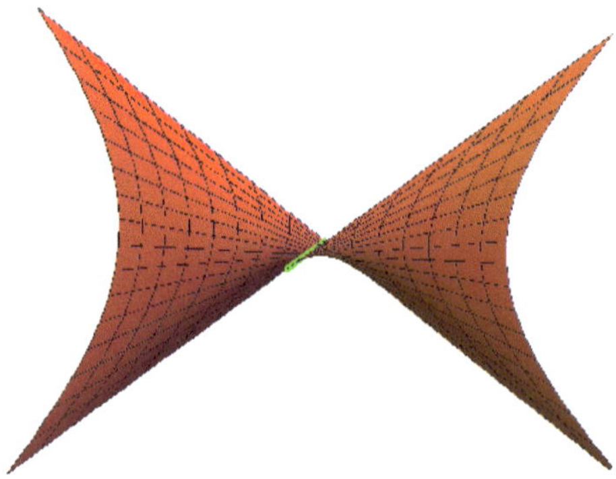

Figure 7. A $\mathbb{T}-$*like* helicoid of the 2nd kind.

For more kinematic analysis of the inflection $\mathbb{T}-$*like* line congruence $(\widehat{x})$, from Equation (30) we can write the equation

$$\widehat{c} : \widehat{x}_1\widehat{\varrho}(1 - \widehat{x}_1^2) + \widehat{x}_3 = 0, \tag{84}$$

which is a curve of third degree. The real part of Equation (84) recognizes a $\mathbb{T}-like$ inflection cone for the real spherical part of $\mathbb{L}_m/\mathbb{L}_f$ and is pointed out by

$$c : x_1\varrho(1 - x_1^2) + x_3 = 0. \tag{85}$$

The mutual lines of the $\mathbb{T}-like$ inflection cone with a real hyperbolic unit sphere concentrated at the head of the cone defines a hyperbolic spherical curve. Furthermore, there is a $\mathbb{T}-like$ plane for each $\mathbb{T}-like$ line, united with each ruling of a $\mathbb{T}-like$ inflection cone, given by the dual part of Equation (84):

$$\pi : \varrho\left(x_1 h + x_1^* - 2x_1^2 x_1^*\right) + x_3^* = 0, \tag{86}$$

where x_1, x_2, and x_3 are the hyperbolic direction cosines of the line $\widehat{\mathbf{x}}$ and x_1^*, x_2^*, and x_3^* are specified by

$$x_1^* = -q_2x_3 + q_3x_2,\ x_2^* = q_3x_1 - q_1x_3,\ x_3^* = q_1x_2 - q_2x_1, \tag{87}$$

where $\mathbf{q}(q_1, q_2, q_3) \in \widehat{\mathbf{x}}$. Equation (84) represents a third-degree equation, it follows that the $\mathbb{T}-like$ line congruence can be traced by all common lines of two cubic $\mathbb{T}-like$ line complexes, as described by Equations (85) and (86). Therefore, the Plückerian coordinates that describe the $\mathbb{T}-like$ lines $\widehat{\mathbf{x}} \in \widehat{c}$ may be expressed by the Equations (21), (85) and (86). In general, these coordinates represent a $\mathbb{T}-like$ ruled surface in the fixed space $\mathbb{L}_f$. However, from Equations (63), (64), (85) and (86), respectively, we obtain

$$c : \varrho \sinh 2\vartheta + 2\sin\varphi = 0, \tag{88}$$

and

$$\pi : \varrho^* \sinh 2\vartheta + 2\varrho\vartheta^* \cosh 2\vartheta + 2\varphi^* \cos\varphi = 0. \tag{89}$$

If the Equation (88) is resolved with respect to ϑ, we have

$$\sinh 2\vartheta = -\left(\frac{2\sin\varphi}{\varrho}\right), \text{ and } \cosh 2\vartheta = \pm\frac{1}{\varrho}\sqrt{\varrho^2 + 4\sin^2\varphi}. \tag{90}$$

Hence, from Equations (89) and (90), we attain

$$\pi : h\sin\varphi \mp \sqrt{\varrho^2 + 4\sin^2\varphi}\,\vartheta^* - \varphi^*\cos\varphi = 0. \tag{91}$$

Equation (91) is linear in $\varphi^*$and ϑ^* of the $\mathbb{T}-like$ line $\widehat{\mathbf{x}}$. Hence, the $\mathbb{T}-like$ lines in a stationary direction within the $\mathbb{L}_m$-space can be found on the $\mathbb{T}-like$ plane denoted as π. As illustrated in Figure 8, the angle φ serves to differentiate the central normal $\widehat{\mathbf{t}}$. Consequently, Equation (91) yields two $\mathbb{T}-like$ lines L^+ and L^- within the $\mathbb{T}-like$ plane π: $Sp\{\widehat{\mathbf{r}}_1, \widehat{\mathbf{t}}\}$ (where L^+ and L^- align with the inflection circle in planar locomotions). Also, if the distance ϑ^* on the central normal $\widehat{\mathbf{t}}$ from the $\mathbb{ISA}$ is taken as the independent parameter, we obtain

$$\pi : \varphi^* = \mp\left(\frac{\sqrt{\varrho^2 + 4\sin^2\varphi}}{\cos\varphi}\right)\vartheta^* + h\tan\varphi. \tag{92}$$

We remark that L^+ (or L^-) will alternate its place if ϑ^* is realized as a various value, but φ =constant. Further, the $\mathbb{T}-like$ plane π is various if φ of L^+ (or L^-) has various value, but ϑ^* =consent. Consequently, the collection of all $\mathbb{T}-like$ lines L^+, and L^- pointed out by Equation (92) is an inflection $\mathbb{T}-like$ congruence for all values of (φ^*, ϑ^*).

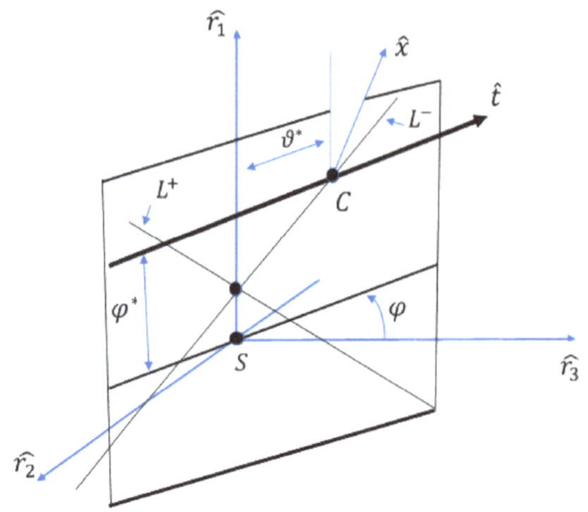

Figure 8. $\mathbb{T}-$*like* inflection line congruence.

However, the ownerships of this inflection $\mathbb{T}-$*like* congruence are clarified as follows: via Figure 8, the two $\mathbb{T}-$*like* lines L^+ and L^- are intersected on the $\mathbb{ISA}$ at distance $h\tan\varphi$. For the orientation $\varphi = 0$, these $\mathbb{T}-$*like* lines passing through the origin ($\mathbf{s} = \mathbf{0}$) and attain the minimal slope is $\pm\varrho$. For $\varphi = \pi/2$, the $\mathbb{T}-$*like* lines are parallel and located on opposite sides of the $\mathbb{ISA}$ at a specific distance $h/\sqrt{\varrho^2+4}$. Furthermore, if the Equation (88) is resolved with respect to φ, we obtain

$$\varphi = -\sin^{-1}\left(\frac{\varrho\sinh 2\vartheta}{2}\right). \tag{93}$$

By substituting Equation (93) into Equation (63), we find

$$x(\vartheta) = \left(\cosh\vartheta, \cos\left[\sin^{-1}\left(\frac{\varrho\sinh 2\vartheta}{2}\right)\right]\sinh\vartheta, -\frac{\varrho\sinh(2\vartheta)}{2}\sinh\vartheta\right). \tag{94}$$

Equation (94) appears the inflection $\mathbb{T}-$*like* curve of the hyperbolic spherical part of the locomotion $\mathbb{L}_m/\mathbb{L}_f$. Further, from the Equations (70), (90) and (94), we obtain

$$(\hat{x}): \left.\begin{aligned} \xi_1(\vartheta,\rho) &= \varphi^* + \rho\cosh\vartheta, \\ \xi_2(\vartheta,\rho) &= \vartheta^*\left(\tfrac{\varrho\sinh 2\vartheta}{2}\right) + \rho\cos\left[\sin^{-1}\left(\tfrac{\varrho\sinh 2\vartheta}{2}\right)\right]\sinh\vartheta, \\ \xi_3(\vartheta,\rho) &= \vartheta^*\cos\left[\sin^{-1}\left(\tfrac{\varrho\sinh 2\vartheta}{2}\right)\right] - \rho\tfrac{\varrho\sinh 2\vartheta}{2}\sinh\vartheta. \end{aligned}\right\} \tag{95}$$

For epitome, via Equations (94) and (95), we have

(1) Hyperbolic spherical inflection curve with its inflection timelike ruled surface: for $\omega = 0.3$, $\vartheta^* = 1$, $\varphi^* = 0$,$-1.3 \leq \vartheta \leq 1.3$, $-5 \leq v \leq 5$ (Figures 9 and 10).
(2) Hyperbolic spherical inflection curve with its inflection $\mathbb{T}-$*like* ruled surface: for $\omega = -0.3$, $\vartheta^* = 1$, $\varphi^* = 0$,$-1.3 \leq \vartheta \leq 1.3$, $-5 \leq v \leq 5$ (Figures 11 and 12).

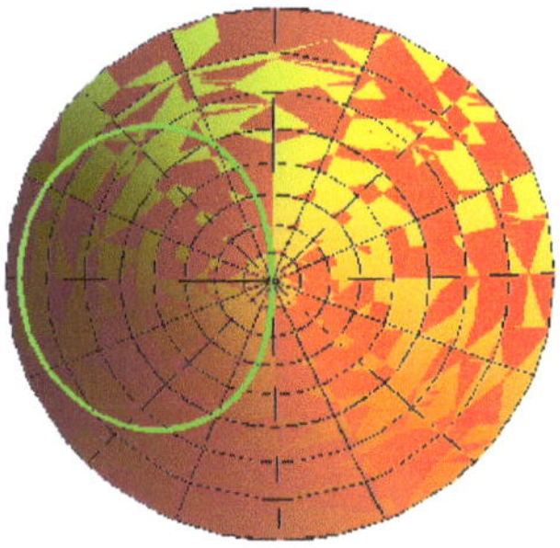

Figure 9. Hyperbolic inflection curve with $\omega = 0.3$.

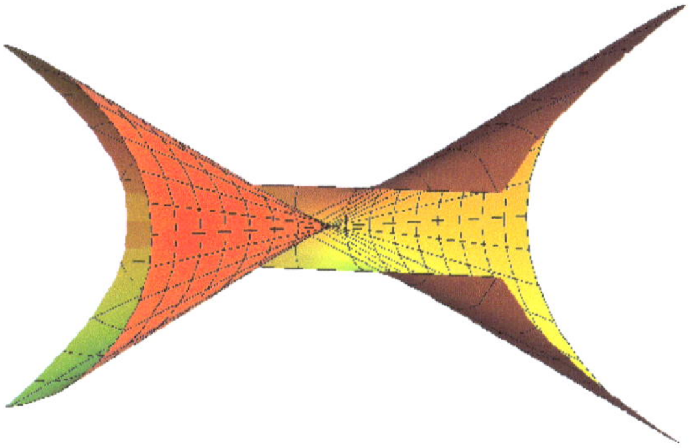

Figure 10. $\mathbb{T}-$*like* Inflection ruled surface.

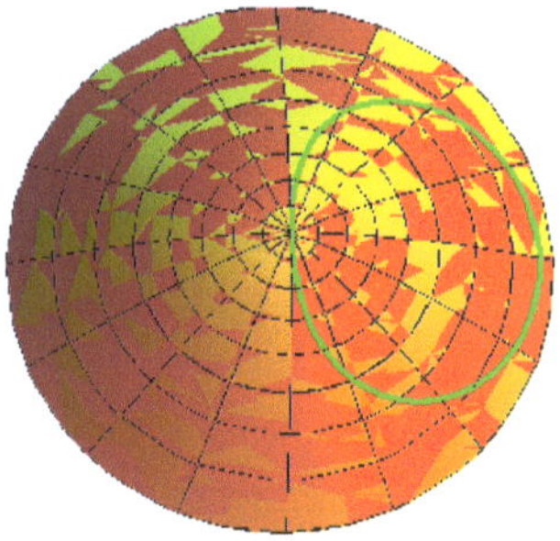

Figure 11. Hyperbolic inflection curve with $\omega = -0.3$.

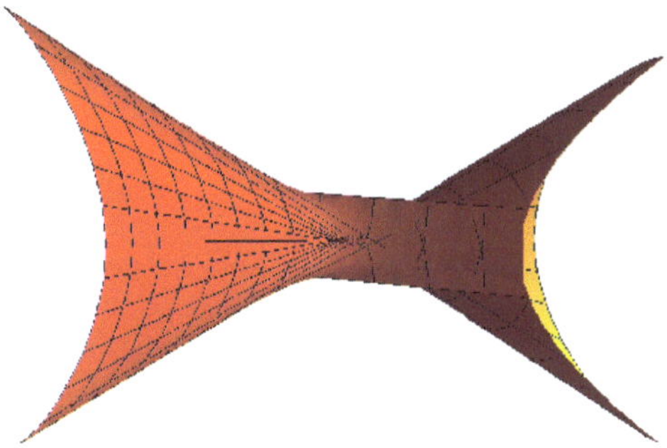

Figure 12. $\mathbb{T}-$*like* Inflection ruled surface.

5. Conclusions

In this paper, the kinematic-geometry of a $\mathbb{T}-like$ trajectory is defined in terms of the axodes invariants of one-parameter hyperbolic spatial locomotion. Then, a new $\mathbb{DF}$ of a $\mathbb{T}-like$ line-trajectory is gained in distinct forms. In symmetry with the plane and spherical locomotions, a new $\mathbb{T}-like$ congruence is pointed and investigated in detail. The main result in this paper is to generalize the $\mathbb{ES}$ formula in the hyperbolic locomotion. We introduced the dual angle, which is represented in Equation (36), and we restricted it to be steady up to 2nd order. Hence, we obtain Equation (39), which introduces the $\mathbb{DA}$. Through this equation, we gave corollary 2 and corollary 3. Also, we reformulated $\mathbb{ES}$ formula for the axodes in a new form given in Equations (47)–(49). Furthermore, in Section 4 of this work, we defined and studied inflection $\mathbb{T}-like$ line congruence, which is the spatial synonym of the inflection circle of planer kinematics. The findings presented in this study have the potential to make significant contributions to the field of spatial locomotion, as well as offer practical applications in the domains of mechanical mathematics and engineering. In our forthcoming research, we intend to explore various applications of the kinematic-geometry of one-parameter hyperbolic spatial locomotion in conjunction with singularity theory, submanifold theory, etc., in [22–25] in order to derive additional novel findings and properties.

Author Contributions: Conceptualization, R.A.A.-B. and A.A.A.; methodology, R.A.A.-B. and A.A.A.; software, R.A.A.-B. and A.A.A.; validation, R.A.A.-B.; formal analysis, R.A.A.-B. and A.A.A.; investigation, R.A.A.-B. and A.A.A.; resources, R.A.A.-B.; data curation, R.A.A.-B. and A.A.A.; writing—original draft preparation, R.A.A.-B. and A.A.A.; writing—review and editing, A.A.A.; visualization, R.A.A.-B. and A.A.A.; supervision, R.A.A.-B.; project administration, R.A.A.-B.; funding acquisition, A.A.A. All authors have read and agreed to the published version of the manuscript.

Funding: This research was funded by Princess Nourah bint Abdulrahman University Researchers Supporting Project number (PNURSP2023R337).

Data Availability Statement: Our manuscript has no associated data.

Acknowledgments: The authors would like to acknowledge the Princess Nourah bint Abdulrahman University Researchers Supporting Project number (PNURSP2023R337), Princess Nourah bint Abdulrahman University, Riyadh, Saudi Arabia.

Conflicts of Interest: The authors declare that there is no conflict of interest regarding the publication of this paper.

References

1. Bottema, O.; Roth, B. *Theoretical Kinematics*; North-Holland Press: New York, NY, USA, 1979.
2. Karger, A.; Novak, J. *Space Kinematics and Lie Groups*; Gordon and Breach Science Publishers: New York, NY, USA, 1985.
3. Pottman, H.; Wallner, J. *Computational Line Geometry*; Springer: Berlin/Heidelberg, Germany, 2001.
4. Stachel, H. On spatial involute gearing, TU Wien, Geometry Preprint No 119. In Proceedings of the 6th International Conference on Applied Informatics, Eger, Hungary, 27–31 January 2004.
5. Dooner, D.; Garcia, R.G.; Martinez, J.M.R. On spatial relations to the Euler–Savary formula. *Mech. Mach. Theory* **2023**, *189*, 105427. [CrossRef]
6. Abdel-Baky, R.A.; Al-Solamy, F.R. A new geometrical approach to one-parameter spatial motion. *J. Eng. Math.* **2008**, *60*, 149–172. [CrossRef]
7. Ayyilidiz, N.; Yalcin, S.N. On instantaneous invariants in dual Lorentzian space kinematics. *Arch. Mech.* **2010**, *62*, 223–238.
8. Figlioini, G.; Stachel, H.; Angeles, J. The computational fundamentals of spatial cycloidal gearing. In *Computational Kinematics: Proceedings of the 5th International Workshop on Computational Kinematics, Duisburg, Germany, 6–8 May 2009*; Kecskeméthy, A., Mü ller, A., Eds.; Springer: Berlin/Heidelberg, Germany, 2009; pp. 375–384.
9. Turhan, T.; Ayyıldız, N. A study on geometry of spatial kinematics in Lorentzian space. *Süleyman Demirel Üniversitesi Fen Bilimleri Enstitüsü Dergisi* **2017**, *21*, 808–811. [CrossRef]
10. Zhang, X.; Zhang, J.; Pang, B.; Zhao, W. An accurate prediction method of cutting forces in 5-axis áank milling of sculptured surface. *Int. J. Mach. Tools Manuf.* **2016**, *104*, 26–36. [CrossRef]
11. Ekinci, Z.; Uğurlu, H.H. Blaschke approach to Euller-Savary formulae. *Konuralp J. Math.* **2016**, *4*, 95–115.
12. Özyilmaz, E. Some results on space-like line congruences and their space-like parameter ruled surface. *Turk. J. Math.* **1999**, *23*, 333–344.

13. Tosun, M.; Gungor, M.A.; Okur, I. On the One-Parameter Lorentzian Spherical Motions and Euler-Savary Formula. *ASME J. Appl. Mech.* **2007**, *74*, 972–977. [CrossRef]
14. Gungor, M.A.; Ersoy, S.; Tosun, M. Dual Lorentzian spherical motions and dual Euler–Savary formula. *Eur. J. Mech. A/Solids* **2009**, *28*, 820–826. [CrossRef]
15. Palavar, S.; Bilici, M. Dual ruled surface constructed by the pole curve of the involute curve. *Int. J. Open Probl. Compt. Math.* **2022**, *15*, 39–53.
16. Rawya, H.A.; Ali, A. Geometry of the line space associated to a given dual ruled surface. *AIMS Math.* **2022**, *7*, 8542–8557. [CrossRef]
17. Saad, M.K.; Ansari, A.Z.; Akram, M.; Alharbi, F. Spacelike surfaces with a common line of curvature in Lorentz-Minkowski 3-space. *Wseas Trans. Math.* **2021**, *20*, 207–217. [CrossRef]
18. Inalcik, A.; Ersoy, S. Ball and Burmester points in Lorentzian sphere kinematics. *Kuwait J. Sci.* **2015**, *42*, 50–63.
19. Alluhaibi, N.S.; Abdel-Baky, R.A.; Naghi, M.F. On the Bertrand offsets of timelike ruled surfaces in Minkowski 3-space. *Symmetry* **2022**, *14*, 673. [CrossRef]
20. Disteli, M. Uber des Analogon der Savaryschen Formel und Konstruktion in der kinematischen Geometrie des Raumes. *Z. Math. Phys.* **1914**, *62*, 261–309.
21. Önder, M.; Uğurlu, H.H.; Caliskan, A. The Euler–Savary analogue equations of a point trajectory in Lorentzian spatial motion. *Proc. Natl. Acad. Sci. India Sect. Phys. Sci.* **2013**, *83*, 163–169. [CrossRef]
22. Nazra, S.; Abdel-Baky, R.A. Singularities of non-lightlike developable surfaces in Minkowski 3-space. *Mediterr. J. Math.* **2023**, *20*, 45. [CrossRef]
23. Li, Y.; Tuncer, O. On (contra) pedals and (anti)orthotomics of frontals in de Sitter 2-space. *Math. Meth. Appl. Sci.* **2023**, *1*, 1–15. [CrossRef]
24. Li, Y.; Aldossary, M.T.; Abdel-Baky, R.A. Spacelike circular surfaces in Minkowski 3-Space. *Symmetry* **2023**, *15*, 173. [CrossRef]
25. Li, Y.; Chen, Z.; Nazra, S.H.; Abdel-Baky, R.A. Singularities for timelike developable surfaces in Minkowski 3- Space. *Symmetry* **2023**, *15*, 277. [CrossRef]

axioms

Article

Surface Pencil Couple with Bertrand Couple as Joint Principal Curves in Galilean 3-Space

Nadia Alluhaibi [1,†] and Rashad A. Abdel-Baky [2,*,†]

[1] Department of Mathematics, Science and Arts College, King Abdulaziz University, Rabigh 21911, Saudi Arabia; nallehaibi@kau.edu.sa
[2] Department of Mathematics, Faculty of Science, University of Assiut, Assiut 71516, Egypt
* Correspondence: baky1960@aun.edu.eg
† These authors contributed equally to this work.

Abstract: A principal curve on a surface plays a paramount role in reasonable implementations. A curve on a surface is a principal curve if its tangents are principal directions. Using the Serret–Frenet frame, the surface pencil couple can be expressed as linear combinations of the components of the local frames in Galilean 3-space $\mathbb{G}_3$. With these parametric representations, a family of surfaces using principal curves (curvature lines) are constructed, and the necessary and sufficient condition for the given Bertrand couple to be the principal curves on these surfaces are derived in our approach. Moreover, the necessary and sufficient condition for the given Bertrand couple to satisfy the principal curves and the geodesic requirements are also analyzed. As implementations of our main consequences, we expound upon some models to confirm the method.

Keywords: sotropic normal; Serret–Frenet formulae; marching-scale functions.

MSC: 53A04; 53A05; 53A17

1. Introduction

Principal curve is designated as one of the significant curves on a surface and it plays a primary role in differential geometry [1–4]. It is an advantageous gadget in surface examination for showing contrast of the principal direction. The harmonic principal curvature and principal curves are considerable and associated with the smooth surfaces. Principal curves can direct the realization of surfaces, quietly utilized in geometric designing, and can constitute consistency, surface polygonization, and surface fulfillment. There exists an enormous literature on the topic, including various monographs, for instance: Martin [5] studied systematic surface patches restricted by principal curves, which are named principal patches. Martin presented that the presence of such patches was conditioned upon confirmed situations corresponding on the patch border curves. Alourdas et al. [6] addressed a mode to confirm a net principal curves on a B-spline surface. Maekawa et al. [7] extended a style to take out the generic characteristics of free-shape parametric surfaces for form inspection. They researched the generic advantage of the umbilics and attitude principal curves that go through an umbilic on a parametric free-shape surface. Che and Paul [8] expanded a manner to resolve and calculate the principal curves and their geometric properties specified on an implicit surface. They also offered a new standard for non-umbilical points and umbilical points on an implicit surface. Zhang et al. [9] proved a planner for calculating and envisaging the principal curves pointed on an implicit surface. Kalogerakis et al. [10] derived a powerful substructure for establishing principal curves by point clouds. Their approach is reasonable for surfaces of random genus, with or without borderlines, and is statistically powerful to use with outliers maintaining surface characteristics. They found the approach to be efficient through an area of synthetic and real-world input data collections with changing amounts of noise and outliers. In practical uses,

Citation: Alluhaibi, N.; Abdel-Baky, R.A. Surface Pencil Couple with Bertrand Couple as Joint Principal Curves in Galilean 3-Space. *Axioms* **2023**, *12*, 1022. https://doi.org/10.3390/axioms12111022

Academic Editors: Mića Stanković, Emil Saucan and Anna Maria Fino

Received: 13 September 2023
Revised: 25 October 2023
Accepted: 28 October 2023
Published: 30 October 2023

however, crucial work has focused on the backward exploration or reverse issue: given a 3D curve, how can we locate those surfaces that are faced with this curve as a distinctive curve, if possible, rather than locating and furnishing curves on analytical curved surfaces? Wang et al. [11] was the first to handle the issue of assembling a surface family with a designated locative geodesic curve, through which every surface can be a candidate for mode style. They demonstrated the necessary and sufficient conditions for the coefficients to be satisfied with both the isoparametric and the geodesic demands. This scheme has been used by many scholars (see, for example, [12–23]).

Galilean geometry is the simplest pattern of a semi-Euclidean geometry for which the isotropic cone reduces to a plane. It is indicated as a bridge from Euclidean geometry to special relativity. The main spine of Galilean geometry is its private gravity, that is, it allows the scientist to search it in detail without using large amounts of time and energy. In other words, the view of Galilean geometry shapes its development with an unpretentious question, and a gross stretch of a modern geometric community is crucial for its efficient comparison with Euclidean geometry. Also, aggregate evolution is practical to supply the scientist with the psychological emphasis of the uniformity of the inspected structure [24,25]. In the 3$\mathbb{D}$ (three-dimensional) Galilean space $\mathbb{G}_3$, there are several studies dealing with $\mathbb{G}_3$, for example, Dede et al. [26,27] resolved tub surfaces, the descriptions of the parallel surfaces. Yuzbasi et al. [28,29] considered a surface family with a curve to be a joint asymptotic curve and geodesic. Jiang et al. [30] considered surface pencil couple with Bertrand couple as joint asymptotic curves. Almoneef and Abdel-Baky [31] designed a surface family with Bertrand curves to be geodesic curves. AL-Jedani and Abdel-Baky [32] considered the surface family and developable surface family with a joint geodesic curve, respectively.

It is known that on any surface there are three main kinds of curves: principal curves (curvature lines), asymptomatic curves, and geodesic curves. In [30], the study focused on how to construct family surfaces via Bertrand curves that are asymptomatic, whereas our work addresses a very new idea related to a differential geometry field based on construction of a family of surfaces using principal curves (curvature lines). Ref. [31] investigated a similar idea but used the geodesic curves instead of principal ones. In addition, a team of researchers, referred to as Li et al. and cited in [32–49],conducted theoretical studies and advancements on soliton theory, submanifold theory, and other related topics. Further motivation can be found in these papers. Their efforts have significantly contributed to the progression of research in these fields.

However, to our knowledge, no additional work has been done to originate surface pencil couples with curve couples that are principal curves. In order to create the surfaces pencil and, specifically ruled ones, we explore Bertrand couples as principal curves and organize a surface pencil couple with a Bertrand couple as joint principal curves. Furthermore, the accessory to the ruled surface pencil is likewise qualified. In addition, some models are exhibited to create the surface pencil, in general, and ruled ones, in particular, with joint Bertrand principal curves.

2. Basic Concepts

The 3$\mathbb{D}$ (3-dimensional) Galilean space $\mathbb{G}_3$ is a Cayley–Klein geometry endued with the projective metric of signature (0, 0, +, +) [48,49]. The absolute figure of $\mathbb{G}_3$ is contingent on the organized triple $\{\pi, \mathfrak{L}, \mathfrak{I}\}$, where π is the (absolute) plane in the real 3$\mathbb{D}$ projective space $\mathbb{P}^3(\mathbb{R})$, $\mathfrak{L}$ is the line (absolute line) in π, and $\mathfrak{L}$ is the steady elliptic involution of points of $\mathfrak{L}$. Homogeneous coordinates in $\mathbb{G}_3$ are developed in such a mode that the absolute plane π is pointed by $z_0 = 0$, the absolute line $\mathfrak{L}$ by $z_0 = z_1 = 0$, and the elliptic involution is pointed by $(0 : 0 : z_2 : z_3) \rightarrow (0 : 0 : z_3 : -z_2)$. A plane is organized Euclidean if it includes $\mathfrak{L}$, otherwise it is organized isotropic, that is, planes z_0=const are Euclidean, and so is the plane π. Other planes are isotropic. Furthermore, an isotropic plane does not involve any isotropic direction. For any $\boldsymbol{v} = (v_1, v_2, v_3)$, and $\boldsymbol{\nu}=(\nu_1, \nu_2, \nu_3) \in \mathbb{G}_3$, their scalar product is

$$<\boldsymbol{v}, \boldsymbol{\nu}>= \begin{cases} v_1\nu_1, \text{ if } v_1 \neq 0 \vee \nu_1 \neq 0, \\ v_2\nu_2 + v_3\nu_3, \text{ if } v_1 = 0 \wedge \nu_1 = 0, \end{cases} \tag{1}$$

and their vector product is

$$\boldsymbol{v} \times \boldsymbol{\nu} = \begin{cases} \begin{vmatrix} \mathbf{x}_1 & \mathbf{x}_2 & \mathbf{x}_3 \\ 0 & v_2 & v_3 \\ 0 & \nu_2 & \nu_3 \end{vmatrix}, \text{ if } v_1 = 0 \wedge \nu_1 = 0. \\ \begin{vmatrix} \mathbf{0} & \mathbf{x}_2 & \mathbf{x}_3 \\ v_1 & v_2 & v_3 \\ \nu_1 & \nu_2 & \nu_3 \end{vmatrix}, \text{ if } v_1 \neq 0 \vee \nu_1 \neq 0, \end{cases} \tag{2}$$

where $\mathbf{x}_1 = (1,0,0)$, $\mathbf{x}_2 = (0,1,0)$, and $\mathbf{x}_3 = (0,0,1)$ are the standard basis vectors in $\mathbb{G}_3$.

A curve $\boldsymbol{\psi}(u) = (\psi_1(u), \psi_2(u), \psi_3(u)); u \in I \subseteq \mathbb{R}$ is a denominated allowable curve if it has no inflection points, that is, $\dot{\boldsymbol{\psi}} \times \ddot{\boldsymbol{\psi}} \neq \mathbf{0}$, and no isotropic tangents $\dot{\psi}_1 \neq 0$. An allowable curve is comparable to a regular curve in Euclidean space. For an allowable curve $\boldsymbol{\psi} : I \subseteq \mathbb{R} \to \mathbb{G}_3$ defined by the Galilean invariant arc-length u, we have:

$$\boldsymbol{\psi}(u) = (u, \psi_2(u), \psi_3(u)). \tag{3}$$

The curvature $\kappa(u)$ and torsion $\tau(u)$ of the curve $\psi(u)$ are

$$\begin{aligned} \kappa(u) &= \left\|\boldsymbol{\psi}''(u)\right\| = \sqrt{(\psi_2''(u))^2 + (\psi_3''(u))^2}, \\ \tau(u) &= \frac{1}{\kappa^2(u)} \det\left(\boldsymbol{\psi}', \boldsymbol{\psi}'', \boldsymbol{\psi}'''\right). \end{aligned} \tag{4}$$

Note that an allowable curve has $\kappa(u) \neq 0$. The Serret–Frenet vectors are:

$$\begin{aligned} \mathbf{g}_1(u) &= \boldsymbol{\psi}'(u) = \left(1, \psi_2'(u), \psi_3'(u)\right), \\ \mathbf{g}_2(u) &= \frac{1}{\kappa(u)}\boldsymbol{\psi}'(u) = \frac{1}{\kappa(u)}\left(0, \psi_2''(u), \psi_3''(u)\right), \\ \mathbf{g}_3(u) &= \frac{1}{\tau(u)}\left(0, \left(\frac{1}{\kappa(u)}\psi_2''(u)\right)', \left(\frac{1}{\kappa(u)}\psi_3''(u)\right)'\right), \end{aligned} \tag{5}$$

where $\mathbf{g}_1(u)$, $\mathbf{g}_2(u)$, and $\mathbf{g}_3(u)$, respectively, are the tangent, principal normal, and binormal vectors. The Serret–Frenet formulae read:

$$\begin{pmatrix} \mathbf{g}_1' \\ \mathbf{g}_2' \\ \mathbf{g}_3' \end{pmatrix} = \begin{pmatrix} 0 & \kappa(u) & 0 \\ 0 & 0 & \tau(u) \\ 0 & -\tau(u) & 0 \end{pmatrix} \begin{pmatrix} \mathbf{g}_1 \\ \mathbf{g}_2 \\ \mathbf{g}_3 \end{pmatrix}. \tag{6}$$

Definition 1 ([31]). *Let $\boldsymbol{\psi}(u)$ and $\widehat{\boldsymbol{\psi}}(u)$ be curves with Galilean invariant arc-length u, $\mathbf{g}_2(u)$ and $\widehat{\mathbf{g}}_2(u)$ are their principal normals, respectively; the set $\{\widehat{\boldsymbol{\psi}}(u), \boldsymbol{\psi}(u)\}$ is an organized Bertrand couple if $\mathbf{g}_2(u)$ and $\widehat{\mathbf{g}}_2(u)$ are linearly related at the conformable points, $\boldsymbol{\psi}(u)$ is organized as the Bertrand mate of $\widehat{\boldsymbol{\psi}}(u)$, and*

$$\widehat{\boldsymbol{\psi}}(u) = \boldsymbol{\psi}(u) + f\mathbf{g}_2(s), \tag{7}$$

where f is a stationary.

We designate a surface S by

$$S : \mathbf{p}(u,t) = (p_1(u,t), p_2(u,t), p_3(u,t)), \ (u,t) \in \mathbb{D} \subseteq \mathbb{R}^2. \tag{8}$$

If $\mathbf{p}_j(u,t) = \frac{\partial \mathbf{p}}{\partial j}$, the isotropic surface normal is

$$\mathbf{u}(u,t) = \mathbf{p}_u \wedge \mathbf{p}_t, \text{ with } <\mathbf{u}, \mathbf{p}_u> = <\mathbf{u}, \mathbf{p}_t> = 0.$$

A curve on a surface S can be principal curve by the status specified by the well-known theorem below [1,2].

Theorem 1. *(Monge's Theorem). A curve on a surface is a principal curve if and only if the surface normals along that curve create a developable surface [1,2].*

3. Main Results

This section presents a new aspect for originating a surface pencil couple with a Bertrand couple as joint principal curves in $\mathbb{G}_3$. For this aim, let $\widehat{\boldsymbol{\psi}}(u)$ be an allowable curve, $\boldsymbol{\psi}(u)$ be its Bertrand mate, and $\{\widehat{\kappa}(u), \widehat{\tau}(u), \widehat{\mathbf{g}}_1(u), \widehat{\mathbf{g}}_2(u), \widehat{\mathbf{g}}_3(u)\}$ is the Serret–Frenet instruments of $\widehat{\boldsymbol{\psi}}(u)$, as in Equation (1). The surface pencil with $\boldsymbol{\psi}(u)$ can be specified by

$$S : \mathbf{p}(u,t) = \boldsymbol{\psi}(u) + \mathfrak{a}(u,t)\mathbf{g}_1(u) + \mathfrak{b}(u,t)\mathbf{g}_2(u) + \mathfrak{c}(u,t)\mathbf{g}_3(u), \tag{9}$$

and the surface pencil with $\widehat{\boldsymbol{\psi}}(u)$ is

$$\widehat{S} : \widehat{\mathbf{p}}(u,t) = \widehat{\boldsymbol{\psi}}(u) + \mathfrak{a}(u,t)\widehat{\mathbf{g}}_1(u) + \mathfrak{b}(u,t)\widehat{\mathbf{g}}_2(u) + \mathfrak{c}(u,t)\widehat{\mathbf{g}}_3(u), \tag{10}$$

where $\mathfrak{a}(u,t)$, $\mathfrak{b}(u,t)$, $\mathfrak{c}(u,t)$ are all C^1 functions, and $0 \leq t \leq T, \ 0 \leq u \leq L$. If the variable t is defined as the time, the functions $\mathfrak{a}(u,t)$, $\mathfrak{b}(u,t)$, and $\mathfrak{c}(u,t)$ can then be explicated as oriented marching distances of a point at the time t in the directions $\widehat{\mathbf{g}}_1$, $\widehat{\mathbf{g}}_2$, and $\widehat{\mathbf{g}}_3$, respectively, and the vector $\widehat{\boldsymbol{\psi}}(u)$ is the initialization of this point.

Our provocation is to confer necessary and sufficient situations for $\psi(u)$ as an isoparametric principal curve on S. First, let us define a unit vector $\mathbf{g}(u)$ orthogonal to the curve $\psi(u)$, that is,

$$\mathbf{g}(u) = \cos\phi\mathbf{g}_2(u) + \sin\phi\mathbf{g}_3(u), \text{ with } \phi = \phi(u). \tag{11}$$

Suppose that the ruled surface

$$\mathbf{y}(u,t) = \boldsymbol{\psi}(u) + t\mathbf{g}(u); \ t \in \mathbb{R}, \tag{12}$$

is a developable one, that is,

$$\det(\boldsymbol{\psi}', \mathbf{g}(u), \mathbf{g}'(u)) = \begin{vmatrix} 1 & 0 & 0 \\ 0 & \cos\phi & \cos\phi \\ 0 & -\phi'\sin\phi - \tau\sin\phi & \phi'\cos\phi + \tau\cos\phi \end{vmatrix} = 0,$$

From which we find

$$\phi'(u) + \tau(u) = 0 \Rightarrow \phi(u) = \phi_0 - \int_{u_0}^{u} \tau(u)du, \tag{13}$$

where $\phi_0 = \phi(u_0)$ and u_0 is the initial value of arc length.

Second, since $\boldsymbol{\psi}(u)$ is an isoparametric curve on S, there exists a value $t = t_0$ such that $\boldsymbol{\psi}(u) = \mathbf{p}(u, t_0)$. Then, we have

$$\begin{aligned} \mathfrak{a}(u,t_0) &= \mathfrak{b}(u,t_0) = \mathfrak{c}(u,t_0) = 0, \\ \frac{\partial \mathfrak{a}(u,t_0)}{\partial u} &= \frac{\partial \mathfrak{b}(u,t_0)}{\partial u} = \frac{\partial \mathfrak{c}(u,t_0)}{\partial u} = 0. \end{aligned}$$

Thus, the isotropic surface normal is

$$\mathbf{u}(u,t_0) := \frac{\partial \mathbf{p}(u,t_0)}{\partial u} \times \frac{\partial \mathbf{p}(u,t_0)}{\partial t} = -\frac{\partial \mathfrak{c}(u,t_0)}{\partial t}\mathbf{g}_2(u) + \frac{\partial \mathfrak{b}(u,t_0)}{\partial t}\mathbf{g}_3(u). \tag{14}$$

Moreover, via Monge's Theorem, $\boldsymbol{\psi}(u)$ is a principal curve on S if and only if $\boldsymbol{u}(u)$ is parallel to $\mathbf{g}(u,t_0)$. Therefore, from Equations (3.3) and (3.6), there exists a function $\chi(u) \neq 0$ such that

$$-\frac{\partial \mathfrak{c}(u,t_0)}{\partial t} = \chi(u)\cos\phi, \quad \frac{\partial \mathfrak{b}(u,t_0)}{\partial t} = \chi(u)\sin\phi, \tag{15}$$

where $\phi(u)$ is designated by Equation (13). The functions $\chi(u)$ and $\phi(u)$ are controlling functions.

Hence, we give the following theorem.

Theorem 2. *The expression $\boldsymbol{\psi}(u)$ is a principal curve on S if and only if*

$$\left.\begin{array}{l} \mathfrak{a}(u,t_0) = \mathfrak{b}(u,t_0) = \mathfrak{c}(u,t_0) = 0, \ \ 0 \le t_0 \le T, \ \ 0 \le u \le L, \\ -\frac{\partial \mathfrak{c}(u,t_0)}{\partial t} = \chi(u)\cos\phi, \ \ \frac{\partial \mathfrak{b}(u,t_0)}{\partial t} = \chi(u)\sin\phi, \ \chi(u) \neq 0, \\ \phi(u) = \phi_0 - \displaystyle\int_{u_0}^{u} \tau(u)du, \ \phi_0 = \phi(u_0), \end{array}\right\} \tag{16}$$

where u_0 is the starting value of the arc length.

Any surface $S : \mathbf{p}(u,t)$ recognized by Equation (9) and identified by Theorem 2 is an element of the surface pencil with $\boldsymbol{\psi}(u)$ as joint principal curve. As reported in [8], for the purpose of resolution and experiment, we also examine the case when $\mathfrak{a}(u,t)$, $\mathfrak{b}(u,t)$, and $\mathfrak{c}(u,t)$ can be realized by

$$\mathfrak{a}(u,t) = \mathfrak{l}(u)\mathfrak{a}(t), \ \mathfrak{b}(u,t) = \mathfrak{m}(u)\mathfrak{b}(t), \ \mathfrak{c}(u,t) = \mathfrak{n}(u)\mathfrak{c}(t). \tag{17}$$

Here $\mathfrak{l}(u)$, $\mathfrak{m}(u)$, $\mathfrak{n}(u)$, $\mathfrak{a}(t)$, $\mathfrak{b}(t)$, and $\mathfrak{c}(t)$ are $\mathfrak{c}^1$ functions that do not identically vanish. Then, from Theorem 2, we gain:

Corollary 1. *The expression $\boldsymbol{\psi}(u)$ is a principal curve on S if and only if*

$$\left.\begin{array}{l} \mathfrak{a}(t_0) = \mathfrak{b}(t_0) = \mathfrak{c}(t_0) = 0, \ \ 0 \le t_0 \le T, \ \ 0 \le u \le L, \\ -\mathfrak{n}(u)\frac{d\mathfrak{c}(t_0)}{dt} = \chi(u)\cos\phi, \ \ \mathfrak{m}(u)\frac{d\mathfrak{b}(t_0)}{dt} = \chi(u)\sin\phi. \\ \phi(u) = \phi_0 - \displaystyle\int_{u_0}^{u} \tau(u)du, \ \phi_0 = \phi(u_0), \end{array}\right\} \tag{18}$$

where u_0 is the starting value of the arc length.

However, we can assume that $\mathfrak{a}(u,t)$, $\mathfrak{b}(u,t)$, and $\mathfrak{c}(u,t)$ are based only on the variable t; that is, $\mathfrak{l}(u) = m(u) = n(u) = 1$. Then, we treat Equation (18) via the dissimilar terms of $\phi(u)$ as follows:

(i) If $\tau(u) \neq 0$, then $\phi(u)$ is a non-stationary function of variable u and Equation (18) can be distinguished by

$$\left.\begin{array}{l} \mathfrak{a}(t_0) = \mathfrak{b}(t_0) = \mathfrak{c}(t_0) = 0, \\ -\frac{d\mathfrak{c}(t_0)}{dt} = \chi(u)\cos\phi, \ \ \frac{d\mathfrak{b}(t_0)}{dt} = \chi(u)\sin\phi. \end{array}\right\} \tag{19}$$

(ii) If $\tau(u) = 0$, that is, the curve is a planar curve, then $\phi(u) = \phi_0$ is a stationary and we have:

(a) If $\phi_0 \neq 0$, Equation (18) can be distinguished by

$$\left.\begin{array}{c} \mathfrak{a}(t_0) = \mathfrak{b}(t_0) = \mathfrak{c}(t_0) = 0, \\ -\frac{d\mathfrak{c}(t_0)}{dt} = \chi(u)\cos\phi_0, \ \frac{d\mathfrak{b}(t_0)}{dt} = \chi(u)\sin\phi_0. \end{array}\right\} \tag{20}$$

(b) If $\phi_0 = 0$, Equation (18) can be distinguished by

$$\left.\begin{array}{c} \mathfrak{a}(t_0) = \mathfrak{b}(t_0) = \mathfrak{c}(t_0) = 0, \\ -\frac{d\mathfrak{c}(t_0)}{dt} = \chi(u), \ \frac{d\mathfrak{b}(t_0)}{dt} = 0, \end{array}\right\} \tag{21}$$

and from Eqsuations (14) and (15), we have $\mathbf{u}(u, t_0) \parallel g_2$. In this situation, the curve $\psi=\psi(u)$ is not only a principal curve but also a geodesic. We also let $\{\widehat{S}, S\}$ to indicate the surface pencil couple with a Bertrand couple $\{\widehat{\boldsymbol{\psi}}(u), \psi(u)\}$ as joint principal curves.

Example 1. *Let $\boldsymbol{\psi}(u)$ be an admissible helix specified by*

$$\boldsymbol{\psi}(u) = (u, \sin u, \cos u), \ 0 \leq u \leq 2\pi.$$

Then,

$$\boldsymbol{\psi}'(u) = (1, \cos u, -\sin u), \ \boldsymbol{\psi}''(u) = (0, -\sin u, -\cos u), \ \boldsymbol{\psi}'''(u) = (0, -\cos u, \sin u).$$

In view of Equations (3)–(5), we gain $\kappa(u) = -\tau(u) = 1$, and

$$\mathbf{g}_1(u) = (1, \cos u, -\sin u), \ \mathbf{g}_2(u) = (0, -\sin u, -\cos u), \ \mathbf{g}_3(u) = (0, \cos u, -\sin u).$$

Then $\phi(u) = -u + \phi_0$. If $\phi_0 = 0$, we gain $\phi(u) = -u$. For

$$\begin{aligned} \mathfrak{l}(u) &= \mathfrak{m}(u) = \mathfrak{n}(u) = 1, \\ \mathfrak{a}(t) &= t, \ \mathfrak{b}(t) = -t\chi(u)\sin u, \ \mathfrak{c}(t) = -t\chi(u)\cos u, \ \chi(u) \neq 0. \end{aligned}$$

The surface pencil S with $\boldsymbol{\psi}(u)$ is

$$S : \mathbf{p}(u,t) = (u, \sin u, \cos u) + t(1, -\chi\sin u, -\chi\cos u) \times \begin{pmatrix} 1 & \cos u & -\sin u \\ 0 & -\sin u & -\cos u \\ 0 & \cos u & -\sin u \end{pmatrix}.$$

The surface pencil $\widehat{S}$ with $\widehat{\boldsymbol{\psi}}(u)$ as joint principal curve is as follows: Let $f = 2$ in Equation (7), we derive $\widehat{\boldsymbol{\psi}}(u) = (u, -\sin u, -\cos u)$. The Serret–Frenet vectors of $\widehat{\boldsymbol{\psi}}(u)$ are

$$\widehat{\mathbf{g}}_1(u) = (1, -\cos u, \sin u), \ \widehat{\mathbf{g}}_2(u) = (0, \sin u, \cos u), \ \widehat{\mathbf{g}}_3(u) = (0, -\cos u, \sin u).$$

Then,

$$\widehat{S} : \widehat{\mathbf{p}}(u,t) = (u, -\sin u, -\cos u) + t(1, -\chi\sin u, -\chi\cos u) \times \begin{pmatrix} 1 & -\cos u & \sin u \\ 0 & \sin u & \cos u \\ 0 & -\cos u & \sin u \end{pmatrix}.$$

Therefore, for $\chi(u) = 1$, $-2 \leq t \leq 2$, $0 \leq u \leq 2\pi$, then $\{\widehat{S}, S\}$ is exhibited in Figure 1, where the blue curve demonstrates $\psi(u)$ and the green curve is $\widehat{\boldsymbol{\psi}}(u)$.

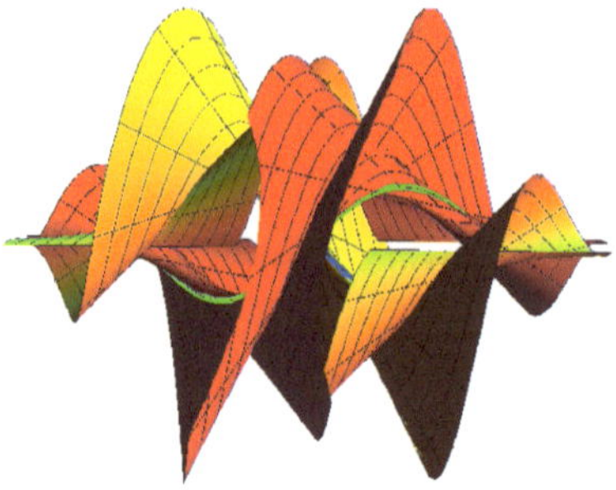

Figure 1. Curves S (yellow) and $\widehat{S}$ (red) with Bertrand couples as joint principal curves.

Example 2. *Suppose we are given an admissible curve by*

$$\boldsymbol{\psi}(u) = (u, 1 + \sin u, \sin u), \ \ 0 \leq u \leq 2\pi.$$

Then,

$$\mathbf{g}_1(u) = (1, \cos u, \cos u), \ \mathbf{g}_2(u) = (0, -\frac{1}{\sqrt{2}}, -\frac{1}{\sqrt{2}}), \ \mathbf{g}_3(u) = (0, \frac{1}{\sqrt{2}}, -\frac{1}{\sqrt{2}}),$$

with $\kappa(u) = \sqrt{2}\sin u$ *and* $\tau(u) = 0$, *which shows that* $\phi(u) = \phi_0$ *is a stationary. For*

$$\begin{aligned} \mathfrak{l}(u) &= \mathfrak{m}(u) = \mathfrak{n}(u) = 1, \\ \mathfrak{a}(t) &= t, \ \mathfrak{b}(t) = t\chi(u)\sin\phi_0, \ \ -\mathfrak{c}(t) = t\chi(u)\cos\phi_0, \ \chi(u) \neq 0. \end{aligned}$$

The surface pencil S *over* $\boldsymbol{\psi}(u)$ *is*

$$S : \mathbf{p}(u,t) = (u, 1 + \sin u, \sin u) + t(1, -\chi(u)\sin\theta_0, \chi\cos\theta_0)\begin{pmatrix} 1 & \cos u & \cos u \\ 0 & -\frac{1}{\sqrt{2}} & -\frac{1}{\sqrt{2}} \\ 0 & \frac{1}{\sqrt{2}} & -\frac{1}{\sqrt{2}} \end{pmatrix}.$$

Similarly, let $f = \sqrt{2}$ *in Equation (7), we acquire* $\widehat{\boldsymbol{\psi}}(u) = (u, \sin u, \sin u - 1)$, *and*

$$\widehat{\mathbf{g}}_1(u) = (1, \cos u, \cos u), \ \widehat{\mathbf{g}}_2(u) = (0, -\frac{1}{\sqrt{2}}, -\frac{1}{\sqrt{2}}), \ \widehat{\mathbf{g}}_3(u) = (0, \frac{1}{\sqrt{2}}, -\frac{1}{\sqrt{2}}),$$

Comparably, we have

$$\widehat{S} : \widehat{\mathbf{p}}(u,t) = (u, \sin u, \sin u - 1) + t(1, -\chi(u)\sin\phi_0, \chi(u)\cos\phi_0) \times \begin{pmatrix} 1 & \cos u & \cos u \\ 0 & -\frac{1}{\sqrt{2}} & -\frac{1}{\sqrt{2}} \\ 0 & \frac{1}{\sqrt{2}} & -\frac{1}{\sqrt{2}} \end{pmatrix}.$$

For $\chi(u) = 1$, $\phi_0 = 0$, $-2.5 \leq t \leq 2.5$, $0 \leq u \leq 2\pi$, *then* $\{\widehat{S}, S\}$ *is exhibited in Figure 2, where the blue curve displays* $\psi(u)$, *and the green curve displays* $\widehat{\boldsymbol{\psi}}(u)$. *Figure 3 specifies the* $\{\widehat{S}, S\}$ *for* $\phi_0 = \pi/2$, $-2.5 \leq t \leq 2.5$, *and* $0 \leq u \leq 2\pi$.

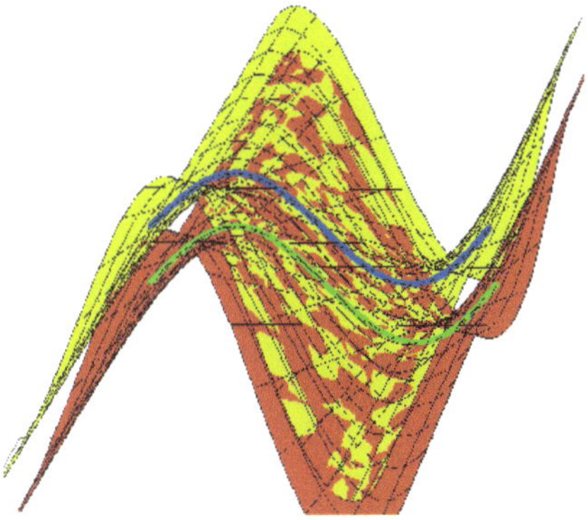

Figure 2. Curves S (yellow) and $\widehat{S}$ (red) with Bertrand couples as joint principal curves, where $\phi = 0$.

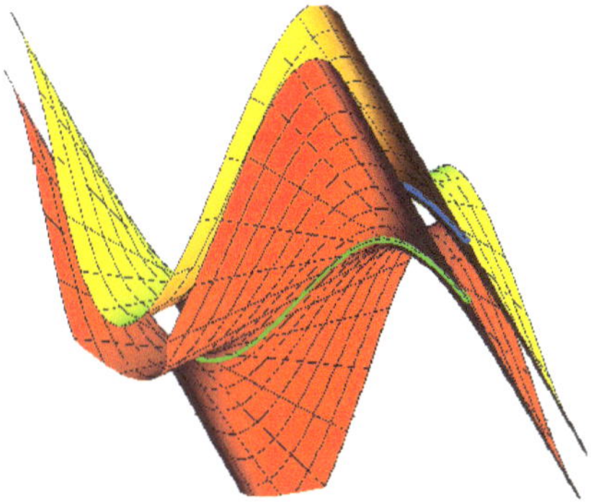

Figure 3. Curves S (yellow) and $\widehat{S}$ (red) with Bertrand couples as joint principal curves, where $\phi = \pi/2$.

Ruled Surface Pencil Couple with Bertrand Couple as Joint Principal Curves

Suppose $S_i : \mathbf{p}_i(u,t)$ is a ruled surface with the directrix $\psi_i(u)$ and $\psi_i(u)$ is also an isoparametric curve of $\mathbf{p}_i(u,t)$; then there exists t_0 such that $\mathbf{p}_i(u,t_0) = \psi_i(u)$. It follows that

$$S_i : \mathbf{p}_i(u,t) - \mathbf{p}_i(u,t_0) = (t - t_0)\mathbf{e}_i(u), 0 \leq u \leq L, \text{ with } t,\ t_0 \in [0,T], \tag{22}$$

where $\mathbf{e}_i(u)(i = 1,2,3)$ explains the orientation of the rulings. In view of Equation (9), we gain

$$(t - t_0)\mathbf{e}_i(u) = \mathfrak{a}(u,t)\mathbf{g}_{1i}(u) + \mathfrak{b}(u,t)\mathbf{g}_{2i}(u) + \mathfrak{c}(u,t)\mathbf{g}_{3i}(u), \tag{23}$$

where $0 \leq u \leq L$, with $t,\ t_0 \in [0,T]$. In fact, Equation (22) is a regulation of equations with the three unknowns $\mathfrak{a}(u,t)$, $\mathfrak{b}(u,t)$, and $\mathfrak{c}(u,t)$. The resolutions can be deduced as

$$\begin{aligned} \mathfrak{a}(u,t) &= (t - t_0) < \mathbf{e}_i(u), \mathbf{g}_{1i}(u) >, \\ \mathfrak{b}(u,t) &= (t - t_0) < \mathbf{e}_i(u), \mathbf{g}_{2i}(u) >, \\ \mathfrak{c}(u,t) &= (t - t_0) < \mathbf{e}_i(u), \mathbf{g}_{3i}(u) > . \end{aligned} \tag{24}$$

In view of Equation (15), if $\psi(u)$ is a principal curve on the surface $\mathbf{p}_i(u,t)$, we get

$$\begin{aligned} \mathfrak{a}(u,t) &= 0, \\ \chi(u)\sin\phi &=< \mathbf{e}_i(u), \mathbf{g}_{2i}(u) >, \\ -\chi(u)\cos\phi &=< \mathbf{e}_i(u), \mathbf{g}_{3i}(u) > . \end{aligned} \tag{25}$$

The above regulations are clearly the necessary and sufficient situations for which $\mathbf{p}_i(u,t)$ is a ruled surface with a directrix $\psi_i(u)$; $i = 1,2,3$.

In Galilean 3-space $\mathbb{G}_3$, there exist only three types of ruled surfaces, specified as follows [22,23]:

Type I. Non-conoidal or conoidal ruled surfaces for which the wrist (striction) curve does not lie in an Euclidean plane;

Type II. Ruled surfaces for which the wrist curve is in an Euclidean plane;

Type III. Conoidal ruled surfaces for an absolute line as the oriented line in infinity.

We immediately research if the curve $\psi_i(u)$ is also a principal curve on these three types:

Type I: $\psi_1(u) = (u, \psi_2(u),\psi_3(u))$ does not lie in an Euclidean plane and $\mathbf{e}_1(u) = (1, e_2(u), e_3(u))$ is non-isotropic. Then,

$$\begin{aligned} \mathbf{g}_{11}(u) &= \left(1, \psi_2'(u), \psi_3'(u)\right), \\ \mathbf{g}_{21}(u) &= \frac{1}{\kappa(u)}\left(0, \psi_2''(u), \psi_3''(u)\right), \\ \mathbf{g}_{31}(u) &= \frac{1}{\kappa(u)}\left(0, -\psi_3''(u), \psi_2''(u)\right), \end{aligned} \tag{26}$$

where $\kappa(u) = \sqrt{(\psi_2''(u))^2 + (\psi_3''(u))^2}$. From Equations (1), (24), and (26), we have:

$$\mathfrak{a}(u,t) = (t - t_0),\ \mathfrak{b}(u,t) = \mathfrak{c}(u,t) = 0, \tag{27}$$

which does not satisfy Theorem 2.

Type II: $\psi_2(u) = (0, \psi_2(u)), z(u))$ lie in an Euclidean plane and $\mathbf{e}_2(u) = (1, e_2(u), e_3(u))$ is non-isotropic. Then,

$$\begin{aligned} \mathbf{g}_{12}(u) &= \left(0, \psi_2'(u), \psi_3'(u)\right), \\ \mathbf{g}_{22}(u) &= \frac{1}{\kappa(u)}\left(0, \psi_2''(u), \psi_3''(u)\right), \\ \mathbf{g}_{32}(u) &= \frac{1}{\kappa(u)}(0,0,0), \end{aligned} \tag{28}$$

where $\kappa(u) = \sqrt{(\psi_2''(u))^2 + (\psi_3''(u))^2}$. From Equations (1), (24), and (28), we gain:

$$\mathfrak{a}(u,t) = \mathfrak{b}(u,t) = \mathfrak{c}(u,t) = 0, \tag{29}$$

which does not satisfy Theorem 2.

Corollary 2. *In $\mathbb{G}_3$, there are no ruled surface pencil couples of Type I and Type II with Bertrand couples as joint principal curves.*

Type III: $\psi_3(u) = (u, \psi_2(u), 0)$ does not lie in an Euclidean plane and $\mathbf{e}_3(u) = (0, e_2(u), e_3(u))$ is non-isotropic. Then,

$$\begin{aligned} \mathbf{g}_{13}(u) &= \left(1, \psi_2'(u), 0\right), \\ \mathbf{g}_{23}(u) &= \frac{1}{\kappa(u)}\left(0, \psi_2''(u)0\right), \\ \mathbf{g}_{33}(u) &= \frac{1}{\kappa(u)}\left(0, 0, \psi_2''(u)\right), \end{aligned} \tag{30}$$

where $\kappa(u) = \sqrt{(\psi_2''(u))^2}$. From Equations (1), (24), and (29), we have:

$$\left.\begin{array}{l} \mathfrak{a}(u,t) = 0,\ \mathfrak{b}(u,t) = \epsilon(t - t_0)e_2(u), \\ \mathfrak{c}(u,t) = \epsilon(t - t_0)e_3(u), \\ e_2(u) \neq 0, e_3(u) \neq 0,\ t_0 \neq 0, \end{array}\right\} \tag{31}$$

where

$$\epsilon = \begin{cases} 1, \text{ if } \psi_2''(u) > 0. \\ -1, \text{ if } \psi_2''(u) < 0. \end{cases} \tag{32}$$

Equation (31) satisfies Theorem 2. Suppose at all points on $\psi_3(u)$, the ruling $\mathbf{e}_3(u) \in up\{\mathbf{g}_{13}(u), \mathbf{g}_{23}(u), \mathbf{g}_{33}(u)\}$, then

$$\mathbf{e}_3(u) = \zeta(u)\mathbf{g}_{13}(u) + \sigma(u)\mathbf{g}_{23}(u) + \mu(u)\mathbf{g}_{33}(u), \tag{33}$$

for some functions $\lambda(u)$, $\sigma(u)$, and $\mu(u)$. Replacing it into Equation (25), we get

$$\sigma(u) = \chi(u)\sin\phi,\ \mu(u) = -\chi(u)\cos\phi. \tag{34}$$

Then,

$$\mathbf{e}_3(u) = \lambda(u)\mathbf{g}_{13}(u) + \chi(u)\sin\phi\mathbf{g}_{23}(u) - \chi(u)\cos\phi\mathbf{g}_{33}(u). \tag{35}$$

Choosing $\mathfrak{a}(u,t) = t\lambda(u)$, $\mathfrak{b}(u,t) = t\chi(u)\sin\phi$, and $\mathfrak{c}(u,t) = -t\chi(u)\cos\phi$, the ruled surface pencil S_3 with $\psi_3(u)$ can be shown by

$$\mathbf{p}_3(u,t) = \boldsymbol{\psi}_3(u) + t\zeta(u)\mathbf{g}_{13}(u) + t\lambda(u)(\sin\phi\mathbf{g}_{23}(u) - \cos\phi\mathbf{g}_{33}(u)),\ 0 \leq u \leq L,\ 0 \leq t \leq T. \tag{36}$$

Then, the ruled surface pencil $\widehat{S}_3$ is

$$\widehat{\mathbf{p}}_3(u,t) = \widehat{\boldsymbol{\psi}}_3(\widehat{u}) + t\lambda(u)\widehat{\mathbf{g}}_{13}(u) + t\chi(u)(\sin\phi\widehat{\mathbf{g}}_{23}(u) - \cos\phi\widehat{\mathbf{g}}_{33}(u)),\ 0 \leq u \leq L,\ 0 \leq t \leq T. \tag{37}$$

The functions $\lambda(u)$ and $\chi(u)$ can control the shape of the surfaces family S_3 and $\widehat{S}_3$.

Example 3. *Via Example 1, we have:*

By taking $\lambda(u) = \chi(u) = u$, then $\{\widehat{S}_3, S_3\}$ with $\{\widehat{\boldsymbol{\psi}}_3(u), \psi_3(u)\}$ as joint Bertrand principal curves is (Figure 4):

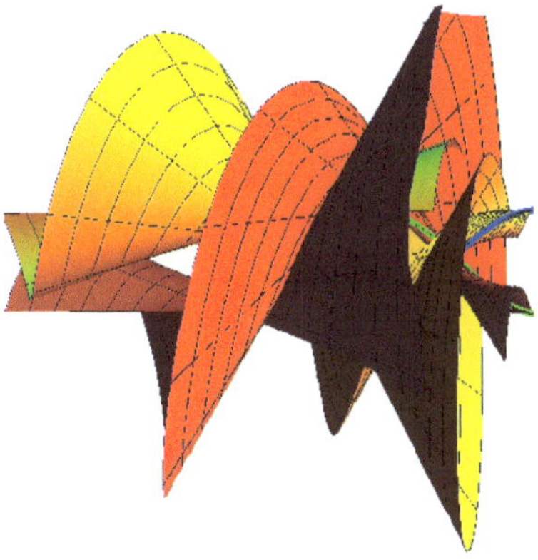

Figure 4. Curves S_3 (yellow) and $\widehat{S}_3$ (red) with Bertrand couples as joint principal curves, where $\lambda(s) = \chi(s) = s$.

$$\begin{cases} \mathbf{p}_3(u,t) = (u(1+t), \sin u + tu(\cos - \cos 2u), \cos u + tu(-\sin u + \sin 2u)), \\ \widehat{\mathbf{p}}_3(u,t) = (u(1+t), -\sin u + tu(-\cos + \cos 2u), -\cos u + tu(\sin u - \sin 2u)), \end{cases}$$

where the blue curve demonstrates $\psi_3(u)$, the green curve is $\widehat{\boldsymbol{\psi}}_3(u)$, $-0.7 \leq t \leq 0.7$, and $0 \leq u \leq 2\pi$.

By taking $\lambda(u) = \chi(u) = \sqrt{u}$, then $\{\widehat{S}_3, S_3\}$ with $\{\widehat{\boldsymbol{\psi}}_3(u), \psi_3(u)\}$ as joint Bertrand principal curves is (Figure 5):

$$\begin{cases} \mathbf{p}_3(u,t) = (\sqrt{u}(\sqrt{u}+t), \sin u + t\sqrt{u}(\cos - \cos 2u), \cos u + t\sqrt{u}(-\sin u + \sin 2u)), \\ \widehat{\mathbf{p}}_3(u,t) = (\sqrt{u}(\sqrt{u}+t), -\sin u + t\sqrt{u}(-\cos + \cos 2u), -\cos \sqrt{u} + tu(\sin u - \sin 2u)), \end{cases}$$

where the blue curve demonstrates $\psi_3(u)$, the green curve is $\widehat{\boldsymbol{\psi}}_3(u)$, $-0.7 \leq t \leq 0.7$, and $0 \leq u \leq 2\pi$.

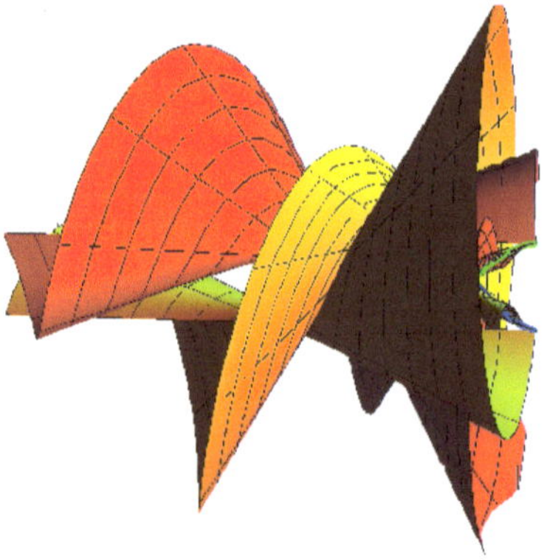

Figure 5. Surface pencil couple S_3 (yellow) and $\widehat{S}_3$ (red).with Bertrand couple as joint principal curves; where $\lambda(u) = \chi(u) = u\sqrt{u}$.

Example 4. *Via Example 2, we have:*

(1) By taking $\phi_0 = 0$, and $\lambda(u) = \chi(u) = u$, then $\{\widehat{S}_3, S_3\}$ with $\{\widehat{\boldsymbol{\psi}}_3(u), \psi_3(u)\}$ as joint Bertrand principal curves is (Figure 6):

$$\begin{cases} \mathbf{p}_3(u,t) = (u(1+t), 1 + \sin u + tu(\cos - \frac{1}{\sqrt{2}}), \sin u + tu(\cos u - \frac{1}{\sqrt{2}})), \\ \widehat{\mathbf{p}}_3(u,t) = u(1+t), \sin u + tu(\cos - \frac{1}{\sqrt{2}}), \sin u - 1 + tu(\cos u - \frac{1}{\sqrt{2}})), \end{cases}$$

where the blue curve demonstrates $\psi_3(u)$, the green curve is $\widehat{\boldsymbol{\psi}}_3(u)$, $-0.2 \leq t \leq 0.2$, and $0 \leq u \leq 2\pi$

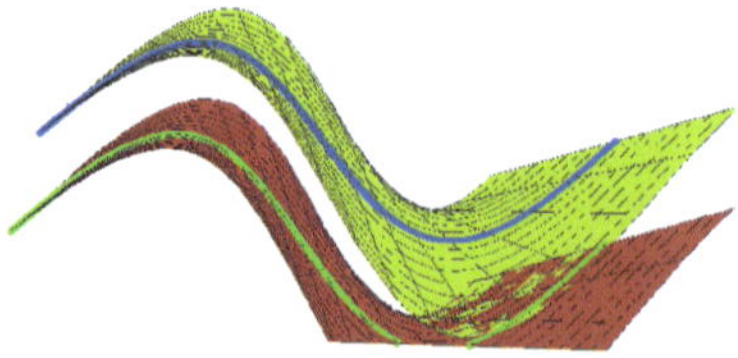

Figure 6. Curves S_3 (yellow) and $\widehat{S}_3$ (red) with Bertrand couples as joint principal curves, where $\lambda(u) = \chi(u) = u$, and $\phi_0 = 0$.

(2) By taking $\phi_0 = \pi/2$, and $\zeta(u) = \chi(u) = u$, then $\{\widehat{S}_3, S_3\}$ with $\{\widehat{\boldsymbol{\psi}}_3(u), \psi_3(u)\}$ as joint Bertrand principal curves is (Figure 7):

$$\begin{cases} \mathbf{y}_3(u,t) = (u(1+t), 1+\sin u + tu(\cos + \frac{1}{\sqrt{2}}), \sin u + tu(\cos u + \frac{1}{\sqrt{2}})), \\ \widehat{\mathbf{y}}_3(u,t) = u(1+t), \sin u + tu(\cos + \frac{1}{\sqrt{2}}), \sin u - 1 + tu(\cos u + \frac{1}{\sqrt{2}})), \end{cases}$$

where the blue curve demonstrates $\psi_3(u)$, the green curve is $\widehat{\boldsymbol{\psi}}_3(u)$, $-0.2 \leq t \leq 0.2$, and $0 \leq u \leq 2\pi$.

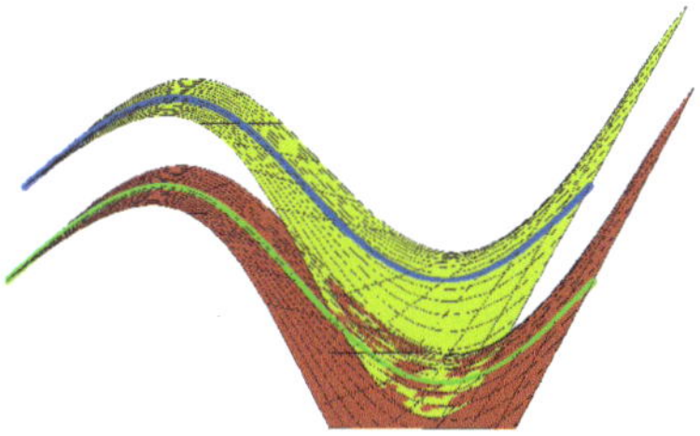

Figure 7. Curves S_3 (yellow) and $\widehat{S}_3$ (red) with Bertrand couples as joint principal curves, where $\lambda(u) = \chi(u) = s$, and $\phi_0 = \pi/2$.

4. Conclusions

This paper studied principal curves and their associated surfaces in Galilean 3-space. Given a 3D curve, we seek surfaces that are faced with this curve as a distinctive curve. The work viewed the Bertrand couples as principal curves and procures a surface pencil couple with a Bertrand couple as joint principal curves. Then, necessary and suffcient conditions for an admissible curve to be an isoparametric principal curve on surface are newly derived in this case. Examples of the surface pencil with principal curves are shown in the figures with a Bertrand couple. Our results in this paper contribute to the work by Jiang et al. [30]. We hope that these outcomes will open modern perceptions for researchers working on geometrical modeling and production evolution procedure in the manufacturing industry. The judgment of stratifying the mechanisms applied here to various spaces such as Lorentz-space, pseudo-Galilean space, and Heisenberg space is already an investigation topic. We will discuss this problem in the future.

Author Contributions: The authors have contributed equally to this work. All authors have read and agreed to the published version of the manuscript.

Funding: This research received no external funding.

Informed Consent Statement: Not applicable.

Data Availability Statement: All of the data are available within the paper.

Conflicts of Interest: The authors declare that there are no conflicts of interest regarding the publication of this paper.

References

1. Do Carmo, M.P. *Differential Geometry of Curves and Surfaces*; Prentice Hall: Englewood Cliffs, NJ, USA, 1976.
2. Spivak, M. *A Comprehensive Introduction to Differential Geometry*, 2nd ed.; Publish or Perish: Houston, TX, USA, 1979.
3. Munchmeyer, F.C. On surface imperfections. In *Mathematics of Surfaces II*; Martin, R., Ed.; Oxford University Press: Oxford, UK, 1987; pp. 459–474.
4. Patrikalakis, N.M.; Maekawa, T. *Shape Interrogation for Computer Aided Design and Manufacturing*; Springer: New York, NY, USA, 2002.
5. Martin, R.R. Principal patches—A new class of surface patches based on differential geometry. In *Eurographics 83, Proceedings of the 4th Ann. European Association for Computer Graphics Conference and Exhibition, Zagreb, Yugoslavia, 31 August–2 September 1983*; Ten Hagen, P.J.W., Ed.; North-Holland: Amsterdam, The Netherlands, 1983; pp. 47–55.
6. Alourdas, P.G.; Hottel, G.R.; Tsohy, S.T. A design and interrogation system for modeling with rational splines. In *Proceedings of the Ninth International Symposium on OMAE, Houston, TX, USA*; AsME: New York, NY, USA, 1990; pp. 555–565. Volume 1, Part B.

7. Maekawa, T.; Wolter, F.-E.; Patrikalakis, N.M. Umbilics and lines of curvature for shape interrogation. *Comput. Aided Geom. Des.* **1996**, *13*, 133–161. [CrossRef]
8. Che, W.J.; Paul, J.C. Lines of curvature and umbilical points for implicit surfaces. *Comput. Aided Geom. Des.* **2007**, *24*, 395–409. [CrossRef]
9. Zhang, X.P.; Che, W.J.; Paul, J.C. Computing lines of curvature for implicit surfaces. *Comput. Geom. Des.* **2009**, *26*, 923–940. [CrossRef]
10. Kalogerakis, E.; Nowrouzezahrai, D.; Simari, P.; Singh, K. Extracting lines of curvature from noisy point clouds. *Comput.-Aided Des.* **2009**, *41*, 282–292. [CrossRef]
11. Wang, G.; Tang, K.; Tai, C.H. Parametric representation of a surface pencil with common spatial geodesic. *Comput.-Aided Des.* **2004**, *36*, 47–59. [CrossRef]
12. Kauap, E.; Akyldz, F.K.; Orbay, K. A generalization of surfaces family with common spatial geodesic. *Appl. Math. Comput.* **2008**, *201*, 781–789.
13. Li, C.Y.; Wang, R.H.; Zhu, C.G. Parametric representation of a surface pencil with a common line of curvature. *Computer Aided Des.* **2011**, *43*, 1110–1117. [CrossRef]
14. Li, C.Y.; Wang, R.H.; Zhu, C.G. An approach for designing a developable surface through a given line of curvature. *Computer Aided Des.* **2013**, *45*, 621–627. [CrossRef]
15. Bayram, E.; Guler, F.; Kauap, E. Parametric representation of a surface pencil with a common asymptotic curve. *Computer Aided Des.* **2012**, *44*, 637–643. [CrossRef]
16. Liu, Y.; Wang, G.J. Designing developable surface pencil through given curve as its common asymptotic curve. *J. Zhejiang Univ.* **2013**, *47*, 1246–1252.
17. Atalay, G.; Kasap, E. Surfaces family with common smarandache geodesic curve. *J. Uci. Artu.* **2017**, *4*, 651–664.
18. Atalay, G.; Kasap, E. Surfaces family with common smarandache geodesic curve according to Bishop frame in Euclidean space. *Math. Sci.* **2016**, *4*, 164–174. [CrossRef]
19. Bayram, E.; Mustafa, B. Surface family with a common involute asymptotic curve. *Int. J. Geom. Methods Mod. Phys.* **2016**, *13*, 447–459.
20. Guler, F.; Bayram, E.; Kasap, E. Offset surface pencil with a common asymptotic curve. *Int. J. Geom. Methods Mod. Phys.* **2018**, *15*, 1850195(1)–1850195(11). [CrossRef]
21. Atalay, G. Surfaces family with a common Mannheim asymptotic curve. *J. Appl. Math. Comput.* **2018**, *2*, 143–154.
22. Atalay, G. Surfaces family with a common Mannheim geodesic curve. *J. Appl. Math. Comput.* **2018**, *2*, 155–165.
23. Abdel-Baky, R.A.; Alluhaibi, N. Surfaces family with a common geodesic curve in Euclidean 3-upace $\mathbb{E}^3$. *Inter. J. Math. Anal.* **2019**, *9*, 433–447.
24. Hobson, M.P.; Efstathiou, G.P.; Lasenby, A.N. *General Relativity An Introduction for Physicists*; Cambridge University Press: Cambridge, UK, 2006.
25. Yaglom, I.M. *A Simple Non-Euclidean Geometry and Its Physical Basis*; Springer: New York, NY, USA, 1979.
26. Dede, D. Tubular surfaces in Galilean space. *Math. Common.* **2013**, *18*, 209–217.
27. Dede, D.; Ekici, E.; Coken, A.C. On the parallel surfaces in Galilean space. *Hacet. J. Math. Stat.* **2013**, *42*, 605–615. [CrossRef]
28. Yuzbası Z.K. On a family of surfaces with joint asymptotic curve in the Galilean space $\mathbb{G}_3$. *J. Nonlinear Sci. Appl.* **2016**, *9*, 518–523. [CrossRef]
29. Yuzbası Z.K.; Bektas, M. On the conutructios of a surface family with joint geodeuic in Galilean space $\mathbb{G}_3$. *Open Phys.* **2016**, *14*, 360–363. [CrossRef]
30. Jiang, X.; Jiang, P.; Meng, J.; Wang, K. Surface pencil couple interpolating Bertrand couple au joint asymptotic curves in Galilean space. *Int. J. Geom. Methods Mod. Phys.* **2021**, *18*, 2150114. [CrossRef]
31. Almoneef, A.A.; Abdel-Baky, R.A. Surface family couple with Bertrand couple au common geodesic curves in Galilean 3-upace $\mathbb{G}_3$. *Mathematics* **2023**, *11*, 2391.
32. Li, Y.; Eren, K.; Ersoy, S. On simultaneous characterizations of partner-ruled surfaces in Minkowski 3-space. *AIMS Math.* **2023**, *8*, 22256–22273. [CrossRef]
33. Li, Y.; Güler, E. A Hypersurfaces of Revolution Family in the Five-Dimensional Pseudo-Euclidean Space $\mathbb{E}^5_2$. *Mathematics* **2023**, *11*, 3427.
34. Li, Y.; Gupta, M.K.; Sharma, S.; Chaubey, S.K. On Ricci Curvature of a Homogeneous Generalized Matsumoto Finsler Space. *Mathematics* **2023**, *11*, 3365. [CrossRef]
35. Li, Y.; Bhattacharyya, S.; Azami, S.; Saha, A.; Hui, S.K. Harnack Estimation for Nonlinear, Weighted, Heat-Type Equation along Geometric Flow and Applications. *Mathematics* **2023**, *11*, 2516. [CrossRef]
36. Li, Y.; Kumara, H.A.; Siddesha, M.S.; Naik, D.M. Characterization of Ricci Almost Soliton on Lorentzian Manifolds. *Symmetry* **2023**, *15*, 1175. [CrossRef]
37. Li, Y.; Mak, M. Framed Natural Mates of Framed Curves in Euclidean 3-Space. *Mathematics* **2023**, 11, 3571.
38. Li, Y.; Mofarreh, F.; Abdel-Baky, R.A. Kinematic-geometry of a line trajectory and the invariants of the axodes. *Demonstr. Math.* **2023**, *56*, 20220252.
39. Li, Y.; Patra, D.; Alluhaibi, N.; Mofarreh, F.; Ali, A. Geometric classifications of k-almost Ricci solitons admitting paracontact metrices. *Open Math.* **2023**, *21*, 20220610. [CrossRef]

40. Li, Y.; Güler, E. Hypersurfaces of revolution family supplying in pseudo-Euclidean space. *AIMS Math.* **2023**, *8*, 24957–24970. [CrossRef]
41. Li, Y.; Eren, K.; Ayvacı, K.H.; Ersoy, S. The developable surfaces with pointwise 1-type Gauss map of Frenet type framed base curves in Euclidean 3-space. *AIMS Math.* **2023**, *8*, 2226–2239. [CrossRef]
42. Li, Y.; Ganguly, D. Kenmotsu Metric as Conformal η-Ricci Soliton. *Mediterr. J. Math.* **2023**, *20*, 193. [CrossRef]
43. Li, Y.; Erdoğdu, M.; Yavuz, A. Differential Geometric Approach of Betchow-Da Rios Soliton Equation. *Hacet. J. Math. Stat.* **2023**, *52*, 114–125. [CrossRef]
44. Li, Y.; Abdel-Salam, A.A.; Saad, M.K. Primitivoids of curves in Minkowski plane. *AIMS Math.* **2023**, *8*, 2386–2406. [CrossRef]
45. Li, J.; Yang, Z.; Li, Y.; Abdel-Baky, R.A.; Saad, M.K. On the Curvatures of Timelike Circular Surfaces in Lorentz-Minkowski Space. *Filomat* **2024**, *38*, 1–15.
46. Li, Y.; Gezer, A.; Karakaş, E. Some notes on the tangent bundle with a Ricci quarter-symmetric metric connection. *AIMS Math.* **2023**, *8*, 17335–17353. [CrossRef]
47. Al-Jedani, A.; Abdel-Baky, R.A. A surface family with a mutual geodesic curve in Galilean 3-upace $\mathbb{G}_3$. *Mathematics* **2023**, *11*, 2971. [CrossRef]
48. Roschel, O. *Die Geometrie des Galileischen Raumes;* Habilitationsschrift: Leoben, Austria, 1984.
49. Divjak, B. Geometrija Pseudogalilejevih Prostora. Ph.D. Thesis, University of Zagreb, Zagreb, Croatia, 1997.

axioms

Article

KCC Theory of the Oregonator Model for Belousov-Zhabotinsky Reaction

M. K. Gupta [1], Abha Sahu [1,*], C. K. Yadav [1], Anjali Goswami [2] and Chetan Swarup [2]

1 Department of Mathematics, Guru Ghasidas Vishwavidyalaya, Bilaspur 495009, India; mkgiaps@gmail.com (M.K.G.); chiranjeev86@gmail.com (C.K.Y.)
2 Department of Basic Science, College of Science and Theoretical Studies, Saudi Electronic University, Riyadh 11673, Saudi Arabia; a.goswami@seu.edu.sa (A.G.); c.swarup@seu.edu.sa (C.S.)
* Correspondence: abhasahu118@gmail.com

Abstract: The behavior of the simplest realistic Oregonator model of the BZ-reaction from the perspective of KCC theory has been investigated. In order to reduce the complexity of the model, we initially transformed the first-order differential equation of the Oregonator model into a system of second-order differential equations. In this approach, we describe the evolution of the Oregonator model in geometric terms, by considering it as a geodesic in a Finsler space. We have found five KCC invariants using the general expression of the nonlinear and Berwald connections. To understand the chaotic behavior of the Oregonator model, the deviation vector and its curvature around equilibrium points are studied. We have obtained the necessary and sufficient conditions for the parameters of the system in order to have the Jacobi stability near the equilibrium points. Further, a comprehensive examination was conducted to compare the linear stability and Jacobi stability of the Oregonator model at its equilibrium points, and We highlight these instances with a few illustrative examples.

Keywords: Oregonator model; KCC theory; Berwald connection; Jacobi stability

MSC: 53B40; 53C22; 53C60

Citation: Gupta, M.K.; Sahu, A.; Yadav, C.K.; Goswami, A.; Swarup, C. KCC Theory of the Oregonator Model for Belousov-Zhabotinsky Reaction. *Axioms* **2023**, *12*, 1133. https://doi.org/10.3390/axioms12121133

Academic Editor: Mića Stanković

Received: 19 October 2023
Revised: 28 November 2023
Accepted: 14 December 2023
Published: 18 December 2023

1. Introduction

In the early 1950's, Belousov studied the behavior of chemical model of the oxidation for organic molecules and found that chemical reaction is also taking place at the end position (equilibrium), but was unable to publish his observation because at that time researchers were convinced that oscillations in homogeneous chemical reactions are not possible. Later, Zhabotinsky [1] confirmed Belousov's discovery and explained that the oscillation is due to the contrast between chemical homogeneous oscillating systems and thermodynamics. Since 1984, oscillating chemical reactions (OCRs) have been recognized, a well-known example is the Belousov–Zhabotinsky (BZ)-reaction [1]. The first mechanism to explain the temporal oscillation of the BZ-reaction was suggested by Field, Koros, and Noyes (FKN) [2]. The FKN mechanism are divided into three subprocesses which are defined according to the factors that control the kinetics of the whole reaction, the concentrations of bromide and cerium ions. OCRs are considered a special case because the oscillating behavior prohibits the second law of thermodynamics which states that *"heat always moves from hotter objects to colder objects, unless energy is supplied to reverse the direction of heat flow"*. There are alot of FKN mechanism like Lotka [3] and Brusselator mechanism [4] which are capable of generating oscillations. The oscillatory BZ-reaction has a simple realistic model called the Oregonator model. The Oregonator is a reduced model of the FKN mechanism [2], containing only a five-steps involving three independent chemical intermidiates that summarises the main features of the BZ reaction. The simplified mechanism is often used to refer the term 'model', instead of attempting to capture the entire chemistry of the process, its goal is to develop a set of differential equations that represents the fundamental features of the original method.

The five fundamental reactions of the Oregonator model are used for the construction of a system of three nonlinear differential equations and the kinetic behavior of the Oregonator can be described by equations [2]

$$\begin{aligned}\frac{dX}{dt} &= k_1AY - k_2XY + k_3AX - 2k_4X^2,\\ \frac{dY}{dt} &= -k_1AY - k_2XY + fk_5BZ,\\ \frac{dZ}{dT} &= k_3AX - k_5BZ,\end{aligned} \tag{1}$$

where $X = HBrO_2, Y = Br^-, Z = Ce(IV), A = BrO_3^-$ and $B = BrMA$ ($k_i's$ are kinetic constants). Here f is a stoichiometric factor and to ensure the existence of oscillations its value has to be in a certain range, i.e., $0.5 < f < 2.4$. The reactions are treated as irreversible and the acidity effects are included in the rate constants. For the sake of simplicity, Cassani et al. [5] rescaled the system of Equation (1) and gave the following system:

$$\begin{cases}\dfrac{dx}{dt} = \dfrac{1}{\epsilon}\left(ax + qay - x^2 - xy\right)\\ \dfrac{dy}{dt} = \dfrac{1}{\delta}(-qay + fbz - xy)\\ \dfrac{dz}{dt} = ax - bz,\end{cases} \tag{2}$$

where $\epsilon = \frac{k_5B_0}{K_3A_0}, \delta = \frac{2k_5k_4B_0}{k_2k_3A_0}, a = \frac{A}{A_0}, b = \frac{B}{B_0}$ and the scaling factor are described on Table 1, with '0' denoting the reference value [6].

Table 1. Scaling factor of the simplified Oregonator model.

Scaling Factor of the Oregonator Model $A_0 = B_0 = 1M$		
X	$x = \frac{X}{X_0}$	$X_0 = \frac{k_3A_0}{2k_4}$
Y	$y = \frac{Y}{Y_0}$	$Y_0 = \frac{k_3A_0}{2k_2}$
Z	$z = \frac{Z}{Z_0}$	$Z_0 = \frac{(k_3A_0)^2}{k_3k_4B_0}$
T	$t = \frac{T}{T_0}$	$\frac{1}{k_5B_0} = 1s$

Mathematical terms for describing the stability of the dynamical system's solution include linear stability and Lyapunov stability. This method yields the Lyapunov exponents, which measure the exponential deviation from the provided trajectories. Since the method of Lyapunov stability is well established, it would be interesting to study the stability of a dynamical system from another viewpoint and comparing the results with corresponding Lyapunov exponents. The KCC theory approach is an alternative method for examining the characteristics of dynamical systems known as geometro-dynamical approach which was first initiated by Kosambi [7], Cartan [8] and Chern [9]. The concept of KCC theory is based on the assumption that the geodesics equation in Finsler space and second-order dynamical systems are topologically equivalent. Antonelli et al. [10,11] initially started the study of Jacobi stability for the geodesic corresponding to a Finslerian metric by deviating the geodesics and using the KCC-covariant derivative for the variation in differential system. The KCC theory is a differential geometric theory for variational equations describing deviations of entire trajectories from neighbouring ones. Each dynamical system in the geometrical description provided by the KCC theory has two types of coefficients of the connection, the first of which is a nonlinear connection and the second of which is a Berwald type connection. With the help of the nonlinear and Berwald connections the five geometrical invariants can be constructed of which the second invariant plays an important role as it gives the Jacobi stability of dynamical system. Jacobi stability analysis for different systems like Lorenz system [12], Chua circuit system [13] and other systems [14–21] have been studied. According to the articles [22,23], one of the geometrical invariants that identifies the beginning of chaos is the deviation vector from the so-called Jacobi equation. Jacobi stability has been analyzed by a large number of authors in the past years as an effective method for predicting chaotic behaviour of the systems [24–26]. Yamasaki and Yajima [27,28] has discussed the KCC stability in the intermediate nonequilibrium region of the Catastrophe and Brusselator model.

In this paper, we discuss the Oregonator model using KCC theory by formulating a set of 2-second order differential equation. Section 2, devotes the basic of KCC theory. In Section 3, we have investigated the general expression and Jacobi stability of the Oregonator model at different equilibrium points. We have analyzed chaotic behavior for the Oregonator model and vector field analysis for deviation vector, in Section 4. Section 5, presents the comparison with the linear stability analysis. At the last section, conclusion is given.

2. KCC Theory and Jacobi Stability

Let us consider a real, smooth n-dimensional manifold M and TM be its tangent bundle, with $(x^1, x^2, \ldots, x^n) = x^i$ and $(\frac{dx^1}{dt}, \frac{dx^2}{dt}, \ldots \frac{dx^n}{dt}) = \frac{dx^i}{dt} = y^i$. Let $(x^i, y^i, t), i = \{1, 2, \ldots, n\}$ be the $(2n+1)$-dimensional coordinate system on a subset Ω of the Euclidean $(2n+1)$-dimensional space $\mathbb{R}^n \times \mathbb{R}^n \times \mathbb{R}^1$, we assume that the time t is an absolute invariant. Consider the following system of second order differential equation (SODE) as

$$\frac{d^2x^i}{dt^2} + 2G^i(x^i, y^i) = 0, \quad i = 1, 2 \tag{3}$$

where G^i are smooth function defined on an open neighborhood of some initial conditions $((x)_0, (y)_0, (t)_0)$ in Ω.

The intrinsic geometric properties of SODE (3), are given by the five different KCC-invariants, under the non-singular coordinate transformations

$$\begin{aligned} \overline{x}^i &= f^i(x^1, x^2, \ldots, x^n), \qquad i = 1, 2, \ldots, n \\ \overline{t} &= t \end{aligned} \tag{4}$$

where f^i are n-smooth functions, possesing derivatives of all orders in their domain of definition. The KCC covariant derivatives of a contravariant vector field $\xi^i(t)$ on the open subset Ω, under the local coordiante system (4), is defined as

$$\frac{D\xi^i}{dt} = \frac{d\xi^i}{dt} + N^i_j \xi^j, \tag{5}$$

where $N^i_j = \frac{\partial G^i}{\partial y^i}$ is the coefficient of the nonlinear connection on TM.

Substituting $\xi^i = y^i$, we get

$$\frac{Dy^i}{dt} = N^i_j y^j - 2G^i = -\varepsilon^i, \tag{6}$$

where the contravariant vector field ε^i is known as the first KCC-invariant. Now, let us assume the transformation of the trajectories $x^i(t)$ into the nearby ones as follows:

$$\overline{x}^i(t) \;=\; x^i(t) + \eta\, \xi^i(t), \quad |\eta| << 1 \tag{7}$$

where η is the very small parameter defined along the trajectory $x^i(t)$ of the SODE (3) and ξ^i is the component of contravariant vector.

Substituting Equation (7) into Equation (3) and taking the limit $\eta \to 0$, we obtain

$$\frac{d^2\xi^i}{dt^2} + 2N^i_j \frac{d\xi^j}{dt} + 2\frac{\partial G^i}{\partial x^j}\xi^j = 0. \tag{8}$$

The above equation represents Jacobi field equation which can be can be reformulate in the covariant form with the use of the KCC-covariant differential as

$$\frac{D^2\xi^i}{dt^2} = P^i_j \xi^j, \tag{9}$$

where P^i_j is a $(1,1)$-type tensor, defined as

$$P^i_j = -2\frac{\partial G^i}{\partial x^j} - 2\,G^i_{jk}G^k + y^k\frac{\partial N^i_k}{\partial x^j} + N^i_k\,N^k_j + \frac{\partial N^i_j}{\partial t}, \tag{10}$$

and the coefficient $G^i_{jk} = \frac{\partial N^i_k}{\partial y^j}$ represents the Berwald connection coefficient. Geometrically, the deviation tensor P^i_j is interpreted as the second KCC-invariant. When the SODE (3) describes about the geodesic equation in Finsler geometry then Equation (9) is called the Jacobi field equation. The torsion tensor, Riemann curvature tensor and Douglas curvature tensor are called the third, fourth and fifth KCC-invariants respectively of the SODE (3) and are defined as

$$P^i_{jk} = \frac{1}{3}\left(\frac{\partial P^i_j}{\partial y^k} - \frac{\partial P^i_k}{\partial y^j}\right), \quad P^i_{jkl} = \frac{\partial P^i_{jk}}{\partial y^l}, \quad D^i_{jkl} = \frac{\partial G^i_{jk}}{\partial y^l}. \tag{11}$$

Jacobi Stability of Dynamical System

The Jacobi stability is a natural generalization of the stability of the geodesic flow on a differentiable manifold endowed with a metric (Finslerian) to the non-metric setting [29]. This kind of stability refers to the focusing tendency of trajectories of systems of ordinary differential equations with respect to nearby trajectories and satisfy the conditions [30].

$$||x^i(t_0) - \overline{x}^i(t_0)|| = 0, \quad ||\dot{x}^i(t_0) - \dot{\overline{x}}^i(t_0)|| \neq 0.$$

Definition 1 ([30]). *The trajectories of SODEs are called Jacobi stable at* $(x^i(t_0), \overline{x}^i(t_0))$ *if and only if the real parts of the all the eigenvalues of second KCC invariants* P^i_j *at point* t_0 *are strictly negative and Jacobi unstable, otherwise.*

The curvature deviation tensor or the second KCC-invariant can be written in a matrix form as

$$P^i_j = \begin{pmatrix} P^1_1 & P^1_2 \\ P^2_1 & P^2_2 \end{pmatrix}$$

where the eigenvalues of the curvature deviation tensor are the solutions of the quadratic equation

$$\lambda^2 - tr(P^i_j)\,\lambda + det(P^i_j) = 0,$$

where

$$tr(P^i_j) = P^1_1 + P^2_2, \quad det(P^i_j) = P^1_1\,P^2_2 - P^1_2\,P^2_1.$$

We use the Routh–Hurwitz criteria [31], to obtain the signs of the eigenvalues of the curvature deviation tensor. According to which, all roots of the 2×2 matrix are negative or have negative real parts if the trace and determinant of the deviation curvature matrix is strictly negative and strictly positive, respectively.

3. Mathematical Model of Oregonator for BZ-Reaction

The Oregonator model uses three independent intermediate, five irreversible reaction step controlled by five kinetic constsnts and a stoichiometric factor. Cassani et al. [5] kept the variables a and b constant and set a unitary value in order to obtain a simplifed version of Equation (2) are as follows:

$$\begin{cases} \frac{dx}{dt} = \frac{1}{\epsilon}\Big(x + qy - x^2 - xy\Big), \\ \frac{dy}{dt} = \frac{1}{\delta}(-qy + fz - xy), \\ \frac{dz}{dt} = x - z. \end{cases} \tag{12}$$

Differentiating first equation of the system (12) with respect to time, we obtain

$$\frac{d^2x}{dt^2} = \frac{1}{\epsilon}\left(\frac{dx}{dt} + q\frac{dy}{dt} - 2x\frac{dx}{dt} - x\frac{dy}{dt} - y\frac{dx}{dt}\right) \tag{13}$$

Second equation of the system (12) can also be written as

$$z = \frac{1}{f}\left(\delta\frac{dy}{dt} + qy + xy\right). \tag{14}$$

Differentiating above equation with respect to $'t'$, we get

$$\frac{dz}{dt} = \frac{1}{f}\left(\delta\frac{d^2y}{dt^2} + q\frac{dy}{dt} + y\frac{dx}{dt} + x\frac{dy}{dt}\right). \tag{15}$$

Using third equation of the system (12) and (14), above equation can be written as

$$\frac{d^2y}{dt^2} + \frac{1}{\delta}\left\{(q + \delta + x)\frac{dy}{dt} + \left(q + x + \frac{dx}{dt}\right)y - fx\right\} = 0 \tag{16}$$

Let us alter the notation to read as

$$x = x^1, \quad \frac{dx}{dt} = y^1, \quad y = x^2, \quad \frac{dy}{dt} = y^2,$$

then from Equations (13) and (16), the system takes the form

$$\begin{aligned} &\frac{d^2x^1}{dt^2} + \frac{1}{\epsilon}\left[\left(2x^1 + x^2 - 1\right)y^1 + \left(x^1 - q\right)y^2\right] = 0 \\ &\frac{d^2x^2}{dt^2} + \frac{1}{\delta}\left[x^2y^1 + \left(q + \delta + x^1\right)y^2 + \left(x^1x^2 + qx^2 - fx^1\right)\right] = 0 \end{aligned} \tag{17}$$

4. Jacobi Stability of Oregonator Model for BZ-Reaction

In this section, we study the dynamical properties of the Oregonator model by using the KCC-theory approach. We will find the non-linear connections, Berwald connections and the deviation curvature tensor for the Oregonator. We also study the eigenvalue of deviation tensor at equilibrium points.

4.1. KCC-Invariants of the Oregonator Model

Now, Equation (3) of Oregonator system can be rewritten as follows:

$$\frac{d^2x^i}{dt^2} + 2G^i(x^i, y^i) = 0, \quad i = 1, 2 \tag{18}$$

where

$$G^1 = \frac{1}{2\epsilon}\left[\left(2x^1 + x^2 - 1\right)y^1 + \left(x^1 - q\right)y^2\right]$$
$$G^2 = \frac{1}{2\delta}\left[x^2y^1 + \left(q + \delta + x^1\right)y^2 + \left(x^1x^2 + qx^2 - fx^1\right)\right] \tag{19}$$

The components of the non-linear connection of Oregonator model can be calculated using $N^i_j = \frac{\partial G^i}{\partial y^j}$, are as follows:

$$N^1_1 = \frac{1}{2\epsilon}\left(2x^1 + x^2 - 1\right), \quad N^1_2 = \frac{1}{2\epsilon}\left(x^1 - q\right)$$
$$N^2_1 = \frac{1}{2\delta}x^2, \qquad N^2_2 = \frac{1}{2\delta}\left(q + \delta + x^1\right) \tag{20}$$

which implies $G^i_{jk} := \frac{\partial N^i_j}{\partial y^k} = 0,\ i, j, k = 1, 2$. Thus, for the Oregonator model the components of Berwald connection vainshes identically. The components of the First KCC-invariant can be obtained using Equation (6) as:

$$\varepsilon^1 = \frac{1}{2\epsilon}\left[\left(2x^1 + x^2 - 1\right)y^1 + \left(x^1 - q\right)y^2\right]$$
$$\varepsilon^2 = \frac{1}{\delta}\left[x^2y^1 + \left(q + \delta + x^1\right)y^2 + 2\left(x^1x^2 + qx^2 - f x^1\right)\right] \tag{21}$$

From Equation (10), the components of the curvature deviation tensor or the second KCC-invariant of the Oregonator model are given by

$$P^1_1 = -\frac{1}{\epsilon}y^1 - \frac{1}{2\epsilon}y^2 + \frac{1}{4\epsilon^2}\left(2x^1 + x^2 - 1\right) + \frac{1}{4\epsilon\delta}\left(x^1 - q\right)x^2$$
$$P^1_2 = -\frac{1}{2\epsilon}y^1 + \frac{1}{4\epsilon^2}\left(2x^1 + x^2 - 1\right)\left(x^1 - q\right) + \frac{1}{4\epsilon\delta}\left(x^1 - q\right)\left(q + \delta + x^1\right)$$
$$P^2_1 = -\frac{1}{\delta}y^2 - \frac{1}{\delta}\left(x^2 - f\right) + \frac{1}{4\epsilon\delta}\left(2x^1 + x^2 - 1\right)x^2 + \frac{1}{\delta^2}\left(q + \delta + x^1\right)x^2$$
$$P^2_2 = -\frac{1}{\delta}x^1 + \frac{1}{2\delta}y^1 + \frac{1}{4\epsilon\delta}\left(x^1 - q\right)x^2 + \frac{1}{4\delta}\left(q + \delta + x^1\right) \tag{22}$$

The third invariant of KCC theory can be interpreted geometrically as torsion tensor and defined as $P^i_{jk} = \frac{1}{3}\left(\frac{\partial P^i_j}{\partial y^k} - \frac{\partial P^i_k}{\partial y^j}\right)$. For Oregonator system

$$P^1_{11} = P^1_{12} = P^1_{21} = P^1_{22} = P^2_{11} = P^2_{22} = 0,$$
$$P^2_{12} = -P^2_{21} = \frac{1}{2\delta}.$$

The fourth and fifth invariant of the Oregonator system for BZ-reaction vanishes identically as P^i_{jk}, G^i_{jk} dosenot contain any term of y^i, for $i = 1, 2$.The time variation of the components of curvature deviation tensor for Oregonator system is represented in Figure 1. The selection of parameters and the initial conditions are purely fictitious and do not necessarily have a geometrical significance.

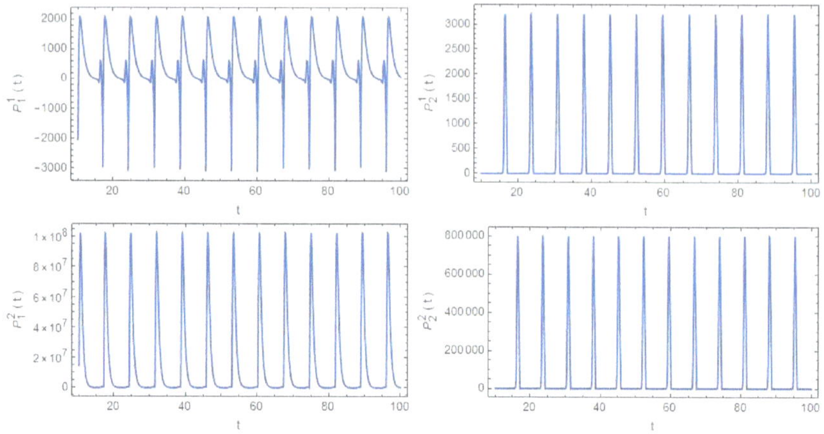

Figure 1. Time variation of components of deviation tensor P_1^1 (left top figure), P_2^1 (right top figure), P_1^2 (left bottom figure) and P_2^1 (right bottom figure) for parameters values $\epsilon = 0.10, \delta = 0.0004, q = 0.0008$ and $f = 1$. The initial conditions for the numerical integration system are $x^1(0) = x^2(0) = x^3(0) = 5$.

4.2. The Jacobi Stability of the Equilibrium Points of the Oregonator Model

The equilibrium points for the Oregonator system (12) is given by

$$S(E_0) = (0,0,0), S(E_1) = \left(\frac{u-v}{2}, \frac{w+v}{4}, \frac{u-v}{2}\right), S(E_2) = \left(\frac{u+v}{2}, \frac{w-v}{4}, \frac{u+v}{2}\right),$$

where

$$u = 1 - f - q, \quad v = \sqrt{(1-f-q)^2 + 4q(q+f)}, \quad w = 1 + 3f + q.$$

With the concern of the system of Equation (17), the equilibrium points are

$$E_0 = (0,0), E_1 = \left(\frac{u-v}{2}, \frac{w+v}{4}\right), E_2 = \left(\frac{u+v}{2}, \frac{w-v}{4}\right).$$

Theorem 1. *For any value of the parameter ϵ, δ, q and f the trivial equilibrium point E_0 of Oregonator model is Jacobi unstable.*

Proof. In view of Equation (21), the first KCC-invariant for the equilibrium point E_0 vanishes identically, i.e., $\varepsilon^1 = \varepsilon^2 = 0$. The deviation curvature matrix at the Equilibrium point E_0 is given by

$$P(E_0) = \begin{pmatrix} 0 & \frac{q}{4\epsilon^2} - \frac{q(q+\delta)}{4\epsilon\delta} \\ \frac{f}{\delta} & \frac{1}{4\delta^2}(q+\delta)^2 \end{pmatrix}$$

and its trace and determinant are $trP_j^i(E_0) = \frac{1}{4\delta^2}(q+\delta)^2$ and $detP_j^i(E_0) = \frac{f}{\delta}\left(\frac{q}{4\epsilon^2} - \frac{q(q+\delta)}{4\epsilon\delta}\right)$ respectively. At Equilibrium point E_0, the characteristic equation of deviation curvature tensor is

$$\lambda^2 - trP_j^i(E_0)\lambda + detP_j^i(E_0) = 0.$$

From Routh-Hurwitz criteria, the eigenvalue of the characteristic equation are negative or have negative real parts if and only if $trP_j^i(E_0) < 0$ and $detP_j^i(E_0) > 0$ holds. Since, trace $\frac{1}{4\delta^2}(q+\delta)^2$ is always positive. Therfore, the system is Jacobi unstable at E_0. $\square$

Now, from Equation (21), the component of the first KCC-invariant and the deviation tensor of Oregonator model at equilibrium point E_1 are

$$\epsilon^1(E_1) = 0, \quad \epsilon^2(E_1) = \frac{1}{4\delta}[(u-v)(w+v) + 4q(w+v) - 4f(u-v)].$$

The components of the second KCC-invariant at the equilibrium point E_1 are given as

$$\begin{aligned}
P_1^1(E_1) &= \frac{1}{64\,\delta\,\epsilon^2}\Big\{(4u-4-3v+w)^2\delta - 2(2q-u+v)(v+w)\epsilon\Big\},\\
P_2^1(E_1) &= -\frac{1}{32\,\delta\,\epsilon^2}\{(2q-u+v)[2(2q+u-v)\epsilon + \delta(4u-4-3v+w+4\epsilon)]\},\\
P_1^2(E_1) &= \frac{1}{64\,\delta\,\epsilon^2}\{(4u-4-3v+w)(v+w)\delta + 2(v+w)(2q+u-v-6\delta)\epsilon + 64f\delta\epsilon\},\\
P_2^2(E_1) &= \frac{1}{32\,\delta\,\epsilon^2}\Big\{-(2q-u+v)(v+w)\delta + 2(2q+u-v-2\delta)^2\epsilon\Big\}.
\end{aligned}$$

Now, from the above deviation tensors at equilibrium point E_1 the trace and determinant are given by

$$\begin{aligned}
trP_j^i(E_1) =& \frac{1}{64\delta^2\epsilon^2}\Big\{(4u-4-3v+w)^2\delta^2 - 4(2q-u+v)(v+w)\delta\epsilon + 4(2q+u-v-2\delta)^2\epsilon^2\Big\},\\
detP_j^i(E_1) =& \frac{1}{2048\delta^3\epsilon^3}\Big\{\Big[(4u-4-3v+w)^2\delta - 2(2q-u+v)(v+w)\epsilon\Big](-(2q-u+v)(v+w)\delta\\
&+2(2q+u-v-2\delta)^2\epsilon) + (2q-u+v)((4u-4-3v+w)(v+w)\delta\\
&+2(v+w)(2q+u-v-6\delta)\epsilon + 64f\delta\,\epsilon)[2(2q+u-v)\epsilon + \delta(4u-4-3v+w+4\epsilon)]\Big\}
\end{aligned}$$

The characteristic equation of the deviation curvature tensor at equilibrium point E_1 can be written as

$$\lambda^2 - trP_j^i(E_1)\lambda + detP_j^i(E_1) = 0$$

In view of Routh-Hurwitz criteria, the eigenvalue of the characteristic equation are negative or have negative real part if and only if $trP_j^i(E_1) < 0$ and $detP_j^i(E_1) > 0$ holds. Thus, we have

Theorem 2. *The equilibrium point E_1 is Jacobi stable if it satisfies simultaneously the constraints*

$$trP_j^i(E_1) < 0, \quad detP_j^i(E_1) > 0,$$

and Jacobi unstable, otherwise.

Now, using Equation (21), the component of the first KCC-invariant at equilibrium point E_2 are

$$\epsilon^1(E_2) = 0, \quad \epsilon^2(E_2) = \frac{1}{4\delta}[(u+v)(w-v) + 4q(w-v) - 4f(u+v)].$$

The components of the second KCC-invariant at the equilibrium point E_2 are given as

$$\begin{aligned}
P_1^1(E_2) &= \frac{1}{64\,\delta\,\epsilon^2}\Big\{(4-4u-5v+w)^2\delta + 2(-2q+u+v)(v-w)\epsilon\Big\},\\
P_2^1(E_2) &= -\frac{1}{32\,\delta\,\epsilon^2}\{(2q-u-v)(2(2q+u+v)\epsilon + \delta(-4+4u+5v-w+4\epsilon))\},\\
P_1^2(E_2) &= \frac{1}{64\,\delta^2\,\epsilon}\{(v-w)(-4+4u+5v-w)\delta + 2(v-w)(2q+u+v-6\delta)\epsilon + 64f\delta\,\epsilon\},\\
P_2^2(E_2) &= -\frac{1}{32\,\delta^2\,\epsilon}\Big\{-(2q-u-v)(v-w)\delta + 2(2q+u+v-2\delta)^2\epsilon\Big\}.
\end{aligned}$$

Now, from the above deviation tensors at equilibrium point E_2 the trace and determinant are given by

$$trP^i_j(E_2) = \frac{1}{64\delta^2\epsilon^2}\Big\{4(2q+u+v-2\delta)^2\epsilon^2 + (4-4u-5v+w)^2\delta^2 - 4\delta\epsilon(2q-u-v)(v-w)\Big\},$$
$$detP^i_j(E_2) = \frac{1}{2048\delta^3\epsilon^3}\Big\{\big[(4-4u-5v+w)^2\delta + 2(u-2q+v)(v-w)\epsilon\big]\big[2(2q+u+v-2\delta)^2\epsilon$$
$$-\ (2q-u-v)(v-w)\delta\big] + (2q-u-v)\big[(v-w)(4u-4+5v-w)\delta$$
$$+\ 2(v-w)(2q+u+v-6\delta)\epsilon + 64f\delta\epsilon\big]\big[2(2q+u+v)\epsilon + \delta(4u-4+5v-w+4\epsilon)\big]\Big\}$$

At equilibrium point E_2, the characteristic equation of the deviation curvature tensor can be written as

$$\lambda^2 - trP^i_j(E_2)\lambda + detP^i_j(E_2) = 0$$

In view of Routh-Hurwitz criteria, the eigenvalue of the characteristic equation are negative or have negative real part if and only if $trP^i_j(E_2) < 0$ and $detP^i_j(E_2) > 0$ holds. Therefore, we obtain

Theorem 3. *The equilibrium point E_2 is Jacobi stable if it satisfies simultaneously the constraints*

$$trP^i_j(E_2) < 0, \quad detP^i_j(E_2) > 0,$$

and Jacobi unstable, otherwise.

Next, we assume different set of parameter for the Oregonator model and calculate the eigenvalue of Jacobi matrix by using MATHEMATICA 12.0.

Example 1. *For $\epsilon = 2$, $\delta = 0.004$, $q = 0.08$, $f = 1$, the stability at equilibrium points are as follows:*

(i) $E_0(0,0)$ *has conjuagte pairs of eigenvalues* $\{0.15625 + 0.784991\,i,\ 0.15625 - 0.784991\,i\}$.
(ii) $E_1(-0.130554, 1.06528)$ *has one positive and one negative eigenvalues* $\{241.46, -0.432151\}$.
(iii) $E_2(0.122554, 0.938723)$ *has one positive and one negative eigenvalues* $\{257.402, -0.42911\}$.

At each of these three points of equilibrium, the system exhibits Jacobi instability.

Example 2. *For $\epsilon = 0.10$, $\delta = 0.004$, $q = 0.0008$, $f = 2.4$, the stability at equilibrium points are as follows:*

- $E_0(0,0)$ *has one positive and one negative eigenvalues* $\{25.4181, -0.258084\}$.
- $E_1(-1.40274, 2.40137)$ *has positive eigenvalues* $\{26327, 394.663\}$.
- $E_2(0.00193, 1.69903)$ *has one positive and one negative eigenvalues* $\{15.4788, -0.682798\}$.

Thus, we can say that the system is Jacobi unstable at each of these three equilibrium points.

5. Chaotic Behavior of Oregonator Model

The trajectory behavior of deviation vector $\xi^i, i = 1,2$ near fixed point is obtained by using the following Equation (8)

$$\frac{d^2\xi^1}{dt^2} + \frac{1}{\epsilon}\Big(2x^1 + x^2 - 1\Big)\frac{d\xi^1}{dt} + \frac{1}{\epsilon}\Big(x^1 - q\Big)\frac{d\xi^2}{dt} + \frac{1}{\epsilon}\Big(2y^1 + y^2\Big)\xi^1 + \frac{1}{\epsilon}y^1\xi^2 = 0,$$
$$\frac{d^2\xi^2}{dt^2} + \frac{1}{\delta}x^2\frac{d\xi^1}{dt} + \frac{1}{\delta}\Big(q + \delta + x^1\Big)\frac{d\xi^2}{dt} + \frac{1}{\delta}\Big(x^2 + y^2 - f\Big)\xi^1 + \frac{1}{\delta}x^1\xi^2 = 0. \tag{23}$$

The value of deviation vector can be obtained from its components ξ^1 and ξ^2 is given as

$$\xi = \sqrt{[\xi^1(t)]^2 + [\xi^2(t)]^2}. \tag{24}$$

In order to obtain the chaotic behaviour of the Oregonator model, we introduce the Lyapunov exponents similar to the instability exponents. The Lyapunov exponents describes the rate of divergence of nearby trajectories, i.e., the presence of chaos in the system as

$$\Delta_i(E) = \lim_{t\to\infty} \frac{1}{t} ln \frac{\xi^i(t)}{\dot{\xi}^i(0)}, \quad i = 1, 2,$$

and

$$\Delta(E) = \lim_{t\to\infty} \frac{1}{t} ln \frac{\xi(t)}{\dot{\xi}(0)},$$

Now, we examine the deviation vectors behaviour close to equilibrium positions. The initial conditions used to integrate the deviation equations are $\xi^1(t) = \xi^2(t) = 0, \dot{\xi}^1(t) = 10^{-10}, \dot{\xi}^2(t) = 10^{-9}$ and values of the parameter are shown in their respective figures.

5.1. *Behavior of the Deviation Vector Near* E_0

The dynamics of deviation vector near equilibrium point E_0 are calculated using Equation (23) as follows:

$$\begin{aligned} \frac{d^2\xi^1}{dt^2} - \frac{1}{\epsilon}\frac{d\xi^1}{dt} - \frac{1}{\epsilon}\frac{d\xi^2}{dt} &= 0 \\ \frac{d^2\xi^2}{dt^2} + \frac{(q+\delta)}{\delta}\frac{d\xi^2}{dt} - \frac{f}{\delta}\xi^1 &= 0 \end{aligned} \tag{25}$$

The behaviour of components of the deviation curvature vector and instability exponents are shown in Figure 2.

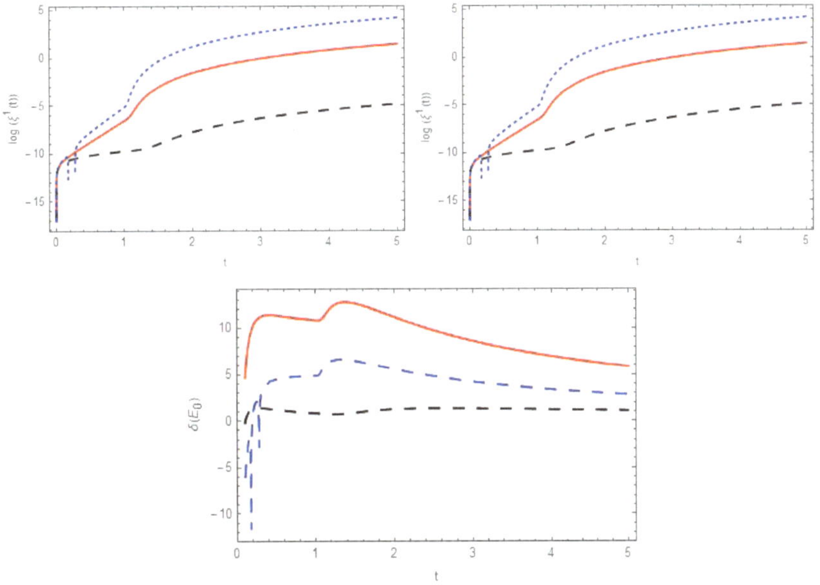

Figure 2. Time variation of the deviation vector components $\xi^1(t)$ (left figure) and $\xi^2(t)$ (right figure), in a logarithmic scale and instability exponents $\delta(E_0)$, near the equilibrium point E_0, for $\epsilon = 0.10, \delta = 0.0004, q = 0.0008, f = 1$ (Solid Red), for $\epsilon = 1, \delta = 4, q = 0.0008, f = 0.5$ (Black), for $\epsilon = 0.10, \delta = 0.004, q = 0.8, f = 2.4$ (Blue). The initial conditions used to integrate the deviation equations are $\xi^1(t) = \xi^2(t) = 0, \dot{\xi}^1(t) = 10^{-10}, \dot{\xi}^2(t) = 10^{-9}$.

5.2. *Behavior of the Deviation Vector Near E_1*

By using Equation (8), the dynamics of deviation vector near the equilibrium point E_1 are obtained as follows:

$$\frac{d^2\xi^1}{dt^2}+\frac{1}{4\epsilon}(4u-3v+w-4)\frac{d\xi^1}{dt}+\frac{1}{2\epsilon}(u-v-2q)\frac{d\xi^2}{dt}=0$$

$$\frac{d^2\xi^2}{dt^2}+\frac{1}{4\delta}(w+v)\frac{d\xi^1}{dt}+\frac{1}{2\delta}(2q+2\delta+u-v)\frac{d\xi^2}{dt}+\frac{1}{4\delta}(w+v-4f)\xi^1+\frac{1}{2\delta}(u-v)\xi^2=0$$

The behaviour of components of the deviation curvature vector and instability exponents are shown in Figure 3.

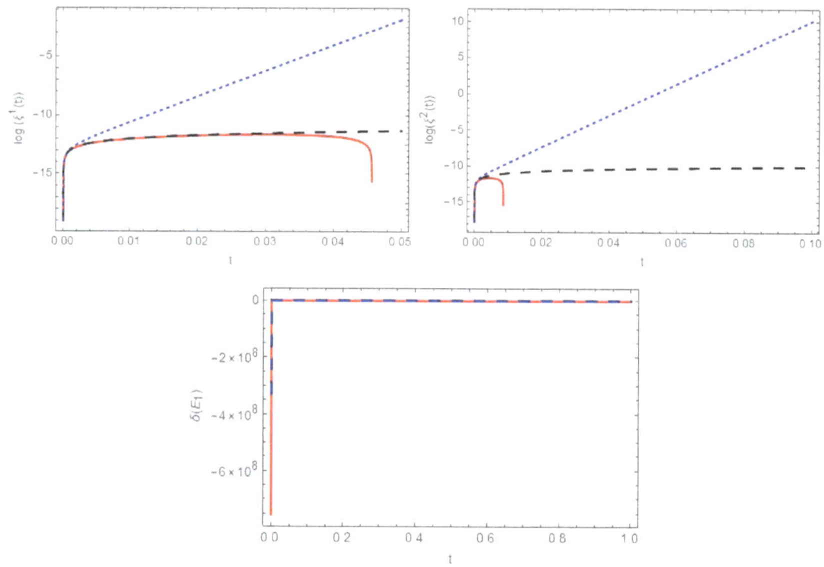

Figure 3. Time variation of the deviation vector components in a logarithmic scale $\xi^1(t)$ (left figure) and $\xi^2(t)$ (right figure), $\delta(E_1)$, near the equilibrium point E_1, for $\epsilon = 0.10$, $\delta = 0.0004$, $q = 0.0008$, $f = 1$ (Solid Red), for $\epsilon = 1$, $\delta = 0.4$, q=0.0008, $f = 0.5$ (Black), for $\epsilon = 0.10$, $\delta = 0.004$, $q = 0.8, f = 2.4$ (Blue). The initial conditions used to integrate the deviation equations are $\xi^1(t) = \xi^2(t) = 0, \dot{\xi}^1(t) = 10^{-10}, \dot{\xi}^2(t) = 10^{-9}$.

5.3. *Behavior of the Deviation Vector Near E_2*

The dynamics of deviation vector near equilibrium point E_2 are calculated using Equation (8) as follows:

$$\frac{d^2\xi^1}{dt^2}+\frac{1}{4\epsilon}(4u+3v+w-4)\frac{d\xi^1}{dt}+\frac{1}{2\epsilon}(u+v-2q)\frac{d\xi^2}{dt}=0$$

$$\frac{d^2\xi^2}{dt^2}+\frac{1}{4\delta}(w-v)\frac{d\xi^1}{dt}+\frac{1}{2\delta}(2q+2\delta+u+v)\frac{d\xi^2}{dt}+\frac{1}{4\delta}(w-v-4f)\xi^1+\frac{1}{2\delta}(u+v)\xi^2=0$$

The behaviour of the components of the deviation curvature vector and instability exponents are shown in Figure 4.

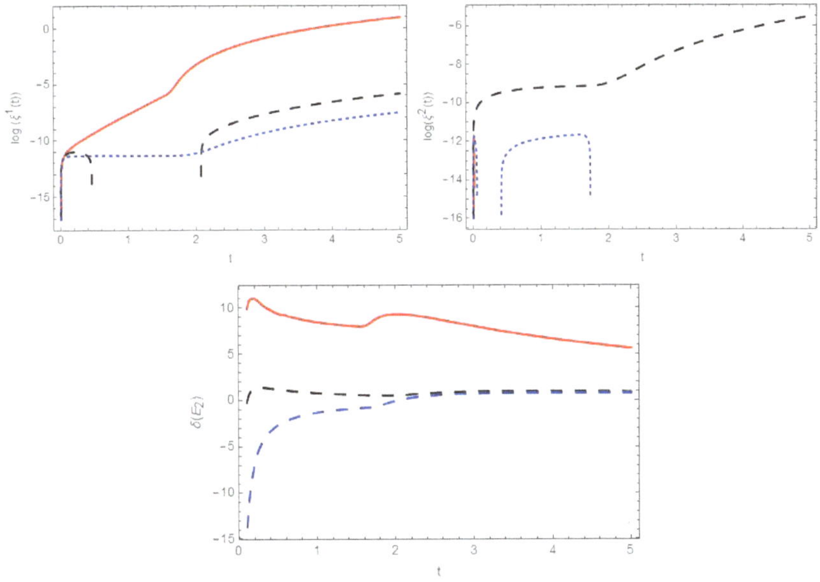

Figure 4. Time variation of the deviation vector components in a logarithmic scale $\xi^1(t)$ (left figure) and $\xi^2(t)$ (right figure), $\delta(E_2)$, near the equilibrium point E_2, for $\epsilon = 0.10$, $\delta = 0.0004$, $q = 0.0008$, $f = 1$ (Solid Red), for $\epsilon = 1, \delta = 4, q = 0.0008, f = 0.5$ (Black), for $\epsilon = 0.10$, $\delta = 0.004$, $q = 0.8$, $f = 2.4$ (Blue). The initial conditions used to integrate the deviation equations are $\xi^1(t) = \xi^2(t) = 0, \dot{\xi}^1(t) = 10^{-10}, \dot{\xi}^2(t) = 10^{-9}$.

5.4. Curavture of the Deviation Tensor

The geometric curvature κ of the curve $\xi(t) = (\xi^1(t), \xi^2(t))$, examines the quantitative explanation of the deviation tensors behaviour which we define, according to the standard approach used in differential geometry of plane curve as [12]

$$\kappa = \frac{\dot{\xi}^1(t)\ddot{\xi}^2(t) - \dot{\xi}^2(t)\ddot{\xi}^1(t)}{\{\,[\dot{\xi}^1(t)]^2 + [\dot{\xi}^2(t)]^2\,\}^{\frac{3}{2}}}$$

The chaotic behavior of the system is described by the curvature of the deviation vector. From Figure 5, we can see that for different value of parameters the curvature of the deviation vector is positive for very brief periods of time, hitting zero at specific times, moving into the region of negative values, and then, after temporarily tending toward zero, returning to positive values before moving back into the region of negative values.

The curvature of the deviation vector of the Oregonator model becomes zero at $\xi^1(t_0) = \xi^2(t_0)$, and the time interval during which it changes sign is an symbol of the development of chaotic behavior.

The behaviour of the curvature deviation tensor are shown in Figure 5.

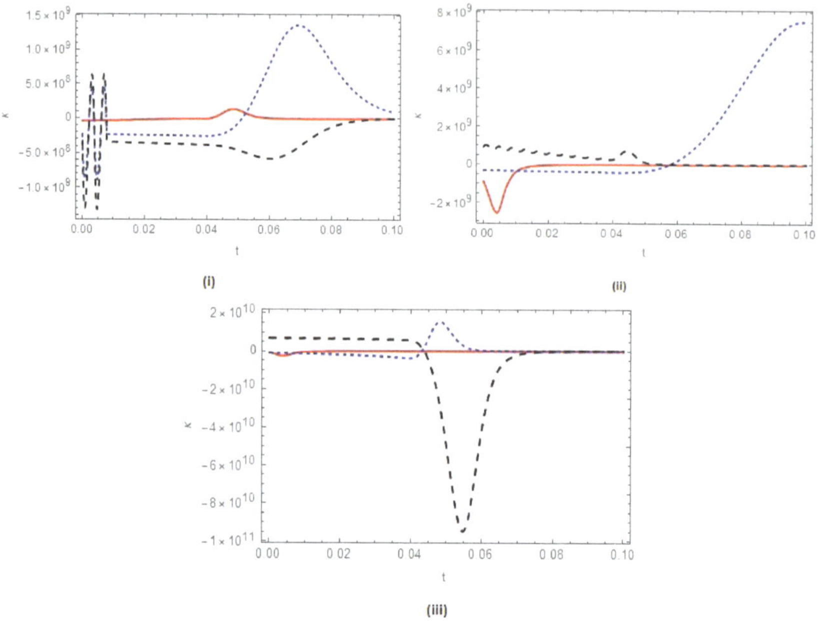

Figure 5. Time variation of curvature κ for E_0 (above left), E_1 (above right), E_2 (below) for parameter values $\epsilon = 0.10, \delta = 0.0004, q = 0.0008, f = 1$ (Red), for $\epsilon = 1, \delta = 0.04, q = 0.008, f = 0.5$ (Blue), (i) for $\epsilon = 0.50, \delta = 0.04, q = 0.008, f = 0.8$ (Black), (ii) for $\epsilon = 0.30, \delta = 0.004, q = 0.08, f = 0.7$ (Black), (iii) for $\epsilon = 0.05, \delta = 0.4, q = 0.008, f = 0.7$ (Black). The initial conditions used to integrate the deviation equations are $\xi^1(t) = \xi^2(t) = 0, \dot{\xi}^1(t) = 10^{-10}, \dot{\xi}^2(t) = 10^{-9}$.

6. Jacobi Stability vs Linear Stability

In this section, we will consider the relationship between Jacobi stability and Linear stability at different equilibrium points. To compare both stability we first find Jacobian matrix for system (12) as follows:

$$J = \begin{pmatrix} \frac{1}{\epsilon}(1-2x-y) & \frac{1}{\epsilon}(q-x) & 0 \\ \frac{1}{\delta}(-y) & \frac{1}{\delta}(-q-x) & \frac{f}{\delta} \\ 1 & 0 & -1 \end{pmatrix}$$

Example 3. *If we take value of the parameters as $\epsilon = 2$, $\delta = 0.004$, $q = 0.08$, $f = 1$ for the Jacobi matrix then the eigenvalues at equilibrium points $S(E_0)$, $S(E_1)$ and $S(E_2)$ are $\{-19.9743, -1.2947, 0.771729$ $\{63.2084, 0.739929, -0.533817\}$ and $\{-84.3937, -0.552687, -0.0408065\}$, repectively.*

At Equilibrium point $S(E_0)$ we have two negative and one positive eigenvalue, for the point $S(E_1)$ we obtain two positive and one negative eigenvalue and all eigenvalues are negative at the equilibrium point $S(E_2)$. Thus, both $S(E_0)$ and $S(E_1)$ points are linearly unstable while $S(E_2)$ is linearly stable. On comparing it with example 1, we found that equilibrium points E_0 and E_1 of the system are both Jacobi and linearly unstable, while the point E_2, is Jacobi unstable but linearly stable.

Now, for the Jacobi matrix if we take a set of parameters $\epsilon = 0.10, \delta = 0.004, q = 0.0008$,

Example 4. *For $f = 2.4$, the eigenvalues at equilibrium points $S(E_0)$, $S(E_1)$, $S(E_2)$ are $\{10.0424, -0.62122 \pm 0.539641\,i\}$, $\{323.185, 40.5648, -0.375287\}$ and $\{-7.44785, -0.557569 \pm 0.57831i\}$ repectively.*

For the first equilibrium point $S(E_0)$, we obtain one real eigenvalue and two complex conjugate and for $S(E_1)$ we get two positive and one negative real eigenvalue. At equilibrium point $S(E_2)$, all three eigenvalues have negative real part. Thus, both $S(E_0)$ and $S(E_1)$ are linearly unstable while $S(E_2)$ is linearly stable. Therefore, on comparing it with example 2, we found that equilibrium points E_0 and E_1 of the system are both Jacobi and linearly unstable, while the point E_2, is Jacobi unstable but linearly stable.

Example 5. *For $f = 1$, the eigenvalues at equilibrium points $S(E_0), S(E_1), S(E_2)$ are $\{10.0178, -0.608883 \pm 0.168974i\}$, $\{3.231 \pm 7.76734i, -0.0573755\}$ and $\{-12.9551, 4.291, 0.0710088\}$ respectively. Equilibrium point $S(E_0)$ has one positive eigenvalue and a conjugate pair of eigenvalues with negative real part. Equilibrium points $S(E_1)$ and $S(E_2)$ has one negative and two positive eigenvalues. Thus, both $S(E_1)$ and $S(E_2)$ are linearly unstable, while $S(E_1)$ is linearly stable.*

Examples 4 and 5, show the value of stoichiometric factor as we can see that the system changes its stability at equilibrium point $S(E_2)$, with the small change in the value of f. Hence, the parameter f is deeply linked to the ocillation rise in the chemical reaction.

7. Conclusions

In the paper, we have looked at the stability analysis of the Oregonator model for BZ-reaction from the perspective of the KCC theory, according to which the geometric properties of the geodesic equations for Finsler space, which are the equivalent of the given system and can be used to deduce the dynamical stability properties of dynamical systems. We first transformed the first order differential equation of the Oregonator model into a system of second order differential equations. We have introduced the instability exponent and the curvature of the deviation vector for the Oregonator model in order to describe the behavior of the trajectories around the equilibrium points. We have also shown the graphical representations of deviation vector components $\xi^1(t)$ and $\xi^2(t)$ in a logrithmic scale and instability exponents near all three equilibrium points for different set of parameters in Figures 2–4. Through the curve $\xi(t)$ transition moment from positive to negative values, it can be directly related to the chaotic behavior of the trajectories. We expressed the geometric quantities of KCC theory in terms of the Jacobian matrix of the linearized system in order to understand the relationship between the Jacobi stability and the linear stability.

This study presents the conditions for Jacobi stability at the equilibrium points of the Oregonator model. It demonstrates that, at the origin the system is always Jacobi unstable, whereas the Jacobi stability of the remaining two equilibrium points is dependent on the particular set of parameter values present in the system. In Mathematical modelling of the BZ reaction the stoichiometric factor has been used as a bifurcation parameter. The phenomenon denotes the transition between the stationary and oscillation during the chemical reaction. The present analysis provides the role of stoichiometric factor as a small increase or decrease in the value leads us to a different stability state of the system. In brief, our analysis of the Belousov–Zhabotinsky (BZ) reaction, a well-established example of an oscillating chemical reaction, using the KCC theory reveals that the system possesses a dual nature (stability and instability) of the system. The analysis emphasizes the coexistence of Jacobi stability and instability. Furthermore, the system exhibits characteristics of both linear stability and instability. This investigation enhances our understanding of the complex dynamics intrinsic in oscillating chemical reactions, thereby making a significant contribution to the wider comprehension of dynamic systems.

Author Contributions: Conceptualization, M.K.G. and A.S.; methodology, M.K.G. and A.S.; software, C.K.Y.; validation, A.G. and C.S.; writing—original draft preparation, A.S.; writing—review and editing, A.S.; visualization, C.K.Y.; supervision, M.K.G. All authors have read and agreed to the published version of the manuscript.

Funding: This research received no external funding.

Data Availability Statement: Data are contained within the article.

Conflicts of Interest: The authors declare no conflict of interest.

References

1. Winfree, A.T. The prehistory of the Belousov-Zhabotinsky oscillator. *J. Chem. Educ.* **1984**, *61*, 661. [CrossRef]
2. Field, R.J.; Noyes, R.M. Oscillations in chemical systems. IV. Limit cycle behavior in a model of a real chemical reaction. *J. Chem. Phys.* **1974**, *60*, 1877–1884. [CrossRef]
3. Lotka, A.J. Undamped oscillations derived from the law of mass action. *J. Am. Chem. Soc.* **1920**, *42*, 1595–1599. [CrossRef]
4. Nicolis, G. Stability and dissipative structures in open systems far from equilibrium. *Adv. Chem. Phys.* **1971**, *19*, 209–324.
5. Cassani, A.; Monteverde, A.; Piumetti, M. Belousov-Zhabotinsky type reaction: The non-linear behavior of chemical systems. *J. Math. Chem.* **2021**, *59*, 792–826. [CrossRef]
6. Pullela, S.R.; Cristancho, D.; He, P.; Luo, D.; Hall, K.R.; Cheng, Z. Temperature dependence of the Oregonator model for the Belousov-Zhabotinsky reaction. *Phys. Chem. Chem. Phys.* **2009**, *11*, 4236–4243. [CrossRef]
7. Kosambi, D.D. Parallelism and path-space. *Math. Z.* **1933**, *37*, 608–818. [CrossRef]
8. Cartan, E. Observations sur le memoire precedent. *Math. Z.* **1933**, *37*, 619–622. [CrossRef]
9. Chern, S.S. Sur la geometrie d'um systemme d'equation differentielles du second ordre. *Bull. Sci. Math.* **1939**, *63*, 206–212.
10. Antonelli, P.L.; Ingarden, R.S.; Matsumoto, M. *The Theories of Sprays and Finsler Spaces with Application in Physics and Biology*; Kluwer Academic Publishers: Dordrecht, The Netherlands, 1993.
11. Antonelli, P.L. Equivalence Problem for Systems of Second Order Ordinary Differential Equations. In *Encyclopedia of Mathematics*; Kluwer Academic Publishers: Dordrecht, The Netherlands, 2000.
12. Harko, T.; Ho, C.Y.; Leung, C.S.; Yip, S. Jacobi stability analysis of Lorenz system. *Int. J. Geom. Methods Mod. Phys.* **2015**, *12*, 1550081. [CrossRef]
13. Gupta, M.K.; Yadav, C.K. Jacobi stability analysis of modified Chua circuit system. *Int. J. Geom. Methods Mod. Phys.* **2017**, *14*, 121–142. [CrossRef]
14. Gupta, M.K.; Yadav, C.K. Jacobi stability analysis of Rikitake system. *Int. J. Geom. Methods Mod. Phys.* **2016**, *13*, 1650098. [CrossRef]
15. Gupta, M.K.; Yadav, C.K. KCC theory and its application in a tumor growth model. *Math. Methods Appl. Sci.* **2017**, *40*, 7470–7487. [CrossRef]
16. Gupta, M.K.; Yadav, C.K. Rabinovich-Fabrikant system in view point of KCC theory in Finsler geometry. *J. Interdiscip. Math.* **2019**, *22*, 219–241. [CrossRef]
17. Gupta, M.K.; Yadav, C.K.; Gupta, A.K. A geometrical study of Wang-Chen system in view of KCC theory. *TWMS J. Appl. Eng. Math.* **2020**, *10*, 1064.
18. Kumar, M.; Mishra, T.N.; Tiwari, B. Stability analysis of Navier–Stokes system. *Int. J. Geom. Methods Mod. Phys.* **2019**, *16*, 1950157. [CrossRef]
19. Munteanu, F. On the Jacobi Stability of Two SIR Epidemic Patterns with Demography. *Symmetry* **2023**, *15*, 1110. [CrossRef]
20. Munteanu, F.; Grin, A.; Musafirov, E.; Pranevich, A.; Şterbeţi, C. About the Jacobi Stability of a Generalized Hopf–Langford System through the Kosambi–Cartan–Chern Geometric Theory. *Symmetry* **2023**, *15*, 598. [CrossRef]
21. Yadav, C. K.; Gupta, M. K. Jacobi stability Analysis of Lu system, *J. Int. Acad. Phys. Sci.* **2019**, *23*, 123–142.
22. Liu, Y.; Huang, Q.; Wei, Z. Dynamics at infinity and Jacobi stability of trajectories for the Yang-Chen system. *Discrete Contin. Dyn. Syst. Ser.* **2021**, *26*, 3357–3380. [CrossRef]
23. Yamasaki, K.; Yajima, T. KCC analysis of a one-dimensional system during catastrophic shift of the Hill function: Douglas tensor in the nonequilibrium region. *Int. J. Bifurc. Chaos* **2020**, *30*, 2030032. [CrossRef]
24. Feng, C.; Qiujian, H.; Liu, Y. Jacobi analysis for an unusual 3D autonomous system. *Int. J. Geom. Methods Mod. Phys.* **2020**, *17*, 2050062. [CrossRef]
25. Liu, Y.; Chen, H.; Lu, X.; Feng, C.; Liu, A. Homoclinic orbits and Jacobi stability on the orbits of Maxwell–Bloch system. *Appl. Anal.* **2022**, *101*, 4377–4396. [CrossRef]
26. Chen, B.; Liu, Y.; Wei, Z.; Feng, C. New insights into a chaotic system with only a Lyapunov stable equilibrium. *Math. Methods Appl. Sci.* **2020**, *43*, 9262–9279. [CrossRef]
27. Yamasaki, K.; Yajima, T. Kosambi–Cartan–Chern Stability in the Intermediate Nonequilibrium Region of the Brusselator Model. *Int. J. Bifurc. Chaos* **2022**, *32*, 2250016. [CrossRef]
28. Yamasaki K.; Yajima T. Kosambi–Cartan–Chern Analysis of the Nonequilibrium Singular Point in One-Dimensional Elementary Catastrophe. *Int. J. Bifurc. Chaos* **2022**, *32*, 2250053. [CrossRef]

29. Boehmer, C.G.; Harko, T.; Sabau, S.V. Jacobi stability analysis of dynamical systems-applications in gravitation and cosmology. *Adv. Theor. Math. Phys.* **2012**, *16*, 1145–1196. [CrossRef]
30. Gupta, M.K.; Yadav, C.K. Jacobi stability analysis of Rossler system. *Int. J. Bifurc. Chaos* **2017**, *27*, 1750056. [CrossRef]
31. Rahman, Q.I.; Schmeisser, G. *Analytic Theory of Polynomials*; London Mathematical Society Monographs, New Series; Oxford University Press: Oxford, UK, 2002; Volume 26.

axioms

MDPI

Article

Differential Cohomology and Gerbes: An Introduction to Higher Differential Geometry

Byungdo Park

Department of Mathematics Education, Chungbuk National University, Cheongju 28644, Republic of Korea; byungdo@chungbuk.ac.kr

Abstract: Differential cohomology is a topic that has been attracting considerable interest. Many interesting applications in mathematics and physics have been known, including the description of WZW terms, string structures, the study of conformal immersions, and classifications of Ramond–Ramond fields, to list a few. Additionally, it is an interesting application of the theory of infinity categories. In this paper, we give an expository account of differential cohomology and the classification of higher line bundles (also known as S^1-banded gerbes) with a connection.We begin with how Čech cohomology is used to classify principal bundles and define their characteristic classes, introduce differential cohomology à la Cheeger and Simons, and introduce S^1-banded gerbes with a connection.

Keywords: Čech cohomology; non-abelian cohomology; characteristic classes; differential characters; differential cohomology; bundle gerbes; Deligne cohomology

MSC: primary 53C08; secondary 55R65; 14F03; 55N05

Citation: Park, B. Differential Cohomology and Gerbes: An Introduction to Higher Differential Geometry. *Axioms* **2024**, *13*, 60. https://doi.org/10.3390/axioms13010060

Academic Editors: Giovanni Calvaruso and Mića Stanković

Received: 26 December 2023
Revised: 11 January 2024
Accepted: 16 January 2024
Published: 19 January 2024

1. Introduction

Higher differential geometry is a study of differential geometry in the context of homotopy theory and higher category theory. It appears in many aspects of differential geometry, such as the theory of the higher analog of line bundles with a connection, considered as sheaves of ∞-groupoids, equivariant refinements, and the theory of orbifolds, and derived geometry (see, for example, [1–3]).

Differential cohomology and the theory of gerbes are topics that have been attracting interest. The idea of differential cohomology is that we can combine data from cohomology groups and differential forms in a homotopy theoretic way. The first construction of differential cohomology was due to Deligne [4] and Cheeger and Simons [5], and numerous applications have been found to date. A few examples include index theorems, the study of conformal immersions, topological quantum field theories, arithmetic Chow groups, and hyperbolic volumes. It is also an interesting application of the theory of infinity categories (see [5–7]). This theory, as generalized by Hopkins and Singer [8], explicitly constructs a differential cohomology theory for any generalized cohomology theory and brings in all of the objects in the category of spectra as topological data for differential extension.

S^1-banded n-gerbes are higher analogs of principal S^1-bundles. Just as line bundles represent, up to isomorphism, the degree-two integral cohomology group of the base space, one-gerbes (or simply gerbes) are geometric objects representing, up to isomorphism, the degree-three integral cohomology, and n-gerbes represent the degree $n+2$ integral cohomology. Endowed with a connection, we have the following pattern of classification:

$$\begin{aligned}
\widehat{H}^1(M) &\cong C^\infty(M, S^1)\\
\widehat{H}^2(M) &\cong \mathrm{Prin}^\nabla_{S^1}(M)/\cong\\
\widehat{H}^3(M) &\cong \mathrm{Grb}^\nabla(M)/\cong\\
\widehat{H}^4(M) &\cong 2\text{-}\mathrm{Grb}^\nabla(M)/\cong
\end{aligned}$$

and so on. It is thus clear that differential cohomology is the proper home for classifying gerbes with a connection and their higher-categorical generalizations.

Historically, gerbes were first conceived by Giraud [9] as sheaves of groupoids (cf. Grothendieck [10]) in the study of non-abelian cohomology. Perhaps the most popular model of gerbes in the literature would be bundle gerbes by Murray and Stevenson [11,12], which has an obvious advantage in that we can remain in the category of smooth manifold while handling it. There are numerous applications of gerbes with a connection, including the description of the Wess–Zumino–Witten terms, string structures, classifications of Ramond–Ramond fields, and topological insulators (see [13–15]). Of course, gerbes banded with other (possibly non-abelian) groups are of interest as well. We do not treat them in this paper, but interested readers should compare Schreiber and Waldorf [16].

The goal of this paper is to give a self-contained expository account of differential cohomology and gerbes and to guide readers to the literature at the forefront of this research. There are several well-written research papers and dissertations in this area from which one can learn about this topic. Nonetheless, there are not many monographs and expository articles for second- or third-year Ph.D. students trying to choose a topic for their dissertations. Perhaps Brylinski [17] is one of such a limited list of compilations. This paper pursues an exposition that is accessible to early-year Ph.D. students and takes the length of three standard 1 h talks. Indeed, this paper is based on the author's notes for a minicourse at the 13th Korea Institute for Advanced Study (KIAS) Winter School on Differential Geometry, intended to accommodate non-experts in the audience.

This paper is organized as follows. In Section 2, we give a gentle introduction to the characteristic classes of complex line bundles and U_1-gerbes. We begin with the principal G-bundles and Čech cohomology with coefficients in G, introduce relevant results from Dixmier and Douady [18], and then give various examples, each of which leads to the construction of a characteristic class. In Section 3, we introduce the differential cohomology group, as in Cheeger–Simons [5], and introduce a classification of complex line bundles with connection by the degree-two differential cohomology group. In Section 4, we introduce bundle gerbes and their Dixmier–Douady classes. After that, we explain what connections, curvings, and three-curvatures are. We then define the Deligne complex and introduce a classification of bundle gerbes with connection by the degree-three differential cohomology group. We also introduce the two-groupoid structure of the category of bundle gerbes with a connection.

2. Čech Cohomology and Characteristic Classes

In this section, we shall review the principal G-bundles and how Čech cohomology can be used to classify them and define their characteristic classes. A good reference for learning more about these topics is Brylinsky [17], which has a broader account.

Definition 1. *Let G be a Lie group. A **principal G-bundle** over a smooth manifold M is a smooth map $\pi\colon P \to M$ and a right G-action on P satisfying:*

(1) π is G-invariant; that is, $\pi(p \cdot g) = \pi(p)$ for all $p \in P$ and $g \in G$.

(2) On each fiber, G acts freely and transitively from the right.

(3) P is locally trivial via G-equivariant trivialization; that is, at every $m \in M$, there exists an open subset $U \subset M$ and a diffeomorphism $\varphi\colon \pi^{-1}(m) \to U \times G$ such that $p \mapsto (\pi(p), \phi(p))$ satisfying $p \cdot g \mapsto (\pi(p), \phi(p) \cdot g)$.

Conditions (1) and (2) mean that the G-orbits are fibers of π. This is equivalent to saying $P \times G \to P \times_M P$, $(p, g) \mapsto (p, p \cdot g)$ is a diffeomorphism; that is, P is a *G-torsor*.

Definition 2. *A **bundle map** of the principal G-bundles from $\pi_1\colon P_1 \to M$ to $\pi_2\colon P_2 \to M$ is a diffeomorphism $f\colon P_1 \to P_2$ that preserves the fiber and G-equivariant; that is, $f(p \cdot g) = f(p) \cdot g$ and $\pi_2 \circ f = \pi_1$.*

The principal G-bundles over M with maps form a groupoid (a category whose morphisms are invertible), and it is denoted by $\mathrm{Prin}_G(M)$. We will also use the notation $\mathrm{Bun}_{\mathbb{C}^n}(M)$ to denote the groupoid of rank n complex vector bundles over M.

Example 1. *Let $G = GL_n(\mathbb{C})$. Consider $\pi\colon P \to M$ and take an associated fiber bundle $E(P) \to M$ with a fiber $\mathbb{C}^n$ defined by $E(P) := (P \times \mathbb{C}^n)/G$ with a diagonal G-action: $(p, v) \mapsto (pg, g^{-1}v)$. The bundle $E(P)$ is a complex vector bundle over M of rank n. On the other hand, let $E \in Bun_{\mathbb{C}^n}(M)$. At each $x \in M$, consider the set $Fr(E)_x$ of all bases of the vector space E_x; equivalently the set of all $\mathbb{C}$-linear maps $p\colon \mathbb{C}^n \to E_x$. Then, the smooth map $\pi\colon Fr(E) \to M$ with $\pi^{-1}(x) = P(E)_x$ and a right G-action on $Fr(E)$ defined by $p \mapsto p \circ g$ is a principal G-bundle over M. It leads to the following equivalence of categories.*

$$Prin_{GL_n(\mathbb{C})}(M) \underset{Fr}{\overset{E}{\rightleftarrows}} Bun_{\mathbb{C}^n}(M)$$

For this reason, in what follows, we do not distinguish a $\mathbb{C}^\times$-, S^1-, or U_1-bundle from a complex line bundle.

Notation 1. *We shall use the notation $U_{i_1\cdots i_n}$ to denote the n-fold intersection $U_{i_1} \cap \cdots \cap U_{i_n}$.*

Definition 3. *Let G be an abelian group, M be a topological space, and $\mathcal{U} = \{U_i\}_{i\in\Lambda}$ be an open cover of M. The set $\check{C}^p(\mathcal{U}; G) = \{f_{i_0\cdots i_p}\colon U_{i_0\cdots i_p} \to G\}_{i_0,\cdots,i_p\in\Lambda}$ inherits the operation from the group G and is termed the degree-p* **Čech cochain group***. Together with the map $\delta_p : \check{C}^p(\mathcal{U}; G) \to \check{C}^{p+1}(\mathcal{U}; G)$, $(f)_{i_0\cdots i_p} \mapsto (\delta f)_{i_0\cdots i_{p+1}} := f_{\widehat{i_0}i_1\cdots i_{p+1}} - f_{i_0\widehat{i_1}\cdots i_{p+1}} + \cdots + (-1)^{p+1}_{i_0 i_1\cdots i_p\widehat{i_{p+1}}}$, the sequence of groups $(\check{C}^\bullet(\mathcal{U}; G), \delta_\bullet)$ is the* **Čech cochain complex***. (It is easy to verify that $\delta^2 = 0$. Here, the hat means an omission.) The cohomology of this complex $\check{H}^\bullet(\mathcal{U}; G) := ker(\delta_\bullet)/Im(\delta_{\bullet-1})$ is the Čech cohomology of M defined on an open cover $\mathcal{U}$.*

Now, if the group G in the definition above is not abelian, in general, the coboundary maps δ are not group homomorphisms, neither $\ker\delta$ nor $\mathrm{Im}\delta$ form a group, and, if we apply δ to a cocycle, we do not obtain $\delta^2 = 1$. We shall see below what goes on starting from the lowest degree:

- $p = 0$: There is no problem. $\check{H}^0(\mathcal{U}; G) = \{f \in \check{C}^0(\mathcal{U}; G) : \delta(f)_{ij} = 0\} = \mathrm{Map}(M, G)$. This is a group under a pointwise group multiplication.
- $p = 1$: Neither $\ker\delta_1$ nor $\mathrm{Im}\delta_0$ form a group. On the set $\ker\delta_1$, we may impose an equivalence relation defined by the action of 0-cochains

$$g_{ij} \sim g'_{ij} \quad \text{if and only if} \quad g'_{ij} = f_i^{-1} g_{ij} f_j.$$

  So, we may define $\check{H}^1(\mathcal{U}; G)$ as the pointed set $\ker\delta_1/\sim$ with a distinguished element of the constant map $g_{ij} \equiv 1$. Notice where set $\check{H}^1(\mathcal{U}; G)$ is precisely the set of isomorphism classes of the principal G-bundles over M defined on the open cover U (see Remark below). For this reason, principal G-bundles are geometric models of a degree-one non-abelian cohomology of M with coefficients in a group G.
- $p \geq 2$: There is no reasonable way to make sense of $\check{H}^p(\mathcal{U}; G)$.

Remark 1. *We shall closely look into how the set $\check{H}^1(M; G)$ classifies the principal G-bundle over M up to isomorphism. Recall that each principal G-bundle is locally trivial and diffeomorphic to $U \times G$ for some open $U \subset M$. This means that if we are given a family of* transition functions *on every double overlap $U_{ij} \in \mathcal{U} = \{U_{ij}\}_{i,j\in\Lambda}$, that is, $\{g_{ij}\colon U_{ij} \to G : i, j \in \Lambda\}$, we can rebuild the principal G-bundle. Since the transition functions satisfy*

$$g_{ij}(x) \cdot g_{jk}(x) \cdot g_{ki}(x) = 1, \text{ for all } x \in U_{ijk} \tag{1}$$

Equation (1) *is called the* cocycle condition *of a principal G-bundle. So, if we have a principal bundle P over M, we have a family of transition functions* $\{g_{ij}\}_{i,j\in\Lambda}$ *satisfying condition* (1) *and vice versa (under a mild condition). Likewise, if we have a bundle map* $f\colon P \to P'$ *covering M, we have a family of functions on open sets in the cover* $\{f_i\}_{i\in\Lambda}$ *satisfying that* $g'_{ij}(x) = f_j^{-1}(x)g_{ij}(x)$ *for all* $x \in U_{ij}$*, and vice versa (under the same mild condition). Here, the mild condition is that the open cover* $\mathcal{U}$ *has to be a good cover. A* ***good cover*** *(also known as Leray's covering) is an open cover of M if all open sets and their intersections are contractible. Such a covering always exists (see [1] (Proposition A.1) and references therein). An open cover* $(\mathcal{V}, \imath)$ *is a* ***refinement*** *of* $\mathcal{U}$ *if* $\imath\colon \mathcal{V} \to \mathcal{U}$ *such that* $V \subseteq \imath(V)$ *for all* $V \in \mathcal{V}$*. A refinement induces a map* $res_{\mathcal{V}\mathcal{U}}\colon \check{H}^1(\mathcal{U};G) \to \check{H}^1(\mathcal{V};G)$*, and it satisfies* $res_{\mathcal{W}\mathcal{U}} = res_{\mathcal{W},\mathcal{V}} \circ res_{\mathcal{V}\mathcal{U}}$*. So, we can define the set* $\check{H}^1(M;G)$ *as a direct limit over refinements of open cover; that is,*

$$\check{H}^1(M;G) = \varinjlim_{\mathcal{U}} \check{H}^1(\mathcal{U};G).$$

If the cover $\mathcal{U}$ *is good, the restriction map* $\check{H}^1(\mathcal{U};G) \xrightarrow{\cong} \check{H}^1(M;G)$ *is an isomorphism. Therefore, we conclude that*

$$\begin{aligned} \pi_0 Prin_G(\mathcal{U}) &\to \check{H}^1(\mathcal{U};G) \\ [P] &\mapsto (g_{ij}). \end{aligned} \tag{2}$$

If we remove the abelian assumption of groups, the long exact sequence induced by a short exact sequence of groups cannot go any further than the degree $p = 1$.

Proposition 1. *Let*

$$1 \longrightarrow K \xrightarrow{\ i\ } \widetilde{G} \xrightarrow{\ j\ } G \longrightarrow 1 \tag{3}$$

be a short exact sequence of groups. We have the following long exact sequence of groups and pointed sets

$$\begin{array}{l} 1 \longrightarrow \check{H}^0(\mathcal{U};K) \xrightarrow{i_*} \check{H}^0(\mathcal{U};\widetilde{G}) \xrightarrow{j_*} \check{H}^0(\mathcal{U};G) \\ \quad\longrightarrow \check{H}^1(\mathcal{U};K) \xrightarrow{i_*} \check{H}^1(\mathcal{U};\widetilde{G}) \xrightarrow{j_*} \check{H}^1(\mathcal{U};G) \end{array}$$

However, in the special case that the second term in the sequence is an abelian group whose image is in the center of the third, we can extend the long exact sequence just one term further. We have the following propositions.

Proposition 2. *If the group K in short exact sequence* (3) *is abelian and* $i(K)$ *belongs to the center of* $\widetilde{G}$*, then the long exact sequence in Proposition 1 extends to* $\check{H}^2(\mathcal{U};K)$*:*

$$\begin{array}{l} 1 \longrightarrow \check{H}^0(\mathcal{U};K) \xrightarrow{i_*} \check{H}^0(\mathcal{U};\widetilde{G}) \xrightarrow{j_*} \check{H}^0(\mathcal{U};G) \\ \quad\longrightarrow \check{H}^1(\mathcal{U};K) \xrightarrow{i_*} \check{H}^1(\mathcal{U};\widetilde{G}) \xrightarrow{j_*} \check{H}^1(\mathcal{U};G) \\ \quad\overset{\alpha}{\dashrightarrow} \check{H}^2(\mathcal{U};K) \end{array}$$

Proposition 3 (Dixmier–Douady [18])**.** *If the sheaf* $\underline{\widetilde{G}}_M$ *is soft, then*

$$\alpha : \check{H}^1(\mathcal{U};G) \to \check{H}^2(\mathcal{U};K)$$

is a bijection.

Proof. See Dixmier–Douady [18] (Lemma 22, p. 278) or Brylinski [17] (Proposition 4.1.8, p. 162). □

In the above, $\underline{G}_M$ is a sheaf such that $\underline{G}_M(U)$ is a group of smooth functions $f: U \to G$ for each open $U \subseteq M$. A sheaf $\underline{G}_M$ is **soft** if $\underline{G}_M(M) \to \underline{G}_M(C)$ is onto for every closed $C \subset M$. Here, we can think of $\underline{G}_M(C) = \lim_U \underline{G}_M(U)$ (since M is paracompact), where the direct limit is taken over all open neighborhoods of C.

Example 2. *(1) Consider a short exact sequence*

$$1 \longrightarrow SO_n \xrightarrow{i} O_n \xrightarrow{det} \mathbb{Z}_2 \longrightarrow 1.$$

The induced map $w_1 \colon \check{H}^1(M; O_n) \to \check{H}^1(M; \mathbb{Z}_2)$ is a correspondence $[P] \in \pi_0 Prin_{O_n}(M) \mapsto w_1([P])$, which is the first Stifel–Whitney class. So, $w_1([P]) = 0$ if and only if P comes from an SO_n-bundle; that is, P is orientable. Equivalently, the obstruction for the transition maps of a Euclidean vector bundle to lift to SO_n is given by the first Stifel–Whitney class.

(2) Consider a short exact sequence

$$1 \longrightarrow \mathbb{Z}_2 \longrightarrow Spin_n \longrightarrow SO_n \longrightarrow 1.$$

The induced map $w_2 \colon \check{H}^1(M; SO_n) \to \check{H}^2(M; \mathbb{Z}_2)$ is a correspondence $[P] \in \pi_0 Prin_{SO_n}(M) \mapsto w_2([P])$, which is the second Stifel–Whitney class. So, $w_2([P]) = 0$ if and only if P comes from a $Spin_n$-bundle. Equivalently, the obstruction for the transition maps of an oriented Euclidean vector bundle to lift to $Spin_n$ is given by the second Stifel–Whitney class. Here, one can think of $Spin_n$ as a double cover of SO_n, which is also a universal cover. For a construction of $Spin_n$ in terms of Clifford algebras, see [19] (Section 1.2).

Remark 2. *The Whitehead tower of O_n is of particular interest. The* ***Whitehead tower*** *of a space X is a factorization of the point inclusion $pt \to X$*

$$pt \simeq \lim_{n\to\infty} X_n \longrightarrow \cdots \longrightarrow X_2 \longrightarrow X_1 \longrightarrow X_0 \simeq X$$

such that each X_n is $(n-1)$-connected (that is, all homotopy groups π_k vanish for $k \le n-1$) and each map $X_n \to X_{n-1}$ is a fibration, which is an isomorphism on all π_k for $k \ge n$. For the space O_n, we have a Whitehead tower as follows:

$$pt \longrightarrow \cdots \longrightarrow FiveBrane_n \longrightarrow String_n \longrightarrow Spin_n \longrightarrow SO_n \longrightarrow O_n$$

Here, $String_n$ is a six-connected cover of $Spin_n$

$$1 \longrightarrow K(\mathbb{Z}, 2) \longrightarrow String_n \longrightarrow Spin_n \longrightarrow 1.$$

and $FiveBrane_n$ is a seven-connected cover of $String_n$

$$1 \longrightarrow K(\mathbb{Z}, 6) \longrightarrow FiveBrane_n \longrightarrow String_n \longrightarrow 1.$$

It is known that the obstruction to lift a $Spin_n$-bundle to a $String_n$-bundle is the first fractional Pontryagin class $\frac{1}{2}p_1$ *and to lift a $String_n$-bundle to a $FiveBrane_n$-bundle is the* second fractional Pontryagin class $\frac{1}{6}p_2$*, and so on (see [20] for more details).*

Example 3. *(3) Consider a short exact sequence*

$$1 \longrightarrow \mathbb{Z} \longrightarrow \mathbb{R} \longrightarrow S^1 \longrightarrow 1.$$

Note that $\underline{\mathbb{R}}_M$ is a soft sheaf (recall the Tietze extension theorem). The induced map $c_1 \colon \check{H}^1(M; S^1) \xrightarrow{\cong} \check{H}^2(M; \mathbb{Z})$ is a correspondence $[L] \in \pi_0 Prin_{S^1}(M) \mapsto c_1([L])$, which is the first Chern class. Note that if group G is abelian, $\mathcal{G}$ is a sheaf of locally constant functions in G, $\check{H}^p(M; \mathcal{G})$, and $H^p(M; G)$, *then the degree-p singular cohomology with coefficients in G is the same. Since group $\mathbb{Z}$ is discrete, we can identify $\check{H}^p(M;\mathbb{Z})$ and $H^p(M;\mathbb{Z})$ for any degree p.*

Proposition 4 (Dixmier–Douady [18]). *Let $\mathcal{H}$ be a complex separable Hilbert space. The sheaf $\underline{U(\mathcal{H})}_M$ is soft.*

Proof. See Dixmier–Douady [18] (Lemma 4, p. 252) or Brylinski [17] (Cor. 4.1.6, p. 162). □

Example 4. *(4) Consider a short exact sequence*

$$1 \longrightarrow U_1 \longrightarrow U(\mathcal{H}) \longrightarrow PU(\mathcal{H}) \longrightarrow 1.$$

Since $U(\mathcal{H})$ is a soft sheaf, the induced map $DD\colon \check{H}^1(M;PU(\mathcal{H})) \overset{\cong}{\to} \check{H}^2(M;S^1) \overset{\cong}{\to} H^3(M;\mathbb{Z})$ is a correspondence $[P] \in \pi_0 Prin_{PU(\mathcal{H})}(M) \mapsto DD([P])$, which is the Dixmier–Douady class of a gerbe.

Definition 4. *A **characteristic class** of a principal G-bundle P over M is an assignment*

$$\begin{aligned} c\colon \pi_0 Prin_G(M) &\to H^\bullet(M;A) \\ [P] &\mapsto c(P) \end{aligned}$$

that is natural; that is, $f^*c(P) = c(\overline{f}^*P)$ for

$$\begin{array}{ccc} P' & \xrightarrow{\overline{f}} & P \\ {\scriptstyle \pi'}\downarrow & \circlearrowleft & \downarrow{\scriptstyle \pi} \\ M' & \xrightarrow{f} & M \end{array}$$

Here, A is an abelian group.

Since $\mathrm{Prin}_G(-) : \mathbf{Man}^{\mathrm{op}} \to \mathbf{Sets}$ is representable by BG, by the Yoneda Lemma (see MacLane [21]), we have the following proposition.

Proposition 5. *An assignment*

$$\{\textit{Characteristic class of principal G-bundles}\} \longrightarrow H^\bullet(BG;A)$$

is one-to-one and onto.

Remark 3. *There is an alternative way to define characteristic classes using a "geometric datum", that is, a connection ∇ on $P \in Prin_G(M)$. This is the Chern–Weil theory. For example, given a line bundle with the connection (L,∇), the first Chern class of ∇ is defined by a Chern–Weil form $\frac{i}{2\pi}curv(\nabla)$. Here, $curv(\nabla)$ is the curvature two-form of the connection ∇. The Chern–Weil theorem shows that the cohomology class of a Chern–Weil form does not depend on the choice of connection. So, $\left[\frac{i}{2\pi}curv(\nabla)\right] \in H^2(M;\mathbb{R})$ is a topological invariant of a line bundle. A priori the class $\left[\frac{i}{2\pi}curv(\nabla)\right]$ is a class in $H^2(M;\mathbb{C})$, but it can be shown that it is actually a class in $H^2(M;\mathbb{R})$. The realification of the first Chern class, Example 3 above, is equal to the first Chern class $\left[\frac{i}{2\pi}curv(\nabla)\right]$ from the Chern–Weil theory. See Morita [22] (Chapter 5) to learn more about Chern–Weil theory of characteristic classes.*

We have seen that, up to isomorphism, complex line bundles are classified by $H^2(M;\mathbb{Z})$ via the first Chern class (Example 3) and principal $PU(\mathcal{H})$-bundles are classified by $H^3(M;\mathbb{Z})$ via the Dixmier–Douady class (Example 4). We can ask the following question: What classifies (higher) line bundles with a connection? For example, if we consider a groupoid $\text{Bun}_{\mathbb{C}}^{\nabla}(M)$ whose objects are line bundles with the connection (L, ∇) and whose morphisms are a bundle isomorphism preserving the connection, what classifies the isomorphism classes of $\text{Bun}_{\mathbb{C}}^{\nabla}(M)$? This question leads us to "differential cohomology". Up to isomorphism, line bundles with a connection are classified by the degree-two differential cohomology $\widehat{H}^2(M)$, gerbes with a connection are classified by $\widehat{H}^3(M)$, two-gerbes with a connection are classified by $\widehat{H}^4(M)$, and so on.

3. Cheeger–Simons Differential Characters

In this section, we introduce a differential extension of the singular cohomology theory $H^*(-;\mathbb{Z})$ on the site of smooth manifolds. Among various known models, we shall introduce the model by Cheeger and Simons [5] which is one of the historical landmarks. Interested readers are referred to the homotopy theoretic model by Hopkins and Singer [8], a spark complex model by Harvey, Lawson, and Zweck [23], and a novel construction using ∞-sheaves of spectra by Bunke, Nikolaus, and Völkl [24].

Notation 2. *We shall define some notations that will be used throughout this section. Let M be a smooth manifold and R be a commutative ring with unity:*

- $C^k(M;R)$: *smooth singular k-cochains in M with coefficients in R.*
- $Z^k(M;R)$: *smooth singular k-cocycles in M with coefficients in R.*
- $\Omega^k(M)$: *differential k-forms on M.*
- $\int : \Omega^k(M) \to C^k(M;\mathbb{R})$ *is a* $\mathbb{R}$*-linear map* $\omega \mapsto \int \omega$*, where* $\int \omega \colon C_k(M;\mathbb{R}) \to \mathbb{R}$ *is a pairing of a singular k-chain and a differential k-form.*
- $\Omega^k_{cl}(M)_{\mathbb{Z}}$: *closed differential k-forms with integral periods; that is,* $\omega \in \Omega^k_{cl}(M)_{\mathbb{Z}}$ *if and only if* $d\omega = 0$ *and* $\int \omega|_{Z_k(M)} \in \mathbb{Z}$.
- $\sim$ *is the natural map* $\mathbb{R} \to \mathbb{R}/\mathbb{Z}$.

A nonvanishing differential form does not take its values in a proper subring $\Lambda \subset \mathbb{R}$. Hence, we have the following:

Proposition 6. *The map*

$$\int : \Omega^k(M) \to C^k(M;\mathbb{R}/\mathbb{Z})$$
$$\omega \mapsto \widetilde{\int \omega}$$

is one-to-one.

Definition 5 (Cheeger and Simons [5])**.** *Let M be a smooth manifold. The group* $\widehat{H}^k(M)$ *of **differential characters** of degree k consists of pairs* (χ, ω)*, where* $\chi \in Hom_{\mathbb{Z}}(Z_{k-1}(M), \mathbb{R}/\mathbb{Z})$ *and* $\omega \in \Omega^k(M)$*, satisfying that*

$$\chi \circ \partial D = \int_D \omega \mod \mathbb{Z}, \text{ for all } D \in C_k(M;\mathbb{Z}),$$

where the group structure is the componentwise addition.

Remark 4. *The degree of the* $\widehat{H}^k(M)$ *in the above definition is different from the one that appears in Cheeger and Simons [5], which defines the same group as degree* $k+1$*. A consequence of adopting their convention would be a mismatch of degree in the group of differential characters and real cohomology, so the forgetful map (see below for a definition) would be* $I \colon \widehat{H}^k(M) \to H^{k+1}(M;\mathbb{R})$*. We stick to our convention for the sake of consistency with the literature from recent years.*

The main goal of this section is to understand the following diagram, known as the *differential cohomology hexagon diagram.*

Proposition 7 (Cheeger and Simons [5]). *The group of differential characters $\widehat{H}^k(M)$ satisfies the following diagram; that is, all squares and triangles are commutative and the diagonal, upper, and lower sequences of the arrows are exact sequences.*

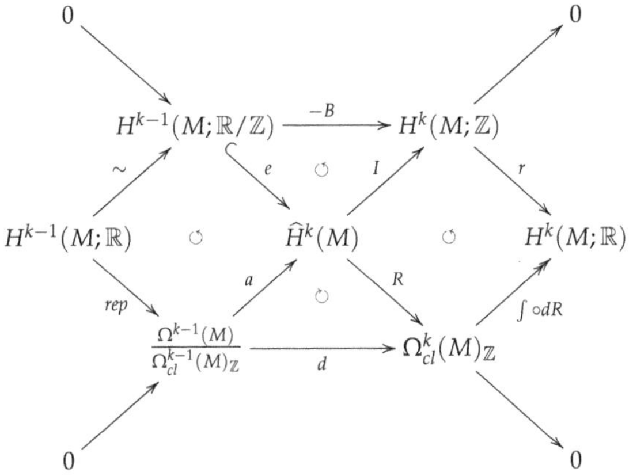

Proof. We shall divide the proof into several parts and enumerate them.

(1) I and R maps: We begin with some algebra facts:

A1. A subgroup of a free abelian group is free.

A2. An abelian group G is **divisible** if, for any $x \in G$ and any $n \in \mathbb{Z}^+$, there exists $y \in G$ such that $x = ny$.

A3. An abelian group G is divisible if and only if the group G is an injective object in the category of abelian groups; if $f\colon A \to G$ and $A \subset B$, there exists a map $\widetilde{f}\colon B \to G$ that satisfies $\widetilde{f}|_A = f$.

Take $(\chi, \omega) \in \widehat{H}^k(M)$ and consider $\chi\colon Z_{k-1}(M) \to \mathbb{R}/\mathbb{Z}$. Since $Z_{k-1}(M)$ is a subgroup of a free abelian group $C_{k-1}(M;\mathbb{Z})$, it is free (A1) and, hence, projective. We have the following commutative diagram:

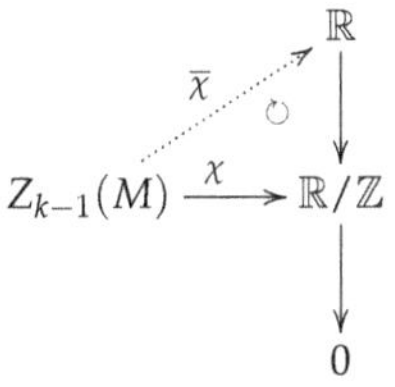

Now, since $\mathbb{R}$ is divisible (A2), it is injective (A3). Hence, $\overline{\chi} : Z_{k-1}(M) \to \mathbb{R}$ lifts to the map T satisfying the following commutative diagram:

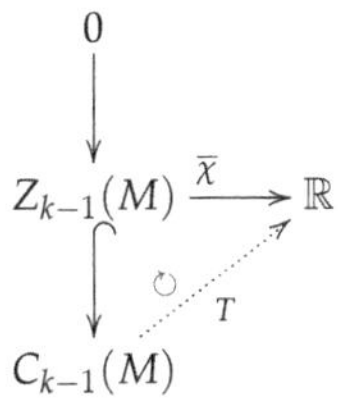

So, $\widetilde{T|_{Z_{k-1}(M)}} = \chi$. It follows that $\widetilde{\delta T} = \delta\widetilde{T} = \widetilde{T} \circ \partial = \int \omega \mod \mathbb{Z}$. Here, the first equality is simply $\sim \circ (T \circ \partial_k) = (\sim \circ T) \circ \partial_k$. Thus, there exists $c \in C^k(M;\mathbb{Z})$ such that

$$\delta T = \int \omega - c. \tag{4}$$

Note that $0 = \delta^2 = \int d\omega - \delta c$, so $\int d\omega = \delta c$. Since a real differential form cannot take its value in a proper subring of $\mathbb{R}$, this means $d\omega \equiv 0 = \delta c$. It is readily seen that ω has an integral period. We define the maps I and R as follows:

$$I : \widehat{H}^k(M) \to H^k(M;\mathbb{R}) \qquad R : \widehat{H}^k(M) \to \Omega^k_{\mathrm{cl}}(M)_{\mathbb{Z}}$$
$$(\chi, \omega) \mapsto [c] \qquad\qquad (\chi, \omega) \mapsto \omega$$

Let us verify that these maps are well defined. Since the choice of lifts is not unique, we have to verify that the above definition does not depend on the choices we made. Suppose T' is another lift satisfying $\delta T' = \int \omega' - c'$. Then, $T' - \widetilde{T|_{Z_{k-1}(M)}} = 0$, so $T' = T + \delta s + d$ for some $d \in C^{k-1}(M;\mathbb{Z})$ and $s \in C^{k-2}(M;\mathbb{R})$. So, $\delta T' = \delta T + 0 + \delta d$ if and only if $\int \omega' - c' = \int \omega - c + \delta d$ if and only if $\int (\omega' - \omega) = c' - c + \delta d$. Again, since the real differential form cannot take its value in a proper subring of $\mathbb{R}$, this means $\omega \equiv \omega'$ and $[c'] = [c]$.

We show that R is surjective. Let r: $H^k(M;\mathbb{Z}) \to H^k(M;\mathbb{R})$ be the realification map (which is from the universal coefficient theorem for cohomology; see [25] (Section 3.1)). Notice that, given $\omega \in \Omega^k_{\mathrm{cl}}(M)_{\mathbb{Z}}$, there exists a $u \in H^k(M;\mathbb{Z})$ such that $r(u) = [\int \omega]$. Since ω has integral periods, $\delta \int \omega = \int \omega \circ \partial \in \mathbb{Z}$ is an integral cochain, and, since ω is closed, $\delta \int \omega = \int d\omega = 0$ (Stokes' theorem). Now, let $u = [c]$ for some $c \in C^k(M;\mathbb{Z})$. Then, $\int \omega - c = \delta\lambda$ for some $\lambda \in C^{k+1}(M;\mathbb{R})$. Define $\chi := \widetilde{\lambda|_{Z_{k-1}(M)}}$. So, R is surjective.

The map I is also surjective. Given any $[c] \in H^k(M;\mathbb{Z})$, $\delta c = 0$ as real cochains. By the de Rham theorem, there exists a $\omega \in \Omega^k_{\mathrm{cl}}(M)$ such that $\int \omega - c = \delta\mu$ for some $\mu \in C^{k-1}(M;\mathbb{R})$. Define $\chi := \widetilde{\mu|_{Z_{k-1}(M)}}$. So, the map I is surjective.

(2) The e map: We define the e map as follows:

$$e : H^{k-1}(M;\mathbb{R}/\mathbb{Z}) \to \widehat{H}^k(M)$$
$$[x] \mapsto (x|_{Z_{k-1}(M)}, 0)$$

The map e is well defined. If we take a different representative $x + \delta y$, the restriction of δy to $Z_{k-1}(M)$ vanishes. The map e is one-to-one: Let $\Lambda \subset \mathbb{R}$ a proper subring. From the universal coefficient theorem, we have $H^k(X;\mathbb{R}/\Lambda) \cong \mathrm{Hom}_{\mathbb{Z}}(H_k(X), \mathbb{R}/\Lambda)$, since $\mathrm{Ext}(H_{n-1}(X), \mathbb{R}/\Lambda) = 0$, from $n(\mathbb{R}/\Lambda) = (n\mathbb{R})/\Lambda = \mathbb{R}/\Lambda$, for any $n \in \mathbb{Z}$. Since $B_k \to Z_k \to H_k \to 0$ is exact if and only if $B_k^* \leftarrow Z_k^* \leftarrow H_k^* \leftarrow 0$ is exact, $\mathrm{Hom}_{\mathbb{Z}}(H_k(X), \mathbb{R}/\Lambda) \hookrightarrow \mathrm{Hom}_{\mathbb{Z}}(Z_k(X), \mathbb{R}/\Lambda)$ is an injection.

(3) The a map: We define the a map as follows:

$$a : \frac{\Omega^{k-1}(M)}{\Omega^{k-1}_{\mathrm{cl}}(M)_{\mathbb{Z}}} \to \widehat{H}^k(M)$$
$$[\alpha] \mapsto (\widetilde{\int \alpha|_{Z_{k-1}(M)}}, d\alpha)$$

It is obvious that the a map is well defined, and the subgroup $\Omega^{k-1}_{\mathrm{cl}}(M)_{\mathbb{Z}}$ is the kernel of the map $\Omega^{k-1}(M) \to \widehat{H}^k(M)$, $\alpha \mapsto (\widetilde{\int \alpha|_{Z_{k-1}(M)}}, d\alpha)$.

(4) Diagonals are exact: First, $\mathrm{Im} e = \ker R$. The inclusion $\subseteq$ is clear. To see $\supseteq$, take (χ, ω) such that $\omega = 0$. Then, $\chi = \widetilde{T|_{Z_{k-1}(M)}}$ satisfying that $\delta T = c$, so T is a $\mathbb{R}/\mathbb{Z}$-valued cocycle, representing a class in $H^{k-1}(M;\mathbb{R}/\mathbb{Z})$, and $T|_{Z_{k-1}(M;\mathbb{R}/\mathbb{Z})} = \chi$.

Now, $\text{Im}\widetilde{a} = \ker I$. Again, the inclusion $\subseteq$ is clear. To see $\supseteq$, take (χ, ω) such that $\chi = \widetilde{T|_{Z_{k-1}(M)}}$ satisfying $\delta T = \int \omega - c$. By assumption, $c = \delta d$ for some $d \in C^{k-1}(M;\mathbb{Z})$. From $\int \omega = \delta(T + d)$, we have $\omega = d\alpha$ for some $\alpha \in \Omega^{k-1}(M)$, and $\int \alpha = T + d + \delta f$ for some $f \in C^{k-2}(M;\mathbb{R})$. Then, δf vanishes when we restrict it to $Z_{k-1}(M)$, and d also vanishes modulo $\mathbb{Z}$. Thus, the preimage of I is $(\widetilde{\int \alpha|_{Z_{k-1}(M)}}, d\alpha)$.

(5) Squares commute: The map rep is defined as follows.

$$\begin{aligned} \text{rep} : H^{k-1}(M;\mathbb{R}) &\to \frac{\Omega^{k-1}(M)}{\Omega^{k-1}_{\text{cl}}(M)_{\mathbb{Z}}} \\ [\beta] &\mapsto \beta + \Omega^{k-1}_{\text{cl}}(M)_{\mathbb{Z}} \end{aligned}$$

which does not depend on the choice of representatives since all exact forms are closed forms with integral periods. From this, it is clear that the square on the left is commutative. Notice that Equation (4) shows the commutativity of the square on the right.

(6) Triangles commute: Two triangle diagrams below commute.

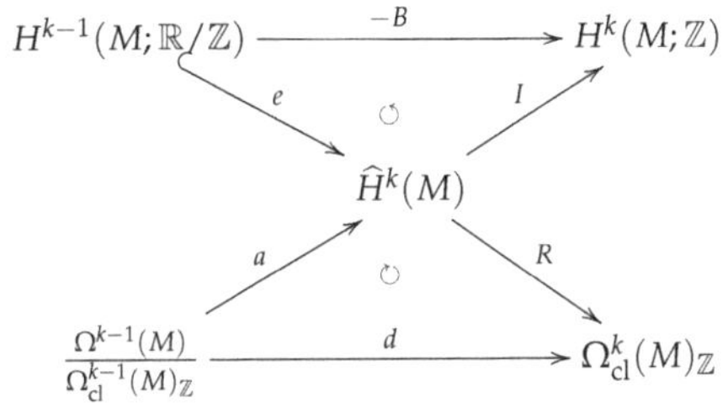

The commutativity of the lower triangle is obvious. Take a $\mathbb{R}/\mathbb{Z}$-valued cocycle x and consider $(x|_{Z_{k-1}(M)}, 0) \in \widehat{H}^k(M)$. There exists $T \in C^{k-1}(M;\mathbb{R})$ such that $x|_{Z_{k-1}(M)} = \widetilde{T|_{Z_{k-1}(M)}}$ satisfying $\delta T = -c$ for some $c \in C^k(M;\mathbb{Z})$, so $I(x|_{Z_{k-1}(M)}, 0) = c = -\delta T = -B([x])$.

(7) Upper and lower sequences are exact: It is readily seen that the following are exact sequences.

$$H^{k-1}(M;\mathbb{R}) \xrightarrow{\sim} H^{k-1}(M;\mathbb{R}/\mathbb{Z}) \xrightarrow{-B} H^k(M;\mathbb{Z}) \xrightarrow{r} H^k(M;\mathbb{R})$$

$$H^{k-1}(M;\mathbb{R}) \xrightarrow{\text{rep}} \frac{\Omega^{k-1}(M)}{\Omega^{k-1}_{\text{cl}}(M)_{\mathbb{Z}}} \xrightarrow{d} \Omega^k_{\text{cl}}(M)_{\mathbb{Z}} \xrightarrow{\int \circ \text{dR}} H^k(M;\mathbb{R})$$

□

Immediately from the definition, $\widehat{H}^0(M) = 0$ and $\widehat{H}^1(M) = C^\infty(M, \mathbb{R}/\mathbb{Z})$. Moreover, note that $\widehat{H}^k(M) = 0$ if $k > \dim(M)$. When $k = 2$, we have the following proposition:

Proposition 8 (Cheeger and Simons [5]). *The following assignment is a one-to-one correspondence:*

$$\begin{aligned} \pi_0 Prin_{S^1,\nabla}(M) &\longrightarrow \widehat{H}^2(M) \\ [(P,\theta)] &\mapsto (\chi, \frac{1}{2\pi} d\theta) \end{aligned}$$

where, for any loop γ in M, χ is defined by the holonomy of the loop γ; that is,

$$\chi(\gamma) := Hol(\gamma)$$

and, for any $D \in C_2(M;\mathbb{Z})$ *bounding* γ,

$$\chi(\partial D) = \frac{1}{2\pi}\int_D d\theta \mod \mathbb{Z}$$

which is extended to all $Z_1(M)$ *by setting* $\chi(x) = \chi(\gamma) + \frac{1}{2\pi}\int d\theta(y)$ *for any* $x = \gamma + \partial y$.

Given $d\theta \in \Omega^2_{\mathrm{cl}}(M)_{\mathbb{Z}}$, as we have seen in the surjectivity of R, there exists $[c] \in H^2(M;\mathbb{Z})$ such that $[\int d\theta] = r([c])$. The class $[c]$ is the characteristic class that classifies P; that is, the first Chern class.

The above proposition addresses the question at the end of Section 2 at least for degree two. What is a higher analog of Proposition 8? How can one define a map? In the following section, we shall see that the isomorphism classes of gerbes with a connection are in one-to-one correspondence with $\widehat{H}^3(M)$, and, to establish the correspondence, one has to construct χ; that is, a holonomy of gerbe.

Remark 5. *Although we do not go into detail, the differential cohomology group* $\widehat{H}^\bullet(M)$ *has a ring structure (see Cheeger and Simons [5] (p. 56, Theorem 1.11)).*

In differential cohomology, the hexagon diagram plays an important role. One uses the hexagon diagram in Proposition 7 to compute differential cohomology groups. Furthermore, it is known that the hexagon diagram uniquely characterizes the differential cohomology. Phrased slightly differently, if there are two $\widehat{H}^k(M)$ *fitting into the middle of the hexagon diagram, then they are naturally isomorphic. This is a theorem of Simons and Sullivan [26] that has been generalized by Bunke and Schick [27] and Stimpson [28] to the uniqueness of the differential extension of all exotic cohomology theories under some mild assumptions.*

4. S^1-Banded Gerbes with Connection

Throughout this section, M is a smooth manifold. In Section 2, we have seen that elements of $H^2(M;\mathbb{Z})$ are represented by complex line bundles, and, in Section 3, differential cohomology classes in $\widehat{H}^2(M;\mathbb{Z})$ are represented by complex line bundles with a connection. What are the corresponding geometric objects representing $H^n(M;\mathbb{Z})$ and $\widehat{H}^n(M;\mathbb{Z})$? The answer is $(n-2)$-gerbes with a connection.

Remark 6. *For a generalized cohomology theory* $E^\bullet$ *and its differential extension* $\widehat{E}^\bullet$, *investigating geometric cocycles representing (differential) cohomology classes is a very interesting research topic that is not fully understood yet. For example, elements of even complex K-group* $K^0(M)$ *are represented by vector bundles over* M *and odd complex K-group* $K^{-1}(M)$ *by* Ω*-vector bundles, but in other interesting generalized cohomology theories, such as elliptic cohomology and topological modular forms, it is largely unknown which geometric objects in the space* M *represent cohomology classes. Moreover, note that this question is closely related to the Stolz–Teichner program [29] wherein they have conjectured a hypothetical equivalence between the totality of supersymmetric field theories of degree* n *over* M *modulo concordance and the group* $E^n(M)$. *There are differential and twisted refinements of this conjecture as well (see, for example, Stoffel [30,31] and references therein).*

Let us observe how a gerbe arises. Consider the short exact sequence of groups

$$1 \longrightarrow U_1 \longrightarrow \widetilde{G} \longrightarrow G \longrightarrow 1.$$

In Example 4 above, we considered the map $DD\colon \check{H}^1(\mathcal{U};G) \to \check{H}^2(\mathcal{U};U_1)$ when $G = PU(\mathcal{H})$. In the proposition below, we shall closely look at how this map is defined.

Proposition 9. *A principal* G*-bundle* P *over* M *lifts to a principal* $\widetilde{G}$*-bundle if and only if the cocycle representing* $DD(P)$ *is trivializable.*

Proof. We look at how the map $DD\colon \check{H}^1(\mathcal{U};G) \to \check{H}^2(\mathcal{U};U_1)$ is defined. Choose a good cover $\mathcal{U}$ on M. Over each U_{ij}, consider the transition map $g_{ij}\colon U_{ij} \to G$ of P. Since $\mathcal{U}$ is a good cover, U_{ij} is contractible. Hence, there is a homotopy between the map g_{ij} and a constant map, which lifts by the homotopy-lifting property, since the map $\widetilde{G} \xrightarrow{r} G$ is a fibration. Let $\widetilde{g} : U_{ij} \to \widetilde{G}$ be a lift of g_{ij}. The cocycle condition $g_{ij}g_{jk}g_{ki} = \lambda_{ijk} \cdot \mathbf{1}_{\widetilde{G}}$, for some $\lambda_{ijk} \in \check{C}^2(\mathcal{U};U_1)$. It is an easy exercise to verify that $\lambda = \{\lambda_{ijk}\}_\Lambda$ is a degree-two Čech cocycle on $\mathcal{U}$ and the class $[\lambda]$ does not depend on the choice of lifting $\widetilde{g}_{ij}$. So, the map DD is a correspondence $[P] \mapsto [\lambda]$, and, using the isomorphism (due to the softness of $\underline{\mathbb{R}}$), it is valued in $H^3(M;\mathbb{Z})$. $\square$

There are several models representing gerbes. The degree-two U_1-valued Čech cocycle λ considered above as an obstruction to lifting a principal G-bundle to a $\widetilde{G}$-bundle is one model, and there are other ways to represent it as a stack. We refer the reader to Giraud [9], Brylinski [17], Behrend and Xu [32], and Moerdijk [33]. In this section, we will specialize in a model called a bundle gerbe by Murray [11], which is presumably the most widely used model in the literature.

Let $\pi\colon Y \to M$ be a surjective submersion. The p-fold *fiber product* of $\pi\colon Y \to M$ is

$$Y^{[p]} := \{(y_1, \cdots, y_n) \in Y^p : \pi(y_1) = \cdots = \pi(y_p) \text{ for } y_i \in Y\}.$$

The projection of $Y^{[p]}$ onto the $(i_1, \cdots, i_k)^{\text{th}}$ copy of $Y^{[k]}$ is $\pi_{i_1 \cdots i_k}\colon Y^{[p]} \to Y^{[k]}$. For example, let $\mathcal{U} = \{U_i\}_{i \in \Lambda}$ be an open cover of M. Then consider

$$Y_{\mathcal{U}} := \{(x,i) \in M \times \Lambda : x \in U_i\} \subset M \times \Lambda.$$

The map $\pi\colon Y_{\mathcal{U}} \to M$ is a surjective submersion, which is an open cover.

Remark 7. *Recall that a fiber product $X \times_M Y$ of $X \xrightarrow{\phi} M \xleftarrow{\pi} Y$ is, in general, not a smooth manifold. If ϕ, π are submersions, then the fiber product is a smooth manifold. So, a surjective submersion is not only a generalization of an open cover; it also lets us stay within the category of smooth manifolds.*

Definition 6 (Murray [11]). *A **bundle gerbe** is a triple $\mathcal{L} = (L, \pi, \mu)$ where:*

(1) $\pi\colon Y \to M$ *is a surjective submersion.*
(2) $L \in Prin_{S^1}(Y^{[2]})$.
(3) $\mu\colon \pi_{12}^*L \otimes \pi_{23}^*L \to \pi_{13}^*L$ *is an S^1-bundle isomorphism.*
(4) μ *is associative over $Y^{[4]}$: that is,*

$$\begin{array}{ccc}
\pi_{12}^*L \otimes \pi_{23}^*L \otimes \pi_{34}^*L & \xrightarrow{\pi_{123}^*\mu \otimes \mathbf{1}} & \pi_{13}^*L \otimes \pi_{34}^*L \\
{\scriptstyle \mathbf{1}\otimes\pi_{234}^*\mu}\downarrow & \circlearrowleft & \downarrow{\scriptstyle \pi_{134}^*\mu} \\
\pi_{12}^*L \otimes \pi_{24}^*L & \xrightarrow{\pi_{124}^*\mu} & \pi_{14}^*L
\end{array}$$

Let us construct the *Dixmier–Douady class*, the characteristic class of a bundle gerbe. Let $\mathcal{L} = (L, \pi, \mu)$ be a bundle gerbe over M. Take a good open cover (cf. Remark 1) $\mathcal{U} = \{U_i\}_{i \in \Lambda}$ of M. Then, local sections on each open set $\sigma_i\colon U_i \to Y$ and on each double intersection $(\sigma_i, \sigma_j)\colon U_{ij} \to Y^{[2]}$ can be defined. We consider the pullback of $L \to Y^{[2]}$ along $(\sigma_i, \sigma_j)\colon U_{ij} \to Y^{[2]}$.

$$\begin{array}{ccc}
(\sigma_i, \sigma_j)^*L & \longrightarrow & L \\
{\scriptstyle s_{ij}}\uparrow\downarrow & \overset{s_{ij}}{\nearrow} & \downarrow \\
U_{ij} & \xrightarrow{(\sigma_i, \sigma_j)} & Y^{[2]}
\end{array}$$

Take a section $s_{ij}\colon U_{ij} \to (\sigma_i, \sigma_j)^*L$, or, equivalently, a map $\mu\colon s_{ij}\colon U_{ij} \to L$. Over triple intersections, we have

$$s_{ij}(x) \otimes s_{jk}(x) \mapsto \lambda_{ijk}(x)s_{ik}(x), \quad x \in U_{ijk}.$$

Here, the associativity of μ implies that λ_{ijk} is a degree-two Čech cocycle in M defined on $\mathcal{U}$.

Definition 7. *Let $\mathcal{L} = (L, \pi, \mu)$ be a bundle gerbe over M. The* ***Dixmier–Douady class*** *$DD(\mathcal{L})$ is the cohomology class $[\lambda] \in \check{H}^2(\mathcal{U}; U_1)$.*

It is not difficult to verify that $DD(\mathcal{L})$ does not depend on the choices we have made.

Let us recall connections and curvatures on a principal G-bundle. A **connection** θ on a principal G-bundle $\pi\colon P \to M$ is a differential one-form on P valued in $\mathfrak{g}$ satisfying that:

(1) $\theta(X^*) = X$ where $X \in \mathfrak{g}$ and $X^*_x := \frac{d}{dt}|_{t=0}x \cdot e^{tX}$ for each $x \in P$.
(2) $R^*_g\theta = \mathrm{Ad}_{g^{-1}} \circ \theta$.

The **curvature** of $(P.\theta)$ is a $\mathfrak{g}$-valued two-form $\mathrm{Curv}(\theta) := d\theta + \frac{1}{2}[\theta, \theta]$ on P.

Now, we define the connection and curving of a bundle gerbe.

Definition 8. *A* ***connection*** *on $\mathcal{L} = (L, \pi, \mu)$ is a connection ∇ on L compatible with μ; that is, $\pi^*_{12}(L, \nabla) \otimes \pi^*_{23}(L, \nabla) \xrightarrow{\mu} \pi^*_{13}(L, \nabla)$ is a connection preserving isomorphism.*

So, a connection on $\mathcal{L}$ has to be an $\mathbb{R}$-valued differential one-form on $Y^{[2]}$.

Definition 9. *A* ***curving*** *B of a bundle gerbe with a connection (L, π, μ, ∇) is a differential two-form on Y satisfying $Curv(\nabla) = \pi_2 B - \pi_1 B$.*

A connection and a curving on a bundle gerbe are called the *connective structure*. By a *bundle gerbe with a connection*, we mean a bundle gerbe with a connective structure.

To work with curvatures and curvings, we need the following proposition.

Proposition 10 (Murray [11])**.** *Let $\pi\colon Y \to M$ be a surjective submersion. The following sequence is a long exact sequence*

$$0 \longrightarrow \Omega^k(M) \xrightarrow{\pi^*} \Omega^k(Y) \xrightarrow{\delta} \Omega^k(Y^{[2]}) \xrightarrow{\delta} \cdots$$

where $\delta = \sum_{k=1}^{p}(-1)^{k-1}\pi^*_{i_1\cdots\hat{i}_k\cdots i_p}$

Proof. See Murray [11] (Section 8). Compare Bott and Tu [34] (Proposition 8.5). □

Note that $0 = d\mathrm{Curv}(\nabla) = d\delta B = \delta dB$ so there exists a unique $H \in \Omega^3(M; \mathbb{R})$ such that $\pi^* H = dB$. The differential form H is closed, so it represents a degree-three real cohomology class in M. Proposition 10 shows that the cohomology class of M does not depend on the choices involved.

Definition 10. *Let $\widehat{\mathcal{L}} = (L, \pi, \mu, \nabla, B)$ be a bundle gerbe with connection. The* ***three-curvature*** *(also known as the three-form flux or the Dixmier–Douady form) of $\widehat{\mathcal{L}}$ is a real differential three-form on M satisfying that $\pi^* H = dB$.*

Remark 8. *In the literature, H is defined as a real-valued differential form in some places and $i\mathbb{R}$-valued differential form in some other places. Recall that, in Definitions 8 and 9, connection forms and curving forms are $\mathbb{R}$-valued, as the Lie algebra of the Lie group S^1 is $\mathbb{R}$. If we consider the Lie group U_1, its Lie algebra is $i\mathbb{R}$ (here $i = \sqrt{-1}$), and we consider differential forms valued in $i\mathbb{R}$.*

It turns out the three-curvature of a gerbe represents the corresponding de Rham cohomology class of the Dixmier–Douady class above.

Proposition 11 (Murray [11]). *Let $\widehat{\mathcal{L}} = (L, \pi, \mu, \nabla, B)$ be a bundle gerbe with a connection. The de Rham cohomology class of its three-curvature form H is equal to the realification of its Dixmier–Douady class $DD(\mathcal{L})$; that is, $r(DD(\mathcal{L})) = [H]_{dR}$, where r is the realification map $r\colon H^3(M;\mathbb{Z}) \to H^3(M;\mathbb{R})$ considered in the proof of Proposition 7.*

Proof. See Murray [11] (Section 11). □

Example 5. *Consider the short exact sequence of groups*

$$1 \longrightarrow U_1 \longrightarrow \widetilde{G} \longrightarrow G \longrightarrow 1.$$

Let $\pi\colon Y \to M$ be a principal G-bundle. There is a natural map $g\colon Y^{[2]} \to \widetilde{G}$ coming from the transitivity of the right G-action. Pull back the fibration $\widetilde{G} \to G$ to obtain a U_1-bundle L over $Y^{[2]}$. Note that the fiber of $(y_1, y_2) \in Y^{[2]}$ is the coset $U_1 g(y_1, y_2)$ in $\widetilde{G}$. So, the multiplication map $\mu\colon \pi_{12}^* L \otimes \pi_{23}^* L \to \pi_{13}^* L$ is defined by the coset multiplication $U_1 g(y_1, y_2) \cdot U_1 g(y_2, y_3) = U_1 g(y_1, y_3)$ and is readily seen to be associative. So, $\mathcal{L} = (L, \pi, \mu)$ is a bundle gerbe over M called the lifting bundle gerbe *of the principal G-bundle $\pi\colon Y \to M$. The Dixmier–Douady class $DD(\mathcal{L})$ is precisely the obstruction for the lifting of the G-valued cocycle to $\widetilde{G}$ considered in Proposition 9.*

Definition 11. *Let $\mathcal{U} = \{U_i\}_{i\in\Lambda}$ be a good cover of M. The* **Deligne complex** *is the double complex $\check{C}^\bullet(\mathcal{U};\Omega^\bullet)$ endowed with total differential $D = d + (-1)^q\delta$ on $\check{C}^p(\mathcal{U};\Omega^q)$ where the Čech differential is δ and the exterior derivative is d; that is,*

$$\begin{array}{ccccccc}
\vdots & & \vdots & & \vdots & & \\
\uparrow{\scriptstyle\delta} & & \uparrow{\scriptstyle -\delta} & & \uparrow{\scriptstyle\delta} & & \\
\check{C}^2(\mathcal{U};\underline{U_1}) & \xrightarrow{d\log} & \check{C}^2(\mathcal{U};\Omega^1) & \xrightarrow{d} & \check{C}^2(\mathcal{U};\Omega^2) & \xrightarrow{d} & \cdots \\
\uparrow{\scriptstyle\delta} & & \uparrow{\scriptstyle -\delta} & & \uparrow{\scriptstyle\delta} & & \\
\check{C}^1(\mathcal{U};\underline{U_1}) & \xrightarrow{d\log} & \check{C}^1(\mathcal{U};\Omega^1) & \xrightarrow{d} & \check{C}^1(\mathcal{U};\Omega^2) & \xrightarrow{d} & \cdots \\
\uparrow{\scriptstyle\delta} & & \uparrow{\scriptstyle -\delta} & & \uparrow{\scriptstyle\delta} & & \\
\check{C}^0(\mathcal{U};\underline{U_1}) & \xrightarrow{d\log} & \check{C}^0(\mathcal{U};\Omega^1) & \xrightarrow{d} & \check{C}^0(\mathcal{U};\Omega^2) & \xrightarrow{d} & \cdots
\end{array}$$

The cohomology of the total complex with the total degree n is the degree n **Deligne cohomology group** *$\check{H}^n_D(\mathcal{U})$ of M defined on $\mathcal{U}$.*

Proposition 12 (Murray [11]). *A bundle gerbe with connection $\widehat{\mathcal{L}} = (L, \pi, \mu, \nabla, B)$ determines a total degree 2 cocycle in the Deligne complex.*

Proof. Recall notations in the paragraph between Definitions 6 and 7. In it, we have obtained a Čech 2-cocycle $\{\lambda_{ijk}\}$. Let us take $A_{ij} = \sigma_{ij}^*\nabla$ and $B_i = \sigma_i^* B$. It is readily seen that the triple $\widehat{\lambda} := (\lambda_{ijk}, A_{ij}, B_i)$ satisfies $D\widehat{\lambda} = 0$ and its cohomology class $[\widehat{\lambda}]_D \in H^2_D(M)$ is independent of the choice of local sections σ_i. □

It is natural to ask if the isomorphic bundle gerbes with connection have Deligne-cohomologous cocycles in the Deligne complex. The answer is yes, but there is a subtlety in isomorphisms of bundle gerbes. One might guess that it is a U_1-bundle isomorphism compatible with the bundle gerbe structure μ, but this is not a notion we want. We will then get non-isomorphic bundle gerbes having the same Dixmier–Douady class. Stevenson [35] and Murray and Stevenson [12] have found that the correct notion of bundle gerbe isomorphism is the "stable isomorphism". We will introduce a version that Waldorf [36] came up with.

Definition 12 (Waldorf [36])**.** *For $\widehat{\mathcal{L}}_i = (L_i, \pi_i, \mu_i, \nabla_i, B_i)$, an **isomorphism** $\widehat{\mathcal{L}}_1 \xrightarrow{\widehat{\mathcal{K}}} \widehat{\mathcal{L}}_2$ is a quadruple $(\zeta, K, \nabla_K, \alpha)$ consists of the following.*

(1) A surjective submersion $\zeta\colon Z \to Y_1 \times_M Y_2$

(2) $(K, \nabla_K) \in Prin^{\nabla}_{S^1}(Z)$ such that $Curv(\nabla_K) = \zeta^*(B_2 - B_1) \in \Omega^2(Z)$.

(3) An isomorphism $\alpha\colon (L_1, \nabla_1) \otimes \zeta_2^*(K, \nabla_K) \to \zeta_1^*(K, \nabla_K) \otimes (L_2, \nabla_2)$ of S^1-bundles with connection over $Z \times_M Z$ compatible with μ_1 and μ_2.

Remark 9. *When $\zeta = \mathbf{1}$, we recover the* stable isomorphism *of Murray and Stevenson [12].*

Proposition 13 (Waldorf [36])**.** *There is an equivalence of groupoids between the* 1*-groupoid of* 1*-morphisms of $Grb(M)$ and the* 1*-groupoid of stable isomorphisms of $Grb_{st}(M)$.*

Definition 13 (Waldorf [36])**.** *A transformation $\widehat{\mathcal{J}}\colon \widehat{\mathcal{K}}_1 \Rightarrow \widehat{\mathcal{K}}_2$, which is an isomorphism between isomorphisms from $\widehat{\mathcal{L}}_1$ to $\widehat{\mathcal{L}}_2$ (that is, a **two-morphism**), is an equivalence class of triples (W, ω, β_W) consisting of the following:*

(1) A surjective submersion $\omega : W \to Z_1 \times_{Y_1 \times_M Y_2} Z_2$.

(2) An isomorphism $\beta_W : (K_1, \nabla_1) \to (K_2, \nabla_2)$ over W compatible with α_1 and α_2.

$$\begin{array}{ccc} L_1 \otimes \omega_2^* K_1 & \xrightarrow{\alpha_1} & \omega_1^* K_1 \otimes L_2 \\ \downarrow {\scriptstyle \mathbf{1} \otimes \omega_2^* \beta_W} & & \downarrow {\scriptstyle \omega_1^* \beta_W \otimes \mathbf{1}} \\ L_1 \otimes \omega_2^* K_2 & \xrightarrow{\alpha_2} & \omega_1^* K_2 \otimes L_2 \end{array}$$

$(W, \omega, \beta_W) \sim (W', \omega', \beta_{W'})$ if there is a smooth manifold X with surjective submersions to W and W' such that the following diagram commutes

$$\begin{array}{ccc} X & \longrightarrow & W \\ \downarrow & & \downarrow {\scriptstyle \omega} \\ W' & \xrightarrow{\omega'} & Z_1 \times_{Y_1 \times_M Y_2} Z_2 \end{array}$$

and β_W and $\beta_{W'}$ coincides if pulled back to X.

Proposition 14 (Stevenson [35])**.** *The category $Grb_{\nabla}(M)$ consisting of bundle gerbes with the connection $\widehat{\mathcal{L}}$ as objects, morphisms as defined in Definition 12, and two-morphisms as defined in Definition 13 is a two-groupoid (that is, a category whose morphisms are invertible and whose morphism between morphisms are invertible).*

Now, we go back to our discussion on Deligne cohomology. Since the cover $\mathcal{U}$ of M is good, we can define the Deligne cohomology group $H^k_D(M)$ as a direct limit over refinements, which is isomorphic to the one defined on $\mathcal{U}$. We have the following result.

Proposition 15 (Murray and Stevenson [12])**.** *Let $\widehat{\mathcal{L}}_i \in Grb_{\nabla}(M)$. $\widehat{\mathcal{L}}_1$ and $\widehat{\mathcal{L}}_2$ are stably isomorphic if and only if they define the same Deligne cohomology class in $\check{H}^2_D(M)$.*

Proof. See Murray and Stevenson [12] (Theorem 4.1). □

Proposition 16 (Esnault [37]). *Let M be a smooth manifold. The following correspondence is an isomorphism:*

$$H^k_D(M) \to \widehat{H}^{k+1}(M)$$

Proof. See Brylinski [17] (Proposition 1.5.7) and references therein. □

Corollary 1. *Let M be a smooth manifold. The following are isomorphic as groups*

$$\pi_0 Grb_\nabla(M) \cong \widehat{H}^3(M).$$

5. Discussion

In this article, we have given an overview of differential cohomology and gerbes. We began with an introduction to characteristic classes and the classification of integral cohomology groups using geometric objects. We then saw differential cohomology and the classification of complex line bundles with connection. Finally, we have seen what a gerbe is and its two-groupoid structure, as well as how gerbes and their higher analogs correspond to cocycles in the Deligne complex.

There are numerous future directions for research based on what we have considered in this paper. We will give three possible directions. First, the G-equivariant differential cohomology has been considered by Redden [2] and Kübel and Thom [38] when the Lie group G is compact. Applications of these constructions have to be developed. Additionally, Redden and the author [39] have established that isomorphism classes of G-equivariant bundle gerbes with a connection are naturally isomorphic to the degree-three differential cohomology of the differential quotient stack. One can expect to establish analogous results for higher gerbes. As a different route, there is an interesting relationship between the arithmetic Chow group of a complex projective variety and its differential cohomology group [6]. One has to cast a light on this result to generalize it as a result over the Deligne–Mumford stacks. Finally, along the vein of the work of Freed and Moore [40] and Gawędzki [15], the theory of differential cohomology and gerbes should be further developed to investigate the topology of matters.

Funding: This research received no funding.

Data Availability Statement: Data are contained within the article.

Acknowledgments: The author thanks Sajjad Lakzian, Insong Choe, and Jaigyoung Choe for giving the motivation to write this paper and its earlier drafts. He also thanks the anonymous referees for helpful comments that improved the readability of the paper.

Conflicts of Interest: The author declares no conflict of interest.

References

1. Fiorenza, D.; Schreiber, U.; Stasheff, J. Čech cocycles for differential characteristic classes: An ∞-Lie theoretic construction. *Adv. Theor. Math. Phys.* **2012**, *16*, 149–250.
2. Redden, C. Differential Borel equivariant cohomology via connections. *N. Y. J. Math.* **2017**, *23*, 441–487.
3. Behrend, K.; Liao, H.-Y.; Xu, P. Derived Differentiable Manifolds. *arXiv* **2021**, arXiv:2006.01376.
4. Deligne, P. Théorie de Hodge. II. *Inst. Hautes Études Sci. Publ. Math.* **1971**, *40*, 5–57.
5. Cheeger, J.; Simons, J. *Differential Characters and Geometric Invariants*; Springer: Berlin/Heidelberg, Germany, 1985; pp. 50–80. [CrossRef]
6. Gillet, H.; Soulé, C. *Arithmetic Chow Groups and Differential Characters*; Springer: Berlin/Heidelberg, Germany, 1989; pp. 29–68.
7. Neumann, W.D. *Realizing Arithmetic Invariants of Hyperbolic 3-Manifolds*; AMS: Providence, RI, USA, 2011; pp. 233–246. [CrossRef]
8. Hopkins, M.J.; Singer, I.M. Quadratic functions in geometry, topology, and M-theory. *J. Differ. Geom.* **2005**, *70*, 329–452.
9. Giraud, J. *Cohomologie non Abélienne*; Die Grundlehren der mathematischen Wissenschaften; Springer: Berlin/Heidelberg, Germany, 1971; Volume Band 179, pp. ix+467.
10. Grothendieck, A.; Raynaud, M. *Revêtements Étales et Groupe Fondamental*; Lecture Notes in Mathematics; Springer: Berlin/Heidelberg, Germany, 1971; Volume 224, pp. xxii+447.
11. Murray, M.K. Bundle gerbes. *J. Lond. Math. Soc.* **1996**, *54*, 403–416. [CrossRef]
12. Murray, M.K.; Stevenson, D. Bundle gerbes: Stable isomorphism and local theory. *J. Lond. Math. Soc.* **2000**, *62*, 925–937. [CrossRef]

13. Berwick-Evans, D.; Pavlov, D. Smooth one-dimensional topological field theories are vector bundles with connection. *Algebr. Geom. Topol.* **2023**, *23*, 3707–3743. [CrossRef]
14. Gawędzki, K.; Reis, N. WZW branes and gerbes. *Rev. Math. Phys.* **2002**, *14*, 1281–1334. [CrossRef]
15. Gawędzki, K. Square root of gerbe holonomy and invariants of time-reversal-symmetric topological insulators. *J. Geom. Phys.* **2017**, *120*, 169–191. [CrossRef]
16. Schreiber, U.; Waldorf, K. Connections on non-abelian gerbes and their holonomy. *Theory Appl. Categ.* **2013**, *28*, 476–540.
17. Brylinski, J.-L. *Loop Spaces, Characteristic Classes and Geometric Quantization*; Modern Birkhäuser Classics; Birkhäuser Boston, Inc.: Boston, MA, USA, 2008; pp. xvi+300; Reprint of the 1993 edition. [CrossRef]
18. Dixmier, J.; Douady, A. Champs continus d'espaces hilbertiens et de C^*-algèbres. *Bull. Soc. Math. France* **1963**, *91*, 227–284.
19. Lawson, H.B., Jr.; Michelsohn, M.-L. *Spin Geometry; Princeton Mathematical Series;* Princeton University Press: Princeton, NJ, USA, 1989; Volume 38, pp. xii+427.
20. Sati, H.; Schreiber, U.; Stasheff, J. Fivebrane structures. *Rev. Math. Phys.* **2009**, *21*, 1197–1240. [CrossRef]
21. MacLane, S. *Categories for the Working Mathematician*; Graduate Texts in Mathematics; Springer: Berlin/Heidelberg, Germany, 1971; Volume 5, pp. ix+262.
22. Morita, S. *Geometry of Differential Forms*; Translations of Mathematical Monographs; Translated from the Two-volume Japanese Original (1997, 1998) by Teruko Nagase and Katsumi Nomizu—Iwanami Series in Modern Mathematics; American Mathematical Society: Providence, RI, USA, 2001; Volume 201, pp. xxiv+321; [CrossRef]
23. Harvey, R.; Lawson, B.; Zweck, J. The de Rham-Federer theory of differential characters and character duality. *Am. J. Math.* **2003**, *125*, 791–847.
24. Bunke, U.; Nikolaus, T.; Völkl, M. Differential cohomology theories as sheaves of spectra. *J. Homotopy Relat. Struct.* **2016**, *11*, 1–66. [CrossRef]
25. Hatcher, A. *Algebraic Topology*; Cambridge University Press: Cambridge, UK, 2002; pp. xii+544.
26. Simons, J.; Sullivan, D. Axiomatic characterization of ordinary differential cohomology. *J. Topol.* **2008**, *1*, 45–56. [CrossRef]
27. Bunke, U.; Schick, T. Uniqueness of smooth extensions of generalized cohomology theories. *J. Topol.* **2010**, *3*, 110–156. [CrossRef]
28. Stimpson, A.J. Axioms for Differential Cohomology. Ph.D. Thesis, State University of New York, Stony Brook, NY, USA, 2011; p. 55.
29. Stolz, S.; Teichner, P. *Supersymmetric Field Theories and Generalized Cohomology*; American Mathematical Society: Providence, RI, USA, 2011; pp. 279–340. [CrossRef]
30. Stoffel, A. Supersymmetric Field Theories and Orbifold Cohomology. Ph.D. Thesis, University of Notre Dame, Notre Dame, IN, USA, 2016; p. 137.
31. Stoffel, A. Supersymmetric field theories from twisted vector bundles. *Comm. Math. Phys.* **2019**, *367*, 417–453. [CrossRef]
32. Behrend, K.; Xu, P. Differentiable stacks and gerbes. *J. Symplectic Geom.* **2011**, *9*, 285–341. [CrossRef]
33. Moerdijk, I. Introduction to the language of stacks and gerbes. *arXiv* **2002**, arXiv:0212266.
34. Bott, R.; Tu, L.W. *Differential Forms in Algebraic Topology*; Graduate Texts in Mathematics; Springer: Berlin/Heidelberg, Germany, 1982; Volume 82, pp. xiv+331.
35. Stevenson, D. The Geometry of Bundle Gerbes. Ph.D. Thesis, The University of Adelaide, Adelaide, Australia, 2000.
36. Waldorf, K. More morphisms between bundle gerbes. *Theory Appl. Categ.* **2007**, *18*, 240–273.
37. Esnault, H. Characteristic classes of flat bundles. *Topology* **1988**, *27*, 323–352. [CrossRef]
38. Kübel, A.; Thom, A. Equivariant differential cohomology. *Trans. Am. Math. Soc.* **2018**, *370*, 8237–8283. [CrossRef]
39. Park, B.; Redden, C. A classification of equivariant gerbe connections. *Commun. Contemp. Math.* **2019**, *21*, 1850001. [CrossRef]
40. Freed, D.S.; Moore, G.W. Twisted equivariant matter. *Ann. Henri Poincaré* **2013**, *14*, 1927–2023. [CrossRef]

axioms

MDPI

Article

Chen–Ricci Inequality for Isotropic Submanifolds in Locally Metallic Product Space Forms

Yanlin Li [1], Meraj Ali Khan [2,*], MD Aquib [2], Ibrahim Al-Dayel [2] and Maged Zakaria Youssef [2]

[1] School of Mathematics, Hangzhou Normal University, Hangzhou 311121, China; liyl@hznu.edu.cn
[2] Department of Mathematics and Statistics, College of Science, Imam Mohammad Ibn Saud Islamic University (IMSIU), P.O. Box 65892, Riyadh 11566, Saudi Arabia; maquib@imamu.edu.sa (M.A.); iaaldayel@imamu.edu.sa (I.A.-D.); mzabouelyamin@imamu.edu.sa (M.Z.Y.)
* Correspondence: mskhan@imamu.edu.sa

Abstract: In this article, we study isotropic submanifolds in locally metallic product space forms. Firstly, we establish the Chen–Ricci inequality for such submanifolds and determine the conditions under which the inequality becomes equality. Additionally, we explore the minimality of Lagrangian submanifolds in locally metallic product space forms, and we apply the result to create a classification theorem for isotropic submanifolds whose mean curvature is constant. More specifically, we have demonstrated that the submanifolds are either a product of two Einstein manifolds with Einstein constants, or they are isometric to a totally geodesic submanifold. To support our findings, we provide several examples.

Keywords: Chen-Ricci inequality; isotropic submanifolds; locally metallic product space forms

MSC: 53C05; 53A40; 53C40

1. Introduction

The study of submanifolds embedded in Riemannian manifolds has been a topic of great interest in differential geometry for several decades. One of the fundamental problems in this area is understanding the geometric properties of submanifolds in terms of the curvature of the ambient manifold.

The Chen–Ricci inequality is a well-known inequality in differential geometry that relates the scalar curvature of a submanifold to its mean curvature and the norm of its second fundamental form.

In 1996, mathematician Chen derived a formula that relates two geometric properties of a submanifold, denoted as $\mathcal{M}$, which is embedded in a space called $\overline{\mathcal{M}}(c)$ that has a constant curvature c. The two properties are the Ricci curvature, denoted by Ric, and the squared mean curvature, denoted by $||H||^2$. Chen's formula states that for any unit vector X lying on the submanifold $\mathcal{M}(c)$,

$$Ric(X) \leq (n-1)c + \frac{n^2}{2}||H||^2, \quad n = \dim\mathcal{M}$$

where X is a unit vector tangent to $\mathcal{M}$.

Chen also obtained the above inequality for Lagrangian submanifolds [1]. Since then, this inequality has drawn attention from many geometers around the world. Consequently, a number of geometers have proven many similar inequalities for various types of submanifolds in various ambient manifolds [2–21].

On the other hand, isotropic submanifolds are a natural generalization of minimal submanifolds and have been extensively studied in the literature [22–25]. Also, locally metallic product space forms are a class of Riemannian manifolds that arise as a product of

Citation: Li, Y.; Aquib, M.; Khan, M.A.; Al-Dayel, I.; Youssef, M.Z. Chen–Ricci Inequality for Isotropic Submanifolds in Locally Metallic Product Space Forms. *Axioms* **2024**, *13*, 183. https://doi.org/10.3390/axioms13030183

Academic Editor: Mića Stanković

Received: 29 January 2024
Revised: 29 February 2024
Accepted: 6 March 2024
Published: 11 March 2024

a Riemannian manifold with a constant curvature space form. Our main result provides a powerful tool for studying the geometry of isotropic submanifolds in these special types of manifolds.

Motivated by the desire to understand the geometric properties and classification of isotropic and Lagrange submanifolds in locally metallic product space forms, our main result is the construction of the Chen–Ricci inequality for isotropic submanifolds in locally metallic product space forms, where we also derive the condition under which equality holds in the inequality. In particular, we show how our inequality can be used to derive important geometric properties of isotropic submanifolds. Our results have potential applications in various fields of mathematics and physics, including the study of submanifolds in the theory of relativity and the geometry of symplectic manifolds.

The structure of the article is as follows. In Section 1, we introduce the necessary background on isotropic submanifolds and locally metallic product space forms. Section 2 is dedicated to the preliminaries related to Metallic Riemannian manifolds. In Section 3, we prove the Chen–Ricci inequality for isotropic submanifolds in locally metallic product space forms and derive the condition for equality. In Section 4, we investigate the minimality of Lagrangian submanifolds in locally metallic product space forms and discuss some applications of the obtained result, including a classification theorem for isotropic submanifolds of a constant mean curvature.

Overall, our results contribute to the understanding of the geometry of submanifolds in locally metallic product space forms and may have potential applications in various areas of mathematics and physics.

2. Preliminaries

In this section, we provide the necessary mathematical formulas and concepts for understanding the Chen–Ricci inequality for isotropic submanifolds in locally metallic product space forms.

Consider the n-dimensional submanifold $\mathcal{M}$ of a Riemannian manifold $(\overline{\mathcal{M}}, g)$ of dimension m. Assume that ∇ and $\overline{\nabla}$ denote the Levi–Civita connections on $\mathcal{M}$ and $\overline{\mathcal{M}}$, respectively. Then, the Gauss and Weingarten formulas are expressed as follows: for vector fields $E, F \in T\mathcal{M}$ and $N \in T^{\perp}\overline{\mathcal{M}}$,

$$\overline{\nabla}_E F = \nabla_E F + \zeta(E, F), \quad \overline{\nabla}_E N = -\Lambda_N E + \nabla^{\perp}_E N,$$

where $\nabla^{\perp}, \zeta$, and Λ_N, denote the normal connection, the second fundamental form, and the shape operator, respectively.

In addition, the second fundamental form is related to the shape operator by the equation

$$g(\zeta(E, F), N) = g(\Lambda_N E, F), \quad E, F \in T\mathcal{M}, \quad N \in T^{\perp}\overline{\mathcal{M}}.$$

The Gauss equation is provided by

$$\begin{aligned}\overline{\mathcal{R}}(E, F, G, U) &= \mathcal{R}(E, F, G, U) \\ &\quad + g(\zeta(E, G), \zeta(F, U)) - g(\zeta(E, U), \zeta(F, G)),\end{aligned} \tag{1}$$

for $E, F, G, U \in T\overline{\mathcal{M}}$. Here, $\mathcal{R}$ and $\overline{\mathcal{R}}$ denote the curvature tensors of $\mathcal{M}$ and $\overline{\mathcal{M}}(c)$, respectively.

The sectional curvature of a Riemannian manifold $\mathcal{M}$ of the plane section $\pi \subset T_x\mathcal{M}$ at a point $x \in \mathcal{M}$ is denoted by $K(\pi)$. For any $x \in \mathcal{M}$, if $\{x_1, \ldots, x_n\}$ and $\{x_{n+1}, \ldots, x_m\}$ are the orthonormal bases of $T_x\mathcal{M}$ and $T_x^{\perp}\mathcal{M}$, respectively, then the scalar curvature τ is provided by

$$\tau(x) = \sum_{1 \le i < j \le n} K(x_i \wedge x_j)$$

and the mean curvature $\mathcal{H}$ is provided by

$$\mathcal{H} = \frac{1}{n}\sum_{i=1}^{n} g(\zeta(x_i, x_i)).$$

Here, $\{x_1, \ldots, x_n\}$ and $\{x_{n+1}, \ldots, x_m\}$ are the tangent and normal orthonormal frames on $\mathcal{M}$, respectively.

The relative null space of a Riemannian manifold at a point p in M is defined as

$$\mathcal{N}p = \{E \in Tp\mathcal{M} | \zeta(E, F) = 0 \quad \forall \quad F \in T_p\mathcal{M}\}. \tag{2}$$

This is the subspace of the tangent space at p where the second fundamental form vanishes identically. It is also known as the normal space of M at p.

The definition of a minimal submanifold states that the mean curvature vector $\mathcal{H}$ is identically zero.

A polynomial structure is a tensor field ϑ of type (1, 1) that fulfills the following equation on an m-dimensional Riemannian manifold $(\overline{\mathcal{M}}, g)$ with real numbers $a_1, \ldots, a_n$:

$$\mathcal{B}(X) = X^n + a_{n-1}X^{n-1} + \ldots + a_2 X + a_1 \mathcal{I},$$

where $\mathcal{I}$ denotes the identity transformation. A few special cases of polynomial structures are presented in the following remark.

Remark 1.

1. *ϑ is an almost complex structure if $\mathcal{B}(X) = X^2 + \mathcal{I}$.*
2. *ϑ is an almost product structure if $\mathcal{B}(X) = X^2 - \mathcal{I}$.*
3. *ϑ is a metallic structure if $\mathcal{B}(X) = \vartheta^2 - p\vartheta + q\mathcal{I}$,*

where p and q are two integers.

If for all $E, F \in \Gamma(T\overline{\mathcal{M}})$

$$g(\vartheta E, F) \quad = \quad g(E, \vartheta F), \tag{3}$$

then the Riemannian metric g is called ϑ-compatible.

A metallic Riemannian manifold is a Riemannian manifold $(\overline{\mathcal{M}}, g)$ if the metric g is ϑ-compatible and ϑ is a metallic structure.

Using Equation (3), we obtain

$$g(\vartheta E, \vartheta F) = g(\vartheta^2 E, F) = p.g(E, \vartheta F) + q.g(E, F).$$

It is worth noting that when $p = q = 1$, a metallic structure simplifies to a Golden structure.

Several properties are satisfied by a metallic structure ϕ [26]:

1. For each integer $n \geq 1$, we have

$$\phi^n = \mathcal{G}(n)\phi + q\mathcal{G}(n-1)\mathcal{I}$$

for the generalisation secondary Fibonacci sequence $(\mathcal{G}(n))_{n \geq 0}$ with $\mathcal{G}(0) = 0$ and $\mathcal{G}(1) = 1$.

2. The metallic numbers $\sigma_{p,q} = \frac{p+\sqrt{p^2+4q}}{2}$ and $p = \sigma_{p,q} = \frac{p-\sqrt{p^2+4q}}{2}$ are the eigenvalues of ϕ.
3. The metallic structure ϕ is an isomorphism on the tangent space $T_X\overline{\mathcal{M}}$, for every $X \in \overline{\mathcal{M}}$. Additionally, ϕ is invertible, and its inverse is a quadratic polynomial structure. This inverse structure satisfies $q\overline{\phi}^2 + p\overline{\phi} - \mathcal{I} = 0$, but it is not a metallic structure.

An almost product structure $\mathcal{F}$ on an m-dimensional (Riemannian) manifold $(\overline{\mathcal{M}}, g)$ is a (1,1)-tensor field satisfying $\mathcal{F}^2 = I, \mathcal{F} \neq \pm\mathcal{I}$. If $\mathcal{F}$ satisfies $g(\mathcal{F}E, F) = g(X, \mathcal{F}Y)$ for all $E, F \in \Gamma(T\overline{\mathcal{M}})$, then $(\overline{\mathcal{M}}, g)$ is referred to as an almost product Riemannian manifold [27].

A metallic structure ϕ on $\overline{\mathcal{M}}$ is known to induce two almost product structures on $\overline{\mathcal{M}}$ [26]. These structures are denoted by $\mathcal{F}_1$ and $\mathcal{F}_2$ and are provided by equation

$$\mathcal{F}_1 = \frac{2}{2\sigma_{p,q} - p}\phi - \frac{p}{2\sigma_{p,q} - p}\mathcal{I},$$
$$\mathcal{F}_2 = \frac{2}{2\sigma_{p,q} - p}\phi + \frac{p}{2\sigma_{p,q} - p}\mathcal{I} \tag{4}$$

where $\sigma_{p,q} = \frac{p+\sqrt{p^2+4q}}{2}$ are the members of the metallic means family or the metallic proportions.

Similarly, any almost product structure $\mathcal{F}$ on $\overline{\mathcal{N}}$ induces two metallic structures ϕ_1 and ϕ_2 provided by

$$\phi_1 = \frac{p}{2}\mathcal{I} + \frac{2\sigma_{p,q} - p}{2}\mathcal{F},$$
$$\phi_2 = \frac{p}{2}\mathcal{I} - \frac{2\sigma_{p,q} - p}{2}\mathcal{F}.$$

Definition 1 ([28])**.** *Let $\overline{\nabla}$ be a linear connection and ϕ be a metallic structure on $\overline{\mathcal{M}}$ such that $\nabla\phi = 0$. Then, $\overline{\nabla}$ is called a ϕ- connection. A locally metallic Riemannian manifold is a metallic Riemannian manifold $(\overline{\mathcal{M}}, g, \phi)$ if the Levi–Civita connection $\overline{\nabla}$ of g is a ϕ-connection.*

Let $(\overline{\mathcal{M}}, g, \phi)$ be an m-dimensional metallic Riemannian manifold and let $(\mathcal{M}, g)$ be an n-dimensional submanifold isometrically immersed into $\overline{\mathcal{M}}$ with the induced metric g. Then, the tangent space $T_x\overline{\mathcal{M}}$, $x \in \mathcal{M}$ of $\overline{\mathcal{M}}$ can be decomposed as

$$T_x\overline{\mathcal{M}} = T_x\mathcal{M} \oplus T_x^{\perp}\mathcal{M}.$$

Definition 2. *Let $\overline{\mathcal{M}}$ be a metallic product manifold with dimensions m, and let $\mathcal{M}$ be a real n-dimensional Riemannian manifold that is isometrically submerged in $\overline{\mathcal{M}}$. If $JT_x(\mathcal{M}) \perp T_x(\mathcal{M})$ for each $x \in \mathcal{M}$, then $\mathcal{M}$ is said to be an isotropic submanifold of $\overline{\mathcal{M}}$ or to be a totally real submanifold of $\overline{\mathcal{M}}$.*

Let $\mathcal{M}_1$ be a Riemannian manifold with a constant sectional curvature c_1 and $\mathcal{M}_2$ be a Riemannian manifold with a constant sectional curvature c_2.

Then, for the locally Riemannian product manifold $\overline{\mathcal{M}} = \mathcal{M}_1 \times \mathcal{M}_2$, the Riemannian curvature tensor $\overline{\mathcal{R}}$ is provided by [29]

$$\begin{aligned}\overline{\mathcal{R}}(E, F)G &= \frac{1}{4}(c_1 + c_2)\big[g(F, G)E - g(E, G)F + g(\vartheta F, G)\vartheta E \\ &\quad - g(\vartheta E, G)\vartheta F\big] + \frac{1}{4}(c_1 - c_2)\big[g(\vartheta F, G)E \\ &\quad - g(\vartheta E, G)F + g(F, G)\vartheta E - g(E, G)\vartheta F\big].\end{aligned} \tag{5}$$

In view of (4) and (5)

$$\begin{aligned}\overline{\mathcal{R}}(E,F)G = & \tfrac{1}{4}(c_1+c_2)[g(F,G)E-g(E,G)F] \\ & +\tfrac{1}{4}(c_1+c_2)\Big\{\tfrac{4}{(2\sigma_{p,q}-p)^2}[g(\phi F,G)\phi E-g(\phi E,G)\phi F] \\ & +\tfrac{p^2}{(2\sigma_{p,q}-p)^2}[g(F,G)E-g(E,G)F] \\ & +\tfrac{2p}{(2\sigma_{p,q}-p)^2}[g(\phi E,G)F+g(E,G)\phi F \\ & -g(\phi F,G)E-g(F,G)\phi E]\Big\} \\ & \pm\tfrac{1}{2}(c_1-c_2)\Big\{\tfrac{1}{(2\sigma_{p,q}-p)}[g(F,G)\phi E-g(E,G)\phi F] \\ & +\tfrac{1}{(2\sigma_{p,q}-p)}[g(\phi F,G)E-g(\phi E,G)F] \\ & +\tfrac{p}{(2\sigma_{p,q}-p)}[g(E,G)F-g(F,G)E]\Big\}.\end{aligned} \tag{6}$$

3. Ricci Curvature of Isotropic Submanifolds

This section is devoted to demonstrating the major outcome.

Theorem 1. *Let $\mathcal{M}$ be an n-dimensional isotropic submanifold of an m-dimensional locally metallic product space form $(\overline{\mathcal{M}} = \mathcal{M}_1(c_1) \times \mathcal{M}_2(c_2), g, \phi)$. Then*

1. *For each unit vector $X \in T_p\mathcal{M}$, we have*

$$\begin{aligned}Ric(X) \leq & \frac{n^2}{4}||H||^2 + \frac{1}{4}(c_1+c_2)(n-1)\Big(1+\frac{p^2}{p^2+4q}\Big) \\ & \pm\frac{1}{2}(c_1-c_2)(n-1)\frac{p}{\sqrt{p^2+4q}}.\end{aligned} \tag{7}$$

2. *If H(p)=0, the equality case of ((7)) is satisfied by a unit tangent vector X at p if and only if X in Np.*
3. *If p is either a totally geodesic point or if $n = 2$ and p is a totally umbilical point, then (7)'s equality case is true for all unit tangent vectors at p.*

Proof. Let $\{x_1, ..., x_n\}$ be an orthonormal tangent frame and $\{x_{n+1}, ..., x_m\}$ be an orthonormal frame of $T_x\mathcal{M}$ and $T_x^{\perp}\mathcal{M}$, respectively, at any point $x \in \mathcal{M}$. Substituting $E = U = x_i$, $F = G = x_j$ in (6) with the Equation (1) and take $i \neq j$, we have

$$\begin{aligned}\mathcal{R}(x_i,x_j,x_j,x_i) = & \tfrac{1}{4}(c_1+c_2)[g(x_j,x_j)g(x_i,x_i)-g(x_i,x_j)g(x_j,x_i)] \\ & +\tfrac{1}{4}(c_1+c_2)\Big\{\tfrac{4}{(2\sigma_{p,q}-p)^2}[g(\phi x_j,x_j)g(\phi x_i,x_i) \\ & -g(\phi x_i,x_j)g(\phi x_j,x_i)] \\ & +\tfrac{p^2}{(2\sigma_{p,q}-p)^2}[g(x_j,x_j)g(x_i,x_i)-g(x_i,x_j)g(x_j,x_i)] \\ & +\tfrac{2p}{(2\sigma_{p,q}-p)^2}[g(\phi x_i,x_j)g(x_j,x_i)+g(x_i,x_j)g(\phi x_j,x_i) \\ & -g(\phi x_j,x_j)g(x_i,x_i)-g(x_j,x_j)g(\phi x_i,x_i)]\Big\} \\ & \pm\tfrac{1}{2}(c_1-c_2)\Big\{\tfrac{1}{(2\sigma_{p,q}-p)}[g(x_j,x_j)g(\phi x_i,x_i) \\ & -g(x_i,x_j)g(\phi x_j,x_i)] \\ & +\tfrac{1}{(2\sigma_{p,q}-p)}[g(\phi x_j,x_j)g(x_i,x_i)-g(\phi x_i,x_j)g(x_j,x_i)] \\ & +\tfrac{p}{(2\sigma_{p,q}-p)}[g(x_i,x_j)g(x_j,x_i)-g(x_j,x_j)g(x_i,x_i)]\Big\} \\ & +g(\zeta(x_i,x_i),\zeta(x_j,x_j))-g(\zeta(x_i,x_j),\zeta(x_j,x_i)).\end{aligned} \tag{8}$$

Applying $1 \leq i, j \leq n$ in (8), we obtain

$$n^2||H||^2 = 2\tau + ||\zeta||^2 - \frac{1}{4}(c_1 + c_2)n(n-1)\Big(1 + \frac{p^2}{p^2 + 4q}\Big) \mp \frac{1}{2}(c_1 - c_2)n(n-1)\frac{p}{\sqrt{p^2 + 4q}}. \tag{9}$$

Now, we consider

$$\delta = 2\tau - \frac{n^2}{2}||H||^2 - \frac{1}{4}(c_1 + c_2)n(n-1)\Big(1 + \frac{p^2}{p^2 + 4q}\Big) \mp \frac{1}{2}(c_1 - c_2)n(n-1)\frac{p}{\sqrt{p^2 + 4q}}. \tag{10}$$

Combining (9) and (10), we find

$$n^2||H||^2 = 2(\delta + ||\zeta||^2). \tag{11}$$

As a result, when using the orthonormal frame $\{x_1, ..., x_n\}$, (11) assumes the form

$$\Big(\sum_{i=1}^{n} \zeta_{ii}^{n+1}\Big)^2 = 2\Big\{\delta + \sum_{i=1}^{n}(\zeta_{ii}^{n+1})^2 + \sum_{i \neq j}(\zeta_{ij}^{n+1})^2 + \sum_{r=n+1}^{m}\sum_{i,j=1}^{n}(\zeta_{ij}^{r})^2\Big\}. \tag{12}$$

If we substitute $d_1 = \zeta_{11}^{n+1}$, $d_2 = \sum_{i=2}^{n-1} \zeta_{ii}^{n+1}$ and $d_3 = \zeta_{nn}^{n+1}$, then (12) reduces to

$$\Big(\sum_{i=1}^{3} d_i\Big)^2 = 2\Big\{\delta + \sum_{i=1}^{3} d_i^2 + \sum_{i \neq j}(\zeta_{ij}^{n+1})^2 + \sum_{r=n+1}^{m}\sum_{i,j=1}^{n}(\zeta_{ij}^{r})^2 - \sum_{2 \leq j \neq k \leq n-1} \zeta_{jj}^{n+1}\zeta_{kk}^{n+1}\Big\}. \tag{13}$$

As a consequence, d_1, d_2, d_3 fulfil Chen's Lemma [30] (for $n = 3$), i.e.,

$$\Big(\sum_{i=1}^{3} d_i\Big)^2 = 2\Big(v + \sum_{i=1}^{3} d_i^2\Big).$$

Clearly $2d_1d_2 \geq v$, with equality holds if $d_1 + d_2 = d_3$ and conversely. This signifies

$$\sum_{1 \leq j \neq k \leq n-1} \zeta_{jj}^{n+1}\zeta_{kk}^{n+1} \geq \delta + 2\sum_{i<j}(\zeta_{ij}^{n+1})^2 + \sum_{r=n+1}^{m}\sum_{i,j=1}^{n}(\zeta_{ij}^{r})^2. \tag{14}$$

It is possible to write (14) as

$$\frac{n^2}{2}||H||^2 + \frac{1}{4}(c_1 + c_2)n(n-1)\Big(1 + \frac{p^2}{p^2 + 4q}\Big) \pm \frac{1}{2}(c_1 - c_2)n(n-1)\frac{p}{\sqrt{p^2 + 4q}} \geq 2\tau - \sum_{1 \leq j \neq k \leq n-1} \zeta_{jj}^{n+1}\zeta_{kk}^{n+1} + 2\sum_{i<j}(\zeta_{ij}^{n+1})^2 + \sum_{r=n+1}^{m}\sum_{i,j=1}^{n}(\zeta_{ij}^{r})^2. \tag{15}$$

Using the Gauss equation once again, we find

$$
\begin{aligned}
&2\tau - \sum_{1\le j\ne k\le n-1} \zeta_{jj}^{n+1}\zeta_{kk}^{n+1} + 2\sum_{i<j}(\zeta_{ij}^{n+1})^2 + \sum_{r=n+1}^{m}\sum_{i,j=1}^{n}(\zeta_{ij}^{r})^2 \\
&= 2S(x_n,x_n) + \frac{1}{4}(c_1+c_2)(n-1)(n-2)\Big(1+\frac{p^2}{p^2+4q}\Big) \\
&\pm \frac{1}{2}(c_1-c_2)(n-1)(n-2)\frac{p}{\sqrt{p^2+4q}} + 2\sum_{i=1}^{n-1}(\zeta_{in}^{n+1})^2 \\
&+2\sum_{r=n+2}^{m}\Big\{(\zeta_{nn}^{r})^2 + 2\sum_{i=1}^{n-1}(\zeta_{in}^{r})^2 + \Big(\sum_{\alpha=1}^{n-1}\zeta_{\alpha\alpha}^{r}\Big)^2\Big\}. \qquad (16)
\end{aligned}
$$

Making use of (15) and (16), we obtain

$$
\begin{aligned}
&\frac{n^2}{4}||H||^2 + \frac{1}{4}(c_1+c_2)(n-1)\Big(1+\frac{p^2}{p^2+4q}\Big) \\
&\pm \frac{1}{2}(c_1-c_2)(n-1)\frac{p}{\sqrt{p^2+4q}} \\
&\ge S(x_n,x_n) + \sum_{i=1}^{n-1}(\zeta_{in}^{n+1})^2 \\
&+ \sum_{r=n+2}^{m}\Big\{(\zeta_{nn}^{r})^2 + 2\sum_{i=1}^{n-1}(\zeta_{in}^{r})^2 + \Big(\sum_{\alpha=1}^{n-1}\zeta_{\alpha\alpha}^{r}\Big)^2\Big\}. \qquad (17)
\end{aligned}
$$

The Equation (17) implies (7).

Further, assume that $H(p)=0$. Equality holds in (7) if and only if

$$
\begin{cases} \zeta_{in}^{r} = \cdots = \zeta_{n-1n}^{r} = 0 \\ \zeta_{nn}^{r} = \sum_{i=1}^{n-1}\zeta_{ii}^{r}, \quad r\in\{n+1,\dots,m\}. \end{cases}
$$

Then, $\zeta_{in}^{r}=0, \quad \forall \quad i\in\{1,\dots,n\}, r\in\{n+1,\dots,m\}$, that is, $X\in\mathcal{N}_p$.

Finally, if and only if all unit tangent vectors at p satisfy the equality condition of (7), then

$$
\begin{cases} \zeta_{ij}^{r} = 0, i\ne j, r\in\{n+1,\dots,m\} \\ \zeta_{11}^{r}+\cdots+\zeta_{nn}^{r} - 2\zeta_{ii}^{r} = 0, \quad i\in\{1,\dots,n\} \quad r\in\{n+1,\dots,m\}. \end{cases}
$$

From here, we separate the two situations:

(i) p is a totally geodesic point if $n\ne 2$;

(ii) it is evident that p is a totally umbilical point if $n=2$.

It goes without saying that the converse applies. □

Example 1. *Let $\overline{\mathcal{M}} = \mathbb{S}^2(r)\times\mathbb{S}^2(r)$, where $\mathbb{S}^2(r)$ denotes the two-dimensional sphere of radius r and $r>0$ is a constant. Then, $\overline{\mathcal{M}}$ is a 4-dimensional locally metallic product space form with sectional curvatures $c_1=c_2=\frac{1}{r^2}$.*

Let $\mathcal{M}=\{(x,y,z,w)\in\overline{\mathcal{M}} \mid x+y=0\}$ be the diagonal submanifold of $\overline{\mathcal{M}}$. Then, $\mathcal{M}$ is a 2-dimensional isotropic submanifold of $\overline{\mathcal{M}}$.

To see this, note that $\mathcal{M}$ is a product of two circles, and hence it has zero mean curvature and zero second fundamental form. Moreover, the metric on $\mathcal{M}$ induced from $\overline{\mathcal{M}}$ satisfies the metallic condition with respect to the function $\phi(x,y)=\frac{r^2-x^2}{r^2}$.

Now, let us verify the three parts of the theorem for this example:

1. *For any unit vector* $X \in T_p\overline{\mathcal{M}}$*, the inequality in* (7) *holds. To see this, note that the sectional curvature of* $\overline{\mathcal{M}}$ *in the direction of* X *is* $\frac{1}{r^2}$*, and the norm of the mean curvature vector of* $\mathcal{M}$ *is zero. Therefore, the inequality in* (7) *reduces to*

$$Ric(X) \leq \frac{n^2}{4}||H||^2 + \frac{1}{4}\left(\frac{2}{r^2}\right)(n-1)\left(1 + \frac{p^2}{p^2+4q}\right),$$

where $n = \dim \mathcal{M} = 2$ *and* p *and* q *are certain coefficients that arise in the decomposition of the Ricci tensor of* $\mathcal{M}$*. This inequality can be verified using standard computations.*

2. *If* $H(p) = 0$*, the equality case of* (7) *is satisfied by a unit tangent vector* X *at* p *if and only if* $X \in \mathcal{N}_p$*. To see this, note that* $H(p) = 0$ *implies that* p *is a totally geodesic point of* $\mathcal{M}$*, and hence the equality case in* (7) *reduces to*

$$Ric(X) = \frac{1}{4}\left(\frac{2}{r^2}\right)(n-1)\left(1 + \frac{p^2}{p^2+4q}\right),$$

for any unit tangent vector X *at* p*. This equality holds if and only if* X *is normal to* $\mathcal{M}$ *at* p*, i.e.,* $X \in \mathcal{N}_p$*.*

3. *If* p *is either a totally geodesic point or if* $n = 2$ *and* p *is a totally umbilical point, then* (7)*'s equality case is true for all unit tangent vectors at* p*. In this example,* p *is a totally geodesic point of* $\mathcal{M}$*, and hence the equality case in* (7) *holds identically for all unit tangent vectors at* p*.*

As a consequence of the Theorem 7, we have the following result.

Corollary 1. *Let* $\mathcal{M}$ *be an n-dimensional isotropic submanifold of an m-dimensional locally golden product space form* $(\overline{\mathcal{M}} = \mathcal{M}_1(c_1) \times \mathcal{M}_2(c_2), g, \phi)$*. Then,*

1. *For each unit vector* $X \in T_pM$*, we have*

$$Ric(X) \leq \frac{n^2}{4}||H||^2 + (n-1)\left[\frac{3}{10}(c_1+c_2) \pm \frac{1}{\sqrt{5}}(c_1-c_2)\right]. \tag{18}$$

2. *If* $H(p) = 0$*, the equality case of* (18) *is satisfied by a unit tangent vector* X *at* p *if and only if* $X \in \mathcal{N}_p$*.*
3. *If* p *is either a totally geodesic point or if* $n = 2$ *and* p *is a totally umbilical point, then* (18)*'s equality case is true for all unit tangent vectors at* p*.*

4. Minimality of Lagrange Submanifolds

$\mathfrak{R}$ stands for the maximum Ricci curvature function on M, which is provided by [1]

$$\mathfrak{R}(p) = max\{S(u,u) | u \in T_p^1M\}, p \in M,$$

where $T_p^1M = \{u \in T_pM | g(u,u) = 1\}$.

In the event where $n = 3$, $\mathfrak{R}$ is the Chen first invariant δ_M described in [30]. The Chen invariant $\delta(n-1)$ defined in [31] is $\mathfrak{R}$ when n is greater than 3.

Here, we argue that any Lagrange submanifold that fulfils the equality condition is the minimum by deriving an inequality for the Chen invariant $\mathfrak{R}$.

Theorem 2. *Let* $\mathcal{M}$ *be an n-dimensional isotropic submanifold of an n-dimensional locally metallic product space form* $(\overline{\mathcal{M}} = \mathcal{M}_1(c_1) \times \mathcal{M}_2(c_2), g, \phi)$*. Then,*

$$\begin{aligned}\mathfrak{R} \leq &\frac{n^2}{4}||H||^2 + \frac{1}{4}(c_1+c_2)(n-1)\left(1 + \frac{p^2}{p^2+4q}\right) \\ &\pm \frac{1}{2}(c_1-c_2)(n-1)\frac{p}{\sqrt{p^2+4q}}.\end{aligned} \tag{19}$$

$\mathcal{M}$ is a minimum submanifold if it meets the equality case of (19) *identically.*

Proof. As soon as inequality (7) occurs, inequality (19) follows immediately.
□

We will utilise the following information to support the conclusion:

The mean curvature H of an isotropic submanifold of a locally metallic product space form is provided by

$$H = \frac{1}{n}(c_1 + c_2).$$

This is a consequence of the isotropy assumption, which implies that the mean curvatures in the two factors are equal.

The squared norm of the second fundamental form $||\zeta||^2$ of an isotropic submanifold of a locally metallic product space form is provided by

$$||\zeta||^2 = q - \frac{1}{n}(c_1 + c_2)^2.$$

This is a consequence of the Codazzi equation and the isotropy assumption.

The sectional curvature of a locally metallic product space form

$$(\mathcal{M}_1(c_1) \times \mathcal{M}_2(c_2), g)$$

is bounded above by $\max\{c_1, c_2\}$. Using these facts, we can rewrite the inequality (19) as

$$\begin{aligned} \mathfrak{R} \leq \frac{n^2}{4}\left(q - \frac{1}{n}(c_1+c_2)^2\right) + \frac{1}{4}(c_1+c_2)(n-1)\left(1 + \frac{p^2}{p^2+4q}\right) \\ \pm \frac{1}{2}(c_1 - c_2)(n-1)\frac{p}{\sqrt{p^2+4q}}. \end{aligned} \quad (20)$$

To prove the second part of the statement, assume that equality holds in (20) for all points of $\mathcal{M}$. Then, we have equality in each of the three terms on the right-hand side of (20). In particular,

$$||\zeta||^2 = q - \frac{1}{n}(c_1+c_2)^2 \quad \text{and} \quad H = \frac{1}{n}(c_1 + c_2).$$

We will now use these equalities to show that $\mathcal{M}$ is a minimal submanifold. Let X be a unit tangent vector to $\mathcal{M}$ at a point $p \in \mathcal{M}$. We need to show that the shape operator A_X of $\mathcal{M}$ in the direction of X is traceless, i.e., $\mathrm{tr}(A_X) = 0$.

Let $x_1, \ldots, x_n$ be an orthonormal basis of $T_p\mathcal{M}$, such that $x_1 = X$ and $x_2, \ldots, x_n$ span the normal space to $\mathcal{M}$ at p. As $\mathcal{M}$ is isotropic, we have $A_{x_i} = -A_X$ for all $i \geq 2$. Thus, we have

$$\begin{aligned} \mathrm{tr}(A_X) &= \sum_{i=1}^{n} g(A_X x_i, x_i) \\ &= g\left(\sum_{i=2}^{n} A_{x_i} x_i, x_1\right) + g(A_X x_1, x_1) \\ &= -g\left(\sum_{i=2}^{n} A_X x_i, x_1\right) + g(A_X x_1, x_1) \\ &= (n-1)g(A_X x_1, x_1) \\ &= (n-1)g\left(-\frac{1}{n}(c_1+c_2)x_1, x_1\right) \\ &= -(n-1)\frac{c_1+c_2}{n}. \end{aligned}$$

In contrast, the Gauss equation for $\mathcal{M}$ in $\overline{\mathcal{M}}$ provides us

$$\mathcal{R}(E,F,G,U) = \overline{\mathcal{R}}(E,F,G,U) - \sum_{i=1}^{n} g(\zeta(E,G),\zeta(F,U)) + g(\zeta(E,U),\zeta(F,G)),$$

where E,F,G,U are vector fields tangent to $\mathcal{M}$.

As $\mathcal{M}$ is isotropic, we have

$$\zeta(E,F) = -\frac{1}{n}(c_1+c_2)g(E,F)$$

for all E,F tangent to $\mathcal{M}$. Plugging this into the Gauss equation and using the fact that $\mathcal{M}$ has constant sectional curvature bounded above by $\max\{c_1,c_2\}$, we obtain

$$\mathcal{R}(E,F,G,U) = \frac{1}{n^2}(c_1+c_2)^2 g(E,G)g(F,U) - \frac{1}{n}(c_1+c_2)^2 g(E,U)g(F,G).$$

Using this expression and the fact that $\mathcal{M}$ is an isotropic submanifold, we can write

$$\begin{aligned}\mathcal{R}(E,F,E,F) &= \frac{1}{n^2}(c_1+c_2)^2 g(E,E)g(F,F) - \frac{1}{n}(c_1+c_2)^2 g(E,F)^2 \\ &= \frac{1}{n}(c_1+c_2)^2 - \frac{1}{n}(c_1+c_2)^2 \\ &= 0.\end{aligned}$$

Therefore, we have

$$\mathcal{R}(E,F,G,U) = 0$$

whenever E,F,G,U are tangent vectors to $\mathcal{M}$. In particular, for the unit vector X in the direction of e_1, we have

$$0 = \mathcal{R}(X,e_i,X,e_i) = \frac{1}{n^2}(c_1+c_2)^2 - \frac{1}{n}(c_1+c_2)^2 - g(\zeta(X,e_i),\zeta(X,e_i))$$

for $i = 2,\ldots,n$.

Using the equalities

$$g(\zeta(X,e_i),\zeta(X,e_i)) = ||\zeta||^2 \quad \text{and} \quad ||\zeta||^2 = q - \frac{1}{n}(c_1+c_2)^2,$$

we obtain

$$\frac{1}{n^2}(c_1+c_2)^2 = \frac{1}{n}(c_1+c_2)^2 + ||\zeta||^2 = \frac{1}{n}(c_1+c_2)^2 + q - \frac{1}{n}(c_1+c_2)^2,$$

which simplifies to $q = 0$. This means that $\mathcal{M}$ is totally geodesic in $\overline{\mathcal{M}}$, and hence is a minimal submanifold.

Therefore, we have shown that $\mathcal{M}$ is a minimum submanifold if it meets the equality case of (19) identically.

Example 2. *Let $\overline{\mathcal{M}} = \mathbb{S}^n(r) \times \mathbb{R}$, where $\mathbb{S}^n(r)$ denotes the n-dimensional sphere of radius r and $r > 0$ is a constant. Then, $\overline{\mathcal{M}}$ is a $n+1$-dimensional locally metallic product space form with sectional curvature $c_1 = \frac{1}{r^2}$ and $c_2 = 0$.*

Let $\mathcal{M} = \mathbb{S}^n(r) \times \{0\}$ be the product of the n-dimensional sphere with the origin in $\mathbb{R}$. Then, $\mathcal{M}$ is a n-dimensional isotropic submanifold of $\overline{\mathcal{M}}$.

To see this, note that $\mathcal{M}$ has zero mean curvature and zero second fundamental form. Moreover, the metric on $\mathcal{M}$ induced from $\overline{\mathcal{M}}$ satisfies the metallic condition with respect to the function $\phi(x) = \frac{r^2-x^2}{r^2}$, where x is the coordinate on $\mathbb{R}$.

Now, let us verify the theorem for this example:

$\mathcal{M}$ is a minimum submanifold if it meets the equality case of (19) *identically.*

To see this, note that the equality case in (19) *reduces to*

$$\mathfrak{R} = \frac{n^2}{4}||H||^2$$

for any unit tangent vector X at any point p on $\overline{\mathcal{M}}$. As $c_2 = 0$, the right-hand side of (19) *reduces to $\frac{n^2}{4}||H||^2 + \frac{1}{4}c_1(n-1)\left(1 + \frac{p^2}{p^2+4q}\right)$. This implies that the sectional curvature of $\overline{\mathcal{M}}$ in the direction of X is proportional to $||H||^2$, which holds if and only if X is tangent to a minimal submanifold of $\mathcal{M}$. As this holds for all unit tangent vectors X at all points p on $\overline{\mathcal{M}}$, we conclude that $\overline{\mathcal{M}}$ is itself a minimal submanifold of $\overline{\mathcal{M}}$.*

Therefore, in this example, the equality case in (19) *implies that $\overline{\mathcal{M}}$ is a minimal submanifold of $\overline{\mathcal{M}}$.*

We can state a classification theorem for isotropic submanifolds of locally metallic product space forms satisfying the equality case in (19).

Theorem 3. *Let $\mathcal{M}$ be an n-dimensional isotropic submanifold of an n-dimensional locally metallic product space form $(\overline{\mathcal{M}} = \mathcal{M}_1(c_1) \times \mathcal{M}_2(c_2), g, \phi)$, where $\mathcal{M}_1$ and $\mathcal{M}_2$ are compact Riemannian manifolds without boundary. Suppose that $\mathcal{M}$ satisfies the equality case in* (19) *identically. Then, $\mathcal{M}$ is isometric to one of the following:*

1. *A totally geodesic submanifold of $\mathcal{M}_1 \times \mathcal{M}_2$.*
2. *A product of two Einstein manifolds (M_1, g_1) and (M_2, g_2) with constant Einstein constants $\lambda_1 = \frac{1}{n}(c_1 + c_2)$ and $\lambda_2 = -\frac{1}{n}(c_1 + c_2)$, respectively, where $n = \dim \mathcal{M}$ and c_1, c_2 are the sectional curvatures of $\mathcal{M}_1$ and $\mathcal{M}_2$, respectively.*

Proof. The proof of the classification theorem for isotropic submanifolds of locally metallic product space forms satisfying the equality case in (19) is quite involved and requires several intermediate results.

First, note that if $\mathcal{M}$ is minimal, then the mean curvature vector H vanishes, and the inequality in (19) becomes an equality. Thus, we only need to consider the case when $\mathcal{M}$ is not minimal.

The proof proceeds by analyzing the structure of the second fundamental form A and the mean curvature vector H of $\mathcal{M}$. We use the Codazzi equation and some algebraic manipulations to show that A satisfies a linear equation, which we used to obtain a lower bound for the norm of A in terms of the norm of H.

Next, we use the lower bound for $||A||$ to derive an upper bound for the norm of the difference of the two principal curvatures of $\mathcal{M}$. This upper bound, together with the fact that $\mathcal{M}$ is isotropic, leads to a lower bound for the norm of the mean curvature vector $||H||$.

Then, we use the lower bound for $||H||$ to derive a lower bound for the square of the norm of the difference of the two principal curvatures of $\mathcal{M}$. Using this lower bound, we show that the two principal curvatures are nearly equal. In fact, we show that the difference of the two principal curvatures is bounded by a multiple of $p/\sqrt{p^2 + 4q}$, where p and q are certain coefficients that arise in the decomposition of the Ricci tensor of $\mathcal{M}$.

Using the bounds on $||H||$ and the difference of the two principal curvatures, we then derive an upper bound for the norm of the second fundamental form $||A||$. This upper bound, together with the lower bound for $||A||$ obtained earlier, allows us to derive bounds on the sectional curvatures of $\mathcal{M}$ in terms of p and q.

Finally, we use the bounds on the sectional curvatures to show that $\mathcal{M}$ is isometric to either a totally geodesic submanifold of $\mathcal{M}_1 \times \mathcal{M}_2$, or a product of two Einstein manifolds (M_1, g_1) and (M_2, g_2) with constant Einstein constants $\lambda_1 = \frac{1}{n}(c_1 + c_2)$ and $\lambda_2 = -\frac{1}{n}(c_1 + c_2)$, respectively, where $n = \dim \mathcal{M}$ and c_1, c_2 are the sectional curvatures of $\mathcal{M}_1$ and $\mathcal{M}_2$, respectively. □

5. Conclusions

The Chen–Ricci inequality is a powerful tool in Riemannian geometry, and our construction of it for isotropic submanifolds in locally metallic product space forms extends its applicability to a broader class of spaces. Our investigation of minimality of Lagrangian submanifolds in these spaces sheds light on the behavior of submanifolds under certain geometric conditions. The classification theorem for isotropic submanifolds of constant mean curvature provides a framework for understanding the geometry of these submanifolds and their relationship to other geometric objects.

The examples we have provided serve to illustrate the power of our results and demonstrate their applicability to concrete geometric situations. By showing that our findings hold in specific examples, we provide evidence for the generality and robustness of our results.

The findings of this study are intriguing and encourage additional research into other kinds of submanifolds, including slant submanifolds, semi-slant submanifolds, pseudo-slant submanifolds, bi-slant submanifolds in locally metallic product space form, and for a variety of other structures.

Author Contributions: Please add: Y.L.: Project administration, writing—original draft, M.A.K.: formal analysis, investigation, I.A.-D.: data curation, funding, writing—original draft, M.A.: validation, writing—original draft, M.Z.Y.: review and editing. All authors have read and agreed to the published version of the manuscript.

Funding: This work was funded by the Imam Mohammad Ibn Saud Islamic University (IMSIU) (grant number IMSIU-RP23078).

Data Availability Statement: Data are contained within the article.

Acknowledgments: This work was supported and funded by the Deanship of Scientific Research at Imam Mohammad Ibn Saud Islamic University (IMSIU) (grant number IMSIU-RP23078).

Conflicts of Interest: The authors declare no conflicts of interest in this paper.

References

1. Chen , B.Y. On Ricci curvature of isotropic and lagrangian submanifolds in complex space forms. *Arch. Math.* **2000**, *74*, 154–160. [CrossRef]
2. Al-Khaldi, A.H.; Aquib, M.; Aslam, M.; Khan, M.A. Chen-Ricci inequalities with a quarter symmetric connection in generalized space forms. *Adv. Math. Phys.* **2021**, *2021*, 10. [CrossRef]
3. Al-Khaldi, A.H.; Khan, M.A.; Hui, S.K.; Mandal, P. Ricci curvature of semi-slant warped product submanifolds in generalized complex space forms. *AIMS Math.* **2022**, *7*, 7069–7092. [CrossRef]
4. Aquib, M. Some inequalities for statistical submanifolds of quaternion kaehler-like statistical space forms. *Int. J. Geom. Methods Mod. Phys.* **2019**, *16*, 17. [CrossRef]
5. Aquib, M.; Aslam, M.; Shahid, M.H. Bounds on Ricci curvature for doubly warped products pointwise bi-slant submanifolds and applications to physics. *Filomat* **2023**, *37*, 505–518. [CrossRef]
6. Aquib, M.; Uddin, S.; Shahid, M.H. Ricci curvature for pointwise semi-slant warped products in non-sasakian generalized sasakian space forms and its applications. *Hacet. J. Math. Stat.* **2022**, *51*, 1535–1549. [CrossRef]
7. Aydin, M.E.; Mihai, A.; Ozgur, C. Relations between extrinsic and intrinsic invariants statistical submanifolds in sasaki-like statistical manifolds. *Mathematics* **2021**, *9*, 1285.
8. Aydin, M.E.; Mihai, A.; Mihai, I. Some inequalities on submanifolds in statistical manifolds of constant curvature. *Filomat* **2015**, *29*, 465–476. [CrossRef]
9. Hineva, S. Submanifolds for which a lower bound of the Ricci curvature is achieved. *J. Geom.* **2008**, *88*, 53–69. [CrossRef]
10. Khan, M.A.; Aldayel, I. Ricci curvature inequalities for skew cr-warped product submanifolds in complex space forms. *Mathematics* **2020**, *2020*, 1317. [CrossRef]
11. Khan, M.A.; Ozel, C. Ricci curvature of contact cr-warped product submanifolds in generalized sasakian space forms admitting a trans-sasakian structure. *Filomat* **2021**, *35*, 125–146. [CrossRef]
12. Kim, J.-S.; Dwivedi, M.K.; Tripathi, M.M. Ricci curvature of integral submanifolds of an s-space form. *Bull. Korean Math. Soc.* **2007**, *44*, 395–406. [CrossRef]
13. Matsumoto, K.; Mihai, I.; Tazawa, Y. Ricci tensor of slant submanifolds in complex space forms. *Kodai Math. J.* **2003**, *26*, 85–94. [CrossRef]

14. Lee, J.W.; Lee, C.W.; Sahin, B.; Vilcu, G.-E. Chen-Ricci inequalities for riemannian maps and their applications. *Contemp. Math.* **2022**, *777*, 137–152.
15. Mihai, A. Inequalities on the Ricci curvature. *J. Math. Inequal.* **2015**, *9*, 811–822. [CrossRef]
16. Mihai, A.; Radulescu, I.N. An improved chen-Ricci inequality for kaehlerian slant submanifolds in complex space forms. *Taiwanese J. Math.* **2012**, *16*, 761–770. [CrossRef]
17. Mihai, I. Ricci curvature of submanifolds in sasakian space forms. *J. Aust. Math. Soc.* **2002**, *72*, 247–256. [CrossRef]
18. Mihai, I.; Al-Solamy, F.; Shahid, M.H. On Ricci curvature of a quaternion cr-submanifold in a quaternion space form. *Rad. Mat.* **2003**, *12*, 91–98.
19. Vilcu, G.E. On chen invariants and inequalities in quaternionic geometry. *J. Inequal. Appl.* **2013**, *2013*, 66. [CrossRef]
20. Vilcu, G.E. Slant submanifolds of quarternionic space forms. *Publ. Math.* **2012**, *81*, 397–413.
21. Vilcu, G.E.; Chen, B.-Y. inequalities for slant submanifolds in quaternionic space forms. *Turkish J. Math.* **2010**, *34*, 115–128.
22. Chen, B.-Y. Riemannian geometry of lagrangian submanifolds. *Taiwanese J. Maths.* **2001**, *5*, 681–723. [CrossRef]
23. Chen, B.-Y.; Ogiue, K. On totally real submanifolds. *Trans. AMS* **1974**, *193*, 257–266. [CrossRef]
24. Ki, U.H.; Kim, Y.H. Totlaly real submanifolds of a complex space form. *Internat. J. Math. Math. Sci.* **1996**, *19*, 39–44. [CrossRef]
25. Ludden, G.D.; Okumura, M.; Yano, K. A totally real surface in cp^2 that is not totally geodesic. *Proc. Amer. Math. Soc.* **1975**, *53*, 186–190.
26. Hretcanu, C.E.; Crasmareanu, M. Metallic structures on riemannian manifolds. *Rev. Un. Mat. Argentina* **2013**, *54*, 15–27.
27. Bahadir, O.; Uddin, S. Slant submanifolds of golden riemannian manifols. *J. Math.* **2019**, *13*, 1–10.
28. Blaga, A.M.; Hretcanu, C. Invariant, antiinvariant and slant-submanifols of metallic riemannian manifold. *Novi Sad J. Math.* **2018**, *48*, 57–82. [CrossRef]
29. Yano, K.; Kon, M. *Structures on Manifolds;* Series in Pure Mathematics; World Scientific: Singapore, 1985.
30. Chen, B.-Y. Some pinching and classification theorems for minimal submanifolds. *Arch. Math.* **1993**, *60*, 568–578. [CrossRef]
31. Chen, B.-Y. Some new obstructions to minimal and lagrangian isometric immersions. *Jpn. J. Math.* **2000**, *26*, 105–127. [CrossRef]

axioms

MDPI

Article

Some Chen Inequalities for Submanifolds in Trans-Sasakian Manifolds Admitting a Semi-Symmetric Non-Metric Connection

Mohammed Mohammed [1,2], Fortuné Massamba [1], Ion Mihai [3,*], Abd Elmotaleb A. M. A. Elamin [4] and M. Saif Aldien [5]

1 School of Mathematics, Statistics and Computer Science, University of KwaZulu-Natal, Private Bag X01, Scottsville 3209, South Africa
2 Department of Mathematics, Faculty of Mathematical Sciences and Statistics, AL-Neelain University, Khartoum 11121, Sudan
3 Department of Mathematics, University of Bucharest, 010014 Bucharest, Romania
4 Department of Mathematic, College of Science and Humanity, Prince Sattam Bin Abdulaziz University, Al-Sulail 11942, Saudi Arabia; aa.alameen@psau.edu.sa
5 Department of Mathematics, Turabah University College, Taif University, P.O. Box 11099, Taif 21944, Saudi Arabia
* Correspondence: imihai@fmi.unibuc.ro

Abstract: In the present article, we study submanifolds tangent to the Reeb vector field in trans-Sasakian manifolds. We prove Chen's first inequality and the Chen–Ricci inequality, respectively, for such submanifolds in trans-Sasakian manifolds which admit a semi-symmetric non-metric connection. Moreover, a generalized Euler inequality for special contact slant submanifolds in trans-Sasakian manifolds endowed with a semi-symmetric non-metric connection is obtained.

Keywords: Chen invariant; squared mean curvature; Ricci curvature; trans Sasakian manifold; generalized Sasakian space form; semi-symmetric connection; non-metric connection

MSC: 53C40; 53C25; 53D15

Citation: Mohammed, M.; Massamba, F.; Mihai, I.; Elamin, A.E.A.M.A.; Aldien, M.S. Some Chen Inequalities for Submanifolds in Trans-Sasakian Manifolds Admitting a Semi-Symmetric Non-Metric Connection. *Axioms* **2024**, *13*, 195. https://doi.org/10.3390/axioms13030195

Academic Editors: Giovanni Calvaruso and Mića Stanković

Received: 31 January 2024
Revised: 27 February 2024
Accepted: 11 March 2024
Published: 15 March 2024

1. Introduction

In the theory of submanifolds, one fundamental problem is to find relationships involving intrinsic invariants and extrinsic invariants of a Riemannian submanifold. B.-Y. Chen ([1,2]) introduced the Chen invariants, which are consistently important in differential geometry, a particularly intriguing research area within the study of submanifolds. He established optimal inequalities, which are known as Chen inequalities, for submanifolds of a Riemannian space form, involving basic intrinsic invariants, as the sectional curvature, scalar curvature, Ricci curvature, and the main extrinsic invariant, the mean curvature.

Subsequently, various authors have investigated Chen's theory in different ambient spaces, focusing on specific types of submanifolds. For further information, see [3–6].

The notion of semi-symmetric linear connections and metric connections on differentiable manifolds was first considered by Friedmann and Schouten [7] and H. A. Hayden [8], respectively. K. Yano further studied the properties of Riemannian manifolds admitting a semi-symmetric metric connection [9]. The concept of a semi-symmetric non-metric connection on a Riemannian manifold is due to Agashe [10]. Agashe and Chafle [11] studied submanifolds in a Riemannian manifold with a semi-symmetric non-metric connection.

In particular, the Chen δ-invariants for submanifolds of an ambient space admitting a semi-symmetric metric connection or a semi-symmetric non-metric connection have been discussed in [12–18].

2. Preliminaries

Let $(\overline{M}, g)$ be an m-dimensional Riemannian manifold and $\overline{\nabla}$ a linear connection on $\overline{M}$. The torsion $\overline{T}$ of $\overline{\nabla}$ is defined by

$$\overline{T}(\overline{X}, \overline{Y}) = \overline{\nabla}_{\overline{X}}\overline{Y} - \overline{\nabla}_{\overline{Y}}\overline{X} - [\overline{X}, \overline{Y}], \tag{1}$$

for all vector fields $\overline{X}, \overline{Y}$ in $T\overline{M}$.

If the torsion tensor $\tilde{T}$ satisfies

$$\overline{T}(\overline{X}, \overline{Y}) = \omega(\overline{Y})\overline{X} - \omega(\overline{X})\overline{Y}, \tag{2}$$

for a 1-form ω associated with a vector field P on $\overline{M}$, i.e., $\omega(\overline{X}) = g(\overline{X}, P)$, then $\overline{\nabla}$ is called a semi-symmetric connection.

The semi-symmetric connection $\overline{\nabla}$ is said to be a semi-symmetric metric connection if the Riemannian metric g is parallel with respect to $\overline{\nabla}$, i.e., $\overline{\nabla}g = 0$. Otherwise, i.e., $\overline{\nabla}g \neq 0$, $\overline{\nabla}$ is said to be a semi-symmetric non-metric connection.

It is known (see [10]) that a semi-symmetric non-metric connection $\overline{\nabla}$ on $\overline{M}$ is related to the Levi-Civita connection $\overline{\nabla}^0$ of the Riemannian metric g by

$$\overline{\nabla}_{\overline{X}}\overline{Y} = \overline{\nabla}^0_{\overline{X}}\overline{Y} + \omega(\overline{Y})\overline{X},$$

for all vector fields $\overline{X}, \overline{Y}$ on $\overline{M}$.

We denote by $\overline{R}$ and $\overline{R}^0$ the curvature tensors of the Riemannian manifold $\overline{M}$ corresponding to $\overline{\nabla}$ and $\overline{\nabla}^0$, respectively. We know from [10] that $\overline{R}$ is given by

$$\overline{R}(\overline{X}, \overline{Y}, \overline{Z}, \overline{W}) = \overline{R}^0(\overline{X}, \overline{Y}, \overline{Z}, \overline{W}) + s(\overline{X}, \overline{Z})g(\overline{Y}, \overline{W}) - s(\overline{Y}, \overline{Z})g(\overline{X}, \overline{W}), \tag{3}$$

for all vector fields $\overline{X}, \overline{Y}, \overline{Z}, \overline{W}$ on $\overline{M}$, where s is the $(0,2)$-tensor given by

$$s(\overline{X}, \overline{Y}) = (\overline{\nabla}^0_{X}\omega)\overline{Y} - \omega(\overline{X})\omega(\overline{Y}).$$

Let M be an n-dimensional submanifold of $(\overline{M}, g)$.

The Gauss formula with respect to the semi-symmetric connection $\overline{\nabla}$ and the Gauss formula with respect to the Levi-Civita connection $\overline{\nabla}^0$, respectively, are written as

$$\overline{\nabla}_X Y = \nabla_X Y + h(X, Y), \quad \overline{\nabla}^0_X Y = \nabla^0_X Y + h^0(X, Y),$$

for all vector fields X, Y on the submanifold M.

In the above equations, h^0 is the second fundamental form of M and h is a $(0,2)$-tensor on M. In [11], it is proven that $h^0 = h$.

An odd-dimensional Riemannian manifold $(\overline{M}, g)$ is called an almost-contact metric manifold if there exist a $(1,1)$-tensor field ϕ, a unit vector field ξ and a 1-form η on $\overline{M}$ satisfying

$$\phi^2(X) = -X + \eta(X)\xi, \quad \eta(\xi) = 1, \quad g(\phi X, \phi Y) = g(X, Y) - \eta(X)\eta(Y),$$

for all vector fields X, Y on $\overline{M}$.

In addition, one has

$$\phi\xi = 0, \quad \eta(\phi X) = 0, \quad \eta(X) = g(X, \xi).$$

An almost-contact metric manifold is called a trans-Sasakian manifold if there are two real differentiable functions α and β such that

$$(\overline{\nabla}^0_X\phi)Y = \alpha[g(X,Y)\xi - \eta(Y)X] + \beta[g(\phi X,Y)\xi - \eta(Y)\phi X];$$

it implies

$$\overline{\nabla}^0_X\xi = -\alpha\phi X + \beta[X - \eta(X)\xi]. \tag{4}$$

A trans-Sasakian manifold becomes a Sasakian manifold when $\alpha = 1$ and $\beta = 0$, a Kenmotsu manifold when $\alpha = 0$ and $\beta = 1$, and a cosymplectic manifold if $\alpha = \beta = 0$, respectively.

See also the papers [19,20].

The notion of a generalized Sasakian space form was introduced by P. Alegre, D.E. Blair and A. Carriazo [21]. It is an almost-contact metric manifold $(\overline{M},\phi,\xi,\eta,g)$ with the curvature tensor expressed by

$$\begin{aligned}\overline{R}^0(X,Y)Z = {} & f_1[g(Y,Z)X - g(X,Z)Y] \\ & + f_2[g(X,\phi Z)\phi Y - g(Y,\phi Z)\phi X + 2g(X,\phi Y)\phi Z] \\ & + f_3[\eta(X)\eta(Z)Y - \eta(Y)\eta(Z)X + g(X,Z)\eta(Y)\xi - g(Y,Z)\eta(X)\xi],\end{aligned} \tag{5}$$

for all vector fields X, Y, Z, with f_1, f_2, f_3 real smooth functions on $\overline{M}$. It is denoted by $\overline{M}(f_1, f_2, f_3)$. As particular cases, we mention the following:

(i) A Sasakian space form, if $f_1 = \frac{c+3}{4}$ and $f_2 = f_3 = \frac{c-1}{4}$;
(ii) A Kenmotsu space form, if $f_1 = \frac{c-3}{4}$ and $f_2 = f_3 = \frac{c+1}{4}$;
(iii) A cosymplectic space form, if $f_1 = f_2 = f_3 = \frac{c}{4}$.

Let $\overline{M}(f_1, f_2, f_3)$ be a $(2m+1)$-dimensional generalized Sasakian space form endowed with a semi-symmetric non-metric connection $\overline{\nabla}$. From (3) and (5), it follows that the curvature tensor $\overline{R}$ of the semi-symmetric non-metric connection $\overline{\nabla}$ has the expression

$$\begin{aligned}\overline{R}(X,Y)Z = {} & f_1[g(Y,Z)X - g(X,Z)Y] \\ & + f_2[g(X,\phi Z)\phi Y - g(Y,\phi Z)\phi X + 2g(X,\phi Y)\phi Z] \\ & + f_3[\eta(X)\eta(Z)Y - \eta(Y)\eta(Z)X + g(X,Z)\eta(Y)\xi - g(Y,Z)\eta(X)\xi] \\ & + s(X,Z)g(Y,W) - s(Y,Z)g(X,W).\end{aligned} \tag{6}$$

The vector field P on M can be written as $P = P^\top + P^\perp$, where $P^\top$ and $P^\perp$ are its tangential and normal components, respectively.

The Gauss equation for the semi-symmetric non-metric connection is (see [11])

$$\begin{aligned}\overline{R}(X,Y,Z,W) = {} & R(X,Y,Z,W) + g(h(X,Z),h(Y,W)) - g(h(X,W),h(Y,Z)) \\ & + g(P^\perp,h(Y,Z))g(X,W) - g(P^\perp,h(X,Z))g(Y,W),\end{aligned} \tag{7}$$

for all vector fields X, Y, Z and W on M, where $R(X,Y)Z = \nabla_X\nabla_Y Z - \nabla_Y\nabla_X Z - \nabla_{[X,Y]}Z$ is the curvature tensor of ∇ and $R(X,Y,Z,W) = g(R(X,Y)Z,W)$.

Because the connection ∇ is not metric, $R(X,Y,Z,W) \neq R(X,Y,W,Z)$; then, we cannot define a sectional curvature on M by the standard definition. We will consider a sectional curvature for a semi-symmetric non-metric connection (for the motivation, see [22]) as follows.

If p is a point in M and $\pi \subset T_pM$ a 2-plane section at p spanned by the orthonormal vectors e_1, e_2, the sectional curvature $K(\pi)$ corresponding to the induced connection ∇ can be defined by

$$K(\pi) = \frac{1}{2}[R(e_1,e_2,e_2,e_1) - R(e_1,e_2,e_1,e_2)]. \tag{8}$$

One can see that this definition does not depend on the orthonormal basis.

The scalar curvature τ of M is defined by

$$\tau(p) = \sum_{1 \leq i < j \leq n} K_{ij}, \tag{9}$$

where K_{ij} denotes the sectional curvature of the 2-plane section spanned by e_i and e_j.

Let M be an $(n+1)$-dimensional submanifold tangent to ξ and $\{e_1, e_2, \ldots, e_n, e_{n+1} = \xi\}$, an orthonormal basis of the tangent space T_pM at $p \in M$; then, from (9), the scalar curvature τ of M at p takes the following form:

$$2\tau = \sum_{1 \leq i \neq j \leq n} K(e_i \wedge e_j) + 2\sum_{i=1}^{n} K(e_i \wedge \xi). \tag{10}$$

Denote by $(\inf K)(p) = \inf\{K(\pi); \pi \subset T_pM, \dim\pi = 2\}$.

B.-Y. Chen defined the invariant δ_M by

$$\delta_M(p) = \tau(p) - \inf K(p). \tag{11}$$

Let L be a k-plane section of T_pM and $X \in L$ a unit vector. For an orthonormal basis $\{e_1 = X, e_2, ..., e_k\}$ of L, the Ricci curvature Ric_L of L at X is defined by

$$\mathrm{Ric}_L(X) = K_{12} + K_{13} + \ldots + K_{1k}. \tag{12}$$

It is called the k-Ricci curvature.

Recall that the mean curvature vector $H(p)$ at $p \in M$ is defined by

$$H(p) = \frac{1}{n+1}\sum_{i=1}^{n+1} h(e_i, e_i). \tag{13}$$

Denoting by $h_{ij}^r = g(h(e_i, e_j), e_r)$, $i, j = 1, \ldots, n+1, r \in \{n+2, \ldots, 2m+1\}$, the squared norm of the second fundamental form h is

$$\|h\|^2 = \sum_{r=n+2}^{2m+1}\sum_{i,j=1}^{n+1} (h_{ij}^r)^2.$$

Obviously, from the definition of the vector field P, one has

$$\omega(H) = \frac{1}{n+1}\sum_{i=1}^{n+1} g(P, h(e_i, e_i)) = g(P^\perp, H). \tag{14}$$

For any $X \in TM$, we can write $\phi X = PX + FX$, where PX and FX are the tangential and the normal parts of ϕX, respectively. Let

$$\|P\|^2 = \sum_{i,j=1}^{n} g^2(Pe_i, e_j).$$

Lemma 1. *Let M be an $(n+1)$-dimensional submanifold tangent to ξ of a $(2m+1)$-dimensional trans-Sasakian manifold $\overline{M}$. Then, one has the following:*
(i) $h(\xi, \xi) = 0$;
(ii) $h(X, \xi) = -\alpha FX$, for any vector field X tangent to M orthogonal to ξ.

Proof. Let $p \in M$ and $X \in T_pM$; then, we have

$$\overline{\nabla}^0_X\xi = -\alpha\phi X + \beta(X - \eta(X)\xi).$$

By the Gauss formula, we get

$$h(X,\xi) = -\alpha FX.$$

Taking $X = \xi$, we obtain (i), and taking X orthogonal to ξ we obtain (ii). □

Lemma 2 ([12]). *Let $f(x_1, x_2, \ldots, x_n)$, $n \geq 3$ be a real function on $\mathbb{R}^n$ defined by*

$$f(x_1, x_2, \ldots, x_n) = (x_1 + x_2) \sum_{i=3}^{n} + \sum_{3 \leq i < j \leq n} x_i x_j.$$

If $x_1 + x_2 + \ldots + x_n = (n-1)a$, then

$$f(x_1, x_2, \ldots, x_n) \leq \frac{(n-1)(n-2)}{2} a^2.$$

The equality holds if and only if $x_1 + x_2 = x_3 = \ldots = x_n = a$.

Lemma 3 ([12]). *Let $f(x_1, x_2, \ldots, x_n)$, $n \geq 3$ be a real function on $\mathbb{R}^n$ defined by*

$$f(x_1, x_2, \ldots, x_n) = x_1 \sum_{i=2}^{n} x_i + \sum_{i=2}^{n} x_i.$$

If $x_1 + x_2 + \ldots + x_n = 2a$, then we have

$$f(x_1, x_2, \ldots, x_n) \leq a^2.$$

The equality holds if and only if $x_1 = x_2 + x_3 + \ldots + x_n = a$.

3. Chen First Inequality

Referring to the work of C. Özgür and A. Mihai [17], they used modifications of the Gauss equation for a semi-symmetric non-metric connection. They subsequently introduced a different concept of sectional curvature by utilizing the modified Gauss equation through the formula $\Omega(X) = s(X,X) + g(P^{\perp}, h(X,X))$. Here, we consider another sectional curvature which was defined above.

In the present section, we obtain Chen's first inequality for submanifolds of trans-Sasakian generalized Sasakian space forms admitting a semi-symmetric non-metric connection.

Theorem 1. *Let M be an $(n+1)$-dimensional $(n \geq 2)$ submanifold tangent to ξ of a (α, β) trans-Sasakian generalized Sasakian space form $\overline{M}(f_1, f_2, f_3)$ admitting a semi-symmetric non-metric connection, $p \in M$ and $\pi \subset T_pM$ a 2-plane section orthogonal to ξ. Then, one has*

$$\begin{aligned} \tau(p) - K(\pi) \leq\ & \frac{(n-2)(n+1)}{2} f_1 + \frac{3}{2} f_2 \{\|P\|^2 - \psi^2(\pi)\} - n f_3 \\ & - \alpha^2 \|F\|^2 - \frac{n}{2} \operatorname{trace} s - \frac{n(n+1)}{2} \omega(H) \\ & + \frac{1}{2} \operatorname{trace}(s_{|\pi}) + \frac{1}{2} g(\operatorname{trace}(h_{|\pi}), P) \\ & + \frac{n^2(n-2)}{2(n-1)} \|H\|^2. \end{aligned} \tag{15}$$

Proof. Let $\overline{M}(f_1, f_2, f_3)$ be a $(2m+1)$-dimensional (α, β) trans-Sasakian generalized Sasakian space form, $\overline{\nabla}$ a semi-symmetric non-metric connection on $\overline{M}(f_1, f_2, f_3)$ and M an $(n+1)$-dimensional submanifold tangent to ξ.

Let $p \in M$, $\pi \subset T_pM$ be a 2-plane section orthogonal to ξ and $\{e_1, \ldots, e_n, e_{n+1} = \xi\}$ be an orthonormal basis of the tangent space T_pM and $\{e_{n+2}, \ldots, e_{2m+1}\}$ an orthonormal basis of the normal space $T^{\perp}{}_pM$, with $Fe_j = \|Fe_j\| e_{n+j+1}, \forall j = 1, \ldots, n$.

We will use formula (10).

If we take $X = W = e_i$, $Y = Z = e_j$, $i, j = 1, \ldots, n$, in the Gauss equation, the scalar curvature τ is expressed by

$$2\tau(p) = \sum_{1 \le i \neq j \le n} \overline{R}(e_i, e_j, e_j, e_i) + 2 \sum_{r=n+2}^{2m+1} \sum_{1 \le i < j \le n} [h^r_{ii} h^r_{jj} - (h^r_{ij})^2] - \sum_{1 \le i \neq j \le n} g(P^{\perp}, h(e_j, e_j)) + 2 \sum_{j=1}^{n} K(\xi \wedge e_j). \quad (16)$$

We calculate $\overline{R}(e_i, e_j, e_j, e_i)$ using formula (6) and put $X = W = e_i$, $Y = Z = e_j$, for $i, j = 1, \ldots n, i \neq j$. We have

$$\overline{R}(e_i, e_j, e_j, e_i) = f_1 + 3f_2 g^2(\phi e_i, e_j) - s(e_j, e_j). \quad (17)$$

Introducing Equation (17) into (16), one has

$$2\tau(p) = n(n-1)f_1 + 3f_2 \sum_{i \neq j} g^2(\phi e_i, e_j) - (n-1)\lambda + 2 \sum_{r=1}^{n} \sum_{1 \le i < j \le n} [h^r_{ii} h^r_{jj} - (h^r_{ij})^2] - (n+1)(n-1)\omega(H) + 2 \sum_{j=1}^{n} K(\xi \wedge e_j), \quad (18)$$

where we denoted $\lambda = \sum_{j=1}^{n} s(e_j, e_j)$.

From our definition of the sectional curvature, we obtain

$$K(\xi \wedge e_j) = \frac{1}{2}[R(\xi, e_j, e_j, \xi) - R(\xi, e_j, \xi, e_j)]. \quad (19)$$

Take $X = W = \xi$, $Y = Z = e_j$, for $j = 1, \ldots, n$, in the Gauss equation. We find

$$R(\xi, e_j, e_j, \xi) = \overline{R}(\xi, e_j, e_j, \xi) + g(h(\xi, \xi), h(e_j, e_j)) - g(h(\xi, e_j), h(\xi, e_j)) - g(P^{\perp}, h(e_j, e_j)) g(\xi, \xi). \quad (20)$$

We can rewrite the last equation as

$$R(\xi, e_j, e_j, \xi) = \overline{R}(\xi, e_j, e_j, \xi) + \sum_{r=n+2}^{2m+1} [h^r_{jj} h^r_{\xi\xi} - (h^r_{j\xi})^2] - g(P^{\perp}, h(e_j, e_j)). \quad (21)$$

By formula (6) we have

$$\overline{R}(\xi, e_j, e_j, \xi) = f_1 - f_3 - s(e_j, e_j), \forall j = 1, \ldots, n. \quad (22)$$

Introducing (22) into (21), one has

$$R(\xi, e_j, e_j, \xi) = f_1 - f_3 - s(e_j, e_j) + \sum_{r=n+2}^{2m+1} [h^r_{jj} h^r_{\xi\xi} - (h^r_{j\xi})^2] - g(P^{\perp}, h(e_j, e_j)). \quad (23)$$

By using Lemma 1, we obtain

$$\sum_{j=1}^{n} \sum_{r=n+2}^{2m+1} (h^r_{j\xi})^2 = \sum_{j=1}^{n} \sum_{r=n+2}^{2m+1} g^2(h(e_j, \xi), e_r) = \alpha^2 \sum_{j=1}^{n} \sum_{r=n+2}^{2m+1} g^2(Fe_j, e_r)$$

$$= \alpha^2 \sum_{j=1}^{n} \|Fe_j\|^2 = \alpha^2 \|F\|^2.$$

Then, Equation (23) can be rewritten as

$$R(\xi, e_j, e_j, \xi) = f_1 - f_3 - s(e_j, e_j) - \alpha^2 \|F\|^2 - g(P^{\perp}, h(e_j, e_j)). \tag{24}$$

Similarly, from the Gauss equation, if we put $X = Z = \xi$, $Y = W = e_j$, for $j = 1, \ldots n$, we have

$$R(\xi, e_j, \xi, e_j) = -f_1 + f_3 + s(\xi, \xi) + \alpha^2 \|Fe_j\|^2. \tag{25}$$

By substituting (24) and (25) in (20), and taking summation, we find

$$\sum_{j=1} K(\xi \wedge e_j) = \frac{1}{2}[2nf_1 - 2nf_3 - 2\alpha^2 \|F\|^2 - \lambda - ns(\xi, \xi) - (n+1)\omega(H)]. \tag{26}$$

If we put (26) in (18), we obtain

$$\begin{aligned} 2\tau(p) = {} & n(n+1)f_1 + 3f_2\|P\|^2 - 2nf_3 \\ & - 2\alpha^2\|F\|^2 - n\lambda - ns(\xi, \xi) - n(n+1)\omega(H) \\ & + 2\sum_{r=n+2}^{2m+1} \sum_{1 \le i < j \le n} [h^r_{ii} h^r_{jj} - (h^r_{ij})^2]. \end{aligned} \tag{27}$$

Let $\pi = \text{span}\{e_1, e_2\}$. In the Gauss equation, we put $X = W = e_1$, $Y = Z = e_2$. Then,

$$\begin{aligned} R(e_1, e_2, e_2, e_1) = {} & f_1 + 3f_2 g^2(\phi e_1, e_2) - s(e_2, e_2) \\ & + \sum_{r=n+1}^{2m+1} [h^r_{11} h^r_{22} - (h^r_{12})^2] - g(P^{\perp}, h(e_2, e_2)). \end{aligned} \tag{28}$$

Similarly, if we put $X = Z = e_1$, $Y = W = e_2$, in the Gauss equation,

$$\begin{aligned} R(e_1, e_2, e_1, e_1) = {} & -f_1 - 3f_2 g^2(\phi e_1, e_2) + s(e_1, e_1) \\ & - \sum_{r=n+2}^{2m+1} [h^r_{11} h^r_{22} - (h^r_{12})^2] + g(P^{\perp}, h(e_1, e_1)). \end{aligned} \tag{29}$$

So from (8), (28) and (29), we have

$$\begin{aligned} K(\pi) = {} & f_1 + 3f_2 g^2(\phi e_1, e_2) \\ & - \frac{1}{2}\texttt{trace}(s_{|\pi}) - \frac{1}{2} g(\texttt{trace}(h_{|\pi}), P)] \\ & + \sum_{r=n+2}^{2m+1} [h^r_{11} h^r_{22} - (h^r_{12})^2]. \end{aligned} \tag{30}$$

We denote $\psi^2(\pi) = g^2(\phi e_1, e_2)$; then

$$\begin{aligned}\tau(p)-K(\pi)=&\frac{(n-2)(n+1)}{2}f_1+\frac{3}{2}f_2\{\|P\|^2-\psi^2(\pi)\}-nf_3\\&-\alpha^2\|F\|^2-\frac{n}{2}\lambda-\frac{n}{2}s(\xi,\xi)-\frac{n(n+1)}{2}\omega(H)\\&+\frac{1}{2}\texttt{trace}(s_{|\pi})+\frac{1}{2}g(\texttt{trace}(h_{|\pi}),P)]\\&+\sum_{r=n+2}^{2m+1}\{(h^r_{11}+h^r_{22})\sum_{3\le i\le n}h^r_{ii}+\sum_{3\le i<j\le n}h^r_{ii}h^r_{jj}\\&-\sum_{3\le j\le n}(h^r_{1j})^2-\sum_{2\le i<j\le n}(h^r_{ij})^2\},\end{aligned}\tag{31}$$

which implies

$$\begin{aligned}\tau(p)-K(\pi)\le&\frac{(n-2)(n+1)}{2}f_1+\frac{3}{2}f_2\{\|P\|^2-\psi^2(\pi)\}-nf_3\\&-\alpha^2\|F\|^2-\frac{n}{2}\lambda-\frac{n}{2}s(\xi,\xi)-\frac{n(n+1)}{2}\omega(H)\\&+\frac{1}{2}\texttt{trace}(s_{|\pi})+\frac{1}{2}g(\texttt{trace}(h_{|\pi}),P)]\\&+\sum_{r=n+2}^{2m+1}\{(h^r_{11}+h^r_{22})\sum_{3\le i\le n}h^r_{ii}+\sum_{3\le i<j\le n}h^r_{ii}h^r_{jj}.\end{aligned}\tag{32}$$

We define the real functions $f_r:\mathbb{R}^n\to\mathbb{R}$ by

$$f_r(h^r_{11},h^r_{22},\dots,h^r_{nn})=(h^r_{11}+h^r_{22})\sum_{3\le i\le n}h^r_{ii}+\sum_{3\le i<j\le n}h^r_{ii}h^r_{jj}$$

We study the problem max f_r, under the condition $h^r_{11}+h^r_{22}+\dots+h^r_{nn}=b^r$, where b^r is a real number.

Lemma 2 implies that the solution $(h^r_{11},h^r_{22},\dots,h^r_{nn})$ must satisfy

$$h^r_{11}+h^r_{22}=h^r_{ii}=\frac{b^r}{(n-1)},\quad i=3,\dots n,$$

which gives

$$f_r\le\frac{(n-2)}{2(n-1)}(b^r)^2.\tag{33}$$

By using (32) and (33), it follows that

$$\begin{aligned}\tau(p)-K(\pi)\le&\frac{(n-2)(n+1)}{2}f_1+\frac{3}{2}f_2\{\|P\|^2-\psi^2(\pi)\}-nf_3\\&-\alpha^2\|F\|^2-\frac{n}{2}\texttt{trace}\,s-\frac{n(n+1)}{2}\omega(H)+\frac{1}{2}\texttt{trace}(s_{|\pi})+\frac{1}{2}g(\texttt{trace}(h_{|\pi}),P)\\&+\frac{n^2(n-2)}{2(n-1)}\|H\|^2.\end{aligned}\tag{34}$$

Then the proof is achieved. □

4. Chen–Ricci Inequality

In [2], B.-Y. Chen established a sharp estimate of the mean curvature in terms of the Ricci curvature for all n-dimensional Riemannian submanifolds in a Riemannian space form $\overline{M}(c)$ of constant sectional curvature c.

$$\text{Ric}\,(X) \le (n-1)c + \frac{n^2}{4}\|H\|^2,$$

It is known as the Chen–Ricci inequality.

One of the present authors [23] derived a Chen–Ricci inequality specifically for submanifolds in Sasakian space forms.

In this section, we obtain a Chen–Ricci inequality for submanifolds tangent to ξ in a trans-Sasakian manifold endowed with a semi-symmetric non-metric connection.

Theorem 2. *Let $\overline{M}(f_1, f_2, f_3)$ be a $(2m+1)$-dimensional (α, β) trans-Sasakian generalized Sasakian space form, $\overline{\nabla}$ a semi-symmetric non-metric connection on it and M an $(n+1)$-dimensional $(n \ge 2)$ submanifold tangent to ξ. Then, we have the following:*

(1) For any unit vector $X \in T_pM$ orthogonal to ξ,

$$\begin{aligned}\text{Ric}(X) \le\; & \frac{n^2}{4}\|H\|^2 + nf_1 + 3f_2\|Pe_1\|^2 - f_3 - \alpha^2\|F\|^2 \\ & - \frac{1}{2}[\text{trace}\, s + (n-1)s(X,X)] \\ & - \frac{1}{2}[(n+1)\omega(H) + (n-1)g(P^{\perp}, h(X,X))]. \end{aligned} \tag{35}$$

(2) If $H(p) = 0$, then a unit tangent vector X at p satisfies the equality case of (35) if and only if $X \in N_p$, where $N_p = \{X \in T_pM | h(X,Y) = 0, \forall Y \in \{\xi\}^{\perp}\}$.

(3) The equality case of (35) holds identically for all unit tangent vectors orthogonal to ξ at p if and only if either

(i) h_p vanishes on $\{\xi\}^{\perp} \times \{\xi\}^{\perp}$ or

(ii) $n = 2$ and $h(X,Y) = g(X,Y)H$, for any $X, Y \in T_pM$ orthogonal to ξ.

Proof.

(1) Let $p \in M$, $X \in T_pM$ a unit tangent vector orthogonal to ξ. Consider an orthonormal basis $\{e_1, \ldots, e_n, e_{n+1} = \xi, e_{n+2}, \ldots, e_{2m+1}\}$ in $T_p\overline{M}(f_1, f_2, f_3)$, with $e_1 = X, e_2, \ldots, e_n$ tangent to M at p.

$$\texttt{Ric}(X) = \sum_{j=2}^{n} K(e_1 \wedge e_j) + K(e_1 \wedge \xi). \tag{36}$$

If we take $X = W = e_1$ and $Y = Z = e_j$ in the Gauss equation, we have

$$\begin{aligned} R(e_1, e_j, e_j, e_1) =\; & f_1 + 3f_2 g^2(\phi e_1, e_j) - s(e_j, e_j) \\ & + \sum_{r=n+2}^{2m+1} [h^r_{11}h^r_{jj} - (h^r_{1j})^2] - g(P^{\perp}, h(e_j, e_j)), \end{aligned} \tag{37}$$

respectively. From the Gauss equation, if we put $X = Z = e_1$, $Y = W = e_j$, we have

$$\begin{aligned} R(e_1, e_j, e_1, e_j) =\; & -f_1 - 3f_2 g^2(\phi e_1, e_j) + s(e_1, e_1) \\ & - \sum_{r=1}^{n} [h^r_{11}h^r_{jj} - (h^r_{1j})^2] + g(P^{\perp}, h(e_1, e_1)). \end{aligned} \tag{38}$$

Similarly to Equation (8), we have

$$K(e_1 \wedge e_j) = \frac{1}{2}[R(e_1, e_j, e_j, e_1) - R(e_1, e_j, e_1, e_j)]. \tag{39}$$

From Equations (37)–(39), we have

$$\begin{aligned} K(e_1 \wedge e_j) = {} & f_1 + 3f_2 g^2(\phi e_1, e_j) - \frac{1}{2}[s(e_j, e_j) + s(e_1, e_1)] \\ & + \sum_{r=n+2}^{2m+1} [h_{11}^r h_{jj}^r - (h_{1j}^r)^2] \\ & - \frac{1}{2}[g(P^\perp, h(e_j, e_j)) + g(P^\perp, h(e_1, e_1))]. \end{aligned} \tag{40}$$

On the other hand, one has

$$\begin{aligned} K(e_1 \wedge \xi) &= \frac{1}{2}[R(\xi, e_1, e_1, \xi) - R(\xi, e_1, \xi, e_1)] \\ &= f_1 - f_3 - \alpha^2 \|Fe_1\|^2 \\ &\quad - \frac{1}{2}[s(e_1, e_1) + g(P^\perp, h(e_1, e_1)) + s(\xi, \xi)]. \end{aligned} \tag{41}$$

By substituting Equations (40) and (41) in (36), we find

$$\begin{aligned} \mathtt{Ric}(X) = {} & nf_1 + 3f_2 \sum_{j=2}^{n} g^2(\phi e_1, e_j) - f_3 \\ & - \frac{1}{2}[\operatorname{trace} s + (n-1)s(X, X)] \\ & - \frac{1}{2}[(n+1)\omega(H) + (n-1)g(P^\perp, h(X, X))] \\ & + \sum_{j=2}^{n} \sum_{r=n+2}^{2m+1} [h_{11}^r h_{jj}^r - (h_{1j}^r)^2] - \alpha^2 \|F\|^2. \end{aligned} \tag{42}$$

Obviously, one has

$$h_{11}^r \left(\sum_{i=2}^{n} h_{ii}^r \right) \leq \frac{1}{4}(h_{11}^r + h_{22}^r + \ldots + h_{nn}^r)^2, \tag{43}$$

and equality holds if and only if

$$h_{11}^r = \sum_{i=2}^{n} h_{ii}^r. \tag{44}$$

From Equations (42) and (43), we have

$$\begin{aligned} \mathtt{Ric}(X) \leq {} & \frac{n^2}{4}\|H\|^2 + nf_1 + 3f_2\|Pe_1\|^2 - f_3 - \alpha^2\|F\|^2 \\ & - \frac{1}{2}[\operatorname{trace} s + (n-1)s(X, X)] \\ & - \frac{1}{2}[(n+1)\omega(H) + (n-1)g(P^\perp, h(X, X))]. \end{aligned} \tag{45}$$

(2) If a unit vector X at p satisfies the equality case of (35), from (42), (43) and (44), one obtains

$$\begin{cases} h_{1i}^r = 0, \ 2 \leq i \leq n, \ \forall r \in \{n+2, \ldots, 2m+1\}, \\ h_{11}^r = h_{22}^r + \ldots + h_{nn}^r, \ \forall r \in \{n+2, \ldots, 2m+1\}. \end{cases}$$

Therefore, because $H(p) = 0$, we have $h^r_{1j} = 0$ for all $j = 1, \ldots, n$, $r \in \{n+2, \ldots, 2m+1\}$; that is, $X \in N_p$.

(3) The equality case of inequality (35) holds for all unit tangent vectors at p if and only if

$$\begin{cases} h^r_{ij} = 0, \quad 1 \leq i \neq j \leq n, \quad r \in \{n+2, \ldots, 2m+1\}, \\ h^r_{11} + \ldots + h^r_{nn} - 2h^r_{ii} = 0, \quad i \in \{1, \ldots, n\}, \quad r \in \{n+2, \ldots, 2m+1\}. \end{cases}$$

There are two cases:

(i) $n \neq 2$, $h^r_{ij} = 0$. It follows that h_p vanishes on $\{\xi\}^\perp \times \{\xi\}^\perp$.

(ii) $n = 2$; then, $h(X,Y) = h(X,Y)H$, for any $X, Y \in \{\xi\}^\perp$.

□

We recall standard definitions of certain classes of submanifolds in trans-Sasakian manifolds.

Let $\overline{M}$ be a trans-Sasakian manifold and M a submanifold of $\overline{M}$ tangent to the Reeb vector field ξ.

According to the behaviour of the tangent spaces of M under the action of ϕ, we distinguish the following classes of submanifolds.

The submanifold M of $\overline{M}$ is an invariant submanifold if all its tangent spaces are invariant by ϕ, i.e., $\phi(T_pM) \subset T_pM, \forall p \in M$.

The submanifold M of $\overline{M}$ is an anti-invariant submanifold if ϕ maps any tangent space into the normal space, i.e., $\phi(T_pM) \subset T_p^\perp M, \forall p \in M$.

The submanifold M is a slant submanifold if for any $p \in M$ and any $X \in T_pM$, linearly independent on ξ, the angle θ between ϕX and T_pM is constant. The angle $\theta \in [0, \frac{\pi}{2}]$ is called the slant angle of M in $\overline{M}$.

We state the corresponding Chen–Ricci inequalities for the above submanifolds.

Corollary 1. *Let $\overline{M}(f_1, f_2, f_3)$ be a $(2m+1)$-dimensional (α, β) trans-Sasakian generalized Sasakian space form, $\overline{\nabla}$ a semi-symmetric non-metric connection on it and M an $(n+1)$-dimensional $(n \geq 2)$ invariant submanifold.*

Then, for each unit vector $X \in T_pM$ orthogonal to ξ, we have

$$\begin{aligned} \mathrm{Ric}(X) \leq & \frac{n^2}{4}\|H\|^2 + nf_1 + 3f_2 - f_3 \\ & - \frac{1}{2}[\mathrm{trace}\, s + (n-1)s(X,X)] \\ & - \frac{1}{2}[(n+1)\omega(H) + (n-1)g(P^\perp, h(X,X))]. \end{aligned} \tag{46}$$

Corollary 2. *Let $\overline{M}(f_1, f_2, f_3)$ be a $(2m+1)$-dimensional (α, β) trans-Sasakian generalized Sasakian space form, $\overline{\nabla}$ a semi-symmetric non-metric connection on it and M an $(n+1)$-dimensional $(n \geq 2)$ anti-invariant submanifold.*

Then, for each unit vector $X \in T_pM$ orthogonal to ξ, we have

$$\begin{aligned} \mathrm{Ric}(X) \leq & \frac{n^2}{4}\|H\|^2 + nf_1 - f_3 - n\alpha^2 \\ & - \frac{1}{2}[\mathrm{trace}\, s + (n-1)s(X,X)] \\ & - \frac{1}{2}(n+1)\omega(H) + (n-1)g(P^\perp, h(X,X))]. \end{aligned} \tag{47}$$

Corollary 3. *Let $\overline{M}(f_1, f_2, f_3)$ be a $(2m+1)$-dimensional (α, β) trans-Sasakian generalized Sasakian space form, $\overline{\nabla}$ a semi-symmetric non-metric connection on it and M an $(n+1)$-dimensional $(n \geq 2)$ slant submanifold.*

Then, for each unit vector $X \in T_pM$ orthogonal to ξ, we have

$$\begin{aligned} \mathrm{Ric}(X) \leq & \frac{n^2}{4}\|H\|^2 + nf_1 + 3f_2\cos^2\theta - f_3 - n\alpha^2\sin^2\theta \\ & - \frac{1}{2}[\mathrm{trace}\, s + (n-1)s(X,X)] \\ & - \frac{1}{2}[(n+1)\omega(H) + (n-1)g(P^{\perp}, h(X,X))]. \end{aligned} \quad (48)$$

5. Generalized Euler Inequality for Special Contact Slant Submanifolds

B.Y. Chen [24] proved a generalized Euler inequality for n-dimensional submanifolds in a Riemannian space form of constant sectional curvature c:

$$\|H\|^2 \geq \frac{2\tau}{n(n-1)} - c,$$

with equality holding identically if and only if the submanifold is totally umbilical.

In this section, we prove a generalized Euler inequality for certain submanifolds in a trans-Sasakian manifold endowed with a semi-symmetric non-metric connection.

In [18], we extended the definition of a special slant submanifold in a Sasakian manifold to trans-Sasakian manifolds.

Let M be a proper slant submanifold ($\theta \neq 0, \frac{\pi}{2}$) of a trans-Sasakian manifold $\overline{M}$. We call M a special contact slant submanifold if

$$(\nabla^0_X P)Y = \cos^2\theta[\alpha(g(X,Y)\xi - \eta(Y)X) + \beta(g(\phi X, Y)\xi - \eta(Y)\phi X0]. \forall X, Y \in \Gamma(TM).$$

Then, the components of the second fundamental form are symmetric, i.e.,

$$h^k_{ij} = h^i_{jk} = h^j_{ik}, \forall i,j,k = 1,\ldots,n.$$

For special contact slant submanifolds, we prove a generalized Euler inequality.

Theorem 3. *Let $\overline{M}(f_1, f_2, f_3)$ be a $(2n+1)$-dimensional (α, β) trans-Sasakian generalized Sasakian space form, $\overline{\nabla}$ a semi-symmetric non-metric connection on it and M an $(n+1)$-dimensional $(n \geq 2)$ special contact slant submanifold. Then,*

$$\begin{aligned} \|H\|^2 \geq & \frac{2(n+2)}{(n-1)(n+1)^2}\tau - \frac{n(n+2)}{n^2-1}f_1 \\ & + 3\frac{n(n+2)}{(n-1)(n+1)^2}f_2\cos^2\theta + \frac{2n(n+2)}{(n-1)(n+1)^2}f_3 + \frac{2n(n+2)}{(n-1)(n+1)^2}\alpha^2\sin^2\theta \\ & - \frac{n(n+2)}{(n-1)(n+1)^2}[\mathrm{trace}\, s + (n+1)\omega(H)]. \end{aligned} \quad (49)$$

Proof. Consider a $(2n+1)$-dimensional (α, β) trans-Sasakian generalized Sasakian space form $\overline{M}(f_1, f_2, f_3)$ with a semi-symmetric non-metric connection $\overline{\nabla}$ and M an $(n+1)$-dimensional special contact slant submanifold.

For any $p \in M$ and $\pi \subset T_pM$, a 2-plane section orthogonal to ξ, let $\{e_1, \ldots, e_n, e_{n+1} = \xi\}$ be an orthonormal basis of the tangent space T_pM and $\{e_{n+2}, \ldots, e_{2n+1}\}$ an orthonormal basis of the normal space $T^{\perp}{}_pM$, with $Fe_j = (\sin\theta)e_{n+j+1}, \forall j = 1, \ldots, n$.

In this case, Equation (27) becomes

$$\begin{aligned} 2\tau(p) = & n(n+1)f_1 + 3nf_2\cos^2\theta - 2nf_3 \\ & - 2n\alpha^2\sin^2\theta - n\mathrm{trace}\, s - n(n+1)\omega(H) \\ & - \|h\|^2 + (n+1)^2\|H\|^2. \end{aligned} \quad (50)$$

On the other hand, we have

$$(n+1)^2\|H\|^2 = \sum_i g(h(e_i,e_i),h(e_i,e_i)) + \sum_{i\neq j} g(h(e_i,e_i),h(e_j,e_j))$$
$$= \sum_{i=1}^{n}[\sum_{j=1}^{n}(h^i_{jj})^2 + 2\sum_{1\le j<k\le n} h^i_{jj}h^i_{kk}]. \tag{51}$$

From Equations (50) and (51), we obtain

$$2\tau(p) = n(n+1)f_1 + 3nf_2\cos^2\theta - 2nf_3$$
$$- 2n\alpha^2\sin^2\theta - n\,\mathrm{trace}\, s - n(n+1)\omega(H)$$
$$+2\sum_i\sum_{j<k} h^i_{jj}h^i_{kk} - 2\sum_{i\neq j}(h^i_{jj})^2 - 6\sum_{i<j<k}(h^k_{ij})^2, \tag{52}$$

Let us now introduce a parameter m given by $m = \frac{n+2}{n-1}$, with $n \geq 2$, for studying the inequality of $\|H\|^2$ by mimicking the technique used in ([25]). Then, we have

$$(n+1)^2\|H\|^2 - m\{2\tau - n(n+1)f_1 + 3n\cos^2\theta - 2nf_3$$
$$- 2n\alpha^2\sin^2\theta - n\,\mathrm{trace}\, s - n(n+1)\omega(H)\}$$
$$= \sum_i(h^i_{ii})^2 + (1+2m)\sum_{i\neq j}(h^i_{jj})^2 + 6m\sum_{i<j<k}(h^k_{ij})^2$$
$$- 2(m-1)\sum_i\sum_{j<k} h^i_{jj}h^i_{kk}$$
$$= \sum_i(h^i_{ii})^2 + 6m\sum_{i<j<k}(h^k_{ij})^2 + (m-1)\sum_i\sum_{j<k}(h^i_{jj} - h^i_{kk})^2$$
$$+ \{1+2m-(n-2)(m-1)\}\sum_{i\neq j}(h^i_{jj})^2 - 2(m-1)\sum_{i\neq j}h^i_{ii}h^i_{jj}$$
$$= 6m\sum_{i<j<k}(h^k_{ij})^2 + (m-1)\sum_{i\neq j,k}\sum_{j<k}(h^i_{jj} - h^i_{kk})^2$$
$$+ \frac{1}{n-1}\sum_{i\neq j}\left\{h^i_{ii} - (n-1)(m-1)h^i_{jj}\right\}^2 \geq 0. \tag{53}$$

It follows that

$$\|H\|^2 \geq \frac{2(n+2)}{(n-1)(n+1)^2}\tau - \frac{n(n+2)}{n^2-1}f_1$$
$$+3\frac{n(n+2)}{(n-1)(n+1)^2}f_2\cos^2\theta + \frac{2n(n+2)}{(n-1)(n+1)^2}f_3 + \frac{2n(n+2)}{(n-1)(n+1)^2}\alpha^2\sin^2\theta$$
$$- \frac{n+2}{(n-1)(n+1)^2}[n\,\mathrm{trace}\, s + n(n+1)\omega(H)]. \tag{54}$$

□

6. Example

We will give an example of a special contact slant submanifold in $\mathbb{R}^5$ with the standard Sasakian strucure, with a semi-symmetric non-metric connection.

Consider on $\mathbb{R}^{2m+1}$ the Sasakian structure $(\mathbb{R}^{2m+1}, \phi_0, \eta, \xi, g)$, given by

$$\eta = \frac{1}{2}\left(dz - \sum_{i=1}^{m} y^i dx^i\right), \quad \xi = 2\frac{\partial}{\partial z},$$
$$g = -\eta \otimes \eta + \frac{1}{4}\sum_{i=1}^{m}(dx^i \otimes dx^i + dy^i \otimes dy^i),$$

$$\phi_0\left(\sum_{i=1}^{m}\left(X_i\frac{\partial}{\partial x^i} + Y_i\frac{\partial}{\partial y^i}\right) + Z\frac{\partial}{\partial z}\right) = \sum_{i=1}^{m}\left(Y_i\frac{\partial}{\partial x^i} - X_i\frac{\partial}{\partial y^i}\right) + \sum_{i=1}^{m} Y_i y^i \frac{\partial}{\partial z},$$

with $\{x^i, y^i, z\}, i = 1, \ldots, m$, the Cartesian coordinates on $\mathbb{R}^{2m+1}$.

A semi-symmetric non-metric connection is given by

$$\nabla_X Y = \nabla^0_X Y + \eta(Y)X.$$

In particular, one derives

$$\phi_0\left(\frac{\partial}{\partial x^i}\right) = -\frac{\partial}{\partial y^i},$$
$$\phi_0\left(\frac{\partial}{\partial x^i}\right) = \frac{\partial}{\partial x^i} + y^i\frac{\partial}{\partial z},$$
$$\phi_0(\xi) = 0.$$

It is known that the ϕ_0-sectional curvature of $\mathbb{R}^{2m+1}$ is -3.

We define a three-dimensional special contact slant submanifold by the equation

$$x(u, v, t) = 2((u + v), k\cos v, v - u, k\sin v, t),$$

in $\mathbb{R}^5$ with the usual Sasakian structure, endowed with the above semi-symmetric non-metric connection.

It is special contact slant submanifold with slant angle $\theta = \cos^{-1}\sqrt{\frac{2}{2+k^2}}$.

An orthonormal frame is given by

$$e_1 = \frac{1}{\sqrt{2}}(1, 0, -1, 0, 0),$$
$$e_2 = \frac{1}{\sqrt{k^2+2}}(1, -k\sin v, 1, k\cos v, 0),$$
$$e_3 = 2(0, 0, 0, 0, 1) = \xi,$$
$$e_4 = \frac{1}{\sin\theta}Fe_1 = e_{1^*},$$
$$e_5 = \frac{1}{\sin\theta}Fe_2 = e_{2^*}.$$

We compute the slant angle and obtain

$$\cos\theta = g(\phi_0 e_2, e_1) = -g(\phi_0 e_1, e_2) = \sqrt{\frac{2}{2+k^2}}$$

Now, we compute the second fundamental form.

Obviously, $h(e_3, e_3) = 0$.

Also, we know from Lemma 1 that $h(e_i, e_3) = -\sin\theta e_{i^*}, \quad i = 1, 2$.

By standard calculations, we obtain

$$h(e_1, e_1) = h(e_1, e_2) = 0$$

and

$$h(e_2, e_2) = \frac{1}{2k^2 + 8}[2(0, -k\cos v, 0, -k\sin v, 0)].$$

Let $\pi = \text{span}\{e_1, e_2\}$. In the Gauss equation, we put $X = W = e_1$, $Y = Z = e_2$. Then,

$$R(e_1, e_2, e_2, e_1) = -3g^2(\phi e_1, e_2) - s(e_2, e_2) + g(h(e_1, e_1), h(e_2, e_2)) - g(h(e_1, e_2), h(e_1, e_2)) - g(P^{\perp}, h(e_2, e_2)).$$

In our case, $s(e_2, e_2) = 0$ and $g(\xi, h(e_2, e_2)) = 0$. Then, $R(e_1, e_2, e_2, e_1) = -\frac{6}{2+k^2}$.

Similarly, $R(e_1, e_2, e_1, e_1) = \frac{6}{k^2+2}$.

Consequently, $K(\pi) = -\frac{6}{2+k^2}$ and $\tau = K(\pi) + 2$.

Also, $H = \frac{1}{3}h(e_2, e_2) \neq 0$, i.e., M is not a minimal submanifold.

7. Conclusions

In this article, we dealt with trans-Sasakian manifolds admitting a semi-symmetric non-metric connection. We considered the sectional curvature defined recently in [22].

We established Chen's first inequality, the Chen–Ricci inequality and the generalized Euler inequality for submanifolds tangent to the Reeb vector field in a trans-Sasakian manifold endowed with a semi-symmetric non-metric connection. Particular cases of such submanifolds were also discussed.

This study can be continued, for instance, to obtain other Chen inequalities or improving the present results for special classes of submanifolds in trans-Sasakian manifolds or in other ambient spaces.

Author Contributions: Conceptualization, F.M.; methodology, A.E.A.M.A.E. and M.M.; software, M.M. and M.S.A.; validation, F.M. and I.M.; formal analysis, F.M. and I.M.; investigation, F.M. and M.M.; resources, I.M.; writing—original draft preparation, M.M. and A.E.A.M.A.E.; writing—review and editing, F.M. and M.M.; visualisation, I.M. and F.M.; supervision, F.M. and I.M.; project administration, F.M. All authors have read and agreed to the published version of the manuscript.

Funding: This research received no external funding.

Institutional Review Board Statement: Not applicable.

Informed Consent Statement: Not applicable.

Data Availability Statement: Data are contained within the article.

Acknowledgments: The researchers would like to acknowledge the Deanship of Scientific Research Taif University for funding this work. The authors are very obliged to the reviewers for their valuable suggestions, which have improved the paper.

Conflicts of Interest: The authors declare no conflicts of interest.

References

1. Chen, B.Y. Some pinching and classification theorems for minimal submanifolds. *Arch. Math.* **1993**, *60*, 568–578. [CrossRef]
2. Chen, B.Y. Relations between Ricci curvature and shape operator for submanifolds with arbitrary codimensions. *Glasg. Math. J.* **1999**, *41*, 33–41. [CrossRef]
3. Aktan, N.; Sarıkaya, M.Z.; Özüsağlam, E. B.Y. Chen's inequality for semi-slant submanifolds in-space forms. *Balk. J. Geom. Appl. (BJGA)* **2008**, *13*, 1–10.
4. Mihai, I.; Presură, I. An improved first Chen inequality for Legendrian submanifolds in Sasakian space forms. *Period. Math. Hung.* **2017**, *74*, 220–226. [CrossRef]
5. Kim, J.S.; Song, Y.M.; Tripathi, M.M. B.-Y. Chen inequalities for submanifolds in generalized complex space forms. *Bull. Korean Math. Soc.* **2003**, *40*, 411–423. [CrossRef]
6. Mohammed, M. Geometric inequalities for bi-slant submanifolds in Kenmotsu space forms. *Rom. J. Math. Comput. Sci* **2022**, *12*, 52–63.
7. Friedmann, A.; Schouten, J.A. Über die Geometrie der halbsymmetrischen Übertragungen. *Math. Z.* **1924**, *21*, 211–223. [CrossRef]
8. Hayden, H. Sub-Spaces of a Space with Torsion. *Proc. Lond. Math. Soc.* **1932**, *2*, 27–50. [CrossRef]
9. Yano, K. On semi symmetric metric connection. *Rev. Roum. Math. Pures Appl.* **1970**, *15*, 1579–1591.

10. Agashe, N.S. A semi-symmetric non-metric connection on a Riemannian manifold. *Indian. J. Pure Appl. Math.* **1992**, *23*, 399–409.
11. Agashe, N.S.; Chafle, M.R. On submanifolds of a Riemannian manifold with a semi-symmetric non-metric connection. *Tensor* **1994**, *55*, 120–130.
12. Zhang, P.; Pan, X.; Zhang, L. Inequalities for submanifolds of a Riemannian manifold of nearly quasi-constant curvature with a semi-symmetric non-metric connection. *Rev. Un. Mat. Argent.* **2015**, *56*, 1–19.
13. He, G.; Liu, H.; Zhang, L. Optimal inequalities for the Casorati curvatures of submanifolds in generalized space forms endowed with semi-symmetric non-metric connections. *Symmetry* **2016**, *8*, 113. [CrossRef]
14. Dogru, Y. Chen Inequalities for Submanifolds of Some Space Forms Endowed With A Semi-Symmetric Non-Metric Connection. *Jordan J. Math. Stat. (JJMS)* **2013**, *6*, 313–339.
15. Mihai, A.; Özgür, C. Chen inequalities for submanifolds of real space forms with a semi-symmetric metric connection. *Taiwan. J. Math.* **2010**, *14*, 1465–1477. [CrossRef]
16. Mihai, A.; Özgür, C. Chen inequalities for submanifolds of complex space forms and Sasakian space forms endowed with semi-symmetric metric connections. *Rocky Mt. J. Math.* **2011**, *41*, 1653–1673. [CrossRef]
17. Özgür, C.; Mihai, A. Chen inequalities for submanifolds of real space forms with a semi-symmetric non-metric connection. *Can. Math. Bull.* **2012**, *55*, 611–622. [CrossRef]
18. Mihai, I.; Mohammed, M. Optimal Inequalities for Submanifolds in Trans-Sasakian Manifolds Endowed with a Semi-Symmetric Metric Connection. *Symmetry* **2023**, *15*, 877. [CrossRef]
19. Rustanov, A.R.; Melekhina, T.L.; Kharitonova, S.V. On the geometry of nearly trans-Sasakian manifolds. *Turk. J. Math.* **2023**, *47*, 1144–1157. [CrossRef]
20. Mikeš, J. On Sasaki spaces and equidistant Kähler spaces. *Sov. Math. Dokl.* **1987**, *34*, 428–431.
21. Alegre, P.; Blair, D.E.; Carriazo, A. Generalized Sasakian-space-forms. *Isr. J. Math.* **2004**, *141*, 157–183. [CrossRef]
22. Mihai, A.; Mihai, I. A note on a well-defined sectional curvature of a semi-symmetric non-metric connection. *Int. Electron. J. Geom.* **2024**, *17*, *to appear*.
23. Mihai, I. Ricci curvature of submanifolds in Sasakian space forms. *J. Aust. Math. Soc.* **2002**, *72*, 247–256. [CrossRef]
24. Chen, B.Y. Mean curvature and shape operator of isometric immersions in real-space-forms. *Glasg. Math. J.* **1996**, *38*, 87–97. [CrossRef]
25. Mihai, A.; Radulescu, I. Scalar and Ricci curvatures of special contact slant submanifolds in Sasakian space forms. *Adv. Geom.* **2014**, *14*, 147–159. [CrossRef]

axioms MDPI

Article

Cross Curvature Solitons of Lorentzian Three-Dimensional Lie Groups

Shahroud Azami [1], Mehdi Jafari [2], Abdul Haseeb [3] and Abdullah Ali H. Ahmadini [3,*]

[1] Department of Pure Mathematics, Faculty of Science, Imam Khomeini International University, Qazvin P.O. Box 34148-96818, Iran; azami@sci.ikiu.ac.ir
[2] Department of Mathematics, Payame Noor University, Tehran P.O. Box 19395-4697, Iran; m.jafarii@pnu.ac.ir
[3] Department of Mathematics, College of Science, Jazan University, P.O. Box 114, Jazan 45142, Saudi Arabia; haseeb@jazanu.edu.sa or malikhaseeb80@gmail.com
* Correspondence: aahmadini@jazanu.edu.sa

Abstract: In this paper, we study left-invariant cross curvature solitons on Lorentzian three-dimensional Lie groups and classify these solitons.

Keywords: cross curvature soliton; lie groups; homogeneous spaces

MSC: 53C50; 53E20; 53C21

1. Introduction

The study of various geometric flows used to improve a given metric for a geometric object has been undertaken by many mathematicians and physicists. Important geometric flows are the Ricci flow, Yamabe flow, mean curvature flow, Ricci-harmonic flow, and cross-curvature flow. These flows are impressive subjects in mathematical physics and geometry. The special solutions for geometric flow are solitons. In fact, solitons are the self-similar solution to flow. R. Hamilton [1] presented the Ricci soliton as $\frac{1}{2}\mathcal{L}_X g + Ric = \lambda g$ for the first time, which is a natural extension of Einstein metrics. After that, many authors generalized this soliton and introduced other solitons corresponding to other geometric flows.

The goal of this study is to discuss three-dimensional homogeneous Lorentzian cross curvature solitons. Three-dimensional locally homogeneous Lorentzian manifolds can fall into one of two categories: they are either locally isometric to a three-dimensional Lie group with a Lorentzian left-invariant metric or locally symmetric.

Suppose that (M, g) is a three-dimensional manifold. We consider the tensor

$$P_{ij} = R_{ij} - \frac{1}{2} R g_{ij}, \tag{1}$$

where R is the scalar curvature and R_{ij} is the Ricci tensor of M. Set

$$P^{ij} = g^{ik} g^{jl} R_{kl} - \frac{1}{2} R g^{ij}. \tag{2}$$

The cross curvature tensor is defined as follows, where V_{ij} is the inverse of P^{ij}:

$$h_{ij} = \frac{\det P^{kl}}{\det g^{kl}} V_{ij}. \tag{3}$$

In the pseudo-Riemannian case, if a vector field X on M and a constant λ exist such that

$$\mathcal{L}_X g + \lambda g = 2h, \tag{4}$$

Citation: Azami, S.; Jafari, M.; Haseeb, A.; Ahmadini, A.A.H. Cross Curvature Solitons of Lorentzian Three-Dimensional Lie Groups. *Axioms* **2024**, *13*, 211. https://doi.org/10.3390/axioms13040211

Academic Editor: Mića Stanković

Received: 7 February 2024
Revised: 3 March 2024
Accepted: 7 March 2024
Published: 25 March 2024

then (M, g) is a cross curvature soliton. We mention that $\mathcal{L}_X g$ indicates the Lie derivative of g with regard to X, and h is the cross curvature tensor of g. A cross curvature soliton is an interesting type of solution to the cross curvature flow. It is actually the self-similar solution of the cross curvature flow [2,3]. A cross curvature soliton is stated as being either expanding, steady, or shrinking if $\lambda > 0$, $\lambda = 0$, or $\lambda < 0$, respectively.

It is worth noting that when $\mathcal{L}_X g$ is equal to zero, a cross curvature soliton is considered trivial. The cross curvature flow, which was introduced by Chow and Hamilton, represented a significant advancement in this field [4]. Buckland's groundbreaking work on the short-term existence of this flow should not be underestimated [5]. Additionally, Cao et al. conducted a comprehensive study on the non-negative cross curvature flow on locally homogeneous Riemannian three-dimensional manifolds, providing valuable insights into the long-term behavior of this flow [6,7]. For further information, please consult [8–10].

Also, other geometric solitons have been studied on locally homogeneous manifolds. For instance, it has been proven that Lie groups with a left-invariant Riemannian metric of dimension of four at most lack non-trivial homogeneous invariant Ricci solitons (see [11–14]), but there are three-dimensional Riemannian homogeneous Ricci solitons [15,16]. Lauret's work established that every algebraic Ricci soliton on a Lie group with left-invariant Riemannian metric is a homogeneous Ricci soliton [17], and Onda later extended this finding to the case of Lie groups with pseudo-Riemannian left-invariant metric [18]. Additionally, Calvaruso and Fino discovered the Ricci solitons on four-dimensional non-reductive homogeneous spaces [19]. Also, for some consequences of Ricci solitons on homogeneous manifolds, refer to [20,21].

The paper is arranged as follows. Section 2 will delve into essential concepts on three-dimensional Lie groups, which will be integral to the paper. Section 3 will present the key findings and their corresponding proofs.

2. Lorentzian Lie Groups in Dimension 3

The Bianchi classification provides a list of all real three-dimensional Lie algebras. This classification contains 11 classes, two of which contain a continuum-sized family of Lie algebras and nine of which contain a single Lie algebra. In the following, we offer a succinct introduction to unimodular and non-unimodular Lie groups in three dimensions. It is important to note that fully connected and simply structured three-dimensional Lorentzian homogeneous manifolds can exhibit either symmetry or a left-invariant Lorentzian metric as a Lie group [22].

2.1. Unimodular Lie Groups

Suppose that $\{e_1, e_2, e_3\}$ is an orthonormal basis of signature $(+\,+\,-)$. We represent the Lorentzian vector product on $\mathbb{R}^3_1$, which is generated by the cross product $\times$, i.e.,

$$e_3 \times e_1 = e_2, \quad e_2 \times e_3 = e_1, \quad e_1 \times e_2 = -e_3.$$

The Lie algebra $\mathfrak{g}$ is defined by the Lie bracket $[\,,\,]$. It is important to note that the algebra is only unimodular if the endomorphism L, which is defined as $[Z, Y] = L(Z \times Y)$, is self-adjoint. Additionally, L is non-unimodular when it is not self-adjoint. By analyzing the various types of L, we can identify four distinct classes of unimodular three-dimensional Lie algebras [23].

Type Ia.

The Lie algebra corresponding to a diagonalizable endomorphism L with three real eigenvalues $\{\alpha, \beta, \gamma\}$ regarding an orthonormal basis $\{e_1, e_2, e_3\}$ of signature $(+\,+\,-)$ is presented by

$$(\mathfrak{g}_{Ia}): \qquad [e_2, e_3] = \alpha e_1, \quad [e_1, e_3] = -\beta e_2, \quad [e_1, e_2] = -\gamma e_3.$$

In this case, Lie groups G admitting a Lie algebra $\mathfrak{g}_{Ia}$ are listed in Table 1.

Table 1. Type Ia. Lie groups G admitting a Lie algebra $\mathfrak{g}_{Ia}$.

G	α	β	γ
$O(1,2)$ or $SL(2,\mathbb{R})$	$+$	$+$	$+$
$O(1,2)$ or $SL(2,\mathbb{R})$	$+$	$-$	$-$
$SO(3)$ or $SU(2)$	$+$	$+$	$-$
$E(2)$	$+$	$+$	0
$E(2)$	$+$	0	$-$
$E(1,1)$	$+$	$-$	0
$E(1,1)$	$+$	0	$+$
H_3	$+$	0	0
H_3	0	0	$-$
$\mathbb{R}\oplus\mathbb{R}\oplus\mathbb{R}$	0	0	0

In this case, the Levi-Civita connection is specified by

$$(\nabla_{e_i}e_j) = \begin{pmatrix} 0 & -\frac{1}{2}(\gamma+\beta-\alpha)e_3 & -\frac{1}{2}(\gamma+\beta-\alpha)e_2 \\ \frac{1}{2}(\alpha-\beta+\gamma)e_3 & 0 & \frac{1}{2}(\gamma-\beta+\alpha)e_1 \\ \frac{1}{2}(\alpha+\beta-\gamma)e_2 & \frac{1}{2}(\gamma-\beta-\alpha)e_1 & 0 \end{pmatrix}.$$

Let ∇ be the Levi-Civita connection; by using the formula $R(X,Y) = \nabla_{[X,Y]} - [\nabla_X, \nabla_Y]$, the only non-vanishing terms of the curvature tensor are presented by

$$\begin{aligned} R_{2332} &= \frac{1}{4}\Big(-\gamma^2-\beta^2+3\alpha^2+2\beta\gamma-2\alpha\gamma-2\alpha\beta\Big), \\ R_{1313} &= \frac{1}{4}\Big(\gamma^2-3\beta^2+\alpha^2+2\beta\gamma-2\alpha\gamma+2\alpha\beta\Big), \\ R_{1221} &= \frac{1}{4}\Big(-3\gamma^2+\beta^2+\alpha^2+2\beta\gamma+2\alpha\gamma-2\alpha\beta\Big), \end{aligned}$$

its Ricci tensor is expressed by

$$\begin{aligned} R_{11} &= -\frac{1}{2}\Big(\alpha^2-(\gamma-\beta)^2\Big), \\ R_{22} &= -\frac{1}{2}\Big(\beta^2-(\gamma-\alpha)^2\Big), \\ R_{33} &= \frac{1}{2}\Big(\gamma^2-(\beta-\alpha)^2\Big), \end{aligned}$$

and other components are 0. The Lie derivative of the metric, $\mathcal{L}_X g$, for an optional left-invariant vector field $X = \sum_{i=1}^{3} x_i e_i$ is given by

$$(\mathcal{L}_X g) = \begin{pmatrix} 0 & (\alpha-\beta)x_3 & (\gamma-\alpha)x_2 \\ (\alpha-\gamma)x_3 & 0 & (\beta-\gamma)x_1 \\ (\gamma-\alpha)x_2 & (\beta-\gamma)x_1 & 0 \end{pmatrix}.$$

Then,

$$R = \frac{1}{2}\Big(\gamma^2+\alpha^2+\beta^2-2\alpha\beta-2\alpha\gamma-2\beta\gamma\Big)$$

and

$$P^{11} = \frac{1}{4}\Big(-3\alpha^2 + \beta^2 + \gamma^2 + 2\alpha\beta + 2\alpha\gamma - 2\beta\gamma \Big),$$
$$P^{22} = \frac{1}{4}\Big(\alpha^2 - 3\beta^2 + \gamma^2 + 2\alpha\beta - 2\alpha\gamma + 2\beta\gamma \Big),$$
$$P^{33} = \frac{1}{4}\Big(-\alpha^2 - \beta^2 + 3\gamma^2 + 2\alpha\beta - 2\alpha\gamma - 2\beta\gamma \Big),$$

and other components of P^{ij} are 0. Throughout the paper, we assume that (P^{ij}) is invertible. Therefore, the only non-vanishing terms of the cross curvature tensor are obtained as follows:

$$h_{11} = -\frac{1}{16}(\gamma^2 - 3\beta^2 + \alpha^2 + 2\beta\gamma - 2\alpha\gamma + 2\alpha\beta)(3\gamma^2 - \beta^2 - \alpha^2 - 2\beta\gamma - 2\alpha\gamma + 2\alpha\beta),$$
$$h_{22} = -\frac{1}{16}(\gamma^2 + \beta^2 - 3\alpha^2 - 2\beta\gamma + 2\alpha\gamma + 2\alpha\beta)(3\gamma^2 - \beta^2 - \alpha^2 - 2\beta\gamma - 2\alpha\gamma + 2\alpha\beta),$$
$$h_{33} = -\frac{1}{16}(\gamma^2 + \beta^2 - 3\alpha^2 - 2\beta\gamma + 2\alpha\gamma + 2\alpha\beta)(\gamma^2 - 3\beta^2 + \alpha^2 + 2\beta\gamma - 2\alpha\gamma + 2\alpha\beta).$$

Type Ib.

Suppose that L has complex eigenvalues $\gamma \pm i\beta$ and one real eigenvalue α. Then, by considering an orthonormal basis $\{e_1, e_2, e_3\}$ of signature $(+ + -)$, we have

$$L = \begin{pmatrix} \alpha & 0 & 0 \\ 0 & \gamma & -\beta \\ 0 & \beta & \gamma \end{pmatrix}, \qquad \beta \neq 0,$$

then the related Lie algebra is provided by

$$(\mathfrak{g}_{Ib}): \qquad [e_2, e_3] = \alpha e_1, \ \ [e_1, e_3] = -\gamma e_2 - \beta e_3, \ \ [e_1, e_2] = \beta e_2 - \gamma e_3.$$

In this case, $G = O(1,2)$ or $G = SL(2, \mathbb{R})$ if $\alpha \neq 0$, while $G = E(1,1)$ if $\alpha = 0$. The Levi-Civita connection is specified by

$$(\nabla_{e_i} e_j) = \begin{pmatrix} 0 & \frac{1}{2}(\alpha - 2\gamma)e_3 & \frac{1}{2}(\alpha - 2\gamma)e_2 \\ \frac{\alpha}{2}e_3 - \beta e_2 & \beta e_1 & \frac{\alpha}{2}e_1 \\ \frac{\alpha}{2}e_2 + \beta e_3 & -\frac{\alpha}{2}e_1 & \beta e_1 \end{pmatrix}.$$

With respect to the basis $\{e_1, e_2, e_3\}$ the only non-vanishing terms of the curvature tensor are described by

$$R_{1231} = -(2\gamma - \alpha)\beta, \quad R_{2332} = \frac{3}{4}\alpha^2 - \alpha\gamma + \beta^2, \quad R_{1313} = R_{1221} = \frac{1}{4}(4\beta^2 + \alpha^2),$$

and its Ricci tensor is expressed by

$$Ric = \begin{pmatrix} -\frac{1}{2}(\alpha^2 + 4\beta^2) & 0 & 0 \\ 0 & \frac{1}{2}\alpha(\alpha - 2\gamma) & -\beta(\alpha - 2\gamma) \\ 0 & \beta(2\gamma - \alpha) & \frac{1}{2}\alpha(2\gamma - \alpha) \end{pmatrix}.$$

For an optional left-invariant vector field $X = \sum_{i=1}^{3} x_i e_i$, we obtain

$$(\mathcal{L}_X g) = \begin{pmatrix} 0 & \beta x_2 + (\alpha-\gamma)x_3 & \beta x_3 + (\gamma-\alpha)x_2 \\ \beta x_2 + (\alpha-\gamma)x_3 & -2\beta x_1 & 0 \\ \beta x_3 + (\gamma-\alpha)x_2 & 0 & -2\beta x_1 \end{pmatrix}.$$

Hence,

$$R = \frac{1}{2}\left(\alpha^2 - 4\beta^2 - 4\alpha\gamma\right)$$

and

$$(P^{ij}) = \begin{pmatrix} \frac{1}{4}(-3\alpha^2 - 4\beta^2 + 4\alpha\gamma) & 0 & 0 \\ 0 & \frac{1}{4}(\alpha^2+4\beta^2) & \beta(\alpha-2\gamma) \\ 0 & \beta(\alpha-2\gamma) & -\frac{1}{4}(\alpha^2+4\beta^2) \end{pmatrix}.$$

Let

$$A_1 = -\left(\frac{1}{16}(\alpha^2+4\beta^2)^2 + \beta^2(\alpha-2\gamma)^2\right)$$

then,

$$(V_{ij}) = \frac{1}{A_1}\begin{pmatrix} -4\frac{\frac{1}{16}(\alpha^2+4\beta^2)^2+\beta^2(\alpha-2\gamma)^2}{-3\alpha^2-4\beta^2+4\alpha\gamma} & 0 & 0 \\ 0 & -\frac{1}{4}(\alpha^2+4\beta^2) & -\beta(\alpha-2\gamma) \\ 0 & \beta(2\gamma-\alpha) & \frac{1}{4}(\alpha^2+4\beta^2) \end{pmatrix}.$$

Therefore, the cross curvature tensor is described by

$$(h_{ij}) = -\frac{1}{4}(-3\alpha^2 - 4\beta^2 + 4\alpha\gamma)\begin{pmatrix} -4\frac{\frac{1}{16}(\alpha^2+4\beta^2)^2+\beta^2(\alpha-2\gamma)^2}{-3\alpha^2-4\beta^2+4\alpha\gamma} & 0 & 0 \\ 0 & -\frac{1}{4}(\alpha^2+4\beta^2) & \beta(2\gamma-\alpha) \\ 0 & \beta(2\gamma-\alpha) & \frac{1}{4}(\alpha^2+4\beta^2) \end{pmatrix}.$$

Type II.

Suppose that the minimal polynomial of L has two roots, α and β, such that $(L-\alpha I)(L-\beta I)^2 = 0$ holds. So, regarding the orthonormal basis $\{e_1, e_2, e_3\}$ of signature $(++-)$ we have

$$L = \begin{pmatrix} \alpha & 0 & 0 \\ 0 & \beta+\frac{1}{2} & -\frac{1}{2} \\ 0 & \frac{1}{2} & \beta-\frac{1}{2} \end{pmatrix},$$

then the related Lie algebra is provided by

$$(\mathfrak{g}_{II}): \qquad [e_2, e_3] = \alpha e_1, \quad [e_1, e_3] = -\frac{1}{2}\beta e_3 - \left(\frac{1}{2}+\beta\right)e_2, \quad [e_1, e_2] = \left(\frac{1}{2}-\beta\right)e_3 + \frac{1}{2}e_2.$$

In this case, Lie groups admitting a Lie algebra $\mathfrak{g}_{II}$ are listed in Table 2.

Table 2. Lie groups admitting a Lie algebra $\mathfrak{g}_{II}$.

G	α	β
$O(1,2)$ or $SL(2,\mathbb{R})$	$\neq 0$	$\neq 0$
$E(1,1)$	0	$\neq 0$
$E(1,1)$	<0	0
$E(2)$	>0	0
H_3	0	0

The Levi-Civita connection in this case is expressed by

$$(\nabla_{e_i}e_j) = \begin{pmatrix} 0 & -\frac{1}{2}(2\beta-\alpha)e_3 & \frac{1}{2}(\alpha-2\beta)e_2 \\ \frac{1}{2}(-1+\alpha)e_3 - \frac{1}{2}e_2 & \frac{1}{2}e_1 & \frac{1}{2}(-1+\alpha)e_1 \\ \frac{1}{2}e_3 + \frac{1}{2}(1+\alpha)e_2 & -\frac{1}{2}(1+\alpha)e_1 & \frac{1}{2}e_1 \end{pmatrix}.$$

With respect to the basis $\{e_1, e_2, e_3\}$, the only non-zero terms of the curvature tensor are described by

$$R_{1231} = \frac{1}{2}\alpha - \beta, \quad R_{2332} = \frac{1}{4}(-4\alpha\beta + 3\alpha^2),$$
$$R_{1313} = \frac{1}{4}(-4\beta + 2\alpha + \alpha^2), \quad R_{1221} = \frac{1}{4}(4\beta - 2\alpha + \alpha^2),$$

and its Ricci tensor is given by

$$Ric = \begin{pmatrix} -\frac{1}{2}\alpha^2 & 0 & 0 \\ 0 & \frac{1}{2}(-2\beta+\alpha)(1+\alpha) & \beta - \frac{1}{2}\alpha \\ 0 & -\frac{1}{2}\alpha + \beta & -\frac{1}{2}(2\beta-\alpha)(1-\alpha) \end{pmatrix}.$$

For $X = \sum_{i=1}^{3} x_i e_i$ as an optional left-invariant vector field, we obtain

$$(\mathcal{L}_X g) = \begin{pmatrix} 0 & \frac{1}{2}(x_2 + (2\alpha - 2\beta - 1)x_3) & \frac{1}{2}(x_3 + (2\beta - 2\alpha - 1)x_2) \\ \frac{1}{2}(x_2 + (2\alpha - 2\beta - 1)x_3) & -x_1 & x_1 \\ \frac{1}{2}(x_3 + (2\beta - 2\alpha - 1)x_2) & x_1 & -x_1 \end{pmatrix}.$$

Thus,

$$R = \frac{1}{2}\alpha^2 - 2\alpha\beta$$

and

$$(P^{ij}) = \begin{pmatrix} -\frac{3}{4}\alpha^2 + \alpha\beta & 0 & 0 \\ 0 & \frac{1}{2}\left(\frac{1}{2}\alpha^2 + \alpha - 2\beta\right) & \frac{1}{2}\alpha - \beta \\ 0 & \frac{1}{2}\alpha - \beta & -\frac{1}{2}\left(\frac{1}{2}\alpha^2 - \alpha + 2\beta\right) \end{pmatrix}.$$

Then,

$$(V_{ij}) = \frac{-16}{\alpha^4}\begin{pmatrix} -\frac{\frac{1}{16}\alpha^4}{-\frac{3}{4}\alpha^2+\alpha\beta} & 0 & 0 \\ 0 & -\frac{1}{2}\left(\frac{1}{2}\alpha^2 - \alpha + 2\beta\right) & -\frac{1}{2}\alpha + \beta \\ 0 & -\frac{1}{2}\alpha + \beta & \frac{1}{2}\left(\frac{1}{2}\alpha^2 + \alpha - 2\beta\right) \end{pmatrix}.$$

Therefore, the cross curvature tensor is described by

$$(h_{ij}) = -(-\frac{3}{4}\alpha^2 + \alpha\beta)\begin{pmatrix} -\frac{\frac{1}{16}\alpha^4}{-\frac{3}{4}\alpha^2+\alpha\beta} & 0 & 0 \\ 0 & -\frac{1}{2}\left(\frac{1}{2}\alpha^2 - \alpha + 2\beta\right) & \beta - \frac{1}{2}\alpha \\ 0 & -\frac{1}{2}\alpha + \beta & \frac{1}{2}\left(\frac{1}{2}\alpha^2 + \alpha - 2\beta\right) \end{pmatrix}.$$

Type III.

Suppose that the minimal polynomial of L has one real root α such that $(L - \alpha I)^3 = 0$ holds. So, regarding the orthonormal basis $\{e_1, e_2, e_3\}$ of signature $(+ + -)$ we have

$$L = \begin{pmatrix} \alpha & \frac{1}{\sqrt{2}} & \frac{1}{\sqrt{2}} \\ \frac{1}{\sqrt{2}} & \alpha & 0 \\ -\frac{1}{\sqrt{2}} & 0 & \alpha \end{pmatrix},$$

then the related Lie algebra is provided by

$$(\mathfrak{g}_{III}) : \begin{cases} [e_1, e_2] = -\frac{1}{\sqrt{2}}e_1 - \alpha e_3, \quad [e_1, e_3] = -\frac{1}{\sqrt{2}}e_1 - \alpha e_2, \\ [e_2, e_3] = \alpha e_1 + \frac{1}{\sqrt{2}}e_2 - \frac{1}{\sqrt{2}}e_3. \end{cases}$$

In this case, $G = O(1,2)$ or $G = SL(2,\mathbb{R})$ if $\alpha \neq 0$ and $G = E(1,1)$ if $\alpha = 0$. The Levi-Civita connection in this case is expressed by

$$(\nabla_{e_i} e_j) = \begin{pmatrix} \frac{1}{\sqrt{2}}e_2 - \frac{1}{\sqrt{2}}e_3 & -\frac{\alpha}{2}e_3 - \frac{1}{\sqrt{2}}e_1 & -\frac{\alpha}{2}e_2 - \frac{1}{\sqrt{2}}e_1 \\ \frac{\alpha}{2}e_3 & \frac{1}{\sqrt{2}}e_3 & \frac{1}{\sqrt{2}}e_2 + \frac{\alpha}{2}e_1 \\ \frac{\alpha}{2}e_2 & \frac{1}{\sqrt{2}}e_3 - \frac{\alpha}{2}e_1 & \frac{1}{\sqrt{2}}e_2 \end{pmatrix}.$$

The only non-zero terms of the curvature tensor are described by

$$R_{1223} = \frac{1}{\sqrt{2}}\alpha, \quad R_{1313} = 1 - \frac{1}{4}\alpha^2, \quad R_{1231} = 1,$$
$$R_{2323} = \frac{1}{4}\alpha^2, \quad R_{1221} = \frac{1}{4}(4 + \alpha^2).$$

and its Ricci tensor is expressed by

$$Ric = \begin{pmatrix} -\frac{1}{2}\alpha^2 & -\frac{1}{\sqrt{2}}\alpha & -\frac{1}{\sqrt{2}}\alpha \\ -\frac{1}{\sqrt{2}}\alpha & -\frac{1}{2}\alpha^2-1 & -1 \\ -\frac{1}{\sqrt{2}}\alpha & -1 & -1+\frac{1}{2}\alpha^2 \end{pmatrix}.$$

For $X = \sum_{i=1}^{3} x_i e_i$ as an optional left-invariant vector field, the Lie derivative of the metric is presented by the following relation.

$$(\mathcal{L}_X g) = \frac{1}{\sqrt{2}} \begin{pmatrix} -2(x_3+x_2) & x_1 & x_1 \\ x_1 & 2x_3 & x_3-x_2 \\ x_1 & -x_2+x_3 & -2x_2 \end{pmatrix}.$$

Hence,

$$R = -\frac{3}{2}\alpha^2$$

and

$$(P^{ij}) = \begin{pmatrix} \frac{1}{4}\alpha^2 & -\frac{1}{\sqrt{2}}\alpha & \frac{1}{\sqrt{2}}\alpha \\ -\frac{1}{\sqrt{2}}\alpha & \frac{1}{4}\alpha^2-1 & 1 \\ \frac{1}{\sqrt{2}}\alpha & 1 & -1-\frac{1}{4}\alpha^2 \end{pmatrix}.$$

Thus,

$$(V_{ij}) = \frac{-16}{\alpha^4} \begin{pmatrix} -\frac{1}{4}\alpha^2 & \frac{-1}{\sqrt{2}}\alpha & \frac{-1}{\sqrt{2}}\alpha \\ \frac{-1}{\sqrt{2}}\alpha & -3-\frac{1}{4}\alpha^2 & -3 \\ \frac{-1}{\sqrt{2}}\alpha & -3 & -3+\frac{1}{4}\alpha^2 \end{pmatrix}.$$

Therefore, the cross curvature tensor is described by

$$(h_{ij}) = -\frac{\alpha^2}{4} \begin{pmatrix} -\frac{1}{4}\alpha^2 & \frac{-1}{\sqrt{2}}\alpha & \frac{-1}{\sqrt{2}}\alpha \\ \frac{-1}{\sqrt{2}}\alpha & -3-\frac{1}{4}\alpha^2 & -3 \\ \frac{-1}{\sqrt{2}}\alpha & -3 & -3+\frac{1}{4}\alpha^2 \end{pmatrix}.$$

2.2. Non-Unimodular Lie Groups

Moving on, we will address the non-unimodular case. We will use the class $\mathfrak{G}$ to represent a set of solvable Lie algebras $\mathfrak{g}$ where, for any $x, y \in \mathfrak{g}$, $[x, y]$ is a linear combination of x and y. According to [24], the Lorentzian non-unimodular Lie algebras with non-constant sectional curvature that do not fall under class $\mathfrak{G}$ can be represented using the following relation in a suitable basis $\mathcal{E} = \{e_1, e_2, e_3\}$,

$$(\mathfrak{g}_{IV}) : [e_2, e_3] = \delta e_2 + \gamma e_1, \quad [e_1, e_3] = \beta e_2 + \alpha e_1, \quad [e_1, e_2] = 0,$$

we have $\delta + \alpha \neq 0$ and one of the next modes is established:

IV.1 $\mathcal{E}$ is orthonormal and $\langle e_3, e_3 \rangle = \langle e_2, e_2 \rangle = -\langle e_1, e_1 \rangle = 1$; also, the constants of structure satisfy $\beta\delta = \alpha\gamma$.

IV.2 $\mathcal{E}$ is orthonormal and $-\langle e_3, e_3 \rangle = \langle e_2, e_2 \rangle = \langle e_1, e_1 \rangle = 1$; also, the constants of structure satisfy $\beta\delta = -\alpha\gamma$.

IV.3 $\mathcal{E}$ is a pseudo-orthonormal basis and

$$\langle \cdot, \cdot \rangle = \begin{pmatrix} 1 & 0 & 0 \\ 0 & 0 & -1 \\ 0 & -1 & 0 \end{pmatrix}$$

also, the constants of structure satisfy $\alpha\gamma = 0$.

Type IV.1.

In this case, the Levi-Civita connection is given by

$$(\nabla_{e_i} e_j) = \begin{pmatrix} \alpha e_3 & -\frac{\beta-\gamma}{2} e_3 & \alpha e_1 + \frac{\beta-\gamma}{2} e_2 \\ -\frac{\beta-\gamma}{2} e_3 & -\delta e_3 & -\frac{\beta-\gamma}{2} e_1 + \delta e_2 \\ -\frac{\beta+\gamma}{2} e_2 & -\frac{\beta+\gamma}{2} e_{31} & 0 \end{pmatrix}.$$

With respect to the basis $\mathcal{E}$, the only non-vanishing terms of the curvature tensor are described by

$$R_{2332} = \frac{1}{4}(-3\gamma^2 + \beta^2 + 2\beta\gamma + 4\delta^2),$$
$$R_{1313} = \frac{1}{4}(-3\beta^2 + 4\alpha^2 + 2\beta\gamma + \gamma^2),$$
$$R_{1212} = \frac{1}{4}(\gamma^2 + \beta^2 - 2\beta\gamma + 4\alpha\delta),$$

its Ricci tensor is expressed by

$$R_{11} = -\frac{1}{2}\big(- \gamma^2 + \beta^2 - 2(\delta\alpha + \alpha^2)\big),$$

$$R_{22} = \frac{1}{2}\big(- \beta^2 + \gamma^2 - 2(\delta^2 + \delta\alpha)\big),$$

$$R_{33} = \frac{1}{2}\big(- 2(\delta^2 + \alpha^2) + (\gamma - \beta)^2\big),$$

and other components are 0. For $X = \sum_{i=1}^{3} x_i e_i$ as an optional left-invariant vector field, the Lie derivative of the metric $\mathcal{L}_X g$ is equal to

$$(\mathcal{L}_X g) = \begin{pmatrix} -2\alpha x_3 & (\beta - \gamma)x_3 & \alpha x_1 + \gamma x_2 \\ (\beta - \gamma)x_3 & 2\delta x_3 & -\beta x_1 - \delta x_2 \\ \alpha x_1 + \gamma x_2 & -\beta x_1 - \delta x_2 & 0 \end{pmatrix}.$$

Then,

$$R = \frac{1}{2}(\beta^2 - \gamma^2 - 4\alpha^2 - 4\delta^2 - 2\beta\gamma - 4\alpha\delta)$$

and

$$P^{11} = \frac{1}{4}\big(- \beta^2 + \gamma^2 - 4\delta^2 - 2\beta\gamma\big),$$

$$P^{22} = \frac{1}{4}(4\alpha^2 - 3\beta^2 + 3\gamma^2 + 2\beta\gamma),$$

$$P^{33} = \frac{1}{4}(3\gamma^2 + \beta^2 + 4\alpha\delta - 2\beta\gamma),$$

and other components are 0. Therefore, the only non-vanishing terms of the cross curvature tensor are described by

$$\begin{aligned} h_{11} &= -\frac{1}{16}\big(4\alpha^2 - 3\beta^2 + 3\gamma^2 + 2\beta\gamma\big)\big(\beta^2 + 3\gamma^2 - 2\beta\gamma + 4\alpha\delta\big), \\ h_{22} &= -\frac{1}{16}\big(- \beta^2 + \gamma^2 - 4\delta^2 - 2\beta\gamma\big)\big(\beta^2 + 3\gamma^2 - 2\beta\gamma + 4\alpha\delta\big), \\ h_{33} &= -\frac{1}{16}\big(- \beta^2 + \gamma^2 - 4\delta^2 - 2\beta\gamma\big)\big(4\alpha^2 - 3\beta^2 + 3\gamma^2 + 2\beta\gamma\big). \end{aligned}$$

Type IV.2.
The Levi-Civita connection of Type IV.2 concerning $\mathcal{E}$ is determined by

$$(\nabla_{e_i} e_j) = \begin{pmatrix} \alpha e_3 & \frac{\beta+\gamma}{2} e_3 & \alpha e_1 + \frac{\beta+\gamma}{2} e_2 \\ \frac{\gamma+\beta}{2} e_3 & \delta e_3 & \delta e_2 + \frac{\gamma+\beta}{2} e_1 \\ -\frac{-\gamma+\beta}{2} e_2 & \frac{-\gamma+\beta}{2} e_1 & 0 \end{pmatrix}.$$

With respect to the basis $\mathcal{E}$, the only non-vanishing terms of the curvature tensor are described by

$$\begin{aligned} R_{2323} &= \frac{1}{4}(-3\gamma^2 + \beta^2 - 2\beta\gamma - 4\delta^2), \\ R_{1331} &= \frac{1}{4}(3\beta^2 + 4\alpha^2 + 2\beta\gamma - \gamma^2), \\ R_{1212} &= -\frac{1}{4}(\gamma + \beta)^2 + \alpha\delta, \end{aligned}$$

and its Ricci tensor is expressed by

$$\begin{aligned} R_{11} &= \frac{1}{2}(-\gamma^2 + \beta^2 + 2(\delta\alpha + \alpha^2)) \\ R_{22} &= \frac{1}{2}(-\beta^2 + \gamma^2 + 2(\delta^2 + \delta\alpha)), \\ R_{33} &= -\frac{1}{2}(2(\delta^2 + \alpha^2) + (\gamma + \beta)^2), \end{aligned}$$

where other components of Ricci tensor are 0. For $X = \sum_{i=1}^{3} x_i e_i$ as an optional left-invariant vector field, we obtain

$$(\mathcal{L}_X g) = \begin{pmatrix} 2\alpha x_3 & (\beta + \gamma)x_3 & -\alpha x_1 - \gamma x_2 \\ (\beta + \gamma)x_3 & 2\delta x_3 & -\beta x_1 - \delta x_2 \\ -\alpha x_1 - \gamma x_2 & -\beta x_1 - \delta x_2 & 0 \end{pmatrix}.$$

Then,

$$R = (\alpha + \delta)^2 + \delta^2 + \alpha^2 + \frac{1}{2}(\gamma + \beta)^2$$

and

$$\begin{aligned} P^{11} &= \frac{1}{4}\big(- 3\gamma^2 + \beta^2 - 4\delta^2 - 2\beta\gamma\big), \\ P^{22} &= \frac{1}{4}\big(- 4\alpha^2 - 3\beta^2 + \gamma^2 - 2\beta\gamma\big), \\ P^{33} &= \frac{1}{4}\big(- \beta^2 - 2\beta\gamma - \gamma^2 + \alpha\gamma\big), \end{aligned}$$

and other components are 0. Therefore, the only non-vanishing terms of the cross curvature tensor are described by

$$\begin{aligned} h_{11} &= -\frac{1}{16}\big(- 4\alpha^2 - 3\beta^2 + \gamma^2 - 2\beta\gamma\big)\big(- \beta^2 - \gamma^2 - 2\beta\gamma + \alpha\gamma\big), \\ h_{22} &= -\frac{1}{16}\big(\beta^2 - 3\gamma^2 - 4\delta^2 - 2\beta\gamma\big)\big(- \beta^2 - \gamma^2 - 2\beta\gamma + \alpha\gamma\big), \\ h_{33} &= -\frac{1}{16}\big(\beta^2 - 3\gamma^2 - 4\delta^2 - 2\beta\gamma\big)\big(- 4\alpha^2 - 3\beta^2 + \gamma^2 - 2\beta\gamma\big). \end{aligned}$$

Type IV.3.
The Levi-Civita connection in this type is specified by

$$(\nabla_{e_i}e_j) = \begin{pmatrix} \alpha e_2 & \frac{\gamma}{2}e_2 & \alpha e_1 - \frac{\gamma}{2}e_3 \\ \frac{\gamma}{2}e_2 & 0 & \frac{\gamma}{2}e_1 \\ -\frac{\gamma}{2}e_3 - \beta e_2 & -\delta e_2 - \frac{\gamma}{2}e_1 & \delta e_3 - \beta e_1 \end{pmatrix}.$$

The only non-zero components of the curvature tensor are given by

$$R_{1213} = \tfrac{1}{4}\gamma^2, \;\; R_{1331} = \alpha^2 - \alpha\delta + \beta\gamma, \;\; R_{2332} = \tfrac{3}{4}\gamma^2,$$

and its Ricci tensor is expressed by

$$Ric = \begin{pmatrix} -\frac{1}{2}\gamma^2 & 0 & 0 \\ 0 & 0 & -\frac{1}{2}\gamma^2 \\ 0 & -\frac{1}{2}\gamma^2 & -(\beta + \alpha^2 - \alpha\delta) \end{pmatrix}.$$

For $X = \sum_{i=1}^{3} x_i e_i$ as an optional left-invariant vector field, we obtain

$$(\mathcal{L}_X g) = \begin{pmatrix} 2\alpha x_3 & \gamma x_3 & -\alpha x_1 - \gamma x_2 - \beta x_3 \\ \gamma x_3 & 0 & -\delta x_3 \\ -\alpha x_1 - \gamma x_2 - \beta x_3 & -\delta x_3 & 2(\beta x_1 + \delta x_2) \end{pmatrix}.$$

Thus, $R = \frac{1}{2}\gamma^2$.

$$(P^{ij}) = \begin{pmatrix} -\frac{3}{4}\gamma^2 & 0 & 0 \\ 0 & -(\alpha^2 - \alpha\delta + \beta) & -\frac{1}{4}\gamma^2 \\ 0 & -\frac{3}{4}\gamma^2 & 0 \end{pmatrix},$$

and

$$(V_{ij}) = \begin{pmatrix} -\frac{4}{3\gamma^2} & 0 & 0 \\ 0 & 0 & -\frac{4}{3\gamma^2} \\ 0 & -\frac{4}{\gamma^2} & \frac{16}{3\gamma^4}(\alpha^2 - \alpha\delta + \beta) \end{pmatrix}.$$

Therefore, the cross curvature tensor is given by

$$(h_{ij}) = \begin{pmatrix} \frac{3}{16}\gamma^4 & 0 & 0 \\ 0 & 0 & \frac{3}{16}\gamma^4 \\ 0 & \frac{9}{16}\gamma^4 & \frac{3}{4}(\alpha\delta - \beta - \alpha^2)\gamma^2 \end{pmatrix}.$$

According to the research conducted by Calvaruso in [25], there has been a significant study of three-dimensional Lorentzian locally conformally flat Lie groups. It has been proposed that these groups possess certain characteristics that are worth exploring further. From [25], we have the following proposition.

Proposition 1. *One of the defining characteristics of a Lorentzian three-dimensional Lie group (G, g) is that it is locally conformally flat if and only if one of the following conditions applies:*

(1) (G, g) is locally symmetric and
- *(1a) of Type Ia with $\gamma = \beta = \alpha$ or any cyclic permutation of $\beta = \alpha, \gamma = 0$*
- *(1b) of Type II with $\beta = \alpha = 0$*
- *(1c) of Type IV.1 with constant sectional curvature, or otherwise $\delta = \gamma = \beta = 0$ and $\alpha \neq 0$, or $\gamma = \alpha = \beta = 0$ and $\delta \neq 0$*
- *(1d) of Type IV.2 with constant sectional curvature, or otherwise $\delta = \gamma = \beta = 0$ and $\alpha \neq 0$, or $\gamma = \beta = \alpha = 0$ and $\delta \neq 0$*
- *(1e) of Type IV.3 and flat, or otherwise $\delta = \gamma = 0$ and $\alpha \neq 0$*
- *(1f) of Type $\mathfrak{S}$ and therefore of constant sectional curvature.*

(2) (G, g) is not locally symmetric and
- *(2a) of Type Ib with $\beta = \pm\sqrt{3}\gamma$ and $\alpha = -2\gamma$*
- *(2b) of Type III with $\alpha = 0$*
- *(2c) of Type IV.3 with $\alpha\delta(\alpha - \delta) \neq 0$ and $\gamma = 0$.*

3. Lorentzian Cross Curvature Solitons on Lorentzian 3-Dimensional Lie Groups

In this section, we will delve into the investigation of left-invariant solutions to (4) on the Lorentzian Lie groups that were examined in Section 2. Our aim is to solve the related equations completely and provide a comprehensive explanation of all left-invariant cross curvature solitons.

Theorem 1. *Suppose that $\mathfrak{g}$ indicate a Lorentzian unimodular three-dimensional Lie algebra of Type Ia. Then, the left-invariant cross curvature soliton on $\mathfrak{g}$ satisfies $\beta = \alpha = \gamma$, $\alpha \neq 0$, and $\lambda = \frac{1}{8}\alpha^2$, for all X. Also, as $\beta = \alpha = \gamma$, all vectors in $\mathfrak{g}$ are Killing.*

Proof. Considering (4), there is a cross curvature soliton of Type Ia if and only if the subsequent system of equations is satisfied:

$$\begin{aligned} &(\beta-\alpha)x_3 = (\alpha-\gamma)x_2 = (\gamma-\beta)x_1 = 0, \\ &-\frac{1}{8}(\gamma^2-3\beta^2+\alpha^2+2\beta\gamma-2\alpha\gamma+2\alpha\beta)(3\gamma^2-\beta^2-\alpha^2-2\beta\gamma-2\alpha\gamma+2\alpha\beta) = \lambda, \\ &-\frac{1}{8}(\gamma^2+\beta^2-3\alpha^2-2\beta\gamma+2\alpha\gamma+2\alpha\beta)(-\alpha^2-\beta^2+3\gamma^2+2\alpha\beta-2\alpha\gamma-2\beta\gamma) = \lambda, \\ &-\frac{1}{8}(\gamma^2+\beta^2-3\alpha^2-2\beta\gamma+2\alpha\gamma+2\alpha\beta)(\gamma^2-3\beta^2+\alpha^2+2\beta\gamma-2\alpha\gamma+2\alpha\beta) = -\lambda. \end{aligned} \tag{5}$$

The first equation in (5) indicates that $\alpha = \beta$ or $x_3 = 0$. We consider $\alpha = \beta$. Then, $(\gamma-\alpha)x_2 = 0$ yields $\gamma = \alpha$ or $x_2 = 0$. If $\gamma = \alpha$, then $\gamma = \beta$. Since the tensor (P^{ij}) is invertible, we conclude $\alpha \neq 0$. Thus, the last three equations in (5) reduce to $\lambda = \frac{1}{8}\alpha^2$. In this case, for any left-invariant vector field X, Equation (4) holds.

Now, we consider $\alpha = \beta$ and $\gamma \neq \alpha$. Then, $\beta \neq \gamma$ and $x_2 = 0$. The equation $(\beta-\gamma)x_1 = 0$ yields $x_1 = 0$. In this case, the last three equations in (5) reduce to

$$\begin{cases} -\frac{1}{8}\gamma^2(3\gamma^2-4\alpha\gamma) = \lambda, \\ -\frac{1}{8}\gamma^4 = -\lambda. \end{cases}$$

Since the tensor (P^{ij}) is invertible, we have $\gamma \neq 0$; this implies that $\gamma = \alpha$, which is a contradiction.

Now, assume that $\alpha \neq \beta$. Then, $x_3 = 0$. From equation $(\gamma-\alpha)x_2 = 0$, we infer $\gamma = \alpha$ or $x_2 = 0$. If $\gamma = \alpha$, then the last three equations in (5) reduce to

$$\begin{cases} -\frac{1}{8}\beta^2(3\beta^2+4\alpha\beta) = \lambda, \\ \frac{1}{8}\beta^4 = \lambda. \end{cases}$$

This system implies that $\alpha = \beta$, which is a contradiction. Hence, this case cannot happen. We suppose that $\beta \neq \alpha$, $\alpha \neq \gamma$ and $x_2 = 0$. From $(\gamma-\beta)x_1 = 0$, we have $\beta = \gamma$ or $x_1 = 0$. Similarly, the case $\beta = \gamma$ cannot occur. Then, we have $\beta \neq \alpha$, $\alpha \neq \gamma$, and $\gamma \neq \beta$. Also, $x_1 = x_2 = x_3 = 0$. In this case, using the last three equations of (5), we obtain

$$\begin{cases} \alpha^2+\beta\gamma = \beta^2+\alpha\gamma, \\ \gamma^2+\alpha\beta = \beta^2+\alpha\gamma. \end{cases}$$

Since $\beta \neq \gamma$, $\alpha \neq \gamma$, and $\alpha \neq \beta$, this system has no solution. □

From Theorem 1 and Proposition 1, we conclude the next result.

Corollary 1. *If a Type Ia Lorentzian unimodular Lie group is a left-invariant cross curvature soliton, then it is locally conformally flat.*

Theorem 2. *A Type Ib Lorentzian unimodular Lie groups does not accept any left-invariant cross curvature soliton.*

Proof. Considering (4), there is a cross curvature soliton of Type Ib if and only if the subsequent system of equations is satisfied:

$$\begin{cases} \frac{1}{8}(\alpha^2+4\beta^2)^2+2\beta^2(\alpha-2\gamma)^2=\lambda, \\ \beta x_2+(\alpha-\gamma)x_3=0, \\ \beta x_3+(\gamma-\alpha)x_2=0, \\ \frac{1}{8}(-3\alpha^2-4\beta^2+4\alpha\gamma)(\alpha^2+4\beta^2)+2\beta x_1=\lambda, \\ \frac{1}{2}(-3\alpha^2-4\beta^2+4\alpha\gamma)\beta(\alpha-2\gamma)=0, \\ -\frac{1}{8}(-3\alpha^2-4\beta^2+4\alpha\gamma)(\alpha^2+4\beta^2)+2\beta x_1=-\lambda. \end{cases} \quad (6)$$

The fourth and the sixth equations of (6) give $4\beta x_1=0$. Since $\beta\neq 0$, we obtain $x_1=0$. The fifth equation of (6) gives $\alpha=2\gamma$ or $-3\alpha^2-4\beta^2+4\alpha\gamma=0$. If $-3\alpha^2-4\beta^2+4\alpha\gamma=0$, the fourth equation indicates that $\lambda=0$. Thus, the foremost equation gives $\beta=0$ and this is a contradiction. If $\alpha=2\gamma$, the first and the fourth equations yield $\frac{1}{8}(\alpha^2+4\beta^2)^2=\lambda$ and $-\frac{1}{8}(\alpha^2+4\beta^2)^2=\lambda$, respectively, which imply $\lambda=0$ and $\beta=0$, which is a contradiction. Hence, the system (6) has no solution. Therefore, no homogeneous cross curvature soliton of Type Ib exists. □

Theorem 3. *Consider the Lorentzian unimodular three-dimensional Lie algebra $\mathfrak{g}_{II}$ of Type II. Then, the left-invariant cross curvature soliton on $\mathfrak{g}_{II}$ satisfies*

$$\alpha=\beta\neq 0, \lambda=\frac{1}{8}\alpha^4, x_1=-\frac{1}{4}\alpha^3, x_2=x_3.$$

Proof. Considering (4), there is a cross curvature soliton of Type II if and only if the subsequent system of equations is satisfied:

$$\begin{cases} \frac{1}{8}\alpha^4=\lambda, \\ x_2+(2\alpha-2\beta-1)x_3=0, \\ x_3+(2\beta-2\alpha-1)x_2=0, \\ (\frac{1}{2}\alpha^2-\alpha+2\beta)(-\frac{3}{4}\alpha^2+\alpha\beta)+x_1=\lambda, \\ (\alpha-2\beta)(-\frac{3}{4}\alpha^2+\alpha\beta)-x_1=0, \\ -(\frac{1}{2}\alpha^2+\alpha-2\beta)(-\frac{3}{4}\alpha^2+\alpha\beta)+x_1=-\lambda. \end{cases} \quad (7)$$

The fourth and the sixth equations of (7) give $\lambda=\frac{1}{2}\alpha^2(-\frac{3}{4}\alpha^2+\alpha\beta)$. Substituting this into the first equation in (7), we obtain $\alpha=\beta$. Since (P^{ij}) is invertible, $\alpha\neq 0$. Then, Equation (7) implies that $x_2=x_3$ and $x_1=-\frac{1}{4}\alpha^3$. □

From Theorem 3 and Proposition 1, we conclude the next result.

Corollary 2. *If a Type II Lorentzian unimodular Lie group is locally conformally flat, it is not necessarily a left-invariant cross curvature soliton.*

Theorem 4. *Consider the Lorentzian unimodular three-dimensional Lie algebra $\mathfrak{g}_{III}$ of Type III. A left-invariant cross curvature soliton on $\mathfrak{g}_{III}$ satisfies*

$$x_1 = \frac{1}{2}\alpha^3, x_2 = -x_3 = -\frac{3\sqrt{2}}{4}\alpha^2, \lambda = \frac{1}{8}\alpha^4, \textit{and } \alpha \neq 0.$$

Proof. In case of Type III, Equation (4) becomes

$$\begin{cases} \frac{1}{8}\alpha^4 + \frac{2}{\sqrt{2}}(x_2 + x_3) = \lambda, \\ \frac{1}{2\sqrt{2}}\alpha^3 - \frac{1}{\sqrt{2}}x_1 = 0, \\ \frac{3}{2}\alpha^2 + \frac{1}{8}\alpha^4 - \frac{2}{\sqrt{2}}x_3 = \lambda, \\ \frac{3}{2}\alpha^2 - \frac{1}{\sqrt{2}}(x_3 - x_2) = 0, \\ \frac{3}{2}\alpha^2 - \frac{1}{8}\alpha^4 + \frac{2}{\sqrt{2}}x_2 = -\lambda. \end{cases} \tag{8}$$

The second equation of (8) implies that $x_1 = \frac{1}{2}\alpha^3$. The first and third equations of (8) give $\frac{3}{2}\alpha^2 - \frac{4}{\sqrt{2}}x_3 - \frac{2}{\sqrt{2}}x_2 = 0$. Also, the first and fifth equations of (8) give $\frac{3}{2}\alpha^2 + \frac{4}{\sqrt{2}}x_2 + \frac{2}{\sqrt{2}}x_3 = 0$. Thus, $x_3 = -x_2 = \frac{3\sqrt{2}}{4}\alpha^2$. □

Theorem 5. *Let $\mathfrak{g}$ indicate a Lorentzian non-unimodular three-dimensional Lie algebra of Type IV.1. Then, the left-invariant cross curvature solitons on $\mathfrak{g}$ satisfy one of the following facts:*

(1) $\beta = \alpha = 0$, $x_2 = x_3 = 0$, *and* $\lambda = \frac{9}{8}\gamma^4$ *for all* x_1, δ, *and* γ *such that* $\gamma \neq 0$ *and* $\delta^2 = \gamma^2$.

(2) $\alpha \neq 0$, $\gamma = \frac{\beta\delta}{\alpha}$, $\lambda = \frac{1}{8}(\alpha^2 + 3\gamma^2 + 2\epsilon\alpha\gamma)$, $x_1 = -\frac{\gamma}{\alpha}x_2$, $\beta = \epsilon\alpha$, *and* $x_3 = 0$ *for all* δ, *and* x_2 *such that and* $\epsilon^2 = 1$.

(3) $\alpha \neq 0$, $\delta = \gamma = x_1 = x_3 = 0$, *and* $\lambda = \frac{1}{8}\beta^4$ *such that* $\alpha^2 = \beta^2$ *for all* x_2.

(4) $\alpha = \delta \neq 0$, $\beta = \gamma$, $\beta^2 \neq \alpha^2$, $x_1 = x_2 = x_3 = 0$, *and* $\lambda = \frac{1}{2}(2\alpha^2 + \beta^2)^2$.

Proof. Equation (4) yields

$$\begin{cases} (\beta - \gamma)x_3 = 0, \\ \alpha x_1 + \gamma x_2 = 0, \\ -\beta x_1 - \delta x_2 = 0, \\ -\frac{1}{8}(4\alpha^2 - 3\beta^2 + 3\gamma^2 + 2\beta\gamma)(\beta^2 + 3\gamma^2 - 2\beta\gamma + 4\alpha\delta) + 2\alpha x_3 = -\lambda, \\ -\frac{1}{8}(\gamma^2 - \beta^2 - 2\beta\gamma - 4\delta^2)(3\gamma^2 + \beta^2 + 4\alpha\delta - 2\beta\gamma) - 2\delta x_3 = \lambda, \\ \frac{1}{8}(\gamma^2 - \beta^2 - 2\beta\gamma - 4\delta^2)(-3\beta^2 + 4\alpha^2 + 2\beta\gamma + 3\gamma^2) = -\lambda. \end{cases} \tag{9}$$

We first analyze the case $\alpha = 0$. In this case, $\delta \neq 0$ and $\beta = 0$. Since (P^{ij}) is invertible, we obtain $\gamma \neq 0$. The first equation of (9) indicates that $x_3 = 0$. By substituting $\alpha = \beta = x_3 = 0$ into the last three equations in (9), we obtain

$$\lambda = \frac{9}{8}\gamma^4, \ \lambda = -\frac{3}{8}\gamma^2(\gamma^2 - 4\delta^2),$$

then $\gamma^2 = \delta^2$. We obtain $x_2 = 0$ from the second equation of (9). Therefore, we have a left-invariant cross curvature soliton (1) in this case.

Now, let $\alpha \neq 0$; then, $\gamma = \frac{\beta\delta}{\alpha}$ and the second equation of (9) indicates $x_1 = -\frac{\gamma}{\alpha}x_2$, while its third equation reduces to $(\beta^2 - \alpha^2)\delta x_2 = 0$. If $\beta^2 = \alpha^2$, then $\beta = \epsilon\alpha$ and $\delta = \epsilon\gamma$, where $\epsilon^2 = 1$. The last three equations of (9) reduce to

$$\begin{cases} -\frac{1}{8}\left(\alpha^2 + 3\gamma^2 + 2\epsilon\alpha\gamma\right)^2 + 2\alpha x_3 = -\lambda, \\ -\frac{1}{8}\left(\alpha^2 + 3\gamma^2 + 2\epsilon\alpha\gamma\right)^2 - 2\delta x_3 = \lambda, \\ \frac{1}{8}\left(\alpha^2 + 3\gamma^2 + 2\epsilon\alpha\gamma\right)^2 = \lambda. \end{cases}$$

Therefore, $\alpha x_3 = 0$; since $\alpha \neq 0$, we obtain $x_3 = 0$ and, in this case, we have a left-invariant cross curvature soliton (2).

If $\delta = 0$, then $\beta\delta - \alpha\gamma = 0$ implies that $\gamma = 0$ and $x_1 = 0$. The last three equations of (9) reduce to

$$\begin{cases} -\frac{1}{8}\beta^2(4\alpha^2 - 3\beta^2) + 2\alpha x_3 = -\lambda, \\ \frac{1}{8}\beta^4 = \lambda, \\ \frac{1}{8}\beta^2(4\alpha^2 - 3\beta^2) = \lambda. \end{cases}$$

Thus, $\alpha x_3 = 0$ and since $\alpha \neq 0$, we obtain $x_3 = 0$. Also, $\alpha^2 = \beta^2$; in this case, we have a left-invariant cross curvature soliton (3).

If $\alpha \neq 0$, $\beta^2 \neq \alpha^2$, $\delta \neq 0$, and $x_2 = 0$, then $x_1 = 0$. The first equation now gives $\beta = \gamma$ or $x_3 = 0$. We assume that $x_3 = 0$, and by using the last three equations of (9), we have

$$\begin{cases} -\frac{1}{8}\left(4\alpha^2 - 3\beta^2 + 3\gamma^2 + 2\beta\gamma\right)\left(\beta^2 + 3\gamma^2 - 2\beta\gamma + 4\alpha\delta\right) = -\lambda, \\ -\frac{1}{8}\left(-\beta^2 + \gamma^2 - 4\delta^2 - 2\beta\gamma\right)\left(\beta^2 + 3\gamma^2 - 2\beta\gamma + 4\alpha\delta\right) = \lambda, \\ -\frac{1}{8}\left(-\beta^2 + \gamma^2 - 4\delta^2 - 2\beta\gamma\right)\left(4\alpha^2 - 3\beta^2 + 3\gamma^2 + 2\beta\gamma\right) = \lambda. \end{cases}$$

Since $\left(P^{ij}\right)$ is invertible, we conclude

$$\begin{cases} \alpha^2 + \beta\gamma = \beta^2 + \alpha\delta, \\ \beta^2 + \delta^2 = \gamma^2 + \alpha^2. \end{cases} \tag{10}$$

Substituting $\gamma = \frac{\beta\delta}{\alpha}$ in (10) and using $\delta + \alpha \neq 0$, we obtain $\delta = \alpha$, $\beta = \gamma$, and $\lambda = \frac{1}{2}(2\alpha^2 + \beta^2)^2$. In this case, we have a left-invariant cross curvature soliton (4).

Now, we consider the case $\alpha \neq 0$, $\beta^2 \neq \alpha^2$, $\delta \neq 0$, $x_2 = 0$, $x_3 \neq 0$, and $\beta = \gamma$. The sixth equation of (9) implies that $\lambda = \frac{1}{2}(2\delta^2 + \beta^2)(2\alpha^2 + \beta^2)$, and by substituting it into the fourth and fifth equations of (9) we obtain $x_3 = \frac{1}{2\alpha}(2\alpha^2 + \beta^2)(\alpha\delta - \delta^2)$ and $x_3 = \frac{1}{2\delta}(2\delta^2 + \beta^2)(\alpha\delta - \alpha^2)$, respectively. Since $x_3 \neq 0$, we obtain $\alpha \neq \delta$; hence,

$$\frac{\delta}{\alpha}(2\alpha^2 + \beta^2) = -\frac{\alpha}{\delta}(2\delta^2 + \beta^2).$$

Thus, $\alpha = 0$, which is a contradiction. □

From Theorem 5 and Proposition 1, we conclude the next result.

Corollary 3. *If a Type IV.1 Lorentzian non-unimodular Lie group is locally conformally flat, then it is not necessarily a left-invariant cross curvature soliton.*

Theorem 6. *Suppose that $\mathfrak{g}$ indicates a Lorentzian non-unimodular three-dimensional Lie algebra of Type IV.2. Then, the left-invariant cross curvature solitons on $\mathfrak{g}$ satisfy one of the following conditions:*

(1) $\beta = \alpha = 0$, $x_2 = x_3 = 0$, $\lambda = \frac{1}{8}\gamma^4$ *for all x_1 and δ such that $\delta \neq 0$ and $\delta^2 = \gamma^2$.*
(2) $\delta = \alpha \neq 0$, $\beta = -4\alpha = -\gamma$, $\lambda = 2\alpha^4$, *and* $x_1 = x_2 = x_3 = 0$.
(3) $\delta = \alpha \neq 0$, $\beta = -\gamma$, $x_1 = x_2 = 0$, $\lambda = 2\beta^4$, *and* $x_3 = -\frac{1}{4}\alpha^2\beta - \alpha^3$.
(4) $\delta \neq 0$, $\alpha \neq 0$, $\gamma = -\beta = -4\delta$, $x_1 = x_2 = 0$, $\lambda = 2\beta^2\delta^2$, *and* $x_3 = -\alpha^2\delta - \alpha\delta^2$.

Proof. Equation (4) becomes

$$\begin{cases} (\beta+\gamma)x_3 = 0, \\ -\alpha x_1 - \gamma x_2 = 0, \\ -\beta x_1 - \delta x_2 = 0, \\ -\frac{1}{8}\big(-4\alpha^2 - 3\beta^2 + \gamma^2 - 2\beta\gamma\big)\big(-\beta^2 - \gamma^2 - 2\beta\gamma + \alpha\gamma\big) - 2\alpha x_3 = \lambda, \\ -\frac{1}{8}\big(\beta^2 - 3\gamma^2 - 4\delta^2 - 2\beta\gamma\big)\big(-\beta^2 - \gamma^2 - 2\beta\gamma + \alpha\gamma\big) - 2\delta x_3 = \lambda, \\ \frac{1}{8}\big(-3\gamma^2 + \beta^2 - 2\beta\gamma - 4\delta^2\big)\big(-3\beta^2 - 4\alpha^2 - 2\beta\gamma + \gamma^2\big) = \lambda. \end{cases} \tag{11}$$

First, we analyze the case $\alpha = 0$. Regarding this matter, $\delta \neq 0$ and $\beta = 0$. Also, we obtain $\gamma \neq 0$ since (P^{ij}) is invertible. The first equation of (11) implies that $x_3 = 0$. By substituting $\alpha = \beta = x_3 = 0$ into the last three equations of (11), we obtain

$$\lambda = \frac{1}{8}\gamma^4, \;\; \lambda = -\frac{1}{8}\gamma^2(3\gamma^2 + 4\delta^2),$$

then $\gamma^2 = \delta^2$. We obtain $x_2 = 0$ from the second equation of (11). Therefore, we have a left-invariant cross curvature soliton (1) in this case.

Now, let $\alpha \neq 0$. Then, $\gamma = -\frac{\beta\delta}{\alpha}$ and the second equation of (11) indicates that $x_1 = -\frac{\gamma}{\alpha}x_2$, while its third equation reduces to $\delta x_2 = 0$. If $\delta = 0$, then $\beta\delta + \alpha\gamma = 0$ implies that $\gamma = 0$ and $x_1 = 0$. The last three equations of (11) reduce to

$$\begin{cases} -\frac{1}{8}\beta^2(4\alpha^2 + 3\beta^2) + 2\alpha x_3 = -\lambda, \\ \frac{1}{8}\beta^4 = \lambda, \\ \frac{1}{8}\beta^2(4\alpha^2 + 3\beta^2) = \lambda. \end{cases}$$

Thus, $4\alpha^2 + 2\beta^2 = 0$ and $\alpha = 0$, which is a contradiction.

If $\alpha \neq 0$, $\delta \neq 0$, and $x_2 = 0$, then $x_1 = 0$. Now, the first equation gives $\beta = -\gamma$ or $x_3 = 0$. We assume that $x_3 = 0$, and by using the last three equations of (11), we have

$$\begin{cases} -\frac{1}{8}\big(-4\alpha^2-3\beta^2+\gamma^2-2\beta\gamma\big)\big(-\beta^2-\gamma^2-2\beta\gamma+\alpha\gamma\big)=\lambda, \\ -\frac{1}{8}\big(\beta^2-3\gamma^2-4\delta^2-2\beta\gamma\big)\big(-\beta^2-\gamma^2-2\beta\gamma+\alpha\gamma\big)=\lambda, \\ -\frac{1}{8}\big(\beta^2-3\gamma^2-4\delta^2-2\beta\gamma\big)\big(-4\alpha^2-3\beta^2+\gamma^2-2\beta\gamma\big)=-\lambda. \end{cases}$$

Since (P^{ij}) is invertible, we conclude

$$\begin{cases} \beta^2+\alpha^2=\gamma^2+\delta^2, \\ \beta^2+\alpha^2+\beta\gamma=\frac{1}{4}\alpha\gamma. \end{cases} \tag{12}$$

Substituting $\gamma=-\frac{\beta\delta}{\alpha}$ in (12) and using $\delta+\alpha\neq 0$, we obtain $\delta=\alpha$, $\gamma=-\beta$, $\gamma=4\alpha$, and $\lambda=2\alpha^4$. In this case, we have a left-invariant cross curvature soliton (2).

Now, we consider the case $\alpha\neq 0$, $\delta\neq 0$, $x_2=0$, $x_3\neq 0$, and $\beta=-\gamma$. Then, $x_1=0$ and the sixth equation of (11) implies that $\lambda=2\alpha^2\delta^2$; substituting it into the fourth and the fifth equations in (11), we obtain $x_3=-\frac{1}{4}\alpha^2\beta-\alpha\delta^2$ and $x_3=-\frac{1}{4}\alpha\beta\delta-\alpha^2\delta$, respectively. We obtain

$$\left(\frac{1}{4}\beta-\delta\right)(\alpha-\delta)=0.$$

If $\alpha=\delta$, then $x_3=-\frac{1}{4}\alpha^2\beta-\alpha^3$; in this case, we have a left-invariant cross curvature soliton satisfying (3).

If $\alpha\neq\delta$ and $\beta=4\delta$, then $x_3=-\alpha^2\delta-\alpha\delta^2$; in this case, we have a left-invariant cross curvature soliton satisfying (4). □

From Theorem 6 and Proposition 1, we conclude the next result.

Corollary 4. *If a Type IV.2 Lorentzian non-unimodular Lie group is locally conformally flat, then it is not necessarily a left-invariant cross curvature soliton.*

Theorem 7. *A Type IV.3 Lorentzian non-unimodular Lie group does not accept any left-invariant cross curvature soliton.*

Proof. Considering (4), there is a cross curvature soliton of Type IV.3 if and only if the subsequent system of equations is satisfied:

$$\begin{cases} \frac{3}{8}\gamma^4-2\alpha x_3=\lambda, \\ \gamma x_3=0, \\ -\alpha x_1-\gamma x_2-\beta x_3=0, \\ \lambda=0, \\ -\alpha x_1-\gamma x_2-\beta x_3+2\delta x_3=-\lambda, \\ -\frac{3}{2}\gamma^2(\alpha^2-\alpha\delta+\beta)-2(\beta x_1+\delta x_2)=0. \end{cases} \tag{13}$$

Since (P^{ij}) is invertible, $\gamma\neq 0$. The condition $\alpha\gamma=0$ yields $\alpha=0$. The first and the fourth equations of (13) imply that $\gamma=0$, which is a contradiction. Therefore, Lorentzian non-unimodular Lie groups do not accept any left-invariant cross curvature soliton. □

4. Conclusions

The main study of the paper is to classify left-invariant cross curvature solitons on Lorentzian three-dimensional Lie groups. Three-dimensional locally homogeneous Lorentzian manifolds are classified into seven classes. The first four classes—Type Ia, Type Ib, Type II, and Type III—are unimodular, and the last three classes—Type IV.1, Type IV.2, and Type IV.3—are non-unimodular. In any of such classes, we obtain the Levi-Civita connection, the Ricci tensor, the Lie derivation of the metric in the direction of the vector field X, and the cross curvature tensor. By solving the cross curvature soliton equation $\mathcal{L}_X g + \lambda g = 2h$, we show that Lorentzian unimodular Lie groups Types Ia, II, III and Lorentzian non-unimodular Lie groups of Types IV.1 and IV.2 admit a left-invariant cross curvature soliton, and Lorentzian unimodular Lie groups of type Ib and Lorentzian non-unimodular Lie groups of type IV.3 do not admit left-invariant cross curvature solitons.

Author Contributions: Conceptualization, S.A., M.J., A.H. and A.A.H.A.; methodology, S.A., M.J., A.H. and A.A.H.A.; investigation, S.A., M.J., A.H. and A.A.H.A.; writing—original draft preparation, S.A., M.J., A.H. and A.A.H.A.; writing—review and editing, S.A., M.J., A.H. and A.A.H.A. All authors have read and agreed to the published version of the manuscript.

Funding: The authors extend their appreciation to the Deputyship for Research & Innovation, Ministry of Education in Saudi Arabia for funding this research work through project number ISP-2024.

Data Availability Statement: Data are contained within the article.

Acknowledgments: The authors are thankful to the reviewer for careful reading of the manuscript and his/her thoughtful comments for the improvement of the paper.

Conflicts of Interest: The authors declare no conflicts of interest.

References

1. Hamilton, R.S. The Ricci flow on surfaces, Mathematics and general relativity. *Am. Math. Soc.* **1988**, *71*, 237–262.
2. Glickenstein, D. Riemannian groupoids and solitons for three-dimensional homogeneous Ricci and cross-curvature flows. *Int. Math. Res. Not.* **2008**, *12*, rnn034. [CrossRef]
3. Ho, P.T.; Shin, J. On the cross curvature flow. *Differ. Geom. Appl.* **2020**, *71*, 101636. [CrossRef]
4. Chow, B.; Hamilton, R.S. The cross curvature flow of 3-dimensional flow of 3-manifolds with negative sectional curvature. *Turkish J. Math.* **2004**, *28*, 1–10.
5. Buckland, J.A. Short-time existence of the solutions to the cross curvature flow on 3-manifolds. *Proc. Am. Math. Soc.* **2006**, *134*, 1803–1807. [CrossRef]
6. Cao, X.; Ni, Y.; Saloff-Coste, L. Cross curvature flow on locally homogeneous three-manifolds, I. *Pac. J. Math.* **2008**, *236*, 263–281.
7. Cao, X.; Saloff-Coste, L. Cross curvature flow on locally homogeneous three-manifolds, II. *Asian J. Math.* **2009**, *13*, 421–458. [CrossRef]
8. Cao, X.; Guckenheimer, J.; Saloff-Coste, L. The backward behavior of the Ricci and cross curvature flows on $SL(2,\mathbb{R})$. *Commun. Anal. Geom.* **2009**, *17*, 777–796. [CrossRef]
9. DeBlois, J.; Knopf, D.; Young, A. Cross curvature flow on a negatively curved solid torus. *Algebr. Geom. Topol.* **2010**, *10*, 343–372. [CrossRef]
10. Ma, L.; Chern, D. Examples for cross curvature flow on 3-manifolds. *Calc. Var. Partial Differ. Equ.* **2006**, *26*, 227–243. [CrossRef]
11. Di Cerbo, F.L. Generic properties of homogeneous Ricci solitons. *Adv. Geom.* **2014**, *14*, 225–237. [CrossRef]
12. Hervik, S. Ricci nilsoliton black holes. *J. Geom. Phys.* **2008**, *58*, 1253–1264. [CrossRef]
13. Klepikov, P.N.; Oskorbin, D.N. Homogeneous invariant Ricci solitons on four-dimensional Lie groups. *Izv. AltGU* **2015**, *85*, 122–129.
14. Payne, T.L. The existence of soliton metrics for nilpotent Lie groups. *Geom. Dedicata* **2010**, *145*, 71–88. [CrossRef]
15. Baird, P.; Danielo, L. Three-dimensional Ricci solitons which project to surfaces. *J. Reine Angew. Math.* **2007**, *608*, 65–91. [CrossRef]
16. Lauret, J. Ricci solitons solvmanifolds. *J. Reine Angew. Math.* **2011**, *650*, 1–21. [CrossRef]
17. Lauret, J. Ricci soliton homogeneous nilmanifolds. *Math. Ann.* **2001**, *319*, 715–733. [CrossRef]
18. Onda, K. Examples of algebraic Ricci solitons in the pseudo-Riemannian case. *Acta Math. Hung.* **2014**, *144*, 247–265. [CrossRef]
19. Calvaruso, G.; Fino, A. Ricci Solitons and geometry of four-dimensional non-reductive homogeneous spaces. *Canad. J. Math.* **2012**, *64*, 778–804. [CrossRef]
20. Jablonski, M. Homogeneous Ricci solitons are algebraic. *Geom. Topol.* **2014**, *18*, 2477–2486. [CrossRef]
21. Arroyo, R.M.; Lafuente, R. Homogeneous Ricci solitons in low dimensions. *Int. Math. Res. Not.* **2015**, *13*, 4901–4932. [CrossRef]

22. Calvaruso, G. Homogeneous structures on three-dimensional homogeneous Lorentzian manifolds. *J. Geom. Phys.* **2007**, *57*, 1279–1291. [CrossRef]
23. García-Río, E.; Haji-Badali, A.; Vázquez-Lorenzo, R. Lorentzian 3-manifolds with special curvature operators. *Class. Quantum Grav.* **2008**, *25*, 015003. [CrossRef]
24. Cordero, L.A.; Parker, P. Left-invariant Lorentzian metrics on 3-dimensional Lie groups. *Rend. Mat.* **1997**, *VII*, 129–155.
25. Calvaruso, G. Einstein-like metrics on three-dimensional homogeneous Lorentzian manifolds. *Geom. Dedicata* **2007**, *127*, 99–119. [CrossRef]

axioms

MDPI

Article

Analyzing the Ricci Tensor for Slant Submanifolds in Locally Metallic Product Space Forms with a Semi-Symmetric Metric Connection

Yanlin Li [1,*], Md Aquib [2], Meraj Ali Khan [2], Ibrahim Al-Dayel [2] and Khalid Masood [2]

1 School of Mathematics, Hangzhou Normal University, Hangzhou 311121, China
2 Department of Mathematics and Statistics, College of Science, Imam Mohammad Ibn Saud Islamic University (IMSIU), Riyadh 11566, Saudi Arabia; maquib@imamu.edu.sa (M.A.); mskhan@imamu.edu.sa (M.A.K.); iaaldayel@imamu.edu.sa (I.A.-D.); kmaali@imamu.edu.sa (K.M.)
* Correspondence: liyl@hznu.edu.cn

Abstract: This article explores the Ricci tensor of slant submanifolds within locally metallic product space forms equipped with a semi-symmetric metric connection (SSMC). Our investigation includes the derivation of the Chen–Ricci inequality and an in-depth analysis of its equality case. More precisely, if the mean curvature vector at a point vanishes, then the equality case of this inequality is achieved by a unit tangent vector at the point if and only if the vector belongs to the normal space. Finally, we have shown that when a point is a totally geodesic point or is totally umbilical with $n = 2$, the equality case of this inequality holds true for all unit tangent vectors at the point, and conversely.

Keywords: Chen–Ricci inequality; isotropic submanifolds; locally metallic product space forms

MSC: 53B50; 53C20; 53C40

Citation: Li, Y.; Aquib, M.; Khan, M.A.; Al-Dayel, I.; Masood, K. Analyzing the Ricci Tensor for Slant Submanifolds in Locally Metallic Product Space Forms with a Semi-Symmetric Metric Connection. *Axioms* **2024**, *13*, 454. https://doi.org/10.3390/axioms13070454

Academic Editor: Mića Stanković

Received: 21 May 2024
Revised: 26 June 2024
Accepted: 1 July 2024
Published: 4 July 2024

1. Introduction

The investigation of submanifolds immersed in Riemannian manifolds has captivated the attention of differential geometry scholars for numerous decades. Within this realm, a fundamental inquiry revolves around comprehending the geometric characteristics of submanifolds in relation to the curvature of the encompassing manifold.

A renowned inequality in differential geometry, known as the Chen–Ricci inequality, establishes a connection between the scalar curvature of a submanifold, its mean curvature, and the norm of its second fundamental form.

In 1996, Chen made a significant breakthrough by formulating an equation that links two fundamental geometric properties of a submanifold, denoted as $\mathcal{M}$, embedded within a space known as $\overline{\mathcal{M}}(c)$ with a constant curvature c. These properties are the Ricci curvature denoted by Ric and the squared mean curvature expressed as $||H||^2$. According to Chen's formula, for any unit vector χ positioned on the submanifold $\mathcal{M}$,

$$Ric(\chi) \leq (k-1)c + \frac{k^2}{2}||H||^2, \quad k = dim\mathcal{M}$$

Chen also established the aforementioned inequality for Lagrangian submanifolds [1]. Since then, this inequality has garnered the interest of geometers worldwide, leading to the proof of similar inequalities by various researchers for diverse types of submanifolds in various ambient manifolds [2–4]. Furthermore, there are some applications in geometric flow and tangent bundles. For example, the study of Harnack estimates [5], Li–Yau-type gradient estimates [6,7], Perelman-type differential Harnack inequalities and Li–Yau-type estimates [8], the new Harnack inequalities of a variety of geometric flows [9], etc. Recent works on differential Harnack inequalities can be found in [10–12]. In Ref. [13], Kumar, R.

et al. considered the problems of NSNMC in the tangent bundle. In Refs. [14,15], Kumar and R. et al. studied the tangent bundles with QSNMC in an LP-Sasakian manifold. The properties, theorems, and results of the curvature tensor and Ricci tensor relevant to QSMC on the tangent bundles were obtained in [16–18]. In the recent years, Li and Khan et al. conducted the research relevant to inequalities [19], solitons [20], submanifolds [21], and classical differential geometry [22] under the viewpoint of submanifold theory, soliton theory, and other related theories [23–25]. The results and methods of those papers motivate us to write the paper.

Simultaneously, a θ-slant submanifold represents a subtype of the submanifold in the domain of differential geometry that generalizes the concept of a slant submanifold. Similar to slant submanifolds, θ-slant submanifolds pertain to submanifolds of Riemannian manifolds that exhibit a tilted or slanted geometry concerning the ambient manifold. Nevertheless, unlike slant submanifolds, which are defined by the angle between the submanifold and a vector distribution in the ambient manifold [26–28], θ-slant submanifolds are defined by a more inclusive angle function θ, which can rely on the submanifold's position within the ambient manifold [29,30]. This broader definition allows for increased flexibility and generality in characterizing the submanifold.

Specifically, a θ-slant submanifold is determined by the prerequisite that its tangent space at each point be slanted in relation to a particular vector distribution in the ambient manifold, wherein the angle of slant is determined by evaluating the angle function θ at that specific point. This angle function enables the capture of various geometric properties of the submanifold, including its curvature or its embedding within the ambient manifold. The classification of minimal surfaces in Euclidean space has historically leveraged slant submanifolds, with Almgren's renowned theorem asserting that any complete, non-flat minimal surface in Euclidean space must be either a plane, a catenoid, or a helicoid. The proof of this theorem employs the theory of slant submanifolds to demonstrate the impossibility of certain types of minimal surfaces.

In the work by Mastsumoto [31], a bound for the Ricci tensor of slant submanifolds in complex space forms was obtained. They also demonstrated that a Kaehlerian slant submanifold satisfying the equality case identically is minimal. Kim et al. [32] derived the Ricci curvature for integral submanifolds of S-space forms and discussed the equality case of the inequality. Additionally, they obtained results for various subclasses, including almost semi-invariant submanifolds, θ-slant submanifolds, anti-invariant submanifolds, and invariant submanifolds.

In 2010, Mihai and Ozgur [33] established an inequality for submanifolds of real space forms with a semi-symmetric connection. They also considered the equality case for this inequality. Mihai and Radulescu improved the inequality for Kaehlerian slant submanifolds in complex space forms [34]. Deng [35] enhanced the Chen–Ricci inequality for Lagrangian submanifolds in complex space forms by utilizing an optimization technique. Mihai [36] improved the Chen–Ricci inequalities for Lagrangian submanifolds of dimension n (where $n \geq 2$) in a $2n$-dimensional complex space form with a semi-symmetric metric connection, as well as for Legendrian submanifolds in a Sasakian space with a semi-symmetric metric connection.

In their work [37], Khan and Ozel established a relationship between the Ricci curvature and the squared norm of the second fundamental form for contact CR-warped product submanifolds in generalized Sasakian space forms admitting a trans-Sasakian structure. Recently, Lee et al. [38] derived Chen–Ricci inequalities for Riemannian maps with different ambient spaces and discussed numerous applications, for which we can refer to [39–41].

Motivated by the above studies, this article centers its focus on θ-slant submanifolds within locally metallic product space forms and explores the Chen–Ricci inequality as it applies to these submanifolds.

Our principal outcome involves the formulation of the Chen–Ricci inequality for θ-slant submanifolds within locally metallic product space forms, along with deriving the conditions under which equality to the inequality is established.

Furthermore, we delve into several applications of our findings, showcasing how our inequality facilitates the derivation of significant geometric properties specific to θ-slant submanifolds.

2. Fundamental Results

In the subsequent section, we present the relevant mathematical formulas and concepts necessary to grasp the Chen–Ricci inequality concerning isotropic submanifolds in locally metallic product space forms.

Let $\overline{\mathcal{M}}$ denote a Riemannian manifold equipped with the linear connection $\overline{\nabla}$. A connection is classified as semi-symmetric if its torsion tensor T satisfies the expression:

$$T(\chi_1, \chi_2) = \pi(\chi_2)\chi_1 - \pi(\chi_1)\chi_2$$

where π is a 1-form. Consequently, $\overline{\nabla}$ is referred to as a semi-symmetric connection. Assuming a Riemannian metric g on $\overline{\mathcal{M}}$, if $\overline{\nabla}g = 0$, then $\overline{\nabla}$ qualifies as a semi-symmetric metric connection on $\overline{\mathcal{M}}$. The mathematical form of this connection is given by:

$$\overline{\nabla}_{\chi_1}\chi_2 = \overline{\tilde{\nabla}}_{\chi_1}\chi_2 + \pi(\chi_2)\chi_1 - g(\chi_1, \chi_2)\Gamma \tag{1}$$

where χ_1 and χ_2 are arbitrary vectors in $\overline{\mathcal{M}}$, $\overline{\tilde{\nabla}}$ represents the Levi–Civita connection with respect to the Riemannian metric g, and Γ is a vector field.

Suppose $\mathcal{M}$ is an m-dimensional submanifold within the Riemannian manifold $\overline{\mathcal{M}}$. Let ∇ denote the semi-symmetric metric connection induced on $\mathcal{M}$, and let $\tilde{\nabla}$ denote the Levi–Civita connection. In this case, the Gauss formulas can be expressed as follows:

$$\overline{\nabla}_{\chi_1}\chi_2 = \nabla_{\chi_1}\chi_2 + \zeta(\chi_1, \chi_2), \quad \chi_1, \chi_2 \in \Gamma(T\mathcal{M}), \tag{2}$$

$$\overline{\tilde{\nabla}}_{\chi_1}\chi_2 = \tilde{\nabla}_{\chi_1}\chi_2 + \tilde{\zeta}(\chi_1, \chi_2), \quad \chi_1, \chi_2 \in \Gamma(T\mathcal{M}), \tag{3}$$

Let $\tilde{\zeta}$ represent the second fundamental form. Additionally, let $\overline{R}$ and $\overline{\tilde{R}}$ denote the curvature tensors of $\overline{\mathcal{M}}$ and $\mathcal{M}$ respectively, with respect to the connections $\overline{\nabla}$ and $\overline{\tilde{\nabla}}$. Similarly, R and $\tilde{R}$ denote the curvature tensors of $\overline{\mathcal{M}}$ and $\mathcal{M}$ respectively, with respect to the connections ∇ and $\tilde{\nabla}$. Given these definitions, we can express the following relations:

$$\begin{aligned}\overline{\tilde{R}}(\chi_1, \chi_2, \chi_3, \chi_4) = {} & \tilde{R}(\chi_1, \chi_2, \chi_3, \chi_4) \\ & + g(\zeta(\chi_1, \chi_3), \zeta(\chi_2, \chi_4)) - g(\zeta(\chi_1, \chi_4), \zeta(\chi_2, \chi_3)),\end{aligned} \tag{4}$$

for $\chi_1, \chi_2, \chi_3, \chi_4 \in T\overline{\mathcal{M}}$. Let us introduce the $(0,2)$ tensors:

$$\beta(\chi_1, \chi_2) = (\overline{\tilde{\nabla}}_{\chi_1}\pi)(\chi_2) - \pi(\chi_1)\pi(\chi_2) + \frac{1}{2}g(\chi_1, \chi_2)\pi(\Gamma).$$

According to Wang [42], the expression for the curvature tensor $\overline{R}$ of the manifold $\overline{\mathcal{M}}$ is as follows:

$$\begin{aligned}\overline{R}(\chi_1, \chi_2, \chi_3, \chi_4) &= \overline{\tilde{R}}(\chi_1, \chi_2, \chi_3, \chi_4) + \beta(\chi_1, \chi_3)g(\chi_2, \chi_4) \\ &- \beta(\chi_2, \chi_3)g(\chi_1, \chi_4) + \beta(\chi_2, \chi_4)g(\chi_1, \chi_3) - \beta(\chi_1, \chi_4)g(\chi_2, \chi_3).\end{aligned} \tag{5}$$

Let us define λ as the trace of β.

Let $\mathcal{M}$ be a Riemannian manifold, and let $\pi \subset T_x\mathcal{M}$ be a plane section at a point $x \in \mathcal{M}$. The sectional curvature of π is denoted by $K(\pi)$. For any $x \in \mathcal{M}$, if $\{\varrho_1, \ldots, \varrho_n\}$

and $\{\varrho_{n+1}, \dots, \varrho_m\}$ are orthonormal bases of $T_x\mathcal{M}$ and $T_x^{\perp}\mathcal{M}$, respectively, then the scalar curvature τ can be expressed as follows:

$$\tau(x) = \sum_{1 \le i < j \le n} K(\varrho_i \wedge \varrho_j). \tag{6}$$

$$\mathcal{H} = \frac{1}{n} \sum_{i=1}^{n} g(\zeta(\varrho_i, \varrho_i)).$$

The orthonormal frames $\{\varrho_1, \dots, \varrho_n\}$ and $\{\varrho_{n+1}, \dots, \varrho_m\}$ represent the tangent and normal spaces, respectively, on the Riemannian manifold $\mathcal{M}$.

The relative null space of the Riemannian manifold at a point x in $\mathcal{M}$ is defined as:

$$\mathcal{N}_x = \{\chi_1 \in T_x\mathcal{M} | \zeta(\chi_1, \chi_2) = 0 \quad \forall \quad \chi_2 \in T_x\mathcal{M}\}. \tag{7}$$

This refers to the subset of the tangent space at point x where the second fundamental form is constantly zero. It is also referred to as the normal space of $\mathcal{M}$ at x.

In the context of minimal submanifolds, it is stated that the mean curvature vector $\mathcal{H}$ is always zero.

On a m-dimensional Riemannian manifold $(\overline{\mathcal{M}}, g)$ with real numbers $a_1, \dots, a_n$, a polynomial structure is defined as a tensor field ϑ of type (1, 1) that satisfies the following equation:

$$\mathcal{B}(\vartheta) \equiv \vartheta^n + a_{n-1}\vartheta^{n-1} + \dots + a_2\vartheta + a_1\mathcal{I} = 0,$$

where $\mathcal{I}$ denotes the identity transformation. The following remark presents a few notable instances of polynomial structures.

Remark 1. *1. ϑ is an almost complex structure if $\mathcal{B}(\vartheta) = \vartheta^2 + \mathcal{I}$.*
2. ϑ is an almost product structure if $\mathcal{B}(\vartheta) = \vartheta^2 - \mathcal{I}$.
3. ϑ is a metallic structure if $\mathcal{B}(\vartheta) = \vartheta^2 - p\vartheta - q\mathcal{I}$,
where p and q are two integers.

If for all $\chi_1, \chi_2 \in \Gamma(T\overline{\mathcal{M}})$

$$g(\vartheta\chi_1, \chi_2) = g(\chi_1, \vartheta\chi_2), \tag{8}$$

then in such a case, the Riemannian metric g is referred to as being compatible with the polynomial structure ϑ.

In the context of Riemannian manifolds, a metallic structure refers to a tensor field ϑ that satisfies two conditions: it is ϑ-compatible with the metric g, and the manifold $(\overline{\mathcal{M}}, g)$ itself is a metallic Riemannian manifold.

By utilizing Equation (8), we derive

$$g(\vartheta\chi_1, \vartheta\chi_2) = g(\vartheta^2\chi_1, \chi_2) = pg(\chi_1, \vartheta\chi_2) + qg(\chi_1, \chi_2).$$

An almost product structure $\mathcal{F}$ defined on an m-dimensional (Riemannian) manifold $(\overline{\mathcal{M}}, g)$ is characterized by being a (1,1)-tensor field that satisfies $\mathcal{F}^2 = I$ and $\mathcal{F} \neq \pm\mathcal{I}$. When $\mathcal{F}$ additionally fulfills the condition $g(\mathcal{F}\chi_1, \chi_2) = g(\chi_1, \mathcal{F}\chi_2)$ for all $\chi_1, \chi_2 \in \Gamma(T\overline{\mathcal{M}})$, the manifold $(\overline{\mathcal{M}}, g)$ is said to be an almost product Riemannian manifold [43].

There exist two almost product structures, denoted as $\mathcal{F}_1$ and $\mathcal{F}_2$, induced by a metallic structure ϕ on $\overline{\mathcal{M}}$ [44]. These structures can be expressed using the following equation:

$$\begin{cases} \mathcal{F}_1 = \frac{2}{2\sigma_{p,q} - p}\phi - \frac{p}{2\sigma_{p,q} - p}\mathcal{I}, \\ \\ \mathcal{F}_2 = \frac{2}{2\sigma_{p,q} - p}\phi + \frac{p}{2\sigma_{p,q} - p}\mathcal{I}, \end{cases}$$

where $\sigma_{p,q} = \frac{p+\sqrt{p^2+4q}}{2}$ are the members of the metallic means family or the metallic proportions.

Likewise, for any given almost product structure $\mathcal{F}$ on $\overline{\mathcal{N}}$, two corresponding metallic structures ϕ_1 and ϕ_2 are induced, and they can be defined as follows:

$$\begin{cases} \phi_1 = \frac{p}{2}\mathcal{I} + \frac{2\sigma_{p,q}-p}{2}\mathcal{F}, \\ \phi_2 = \frac{p}{2}\mathcal{I} - \frac{2\sigma_{p,q}-p}{2}\mathcal{F}. \end{cases}$$

Definition 1 ([45]). *Consider a metallic structure ϕ on $\overline{\mathcal{M}}$ and a linear connection $\overline{\nabla}$ such that $\nabla\phi = 0$. In this case, $\overline{\nabla}$ is referred to as a ϕ connection. A locally metallic Riemannian manifold is defined as a metallic Riemannian manifold $(\overline{\mathcal{M}}, g, \phi)$, wherein the Levi–Civita connection $\overline{\nabla}$ associated with the metric g serves as a ϕ connection.*

Here, we recall the following.

Remark 2. *It is essential to bear in mind that the metallic family includes various members, which are categorized as follows [44]:*

1. *The golden structure, when $p = q = 1$.*
2. *The copper structure, when $p = 1$ and $q = 2$.*
3. *The nickel structure, when $p = 1$ and $q = 3$.*
4. *The silver structure, when $p = 2$ and $q = 1$.*
5. *The bronze structure, when $p = 3$ and $q = 1$.*
6. *The subtle structure, when $p = 4$ and $q = 1$, and so on.*

Suppose we have an m-dimensional metallic Riemannian manifold $(\overline{\mathcal{M}}, g, \phi)$ and an n-dimensional submanifold $(\mathcal{M}, g)$ that is isometrically immersed into $\overline{\mathcal{M}}$ with the induced metric g. For any $x \in \mathcal{M}$, the tangent space $T_x\overline{\mathcal{M}}$ of $\overline{\mathcal{M}}$ at x can be expressed as the direct sum of $T_x\mathcal{M}$ and $T_x^{\perp}\mathcal{M}$, where $T_x\mathcal{M}$ is the tangent space of $\mathcal{M}$ at x, and $T_x^{\perp}\mathcal{M}$ is the orthogonal complement of $T_x\mathcal{M}$ in $T_x\overline{\mathcal{M}}$.

In an almost Hermitian manifold $\overline{\mathcal{M}}$, a submanifold $\mathcal{M}$ is considered to be a slant submanifold if the angle between $J\mathcal{M}$ and $T_x\mathcal{M}$ remains constant for any $x \in \mathcal{M}$ and a non-zero vector $X \in T_x\mathcal{M}$. The slant angle of $\mathcal{M}$ in $\overline{\mathcal{M}}$ is denoted by θ and takes values in the interval $[0, \frac{\pi}{2}]$.

Moreover, if $\mathcal{M}$ is a slant submanifold of a metallic Riemannian manifold $(\overline{\mathcal{M}}, g, \phi)$ with a slant angle θ, then according to [45]:

$$g(T\chi_1, T\chi_2) = \cos^2\theta[pg(\chi_1, T\chi_2) + qg(\chi_1, \chi_2)]$$

and

$$g(N\chi_1, N\chi_2) = \sin^2\theta[pg(\chi_1, T\chi_2) + qg(\chi_1, \chi_2)],$$

$\forall \chi_1, \chi_2 \in \Gamma(T\mathcal{M})$.

Additionally,

$$T^2 = \cos^2\theta(pT + q\mathcal{I}),$$

Here, $\mathcal{I}$ denotes the identity operator acting on $\Gamma(T\mathcal{M})$, the space of smooth sections of the tangent bundle of $\mathcal{M}$, and

$$\nabla T^2 = p\cos^2\theta\nabla T.$$

Consider $\mathcal{M}_1$, a Riemannian manifold with constant sectional curvature c_1, and $\mathcal{M}_2$, a Riemannian manifold with constant sectional curvature c_2. In this case, the Riemannian curvature tensor $\overline{R}$ of the locally Riemannian product manifold can be expressed as follows.

$\overline{\mathcal{M}} = \mathcal{M}_1 \times \mathcal{M}_2$ is given by [43]

$$\begin{aligned}
\tilde{\overline{R}}(\chi_1,\chi_2)\chi_3 \;=\; & \frac{1}{4}(c_1+c_2)[g(\chi_2,\chi_3)\chi_1 - g(\chi_1,\chi_3)\chi_2] \\
& +\frac{1}{4}(c_1+c_2)\left\{\frac{4}{(2\sigma_{p,q}-p)^2}[g(\phi\chi_2,\chi_3)\phi\chi_1 - g(\phi\chi_1,\chi_3)\phi\chi_2]\right. \\
& +\frac{p^2}{(2\sigma_{p,q}-p)^2}[g(\chi_2,\chi_3)X - g(\chi_1,\chi_3)\chi_2] \\
& +\frac{2p}{(2\sigma_{p,q}-p)^2}[g(\phi\chi_1,\chi_3)\chi_2 + g(\chi_1,\chi_3)\phi\chi_2 \\
& \left. -g(\phi\chi_2,\chi_3)\chi_1 - g(\chi_1,\chi_3)\phi\chi_1]\right\} \\
& \pm\frac{1}{2}(c_1-c_2)\left\{\frac{1}{(2\sigma_{p,q}-p)}[g(\chi_2,\chi_3)\phi\chi_1 - g(\chi_1,\chi_3)\phi\chi_2]\right. \\
& +\frac{1}{(2\sigma_{p,q}-p)}[g(\phi\chi_2,\chi_3)\chi_1 - g(\phi\chi_1,\chi_3)\chi_2] \\
& \left. +\frac{p}{(2\sigma_{p,q}-p)}[g(\chi_1,\chi_3)\chi_2 - g(\chi_2,\chi_3)\chi_1]\right\}. \qquad (9)
\end{aligned}$$

From (5) and (9), we have

$$\begin{aligned}
\overline{R}(\chi_1,\chi_2,\chi_3,\chi_4) = & \frac{1}{4}(c_1+c_2)[g(\chi_2,\chi_3)g(\chi_1,\chi_4) - g(\chi_1,\chi_3)g(\chi_2,\chi_4)] \\
& +\frac{1}{4}(c_1+c_2)\left\{\frac{4}{(2\sigma_{p,q}-p)^2}[g(\phi\chi_2,\chi_3)g(\phi\chi_1,\chi_4) - g(\phi\chi_1,\chi_3)g(\phi\chi_2,\chi_4)]\right. \\
& +\frac{p^2}{(2\sigma_{p,q}-p)^2}[g(\chi_2,\chi_3)g(\chi_1,\chi_4) - g(\chi_1,\chi_3)g(\chi_2,\chi_4)] \\
& +\frac{2p}{(2\sigma_{p,q}-p)^2}[g(\phi\chi_1,\chi_3)g(\chi_2,\chi_4) + g(\chi_1,\chi_3)g(\phi\chi_2,\chi_4) \\
& \left. - g(\phi\chi_2,\chi_3)g(\chi_1,\chi_4) - g(\chi_2,\chi_3)g(\phi\chi_1,\chi_4)]\right\} \\
& \pm\frac{1}{2}(c_1-c_2)\left\{\frac{1}{(2\sigma_{p,q}-p)}[g(\chi_2,\chi_3)g(\phi\chi_1,\chi_4) - g(\chi_1,\chi_3)g(\phi\chi_2,\chi_4)]\right. \\
& +\frac{1}{(2\sigma_{p,q}-p)}[g(\phi\chi_2,\chi_3)g(\chi_1,\chi_4) - g(\phi\chi_1,\chi_3)g(\chi_2,\chi_4)] \\
& \left. +\frac{p}{(2\sigma_{p,q}-p)}[g(\chi_1,\chi_3)g(\chi_2,\chi_4) - g(\chi_2,\chi_3)g(\chi_1,\chi_4)]\right\} \\
& +\beta(\chi_1,\chi_3)g(\chi_2,\chi_4) - \beta(\chi_2,\chi_3)g(\chi_1,\chi_4) \\
& +\beta(\chi_2,\chi_4)g(\chi_1,\chi_3) - \beta(\chi_1,\chi_4)g(\chi_2,\chi_3). \qquad (10)
\end{aligned}$$

3. Ricci Tensor Analysis with Semi-Symmetric Metric Connection

The primary objective of this section is to introduce and analyze the principal outcome.

Theorem 1. *Suppose we have a submanifold $\mathcal{M}$ of dimension n that is slanted at an angle of θ in a locally metallic product space form $\overline{\mathcal{M}} = \mathcal{M}_1(c_1) \times \mathcal{M}_2(c_2)$ with SSMC.*

Given a point p on $\mathcal{M}$ and a unit vector X in the tangent space $T_p\mathcal{M}$, the following inequality holds:

$$Ric(X) \leq \frac{n^2}{4}||H||^2 \pm \frac{1}{2}\frac{c_1 - c_2}{\sqrt{p^2+4q}}[2\,\mathrm{tr}\,\phi - p(n-1)] - (n-2)\beta(X,X) - \lambda$$
$$+ \frac{1}{2}\frac{c_1+c_2}{p^2+4q}(n-1)\left[p^2 + 2q - \frac{1}{n-1}(p\,\mathrm{tr}\,\phi + q\cos^2\theta)\right]. \tag{11}$$

Moreover, if $H(p) = 0$, then the equality case of this inequality is achieved by a unit tangent vector X at p if and only if X belongs to the normal space $\mathcal{N}_p$. Finally, when p is a totally geodesic point or is totally umbilical with $n = 2$, the equality case of this inequality holds true for all unit tangent vectors at p, and conversely.

Proof. Let $\{\varrho_1, \ldots, \varrho_n\}$ be an orthonormal tangent frame and $\{\varrho_{n+1}, \ldots, \varrho_m\}$ be an orthonormal frame of $T_x\mathcal{M}$ and $T_x^{\perp}\mathcal{M}$, respectively at any point $x \in \mathcal{M}$. Substituting $\chi_1 = \chi_4 = \varrho_i, \chi_2 = \chi_3 = \varrho_j$ in (10) with the Equation (4) and take $i \neq j$, we obtain

$$\begin{aligned}
R(\varrho_i,\varrho_j,\varrho_j,\varrho_i) &= \frac{1}{4}(c_1+c_2)[g(\varrho_j,\varrho_j)g(\varrho_i,\varrho_i) - g(\varrho_i,\varrho_j)g(\varrho_j,\varrho_i)] \\
&+ \frac{1}{4}(c_1+c_2)\bigg\{\frac{4}{(2\sigma_{p,q}-p)^2}[g(\phi\varrho_j,\varrho_j)g(\phi\varrho_i,\varrho_i) \\
&- g(\phi\varrho_i,\varrho_j)g(\phi\varrho_j,\varrho_i)] \\
&+ \frac{p^2}{(2\sigma_{p,q}-p)^2}[g(\varrho_j,\varrho_j)g(\varrho_i,\varrho_i) - g(\varrho_i,\varrho_j)g(\varrho_j,\varrho_i)] \\
&+ \frac{2p}{(2\sigma_{p,q}-p)^2}[g(\phi\varrho_i,\varrho_j)g(\varrho_j,\varrho_i) + g(\varrho_i,\varrho_j)g(\phi\varrho_j,\varrho_i) \\
&- g(\phi\varrho_j,\varrho_j)g(\varrho_i,\varrho_i) - g(\varrho_j,\varrho_j)g(\phi\varrho_i,\varrho_i)]\bigg\} \\
&\pm \frac{1}{2}(c_1-c_2)\bigg\{\frac{1}{(2\sigma_{p,q}-p)}[g(\varrho_j,\varrho_j)g(\phi\varrho_i,\varrho_i) \\
&- g(\varrho_i,\varrho_j)g(\phi\varrho_j,\varrho_i)] \\
&+ \frac{1}{(2\sigma_{p,q}-p)}[g(\phi\varrho_j,\varrho_j)g(\varrho_i,\varrho_i) - g(\phi\varrho_i,\varrho_j)g(\varrho_j,\varrho_i)] \\
&+ \frac{p}{(2\sigma_{p,q}-p)}[g(\varrho_i,\varrho_j)g(\varrho_j,\varrho_i) - g(\varrho_j,\varrho_j)g(\varrho_i,\varrho_i)]\bigg\} \\
&+ \beta(\varrho_i,\varrho_j)g(\varrho_j,\varrho_i) - \beta(\varrho_j,\varrho_j)g(\varrho_i,\varrho_i) \\
&+ \beta(\varrho_j,\varrho_i)g(\varrho_i,\varrho_j) - \beta(\varrho_i,\varrho_i)g(\varrho_j,\varrho_j) \\
&+ g(\zeta(\varrho_i,\varrho_i),\zeta(\varrho_j,\varrho_j)) - g(\zeta(\varrho_i,\varrho_j),\zeta(\varrho_j,\varrho_i)).
\end{aligned} \tag{12}$$

Using $1 \leq i,j \leq n$ in (12), we find

$$2\tau(x) = \frac{1}{4}(c_1+c_2)\frac{n(n-1)}{p^2+4q}\bigg\{2p^2 + 4q + \frac{4}{n(n-1)}[tr^2\phi - \cos^2\theta(p.trT + nq)]$$
$$- \frac{4p}{n}tr\phi\bigg\} - 2(n-1)\lambda$$
$$\pm \frac{1}{4}\frac{(n-1)}{\sqrt{p^2+4q}}(c_1-c_2)(4tr\phi - 2np) + n^2||H||^2 - ||\zeta||^2. \tag{13}$$

Now, we consider

$$\delta = 2\tau - \frac{n^2}{2}||\mathcal{H}||^2 \mp \frac{1}{4}\frac{(n-1)}{\sqrt{p^2+4q}}(c_1-c_2)(4tr\phi - 2np) + 2(n-1)\lambda$$
$$-\frac{1}{4}(c_1-c_2)\frac{n(n-1)}{p^2+4q}\left\{2p^2+4q+\frac{4}{n(n-1)}\left[tr^2\phi - \cos^2\theta(ptrT+nq)\right] - \frac{4p}{n}tr\phi\right\}. \quad (14)$$

Combining (13) and (14), we obtain

$$n^2||H||^2 = 2(\delta + ||\zeta||^2). \quad (15)$$

Consequently, when employing the orthonormal frame $\{\varrho_1, \ldots, \varrho_n\}$, Equation (15) takes on the subsequent expression:

$$\left(\sum_{i=1}^{n}\zeta_{ii}^{n+1}\right)^2 = 2\left\{\delta + \sum_{i=1}^{n}(\zeta_{ii}^{n+1})^2 + \sum_{i\neq j}(\zeta_{ij}^{n+1})^2 + \sum_{r=n+1}^{m}\sum_{i,j=1}^{n}(\zeta_{ij}^{r})^2\right\}. \quad (16)$$

If we substitute $d_1 = \zeta_{11}^{n+1}$, $d_2 = \sum_{i=2}^{n-1}\zeta_{ii}^{n+1}$ and $d_3 = \zeta_{nn}^{n+1}$, then (16) reduces to

$$\left(\sum_{i=1}^{3}d_i\right)^2 = 2\left\{\delta + \sum_{i=1}^{3}d_i^2 + \sum_{i\neq j}(\zeta_{ij}^{n+1})^2 + \sum_{r=n+1}^{m}\sum_{i,j=1}^{n}(\zeta_{ij}^{r})^2 - \sum_{2\leq j\neq k\leq n-1}\zeta_{jj}^{n+1}\zeta_{kk}^{n+1}\right\}. \quad (17)$$

As a result, d_1, d_2, d_3 fulfill Chen's Lemma [41], that is,

$$\left(\sum_{i=1}^{3}d_i\right)^2 = 2\left(v + \sum_{i=1}^{3}d_i^2\right).$$

Clearly, $2d_1d_2 \geq v$ with equality holds if $d_1 + d_2 = d_3$ and vice versa. This signifies

$$\sum_{1\leq j\neq k\leq n-1}\zeta_{jj}^{n+1}\zeta_{kk}^{n+1} \geq \delta + 2\sum_{i<j}(\zeta_{ij}^{n+1})^2 + \sum_{r=n+1}^{m}\sum_{i,j=1}^{n}(\zeta_{ij}^{r})^2. \quad (18)$$

It is possible to write (18) as

$$\frac{n^2}{2}||\mathcal{H}||^2 \pm \frac{1}{4}\frac{(n-1)}{\sqrt{p^2+4q}}(c_1-c_2)(4tr\phi - 2np) - 2(n-1)\lambda$$
$$+\frac{1}{4}(c_1-c_2)\frac{n(n-1)}{p^2+4q}\left\{2p^2+4q+\frac{4}{n(n-1)}\left[tr^2\phi - \cos^2\theta(ptrT+nq)\right] - \frac{4p}{n}tr\phi\right\}$$
$$\geq 2\tau - \sum_{1\leq j\neq k\leq n-1}\zeta_{jj}^{n+1}\zeta_{kk}^{n+1} + 2\sum_{i<j}(\zeta_{ij}^{n+1})^2 + \sum_{r=n+1}^{m}\sum_{i,j=1}^{n}(\zeta_{ij}^{r})^2. \quad (19)$$

Invoking the Gauss equation once again and making use of (19), we obtain

$$Ric(X) \leq \frac{n^2}{4}||H||^2 \pm \frac{1}{2}\frac{(c_1-c_2)}{\sqrt{p^2+4q}}\left[2tr\phi - p(n-1)\right] - (n-2)\beta(X,X) - \lambda$$
$$+\frac{1}{2}\frac{(c_1+c_2)}{p^2+4q}(n-1)\left[p^2+2q-\frac{1}{n-1}(ptr\phi + q\cos^2\theta)\right]. \quad (20)$$

Hence, we have obtained the required inequality (1).

Further, assume that $H(p) = 0$. Equality holds in (1) if and only if

$$\begin{cases} \zeta_{1n}^r = \cdots = \zeta_{n-1n}^r = 0 \\ \zeta_{nn}^r = \sum_{i=1}^{n-1} \zeta_{ii}^r, \quad r \in \{n+1, \ldots, m\}. \end{cases} \tag{21}$$

Then,

$$\zeta_{in}^r = 0,$$

for all $i \in \{1, \ldots, n\}$, and $r \in \{n+1, \ldots, m\}$, i.e., $X \in \mathcal{N}_p$.

In conclusion, the equality condition of (1) holds for all unit tangent vectors at p if and only if

$$\begin{cases} \zeta_{ij}^r = 0, i \neq j, r \in \{n+1, \ldots, m\} \\ \zeta_{11}^r + \cdots + \zeta_{nn}^r - 2\zeta_{ii}^r = 0, \quad i \in \{1, \ldots, n\} \quad r \in \{n+1, \ldots, m\}. \end{cases} \tag{22}$$

From here, we separate the two situations:

(i) p is a totally geodesic point if $n \neq 2$;

(ii) It is evident that p is a totally umbilical point if $n = 2$.

It goes without saying that the converse applies. □

4. Some Applications

We can have two different approaches to see the various applications: either by considering particular classes of locally metallic product space forms, or by considering particular classes of θ-slant submanifolds.

4.1. Application by Considering Particular Classes of θ-Slant Submanifolds

Two specific classes of θ-slant submanifolds, namely, invariant and anti-invariant submanifolds, were introduced in [45] for metallic Riemannian manifolds. With the help of the definitions of these submanifolds in Theorem 1, we obtain the following results.

Corollary 1. *Suppose we have a submanifold $\mathcal{M}$ of dimension n that is invariant in a locally metallic product space form $\overline{\mathcal{M}} = \mathcal{M}_1(c_1) \times \mathcal{M}_2(c_2)$ with SSMC.*

For any unit vector X in the tangent space $T_p\mathcal{M}$ at a point p on $\mathcal{M}$, the following inequality holds:

$$\begin{aligned} Ric(X) \leq{} & \frac{n^2}{4}||H||^2 \pm \frac{1}{2}\frac{(c_1 - c_2)}{\sqrt{p^2 + 4q}}[2tr\phi - p(n-1)] - (n-2)\beta(X, X) - \lambda \\ & + \frac{1}{2}\frac{(c_1 + c_2)}{p^2 + 4q}(n-1)\left[p^2 + 2q - \frac{1}{n-1}(ptr\phi + q)\right]. \end{aligned} \tag{23}$$

Moreover, if $H(p) = 0$, then the equality case of this inequality is achieved by a unit tangent vector X at p if and only if X belongs to the normal space $\mathcal{N}_p$. Finally, when p is a totally geodesic point or is totally umbilical with $n = 2$, the equality case of this inequality holds true for all unit tangent vectors at p, and vice versa.

Proof. The result is directly obtained by taking $\theta = 0$ in Theorem 1. □

Corollary 2. *Suppose we have a submanifold $\mathcal{M}$ of dimension n that is anti-invariant in a locally metallic product space form $\overline{\mathcal{M}} = \mathcal{M}_1(c_1) \times \mathcal{M}_2(c_2)$ with SSMC.*

For any unit vector X in the tangent space $T_p\mathcal{M}$ at a point p on $\mathcal{M}$, the following inequality holds:

$$Ric(X) \leq \frac{n^2}{4}||H||^2 \pm \frac{1}{2}\frac{(c_1 - c_2)}{\sqrt{p^2 + 4q}}\big[2tr\phi - p(n-1)\big] - (n-2)\beta(X,X) - \lambda$$
$$+ \frac{1}{2}\frac{(c_1 + c_2)}{p^2 + 4q}(n-1)\big[p^2 + 2q - \frac{1}{n-1}(ptr\phi)\big]. \tag{24}$$

Moreover, if $H(p) = 0$, then the equality case of this inequality is achieved by a unit tangent vector X at p if and only if X belongs to the normal space $\mathcal{N}_p$. Finally, when p is a totally geodesic point or is totally umbilical with $n = 2$, the equality case of this inequality holds true for all unit tangent vectors at p, and vice versa.

Proof. The result is directly obtained by taking $\theta = \frac{\pi}{2}$ in Theorem 1. □

4.2. Application by Considering Particular Classes of Locally Metallic Product Space Forms

As a consequence of Theorem 1 and together with Remark 2 (1), we obtained the following results.

Corollary 3. *Suppose we have a submanifold $\mathcal{M}$ of dimension n that is slanted at an angle of θ in a locally golden product space form $\overline{\mathcal{M}} = \mathcal{M}_1(c_1) \times \mathcal{M}_2(c_2)$ with SSMC.*

For any unit vector X in the tangent space $T_p\mathcal{M}$ at a point p on $\mathcal{M}$, the following inequality holds:

$$Ric(X) \leq \frac{n^2}{4}||H||^2 \pm \frac{1}{2\sqrt{5}}(c_1 - c_2)\big[2.tr\phi - (n-1)\big] - (n-2)\beta(X,X) - \lambda$$
$$+ \frac{1}{10}(c_1 + c_2)(n-1)\big[3 - \frac{1}{n-1}(tr\phi + \cos^2\theta)\big]. \tag{25}$$

Moreover, if $H(p) = 0$, then the equality case of this inequality is achieved by a unit tangent vector X at p if and only if X belongs to the normal space $\mathcal{N}_p$. Finally, when p is a totally geodesic point or is totally umbilical with $n = 2$, the equality case of this inequality holds true for all unit tangent vectors at p, and vice versa.

Proof. The result is directly obtained by taking $p = q = 1$ in Theorem 1. □

Corollary 4. *Suppose we have a submanifold $\mathcal{M}$ of dimension n that is invariant in a locally golden product space form $\overline{\mathcal{M}} = \mathcal{M}_1(c_1) \times \mathcal{M}_2(c_2)$.*

For any unit vector X in the tangent space $T_p\mathcal{M}$ at a point p on $\mathcal{M}$, the following inequality holds:

$$Ric(X) \leq \frac{n^2}{4}||H||^2 \pm \frac{1}{2\sqrt{5}}(c_1 - c_2)\big[2tr\phi - (n-1)\big]$$
$$+ \frac{1}{10}(c_1 + c_2)\big[3n - 4 - tr\phi\big] - (n-2)\beta(X,X) - \lambda. \tag{26}$$

Moreover, if $H(p) = 0$, then the equality case of this inequality is achieved by a unit tangent vector X at p if and only if X belongs to the normal space $\mathcal{N}_p$. Finally, when p is a totally geodesic point or is totally umbilical with $n = 2$, the equality case of this inequality holds true for all unit tangent vectors at p, and vice versa.

Proof. The result is directly obtained by taking $\theta = 0$ and $p = q = 1$ in Theorem 1. □

Corollary 5. *Suppose we have a submanifold $\mathcal{M}$ of dimension n that is anti-invariant in a locally golden product space form $\overline{\mathcal{M}} = \mathcal{M}_1(c_1) \times \mathcal{M}_2(c_2)$ with SSMC.*

For any unit vector X in the tangent space $T_p\mathcal{M}$ at a point p on $\mathcal{M}$, the following inequality holds:

$$Ric(X) \leq \frac{n^2}{4}||H||^2 \pm \frac{1}{2\sqrt{5}}(c_1 - c_2)\left[2tr\phi - (n-1)\right] + \frac{1}{10}(c_1 + c_2)\left[3n - 3 - tr\phi\right] - (n-2)\beta(X,X) - \lambda. \tag{27}$$

Moreover, if $H(p) = 0$, then the equality case of this inequality is achieved by a unit tangent vector X at p if and only if X belongs to the normal space $\mathcal{N}_p$. Finally, when p is a totally geodesic point or is totally umbilical with $n = 2$, the equality case of this inequality holds true for all unit tangent vectors at p, and vice versa.

Proof. The result is directly obtained by taking $\theta = \frac{\pi}{2}$ and $p = q = 1$ in Theorem 1. □

Remark 3. *Similar results can also be obtained for other particular classes such as copper, silver, nickel, bronze, etc., by providing different particular values to p and q with the help of Remark 2.*

5. Conclusions

This article has not only explored the Ricci tensor of slant submanifolds within locally metallic product space forms equipped with a semi-symmetric metric connection but has also contributed to our understanding of these mathematical constructs. The derivation of the Chen–Ricci inequality, the analysis of its equality case, and the applications arising from our findings collectively demonstrate the significance and relevance of this research.

Author Contributions: Conceptualization, Y.L., M.A., M.A.K., I.A.-D. and K.M.; methodology, Y.L., M.A., M.A.K., I.A.-D. and K.M.; investigation, Y.L., M.A., M.A.K., I.A.-D. and K.M.; writing—original draft preparation, Y.L., M.A., M.A.K., I.A.-D. and K.M.; writing—review and editing, Y.L., M.A., M.A.K., I.A.-D. and K.M. All authors have read and agreed to the published version of the manuscript.

Funding: This work was supported and funded by the Deanship of Scientific Research at Imam Mohammad Ibn Saud Islamic University (IMSIU) (grant number IMSIU-RG23036).

Data Availability Statement: Data are contained within the article.

Acknowledgments: This work was supported and funded by the Deanship of Scientific Research at Imam Mohammad Ibn Saud Islamic University (IMSIU) (grant number IMSIU-RG23036). The authors are thankful to the reviewers for their worthy suggestions.

Conflicts of Interest: There are no conflicts of interest.

References

1. Chen, B.Y. On Ricci curvature of isotropic and Lagrangian submanifolds in complex space forms. *Arch. Math.* **2000**, *74*, 154–160. [CrossRef]
2. Hineva, S. Submanifolds for which a lower bound of the Ricci curvature is achieved. *J. Geom.* **2008**, *88*, 53–69. [CrossRef]
3. Mihai, I. Ricci curvature of submanifolds in Sasakian space forms. *J. Aust. Math. Soc.* **2002**, *72*, 247–256. [CrossRef]
4. Mihai, I.; Al-Solamy, F.; Shahid, M.H. On Ricci curvature of a quaternion CR-submanifold in a quaternion space form. *Rad. Mat.* **2003**, *12*, 91–98.
5. Guo, H.; He, T. Harnack estimates for geometric flows, applications to Ricci flow coupled with harmonic map flow. *Geom. Dedicata* **2014**, *169*, 411–418. [CrossRef]
6. Abolarinwa, A. Gradient estimates for heat-type equations on evolving manifolds. *J. Nonlinear Evol. Equ. Appl.* **2015**, *1*, 1–19.
7. Ma, L. Gradient estimates for a simple elliptic equation on non-compact Riemannian manifolds. *J. Funct. Anal.* **2006**, *241*, 374–382.
8. Abolarinwa, A. Harnack estimates for heat equations with potentials on evolving manifolds. *Mediterr. J. Math.* **2016**, *13*, 3185–3204. [CrossRef]
9. Guo, H.; Ishida, M. Harnack estimates for nonlinear backward heat equations in geometric flows. *J. Func. Anal.* **2014**, *267*, 2638–2662. [CrossRef]
10. Abolarinwa, A. Differential Harnack inequalities for nonlinear parabolic equation on time–dependent metrics. *Adv. Theor. Appl. Math.* **2014**, *9*, 155–166.

11. Cao, X.; Hamilton, R. Differential Harnack estimates for time-dependent heat equations with potentials. *Geom. Funct. Anal.* **2009**, *19*, 989–1000. [CrossRef]
12. Fang, S. Differential Harnack inequalities for heat equations with potentials under the geometric flow. *Arch. Math.* **2013**, *100*, 179–189. [CrossRef]
13. Kumar, R.; Colney, L.; Khan, M.N.I. Proposed theorems on the lifts of Kenmotsu manifolds admitting a non-symmetric non-metric connection (NSNMC) in the tangent bundle. *Symmetry* **2023**, *15*, 2037. [CrossRef]
14. Kumar, R.; Colney, L.; Shenawy, S.; Turki, N.B. Tangent bundles endowed with quarter-symmetric non-metric connection (QSNMC) in a Lorentzian Para-Sasakian manifold. *Mathematics* **2023**, *11*, 4163. [CrossRef]
15. Khan, M.N.I. Liftings from a para-sasakian manifold to its tangent bundles. *Filomat* **2023**, *37*, 6727–6740. [CrossRef]
16. Khan, M.N.I.; Mofarreh, F.; Haseeb, A.; Saxena, M. Certain results on the lifts from an LP-Sasakian manifold to its tangent bundles associated with a quarter-symmetric metric connection. *Symmetry* **2023**, *15*, 1553. [CrossRef]
17. Khan, M.N.I.; De, U.C.; Velimirovic, L.S. Lifts of a quarter-symmetric metric connection from a Sasakian manifold to its tangent bundle. *Mathematics* **2023**, *11*, 53. [CrossRef]
18. Khan, M.N.I.; Mofarreh, F.; Haseeb, A. Tangent bundles of P-Sasakian manifolds endowed with a quarter-symmetric metric connection. *Symmetry* **2023**, *15*, 753. [CrossRef]
19. Li, Y.; Aquib, M.; Khan, M.A.; Al-Dayel, I.; Youssef, M.Z. Chen-Ricci Inequality for Isotropic Submanifolds in Locally Metallic Product Space Forms. *Axioms* **2024**, *13*, 183. [CrossRef]
20. Li, Y.; Siddiqi, M.; Khan, M.; Al-Dayel, I.; Youssef, M. Solitonic effect on relativistic string cloud spacetime attached with strange quark matter. *AIMS Math.* **2024**, *9*, 14487–14503. [CrossRef]
21. Li, Y.; Mofarreh, F.; Abolarinwa, A.; Alshehri, N.; Ali, A. Bounds for Eigenvalues of q-Laplacian on Contact Submanifolds of Sasakian Space Forms. *Mathematics* **2023**, *11*, 4717. [CrossRef]
22. Li, J.; Yang, Z.; Li, Y.; Abdel-Baky, R.A.; Saad, M.K. On the Curvatures of Timelike Circular Surfaces in Lorentz-Minkowski Space. *Filomat* **2024**, *38*, 1–15.
23. Li, Y.; Güler, E. Twisted Hypersurfaces in Euclidean 5-Space. *Mathematics* **2023**, *11*, 4612. [CrossRef]
24. Li, Y.; Jiang, X.; Wang, Z. Singularity properties of Lorentzian Darboux surfaces in Lorentz–Minkowski spacetime. *Res. Math. Sci.* **2024**, *11*, 7. [CrossRef]
25. Li, Y.; Güler, E. Toda, M. Family of right conoid hypersurfaces with light-like axis in Minkowski four-space. *AIMS Math.* **2024**, *9*, 18732–18745. [CrossRef]
26. Cabrezo, J.L.; Carriazo, A.; Fernandez, L.M. Slant submanifolds in Sasakian manifolds. *Glass Math. J.* **2000**, *42*, 125–138. [CrossRef]
27. Carriazo, A. *New Development in Slant Submanifold Theory, Applicable Mathematics in the GoldenAge*; Mishra, J.C., Ed.; Narosa Publishing House: Delhi, India, 2020; pp. 339–356.
28. Vilcu, G.E. B.-Y. Chen inequalities for slant submanifolds in quaternionic space forms. *Turkish J. Math.* **2010**, *34*, 115–128. [CrossRef]
29. Vilcu, G.E. On Chen invariants and inequalities in quaternionic geometry. *J. Inequal. Appl.* **2013**, *2013*, 66. [CrossRef]
30. Vilcu, G.E. Slant submanifolds of quarternionic space forms. *Publ. Math.* **2012**, *81*, 397–413.
31. Matsumoto, K.; Mihai, I.; Tazawa, Y. Ricci tensor of slant submanifolds in complex space forms. *Kodai Math. J.* **2003**, *26*, 85–94. [CrossRef]
32. Kim, J.-S.; Dwivedi, M.K.; Tripathi, M.M. Ricci curvature of integral submanifolds of an S-space form. *Bull. Korean Math. Soc.* **2007**, *44*, 395–406. [CrossRef]
33. Mihai, A.; Ozgur, C. Chen inequalities for submanifolds of real space forms with a semi-symmetric metric connection. *Taiwan J. Math.* **2010**, *14*, 1465–1477. [CrossRef]
34. Mihai, A.; Radulescu, I.N. An improved Chen-Ricci inequality for Kaehlerian slant submanifolds in complex space forms. *Taiwanese J. Math.* **2012**, *16*, 761–770. [CrossRef]
35. Deng, S. An Improved Chen-Ricci Inequality. *Int. Elect. J. Geom.* **2009**, *2*, 39–45.
36. Mihai, A. Inequalities on the Ricci curvature. *J. Math. Inequal.* **2015**, *9*, 811–822. [CrossRef]
37. Khan, M.A.; Ozel, C. Ricci curvature of contact CR-warped product submanifolds in generalized sasakian space forms admitting a trans-Sasakian structure. *Filomat* **2021**, *35*, 125–146. [CrossRef]
38. Lee, J.W.; Lee, C.W.; Sahin, B.; Vilcu, G.-E. Chen-Ricci inequalities for Riemannian maps and their applications. *Contemp. Math.* **2022**, *777*, 137–152.
39. Aydin, M.E.; Mihai, A.; Ozgur, C. Relations between extrinsic and intrinsic invariants statistical submanifolds in Sasaki-like statistical manifolds. *Mathematics* **2021**, *9*, 1285. [CrossRef]
40. Aydin, M.E.; Mihai, A.; Mihai, I. Some inequalities on submanifolds in statistical manifolds of constant curvature. *Filomat* **2015**, *29*, 465–476. [CrossRef]
41. Chen, B.-Y. Some pinching and classification theorems for minimal submanifolds. *Arch. Math.* **1993**, *60*, 568–578. [CrossRef]
42. Wang, Y. Chen inequalities for submanifolds of complex space forms and Sasakian space forms with quarter symmetric connections. *Int. J. Geom. Methods Mod. Phys.* **2019**, *16*, 1950118. [CrossRef]
43. Yano, K.; Kon, M. *Structures on Manifolds: Series in Pure Mathematics*; World Scientific: Singapore, 1985.

44. Hretcanu, C.E.; Crasmareanu, M. Metallic structures on Riemannian manifolds. *Rev. Un. Mat. Argentina* **2013**, *54*, 15–27.
45. Blaga, A.M.; Hretcanu, C. Invariant, anti-invariant and slant-submanifols of metallic riemannian manifold. *Novi Sad J. Math.* **2018**, *48*, 57–82. [CrossRef]

axioms

MDPI

Article

Geometric Inequalities of Slant Submanifolds in Locally Metallic Product Space Forms

Yanlin Li [1], Md Aquib [2], Meraj Ali Khan [2,*], Ibrahim Al-Dayel [2] and Maged Zakaria Youssef [2]

[1] School of Mathematics, Hangzhou Normal University, Hangzhou 311121, China; liyl@hznu.edu.cn
[2] Department of Mathematics and Statistics, College of Science, Imam Mohammad Ibn Saud Islamic University (IMSIU), P.O. Box 65892, Riyadh 11566, Saudi Arabia; maquib@imamu.edu.sa (M.A.); iaaldayel@imamu.edu.sa (I.A.-D.); mzabouelyamin@imamu.edu.sa (M.Z.Y.)
* Correspondence: mskhan@imamu.edu.sa

Abstract: In this particular article, our focus revolves around the establishment of a geometric inequality, commonly referred to as Chen's inequality. We specifically apply this inequality to assess the square norm of the mean curvature vector and the warping function of warped product slant submanifolds. Our investigation takes place within the context of locally metallic product space forms with quarter-symmetric metric connections. Additionally, we delve into the condition that determines when equality is achieved within the inequality. Furthermore, we explore a number of implications of our findings.

Keywords: geometric inequalities; metallic product space; θ-slant submanifolds; Chen's inequality

MSC: 53B50; 53C20; 53C40

check for updates

Citation: Li, Y.; Aquib, M.; Khan, M.A.; Al-Dayel, I.; Youssef, M.Z. Geometric Inequalities of Slant Submanifolds in Locally Metallic Product Space Forms. *Axioms* **2024**, *13*, 486. https://doi.org/10.3390/axioms13070486

Academic Editors: Mića Stanković and Giovanni Calvaruso

Received: 28 April 2024
Revised: 3 July 2024
Accepted: 15 July 2024
Published: 19 July 2024

1. Introduction

The theory of product manifolds encompasses significant implications in both physics and geometry, particularly in the realm of Hermitian geometry. In physics, Einstein's general theory of relativity describes space time as a product of three-dimensional space and one-dimensional time, each possessing its own metrics that determine the overall topology. Various theories such as Kaluza–Klein, brane theory, and gauge theory have intriguing applications involving product manifolds.

Modern physics relies on gauge theories, which are based on the geometric framework given by moduli spaces. These moduli spaces enable the categorization and exploration of characteristics of bundle configurations on compact Riemann surfaces or algebraic curves [1,2]. Notably, at the forefront of the construction of novel gauge theories are subvarieties of the moduli space of primary bundles with exceptional structure groups. Gauge theories are further illuminated by investigating the stratifications and fixed points in the moduli space of principal and Higgs bundles [3–6].

Moreover, the connection between Riemannian surfaces and gauge theories extends beyond the study of bundles. The moduli space of vector bundles over a compact Riemann surface or algebraic curve provides valuable insights into the formulation of gauge theories, offering a geometric understanding of the topological and geometric properties inherent in these theories [7,8].

A significant development in the study of manifolds with negative sectional curvature, referred to as warped product manifolds, was introduced by R. L. Bishop et al. in 1969 [9]. These generalized Riemannian product manifolds have found prominence in differential geometry and physics, particularly in general relativity [10,11]. Warped products have been widely used to examine energy, angles, and lengths through the lens of the second fundamental form. From a mathematical perspective, warped product manifolds extend the concept of Riemann product manifolds and provide examples of manifolds with strictly negative curvature. Notably, the best relativistic representation of Schwarzschild space time, which describes the region surrounding a massive star or black hole, can be expressed as a warped product [11]. Moreover, these manifolds have practical applications in modeling bodies with significant gravitational fields from a mechanical standpoint.

From a mathematical standpoint, warped product manifolds, a generalization of the Riemann product manifold [12–14], also give instances of manifolds with strictly negative curvature. A warped product, for example, is supplied as the best relativistic representation of the Schwarzschid space time, which describes the outer space around a massive star or black hole. From a mechanical aspect, they may also be employed to simulate bodies with massive gravitational fields.

The construction of warped product manifolds is defined as follows:

Let us consider a Riemannian manifold N_T of dimension d_1 with Riemannian metric g_1, N_θ of dimension d_2 with Riemannian metric g_2, and let f be positive differentiable functions on N_T. Consider the warped product $N_T \times N_\theta$ with its projections $\iota_1 : N_T \times N_\theta \to N_T$ and $\iota_2 : N_T \times N_\theta \to N_\theta$. Then, their warped product manifold $\mathcal{M} = N_T \times_f N_\theta$ is the product manifold equipped with the structure

$$g(X,Y) = g_1(\iota_{1*}X, \iota_{1*}Y) + (f \circ \iota_1)^2 g_2(\iota_{2*}X, \iota_{2*}Y),$$

for any vector fields X, Y on $\mathcal{M}$, where $*$ denotes the symbol for tangent maps. The function f is called the warping function of the warped product [15–17]. This concept has been extensively explored, leading to numerous research articles in the field of complex geometry [18,19] and contact geometry [20–22].

However, despite the extensive exploration of warped product manifolds, the immersibility/non-immersibility of Riemannian manifolds in space forms remains a fundamental problem in submanifold theory. In this regard, the groundbreaking work of Chen and his introduction of new Riemannian invariants, notably Chen's inequality, established an optimal relationship between extrinsic and intrinsic invariants on submanifolds.

Motivated by these considerations, the objective of this article is twofold: first, to derive Chen's inequality for warped product submanifolds in locally metallic product space forms with a quarter-symmetric metric connection, and secondly, to explore a few applications of the obtained result.

2. Preliminaries

Let $\overline{\mathcal{M}}$ be a Riemannian manifold endowed with the linear connection $\overline{\nabla}$. A connection is deemed semi-symmetric if its torsion tensor T satisfies the elegant expression

$$T(U,V) = \pi(V)U - \pi(U)V$$

where π is a one form. Consequently, $\overline{\nabla}$ is referred to as a semi-symmetric connection. Assuming a Riemannian metric g on $\overline{\mathcal{M}}$, if $\overline{\nabla} g = 0$, then $\overline{\nabla}$ qualifies as a semi-symmetric metric connection on $\overline{\mathcal{M}}$. The mathematical form of this connection is given by

$$\overline{\nabla}_U V = \tilde{\overline{\nabla}}_U V + \pi(V)U - g(U,V)\Gamma \tag{1}$$

where U and V are arbitrary vectors in $\overline{\mathcal{M}}$, $\tilde{\overline{\nabla}}$ represents the Levi-Civita connection with respect to the Riemannian metric g, and Γ is a vector field.

Furthermore, if $\overline{\nabla}$ satisfies the condition

$$\overline{\nabla}_U V = \tilde{\overline{\nabla}}_U V + \pi(V)U, \tag{2}$$

then it is termed a semi-symmetric non-metric connection.

Additionally, a linear connection $\overline{\nabla}$ on a Riemannian manifold $\overline{\mathcal{M}}$ with metric g is classified as a quarter-symmetric connection if its torsion tensor T is given by

$$T(U,V) = \overline{\nabla}_U V - \overline{\nabla}_V U - [U,V]$$

which satisfies the condition

$$T(U,V) = \pi(V)\phi U - \pi(U)\phi V$$

where $\pi(U) = g(U,\Gamma)$ and ϕ is a (1,1) tensor field.

Consequently, a special quarter-symmetric connection can be defined as follows

$$\overline{\nabla}_U V = \tilde{\overline{\nabla}}_U V + \psi_1 \pi(V) U - \psi_2 g(U,V)\Gamma \tag{3}$$

where ψ_1 and ψ_2 are real constants.

Remarkably, from Equations (1)–(3), it is evident that [23]

1. If $\psi_1 = \psi_2 = 1$, a quarter-symmetric connection reduces to a semi-symmetric metric connection.
2. If $\psi_1 = 1$ and $\psi_2 = 0$, a quarter-symmetric connection becomes a semi-symmetric non-metric connection.

It is worth mentioning that the quarter-symmetric connections generalize several well-known connections.

Moving on, the curvature tensor $\overline{R}$ associated with $\overline{\nabla}$ is expressed as

$$\overline{R}(U,V)Z = \overline{\nabla}_U \overline{\nabla}_V Z - \overline{\nabla}_V \overline{\nabla}_U Z - \overline{\nabla}_{[U,V]} Z. \tag{4}$$

Similarly, the curvature tensor $\tilde{\bar{R}}$ can also be defined.

Let us introduce the $(0,2)$ tensors

$$\beta_1(U,V) = (\tilde{\overline{\nabla}}_U \pi)(V) - \psi_1 \pi(U)\pi(V) + \frac{\psi_2}{2} g(U,V)\pi(\Gamma),$$

and

$$\beta_2(U,V) = \frac{\pi(\Gamma)}{2} g(U,V) + \pi(U)\pi(V).$$

The curvature tensor $\bar{R}$ of the manifold $\overline{\mathcal{M}}$ is then given by [24]

$$\begin{aligned} \bar{R}(U,V,Z,W) &= \tilde{\bar{R}}(U,V,Z,W) + \psi_1 \beta_1(U,Z) g(V,W) \\ &- \psi_1 \beta_1(V,Z) g(U,W) + \psi_2 \beta_1(V,W) g(U,Z) \\ &- \psi_2 \beta_1(U,W) g(V,Z) + \psi_2(\psi_1 - \psi_2) g(U,Z) \beta_2(V,W) \\ &- \psi_2(\psi_1 - \psi_2) g(V,Z) \beta_2(U,W). \end{aligned} \tag{5}$$

Moreover, let us define λ as the trace of β_1 and μ as the trace of β_2.

Let $\mathcal{M}$ be an m-dimensional submanifold in a Riemannian manifold $\overline{\mathcal{M}}$. Let ∇ and $\tilde{\nabla}$ be the induced quarter-symmetric metric connection and Levi-Civita connection, respectively, on $\mathcal{M}$. Then, the Gauss formulas are

$$\overline{\nabla}_U V = \nabla_U V + \zeta(U,V), \quad U,V \in \Gamma(T\mathcal{M}), \tag{6}$$

$$\tilde{\overline{\nabla}}_U V = \tilde{\nabla}_U V + \tilde{\zeta}(U,V), \quad U,V \in \Gamma(T\mathcal{M}), \tag{7}$$

where $\tilde{\zeta}$ is the second fundamental form that satisfies the relation

$$\zeta(U,V) = \tilde{\zeta}(U,V) - \psi_2 g(U,V)\Gamma^{\perp},$$

where $\Gamma^{\perp}$ is the normal component of the vector field Γ on $\mathcal{M}$.

Moreover, the equation of Gauss is defined by [24]

$$\begin{aligned} \bar{R}(U,V,Z,W) &= R(U,V,Z,W) \\ &- g(\zeta(U,W),\zeta(V,Z)) + g(\zeta(V,W),\zeta(U,Z)) \end{aligned}$$

$$+ \quad (\psi_1 - \psi_2)g(\zeta(V,Z),\Gamma^{\perp})g(U,W)$$
$$+ \quad (\psi_2 - \psi_1)g(\zeta(U,Z),\Gamma^{\perp})g(V,W). \tag{8}$$

Let $K(\pi)$ denote the sectional curvature of a Riemannian manifold $\mathcal{M}$ of the plane section $\pi \subset T_x\mathcal{M}$ at a point $x \in \mathcal{M}$. If $\{e_1, \ldots, e_n\}$ is the orthonormal basis of $\mathcal{T}_x\mathcal{M}$ and $\{e_{n+1}, \ldots, e_m\}$ is the orthonormal basis of $\mathcal{T}_x^{\perp}\mathcal{M}$ at any $x \in \mathcal{M}$, then

$$\tau(x) = \sum_{1 \le i < j \le n} K(e_i \wedge e_j), \tag{9}$$

where τ is the scalar curvature.

Let $\{e_1, ..., e_n\}$ and $\{e_{n+1}, ..., e_m\}$ be the tangent and normal orthonormal frames on $\mathcal{M}$, respectively. Then,

$$\mathcal{H} = \frac{1}{n}\sum_{i=1}^{n} g(\zeta(e_i, e_i).$$

is known as the mean curvature vector field.

A tensor field ϑ of type (1, 1) earns the title of a polynomial structure when it satisfies the following remarkable equation on an m-dimensional Riemannian manifold $(\overline{\mathcal{M}}, g)$, adorned with real numbers $b_1, .., b_n$:

$$\mathcal{B}(X) = X^n + b_{n-1}X^{n-1} + ... + b_2X + b_1\mathcal{I} = 0$$

Here, $\mathcal{I}$ represents the identity transformation [25,26].

Remark 1. *Behold the following revelations:*

1. *When* $\mathcal{B}(X) = X^2 + \mathcal{I}$, ϑ *unveils itself as an almost complex structure.*
2. *When* $\mathcal{B}(X) = X^2 - \mathcal{I}$, ϑ *emerges as an almost product structure.*
3. *When* $\mathcal{B}(X) = \vartheta^2 - p\vartheta + q\mathcal{I}$, ϑ *takes on the form of a metallic structure.*

In this case, p and q are two integers.

If

$$g(\vartheta X, Y) = g(X, \vartheta Y), \quad \forall X, Y \in \Gamma(\mathcal{T}\overline{\mathcal{M}}), \tag{10}$$

then the Riemannian metric g is bestowed with the grand title of being ϑ-compatible.

Imagine a scenario where g is ϑ-compatible and ϑ assumes the form of a metallic structure on the Riemannian manifold $\overline{\mathcal{M}}$. In this wondrous situation, we refer to $(\overline{\mathcal{M}}, g)$ as a metallic Riemannian manifold.

Exploiting the power of Equation (10), we can unfold the following revelation:

$$g(\vartheta X, \vartheta Y) = g(\vartheta^2 X, Y) = p.g(X, \vartheta Y) + q.g(X, Y). \tag{11}$$

It is worth mentioning that when we set $p = q = 1$ in (11), a metallic structure magically transforms into a golden structure.

The esteemed members of the metallic family are elegantly categorized as follows [27]:

1. The golden structure $\vartheta = \frac{1+\sqrt{5}}{2}$ for $p = q = 1$, entwined with the ratio of two consecutive classical Fibonacci numbers.
2. The copper structure $\kappa_{1,2} = 2$ with $p = 1$ and $q = 2$.
3. The nickel structure $\kappa_{1,3} = \frac{1+\sqrt{13}}{2}$ if $p = 1$ and $q = 3$.
4. The silver structure $\kappa_{2,1} = 1 + \sqrt{2}$ if $p = 2$ and $q = 1$, enchanted by the ratio of two consecutive Pell numbers.
5. The bronze structure $\kappa_{3,1} = \frac{3+\sqrt{13}}{2}$ with $p = 3$ and $q = 1$.
6. The subtle structure $\kappa_{4,1} = 2 + \sqrt{5}$ if $p = 4$ and $q = 1$, and so forth.

Let $(\overline{\mathcal{M}}, g)$ be an m-dimensional Riemannian manifold and let ϑ be a (1,1)-tensor field on $\overline{\mathcal{M}}$ such that $\vartheta^2 = \mathcal{I}$, $\vartheta \neq \pm\mathcal{I}$; then, ϑ is called an almost product structure. The structure ϑ with

$$g(\vartheta X, Y) = g(X, \vartheta Y), \quad \forall X, Y \in \Gamma(\mathcal{T}\overline{\mathcal{M}})$$

is known as an almost product Riemannian manifold [26].

Any metallic structure ϕ on $\overline{\mathcal{M}}$ is known to induce two almost product structures ϕ on $\overline{\mathcal{M}}$ [27]:

$$\vartheta_1 = \frac{2}{2\sigma_{p,q} - p}\phi - \frac{p}{2\sigma_{p,q} - p}\mathcal{I}, \tag{12}$$

$$\vartheta_2 = \frac{2}{2\sigma_{p,q} - p}\phi + \frac{p}{2\sigma_{p,q} - p}\mathcal{I} \tag{13}$$

where $\sigma_{p,q} = \frac{p+\sqrt{p^2+4q}}{2}$.

Also, an almost product structure ϑ on $\overline{\mathcal{M}}$ induces two metallic structures:

$$\phi_1 = \frac{p}{2}\mathcal{I} + \frac{2\sigma_{p,q} - p}{2}\vartheta,$$

$$\phi_2 = \frac{p}{2}\mathcal{I} - \frac{2\sigma_{p,q} - p}{2}\vartheta.$$

Definition 1 ([28]). *(i)* *Let $\overline{\nabla}$ be a linear connection and ϕ be a metallic structure on $\overline{\mathcal{M}}$ such that $\nabla\phi = 0$. Then, $\overline{\nabla}$ is called a ϕ-connection.*

(ii) *A locally metallic Riemannian manifold is a metallic Riemannian manifold $(\overline{\mathcal{M}}, g, \phi)$ if the Levi-Civita connection $\overline{\nabla}$ of g is a ϕ-connection.*

Consider an almost Hermitian manifold $\overline{\mathcal{M}}$ and a submanifold $\mathcal{M}$ embedded within it. We refer to $\mathcal{M}$ as a slant submanifold if, for any point x on $\mathcal{M}$ and any non-zero vector X in the tangent space $\mathcal{T}_x\mathcal{M}$, the angle between the tangent space $J\mathcal{M}$ and $\mathcal{T}_x\mathcal{M}$ remains constant. In other words, this angle does not vary based on the specific choice of x and X on $\mathcal{M}$. The constant angle is known as the slant angle θ, which lies in the range $[0, \frac{\pi}{2}]$ and characterizes the slant submanifold within $\overline{\mathcal{M}}$.

Moreover, if $\mathcal{M}$ is a slant submanifold of a metallic Riemannian manifold $(\overline{\mathcal{M}}, g, \phi)$ with a slant angle θ, the following relationships hold [28]:

$$g(\mathcal{T}X, \mathcal{T}Y) = \cos^2\theta[pg(X, \mathcal{T}Y) + qg(X, Y)],$$

and

$$g(NX, NY) = \sin^2\theta[pg(X, \mathcal{T}Y) + qg(X, Y)],$$

for all $X, Y \in \Gamma(\mathcal{T}\mathcal{M})$.

Furthermore, we have the additional relations

$$\mathcal{T}^2 = \cos^2\theta(p\mathcal{T} + q\mathcal{I}),$$

where $\mathcal{I}$ represents the identity operator on $\Gamma(\mathcal{T}\mathcal{M})$ and

$$\nabla\mathcal{T}^2 = pcos^2\theta.\nabla\mathcal{T}.$$

These expressions provide valuable insights into the geometric properties of slant submanifolds and their relationships within the broader context of metallic Riemannian manifolds.

Also, let $\mathcal{M}_1$ be a Riemannian manifold with constant sectional curvature c_1 and $\mathcal{M}_2$ be a Riemannian manifold with constant sectional curvature c_2. Then, the Riemannian curvature tensor $\overline{\mathcal{R}}$ of the locally Riemannian product manifold $\overline{\mathcal{M}} = \mathcal{M}_1 \times \mathcal{M}_2$ is given by [29]

$$\begin{aligned}
\tilde{\bar{\mathcal{R}}}(X,Y)Z &= \frac{1}{4}(c_1+c_2)\left[g(Y,Z)X - g(X,Z)Y\right] \\
&+\frac{1}{4}(c_1+c_2)\left\{\frac{4}{(2\sigma_{p,q}-p)^2}\left[g(\phi Y,Z)\phi X - g(\phi X,Z)\phi Y\right]\right. \\
&+\frac{p^2}{(2\sigma_{p,q}-p)^2}\left[g(Y,Z)X - g(X,Z)Y\right] \\
&+\frac{2p}{(2\sigma_{p,q}-p)^2}\left[g(\phi X,Z)Y + g(X,Z)\phi Y\right. \\
&\left.\left.-g(\phi Y,Z)X - g(Y,Z)\phi X\right]\right\} \\
&\pm\frac{1}{2}(c_1-c_2)\left\{\frac{1}{(2\sigma_{p,q}-p)}\left[g(Y,Z)\phi X - g(X,Z)\phi Y\right]\right. \\
&+\frac{1}{(2\sigma_{p,q}-p)}\left[g(\phi Y,Z)X - g(\phi X,Z)Y\right] \\
&\left.+\frac{p}{(2\sigma_{p,q}-p)}\left[g(X,Z)Y - g(Y,Z)X\right]\right\}.
\end{aligned} \tag{14}$$

From (5) and (14), we have

$$\begin{aligned}
\bar{R}(X,Y,Z,W) = &\frac{1}{4}(c_1+c_2)\left[g(Y,Z)g(X,W) - g(X,Z)g(Y,W)\right] \\
&+\frac{1}{4}(c_1+c_2)\left\{\frac{4}{(2\sigma_{p,q}-p)^2}\left[g(\phi Y,Z)g(\phi X,W) - g(\phi X,Z)g(\phi Y,W)\right]\right. \\
&+\frac{p^2}{(2\sigma_{p,q}-p)^2}\left[g(Y,Z)g(X,W) - g(X,Z)g(Y,W)\right] \\
&+\frac{2p}{(2\sigma_{p,q}-p)^2}\left[g(\phi X,Z)g(Y,W) + g(X,Z)g(\phi Y,W)\right. \\
&\left.\left.-g(\phi Y,Z)g(X,W) - g(Y,Z)g(\phi X,W)\right]\right\} \\
&\pm\frac{1}{2}(c_1-c_2)\left\{\frac{1}{(2\sigma_{p,q}-p)}\left[g(Y,Z)g(\phi X,W) - g(X,Z)g(\phi Y,W)\right]\right. \\
&+\frac{1}{(2\sigma_{p,q}-p)}\left[g(\phi Y,Z)g(X,W) - g(\phi X,Z)g(Y,W)\right] \\
&\left.+\frac{p}{(2\sigma_{p,q}-p)}\left[g(X,Z)g(Y,W) - g(Y,Z)g(X,W)\right]\right\} \\
&+\psi_1\beta_1(X,Z)g(Y,W) - \psi_1\beta_1(Y,Z)g(X,W) + \psi_2\beta_1(Y,W)g(X,Z) \\
&-\psi_2\beta_1(X,W)g(Y,Z) + \psi_2(\psi_1-\psi_2)g(X,Z)\beta_2(Y,W) \\
&-\psi_2(\psi_1-\psi_2)g(Y,Z)\beta_2(X,W).
\end{aligned} \tag{15}$$

3. Unveiling the Pinching Phenomenon: Main Result

The proof of the major finding is the focus of this section.

Theorem 1. *Let $\mathcal{M}$ be an n-dimensional warped product θ-slant submanifold of an m-dimensional locally metallic product space form $(\overline{\mathcal{M}} = \mathcal{M}_1(c_1) \times \mathcal{M}_2(c_2), g, \phi)$ with quarter-symmetric metric connections. Then,*

$$\begin{aligned} n_2\frac{\Delta f}{f} \leq\ & \frac{n^2(n-2)}{2(n-1)}||\mathcal{H}||^2 \\ & + \frac{1}{8}(c_1+c_2)\frac{1}{p^2+4q}\Big\{2n_1n_2(p^2+4q) - 4\big[tr^2\phi + tr\phi \\ & - cos^2\theta(p.trT)\big]\Big\} \pm \frac{1}{8}\frac{(c_1-c_2)}{\sqrt{(p^2+4q)}}\{4pn_1n_2 - 4tr\phi\} \\ & - (\psi_1+\psi_2)\Big[\lambda(n-1) + \lambda|_{M_1}(n_1-1) + \lambda|_{M_2}(n_2-1)\Big] \\ & - \psi_2(\psi_1-\psi_2)\Big[\mu(n-1) + \mu|_{M_1}(n_1-1) + \mu|_{M_2}(n_2-1)\Big] \\ & + (\psi_1-\psi_2)\Big[n(n-1)\pi(H) + n_1(n_1-1)\pi(H_1) + n_2(n_2-1)\pi(H_2)\Big], \end{aligned} \tag{16}$$

where Δ is the Laplacian operator on $\mathcal{M}_1$. The equality case holds in (16) if and only if $\mathcal{M}$ is a mixed totally geodesic isometric immersion and the following satisfies

$$\frac{\mathcal{H}_1}{\mathcal{H}_2} = \frac{n_1}{n_2}, \tag{17}$$

where $\mathcal{H}_1$ and $\mathcal{H}_2$ are the mean curvature vectors along $\mathcal{M}_1^{n_1}$ and $\mathcal{M}_2^{n_2}$, respectively.

Proof. Let $\{e_1, ..., e_n\}$ be an orthonormal tangent frame and $\{e_{n+1}, ..., e_m\}$ be an orthonormal frame of $T_x\mathcal{M}$ and $T_x^\perp\mathcal{M}$, respectively, at any point $x \in \mathcal{M}$. Putting $X = W = e_i$, $Y = Z = e_j$ in (15) with Equation (8) and take $i \neq j$, we have

$$\begin{aligned} \mathcal{R}(e_i,e_j,e_j,e_i) =\ & \frac{1}{4}(c_1+c_2)[g(e_j,e_j)g(e_i,e_i) - g(e_i,e_j)g(e_j,e_i)] \\ + & \frac{1}{4}(c_1+c_2)\Big\{\frac{4}{(2\sigma_{p,q}-p)^2}[g(\phi e_j,e_j)g(\phi e_i,e_i) \\ - & g(\phi e_i,e_j)g(\phi e_j,e_i)] \\ + & \frac{p^2}{(2\sigma_{p,q}-p)^2}[g(e_j,e_j)g(e_i,e_i) - g(e_i,e_j)g(e_j,e_i)] \\ + & \frac{2p}{(2\sigma_{p,q}-p)^2}[g(\phi e_i,e_j)g(e_j,e_i) + g(e_i,e_j)g(\phi e_j,e_i) \\ - & g(\phi e_j,e_j)g(e_i,e_i) - g(e_j,e_j)g(\phi e_i,e_i)]\Big\} \\ \pm & \frac{1}{2}(c_1-c_2)\Big\{\frac{1}{(2\sigma_{p,q}-p)}[g(e_j,e_j)g(\phi e_i,e_i) \\ - & g(e_i,e_j)g(\phi e_j,e_i)] \\ + & \frac{1}{(2\sigma_{p,q}-p)}[g(\phi e_j,e_j)g(e_i,e_i) - g(\phi e_i,e_j)g(e_j,e_i)] \\ + & \frac{p}{(2\sigma_{p,q}-p)}[g(e_i,e_j)g(e_j,e_i) - g(e_j,e_j)g(e_i,e_i)]\Big\} \\ + & \psi_1\beta_1(e_i,e_j)g(e_j,e_i) - \psi_1\beta_1(e_j,e_j)g(e_i,e_i) \\ + & \psi_2\beta_1(e_j,e_i)g(e_i,e_j) - \psi_2\beta_1(e_i,e_i)g(e_j,e_j) \\ + & \psi_2(\psi_1-\psi_2)g(e_i,e_j)\beta_2(e_j,e_i) \\ - & \psi_2(\psi_1-\psi_2)g(e_j,e_j)\beta_2(e_i,e_i) \\ + & g(\zeta(e_i,e_i),\zeta(e_j,e_j)) - g(\zeta(e_j,e_i),\zeta(e_i,e_j)) \end{aligned}$$

$$
\begin{aligned}
&- \quad (\psi_1 - \psi_2) g(\zeta(e_j, e_j), \Gamma^{\perp}) g(e_i, e_i) \\
&- \quad (\psi_2 - \psi_1) g(\zeta(e_i, e_j), \Gamma^{\perp}) g(e_j, e_i).
\end{aligned} \tag{18}
$$

Applying $1 \leq i, j \leq n$ in (18), we obtain

$$
\begin{aligned}
2\tau(x) = &\frac{1}{4}(c_1 + c_2)\frac{n(n-1)}{p^2 + 4q}\Big\{2p^2 + 4q + \frac{4}{n(n-1)}\left[tr^2\phi - cos^2\theta(p.trT + nq)\right] \\
&- \frac{4p}{n}tr\phi\Big\} \\
&\pm \frac{1}{4}\frac{(n-1)}{\sqrt{p^2 + 4q}}(c_1 - c_2)(4tr\phi - 2np) + n^2||H||^2 - ||\zeta||^2 \\
&- (\psi_1 + \psi_2)\lambda(n-1) - \psi_2(\psi_1 - \psi_2)\mu(n-1) + n(n-1)(\psi_1 - \psi_2)\pi(H).
\end{aligned} \tag{19}
$$

We take

$$
\begin{aligned}
\delta \quad = \quad & 2\tau - \frac{1}{4}(c_1 + c_2)\frac{n(n-1)}{p^2 + 4q}\Big\{2p^2 + 4q \\
&+ \frac{4}{n(n-1)}\left[tr^2\phi - cos^2\theta(p.trT + nq)\right] - \frac{4p}{n}tr\phi\Big\} \\
&\mp \frac{1}{4}\frac{(n-1)}{\sqrt{p^2 + 4q}}(c_1 - c_2)(4tr\phi - 2np) - \frac{n^2(n-2)}{(n-1)}||\mathcal{H}||^2 \\
&+ (\psi_1 + \psi_2)\lambda(n-1) + \psi_2(\psi_1 - \psi_2)\mu(n-1) - n(n-1)(\psi_1 - \psi_2)\pi(H).
\end{aligned} \tag{20}
$$

Then, from (19) and (20), we have

$$
n^2||\mathcal{H}||^2 = (n-1)(\delta + ||\zeta||^2). \tag{21}
$$

As a result, when using the orthonormal frame $\{e_1, ..., e_n\}$, (21) assumes the following form:

$$
\left(\sum_{i=1}^{n} \zeta_{ii}^{n+1}\right)^2 = (n-1)\Big\{\delta + \sum_{i=1}^{n}(\zeta_{ii}^{n+1})^2 + \sum_{i \neq j}(\zeta_{ij}^{n+1})^2 + \sum_{r=n+1}^{m}\sum_{i,j=1}^{n}(\zeta_{ij}^{r})^2\Big\}. \tag{22}
$$

If we substitute $a_1 = \zeta_{11}^{n+1}$, $a_2 = \sum_{i=2}^{n_1} \zeta_{ii}^{n+1}$, and $a_3 = \sum_{t=n_1+1}^{n} \zeta_{tt}^{n+1}$, then (22) reduces to

$$
\begin{aligned}
\left(\sum_{i=1}^{n} a_i\right)^2 \quad = \quad & (n-1)\Big\{\delta + \sum_{i=1}^{n} a_i^2 + \sum_{i \neq j \leq n}(\zeta_{ij}^{n+1})^2 + \sum_{r=n+1}^{m}\sum_{i,j=1}^{n}(\zeta_{ij}^{r})^2 \\
&- \sum_{2 \leq j \neq k \leq n_1} \zeta_{jj}^{n+1}\zeta_{kk}^{n+1} - \sum_{n_1+1 \leq s \neq t \leq n} \zeta_{ss}^{n+1}\zeta_{tt}^{n+1}\Big\}.
\end{aligned} \tag{23}
$$

As a result, a_1, a_2, a_3 fulfill Chen's Lemma (for $n = 3$), i.e.,

$$
\left(\sum_{i=1}^{3} a_i\right)^2 = 2\left(b + \sum_{i=1}^{3} a_i^2\right).
$$

Clearly, $2a_1a_2 \geq b$ with equality holds if $a_1 + a_2 = a_3$, and conversely, this signifies

$$
\begin{aligned}
&\sum_{1 \leq j < k \leq n_1} \zeta_{jj}^{n+1}\zeta_{kk}^{n+1} + \sum_{n_1+1 \leq s < t \leq n} \zeta_{ss}^{n+1}\zeta_{tt}^{n+1} \\
&\qquad\qquad \geq \frac{\delta}{2} + \sum_{1 \leq \alpha_3 < \beta_3 \leq n}(\zeta_{\alpha_3\beta_3}^{n+1})^2 + \sum_{r=n+1}^{p+q}\sum_{\alpha_3\beta_3=1}^{n}(\zeta_{\alpha_3\beta_3}^{r})^2
\end{aligned}
$$

(24)

and equality holds if and only if

$$\sum_{i=1}^{n_1} \zeta_{ii}^{n+1} = \sum_{t=n_1+1}^{n} \zeta_{tt}^{n+1}. \tag{25}$$

Again taking into consideration Equation (3.3) in [15], we arrive at the following conclusion:

$$n_2 \frac{\Delta f}{f} = \tau - \sum_{1 \le j < k \le n_1} \kappa(e_j \wedge e_k) - \sum_{n_1+1 \le s < t \le n} \kappa(e_s \wedge e_t). \tag{26}$$

Then, from (24) and (26), we compute

$$\begin{aligned} n_2 \frac{\Delta f}{f} \le{}& \tau - \frac{1}{8}(c_1 + c_2)\frac{1}{p^2+4q}\Big\{(n(n-1) - 2n_1n_2)(p^2+4q) + 8\big[tr^2\phi \\ & - 4cos^2\theta(2p.trT + nq) - 4p(n-2)tr\phi\big]\Big\} \\ & \mp \frac{1}{8}\frac{(c_1 - c_2)}{(\sqrt{p^2+4q})}\{4tr\phi(n-2) - 2pn(n-1) - 4pn_1n_2)\} - \frac{\delta}{2} \\ & - (\psi_1 + \psi_2)\Big[\lambda|_{M_1}(n_1 - 1) + \lambda|_{M_2}(n_2 - 1)\Big] \\ & - \psi_2(\psi_1 - \psi_2)\Big[\mu|_{M_1}(n_1 - 1) + \mu|_{M_2}(n_2 - 1)\Big] \\ & + (\psi_1 - \psi_2)\Big[n_1(n_1 - 1)\pi(H_1) + n_2(n_2 - 1)\pi(H_2)\Big]. \end{aligned} \tag{27}$$

Using (20) in the above equation, we obtain

$$\begin{aligned} n_2 \frac{\Delta f}{f} \le{}& \frac{n^2(n-2)}{2(n-1)}||\mathcal{H}||^2 \\ & + \frac{1}{8}(c_1 + c_2)\frac{1}{p^2+4q}\Big\{2n_1n_2(p^2+4q) - 4\big[tr^2\phi + tr\phi \\ & - cos^2\theta(p.trT)\big]\Big\} \pm \frac{1}{8}\frac{(c_1 - c_2)}{\sqrt{(p^2+4q)}}\{4pn_1n_2 - 4tr\phi\} \\ & - (\psi_1 + \psi_2)\Big[\lambda(n-1) + \lambda|_{M_1}(n_1 - 1) + \lambda|_{M_2}(n_2 - 1)\Big] \\ & - \psi_2(\psi_1 - \psi_2)\Big[\mu(n-1) + \mu|_{M_1}(n_1 - 1) + \mu|_{M_2}(n_2 - 1)\Big] \\ & + (\psi_1 - \psi_2)\Big[n(n-1)\pi(H) + n_1(n_1 - 1)\pi(H_1) + n_2(n_2 - 1)\pi(H_2)\Big], \end{aligned} \tag{28}$$

which implies the required inequality.

We deduce from (24) and (25) that the equality in (16) holds if and only if

$$\sum_{r=n+1}^{m} \sum_{i=1}^{n_1} \zeta_{ii}^{r} = \sum_{r=n+1}^{2m} \sum_{t=n_1+1}^{n} \zeta_{tt}^{r} = 0. \tag{29}$$

Moreover, from (25), we obtain

$$\zeta_{jt} = 0, \ \forall\, 1 \le j \le n_1, n+1 \le t \le n, n+1 \le r \le m. \tag{30}$$

This shows that (30) is equivalent to the mixed total geodesicness of the doubly warped product $\mathcal{M} = \mathcal{M}_1(c_1) \times \mathcal{M}_2(c_2)$ and (25) and (29) imply $n_1\mathcal{H}_1 = n_2\mathcal{H}_2$. □

4. Some Applications of the Result

The significance and applicability of the findings can be observed from three distinct perspectives. Firstly, they can be regarded as specific instances within the realm of quarter-symmetric connections, shedding light on the broader understanding of this field. Secondly, the results can be viewed as particular cases within the framework of slant submanifolds, contributing to the knowledge and characterization of these geometric structures. Lastly, they hold relevance as specific instances within the domain of metallic space forms, providing valuable insights into the properties and behavior of such spaces. The multifaceted nature of these applications underscores the depth and breadth of the implications derived from this research, making it a compelling contribution to the respective fields and offering new avenues for exploration and discovery.

4.1. Results on Specific Instances within the Realm of Quarter-Symmetric Connection

It is known that a quarter-symmetric connection becomes a semi-symmetric metric connection with $\psi_1 = 1$ and $\psi_2 = 1$. Taking this into consideration together with Theorem 1, we have the following result:

Corollary 1. *Let $\mathcal{M}$ be an n-dimensional warped product θ-slant submanifold of an m-dimensional locally metallic product space form $(\overline{\mathcal{M}} = \mathcal{M}_1(c_1) \times \mathcal{M}_2(c_2), g, \phi)$ with semi-symmetric metric connections. Then,*

$$\begin{aligned} n_2\frac{\Delta f}{f} \leq\ & \frac{n^2(n-2)}{2(n-1)}||\mathcal{H}||^2 \\ & + \frac{1}{8}(c_1+c_2)\frac{1}{p^2+4q}\Big\{2n_1n_2(p^2+4q) - 4\big[tr^2\phi + tr\phi \\ & - cos^2\theta(p.trT)\big]\Big\} \pm \frac{1}{8}\frac{(c_1-c_2)}{\sqrt{(p^2+4q)}}\{4pn_1n_2 - 4tr\phi\} \\ & - 2\Big[\lambda(n-1) + \lambda|_{M_1}(n_1-1) + \lambda|_{M_2}(n_2-1)\Big]. \end{aligned} \tag{31}$$

The equality in (31) *is attained if and only if $\mathcal{M}$ is a mixed totally geodesic isometric immersion and meets the condition* (17).

We also know that a quarter-symmetric connection becomes a semi-symmetric non-metric connection with $\psi_1 = 1$ and $\psi_2 = 0$. Taking this into consideration together with Theorem 1, we have the following result.

Corollary 2. *Let $\mathcal{M}$ be an n-dimensional warped product θ-slant submanifold of an m-dimensional locally metallic product space form $(\overline{\mathcal{M}} = \mathcal{M}_1(c_1) \times \mathcal{M}_2(c_2), g, \phi)$ with a semi-symmetric non-metric connection. Then,*

$$\begin{aligned} n_2\frac{\Delta f}{f} \leq\ & \frac{n^2(n-2)}{2(n-1)}||\mathcal{H}||^2 \\ & + \frac{1}{8}(c_1+c_2)\frac{1}{p^2+4q}\Big\{2n_1n_2(p^2+4q) - 4\big[tr^2\phi + tr\phi \\ & - cos^2\theta(p.trT)\big]\Big\} \pm \frac{1}{8}\frac{(c_1-c_2)}{\sqrt{(p^2+4q)}}\{4pn_1n_2 - 4tr\phi\} \\ & - \Big[\lambda(n-1) + \lambda|_{M_1}(n_1-1) + \lambda|_{M_2}(n_2-1)\Big] \\ & + \Big[n(n-1)\pi(H) + n_1(n_1-1)\pi(H_1) + n_2(n_2-1)\pi(H_2)\Big]. \end{aligned} \tag{32}$$

The equality in (32) *is attained if and only if $\mathcal{M}$ is a mixed totally geodesic isometric immersion and meets the condition* (17).

4.2. *Results on Specific Instances within the Realm of θ-Slant Submanifold*

We know that the particular classes of the θ-slant submanifold are either invariant or anti-invariant with $\theta = 0$ or $\theta = \frac{\pi}{2}$, respectively. Thus, we have the following result as a consequence of Theorem 1.

Corollary 3. *Let $\mathcal{M}$ be an n-dimensional warped product invariant submanifold of an m-dimensional locally metallic product space form $(\overline{\mathcal{M}} = \mathcal{M}_1(c_1) \times \mathcal{M}_2(c_2), g, \phi)$ with quarter-symmetric metric connections. Then,*

$$\begin{aligned} n_2\frac{\Delta f}{f} \leq & \frac{n^2(n-2)}{2(n-1)}||\mathcal{H}||^2 \\ & + \frac{1}{8}(c_1+c_2)\frac{1}{p^2+4q}\Big\{2n_1n_2(p^2+4q) - 4\big[tr^2\phi + tr\phi - p.trT\big]\Big\} \\ & \pm \frac{1}{8}\frac{(c_1-c_2)}{\sqrt{(p^2+4q)}}\{4pn_1n_2 - 4tr\phi\} \\ & - (\psi_1+\psi_2)\Big[\lambda(n-1) + \lambda|_{M_1}(n_1-1) + \lambda|_{M_2}(n_2-1)\Big] \\ & - \psi_2(\psi_1-\psi_2)\Big[\mu(n-1) + \mu|_{M_1}(n_1-1) + \mu|_{M_2}(n_2-1)\Big] \\ & + (\psi_1-\psi_2)\Big[n(n-1)\pi(H) + n_1(n_1-1)\pi(H_1) + n_2(n_2-1)\pi(H_2)\Big]. \end{aligned} \tag{33}$$

The equality in (33) *is attained if and only if $\mathcal{M}$ is a mixed totally geodesic isometric immersion and meets the condition* (17).

Corollary 4. *Let $\mathcal{M}$ be an n-dimensional warped product anti-invariant submanifold of an m-dimensional locally metallic product space form $(\overline{\mathcal{M}} = \mathcal{M}_1(c_1) \times \mathcal{M}_2(c_2), g, \phi)$ with quarter-symmetric metric connections. Then,*

$$\begin{aligned} n_2\frac{\Delta f}{f} \leq & \frac{n^2(n-2)}{2(n-1)}||\mathcal{H}||^2 \\ & + \frac{1}{8}(c_1+c_2)\frac{1}{p^2+4q}\Big\{2n_1n_2(p^2+4q) - 4\big[tr^2\phi + tr\phi\big]\Big\} \\ & \pm \frac{1}{8}\frac{(c_1-c_2)}{\sqrt{(p^2+4q)}}\{4pn_1n_2 - 4tr\phi\} \\ & - (\psi_1+\psi_2)\Big[\lambda(n-1) + \lambda|_{M_1}(n_1-1) + \lambda|_{M_2}(n_2-1)\Big] \\ & - \psi_2(\psi_1-\psi_2)\Big[\mu(n-1) + \mu|_{M_1}(n_1-1) + \mu|_{M_2}(n_2-1)\Big] \\ & + (\psi_1-\psi_2)\Big[n(n-1)\pi(H) + n_1(n_1-1)\pi(H_1) + n_2(n_2-1)\pi(H_2)\Big]. \end{aligned} \tag{34}$$

The equality in (34) *is attained if and only if $\mathcal{M}$ is a mixed totally geodesic isometric immersion and meets the condition* (17).

Further, from Corollary 1, we mind the following results.

Corollary 5. *Let $\mathcal{M}$ be an n-dimensional warped product invariant submanifold of an m-dimensional locally metallic product space form $(\overline{\mathcal{M}} = \mathcal{M}_1(c_1) \times \mathcal{M}_2(c_2), g, \phi)$ with semi-symmetric metric connections. Then,*

$$\begin{aligned} n_2\frac{\Delta f}{f} \leq & \frac{n^2(n-2)}{2(n-1)}||\mathcal{H}||^2 \\ & + \frac{1}{8}(c_1+c_2)\frac{1}{p^2+4q}\Big\{2n_1n_2(p^2+4q) - 4\big[tr^2\phi + tr\phi - p.trT\big]\Big\} \end{aligned}$$

$$\pm \frac{1}{8}\frac{(c_1-c_2)}{\sqrt{(p^2+4q)}}\{4pn_1n_2-4tr\phi\}$$
$$-2\Big[\lambda(n-1)+\lambda|_{M_1}(n_1-1)+\lambda|_{M_2}(n_2-1)\Big]. \quad (35)$$

The equality in (35) *is attained if and only if* $\mathcal{M}$ *is a mixed totally geodesic isometric immersion and meets the condition* (17).

Corollary 6. *Let* $\mathcal{M}$ *be an n-dimensional warped product anti-invariant submanifold of an m-dimensional locally metallic product space form* $(\overline{\mathcal{M}} = \mathcal{M}_1(c_1) \times \mathcal{M}_2(c_2), g, \phi)$ *with semi-symmetric metric connections. Then,*

$$n_2\frac{\Delta f}{f} \leq \frac{n^2(n-2)}{2(n-1)}||\mathcal{H}||^2$$
$$+\frac{1}{8}(c_1+c_2)\frac{1}{p^2+4q}\Big\{2n_1n_2(p^2+4q)-4\big[tr^2\phi+tr\phi\big]\Big\}$$
$$\pm\frac{1}{8}\frac{(c_1-c_2)}{\sqrt{(p^2+4q)}}\{4pn_1n_2-4tr\phi\}$$
$$-2\Big[\lambda(n-1)+\lambda|_{M_1}(n_1-1)+\lambda|_{M_2}(n_2-1)\Big]. \quad (36)$$

The equality in (36) *is attained if and only if* $\mathcal{M}$ *is a mixed totally geodesic isometric immersion and meets the condition* (17).

Moreover, from Corollary 2, we obtain the following results.

Corollary 7. *Let* $\mathcal{M}$ *be an n-dimensional warped product invariant submanifold of an m-dimensional locally metallic product space form* $(\overline{\mathcal{M}} = \mathcal{M}_1(c_1) \times \mathcal{M}_2(c_2), g, \phi)$ *with a semi-symmetric non-metric connection. Then,*

$$n_2\frac{\Delta f}{f} \leq \frac{n^2(n-2)}{2(n-1)}||\mathcal{H}||^2$$
$$+\frac{1}{8}(c_1+c_2)\frac{1}{p^2+4q}\Big\{2n_1n_2(p^2+4q)-4\big[tr^2\phi+tr\phi-p.trT\big]\Big\}$$
$$\pm\frac{1}{8}\frac{(c_1-c_2)}{\sqrt{(p^2+4q)}}\{4pn_1n_2-4tr\phi\}$$
$$-\Big[\lambda(n-1)+\lambda|_{M_1}(n_1-1)+\lambda|_{M_2}(n_2-1)\Big]$$
$$+\Big[n(n-1)\pi(H)+n_1(n_1-1)\pi(H_1)+n_2(n_2-1)\pi(H_2)\Big]. \quad (37)$$

The equality in (37) *is attained if and only if* $\mathcal{M}$ *is a mixed totally geodesic isometric immersion and meets the condition* (17).

Corollary 8. *Let* $\mathcal{M}$ *be an n-dimensional warped product anti-invariant submanifold of an m-dimensional locally metallic product space form* $(\overline{\mathcal{M}} = \mathcal{M}_1(c_1) \times \mathcal{M}_2(c_2), g, \phi)$ *with a semi-symmetric non-metric connection. Then,*

$$n_2\frac{\Delta f}{f} \leq \frac{n^2(n-2)}{2(n-1)}||\mathcal{H}||^2$$
$$+\frac{1}{8}(c_1+c_2)\frac{1}{p^2+4q}\Big\{2n_1n_2(p^2+4q)-4\big[tr^2\phi+tr\phi\big]\Big\}$$
$$\pm\frac{1}{8}\frac{(c_1-c_2)}{\sqrt{(p^2+4q)}}\{4pn_1n_2-4tr\phi\}$$

$$- \Big[\lambda(n-1) + \lambda|_{M_1}(n_1-1) + \lambda|_{M_2}(n_2-1)\Big]$$
$$+ \Big[n(n-1)\pi(H) + n_1(n_1-1)\pi(H_1) + n_2(n_2-1)\pi(H_2)\Big]. \quad (38)$$

The equality in (38) *is attained if and only if $\mathcal{M}$ is a mixed totally geodesic isometric immersion and meets the condition* (17).

4.3. Results on Specific Instances within the Realm of Metallic Product Space

A metallic structure can be characterized as a golden structure, copper structure, nickel structure, silver structure, bronze structure, subtle structure, and so on for providing different particular values to p and q. For instance, the metallic structure implies a golden structure when $p = 1$ and $q = 1$. Hence, from Theorem 1, we obtain the following results.

Corollary 9. *Let $\mathcal{M}$ be an n-dimensional warped product θ-slant submanifold of an m-dimensional locally golden product space form $(\overline{\mathcal{M}} = \mathcal{M}_1(c_1) \times \mathcal{M}_2(c_2), g, \phi)$ with quarter-symmetric metric connections. Then,*

$$n_2\frac{\Delta f}{f} \le \frac{n^2(n-2)}{2(n-1)}||\mathcal{H}||^2$$
$$+ \frac{1}{40}(c_1+c_2)\Big\{10n_1n_2 - 4\big[tr^2\phi + tr\phi - cos^2\theta(trT)\big]\Big\}$$
$$\pm \frac{1}{8\sqrt{5}}(c_1-c_2)\{4n_1n_2 - 4tr\phi\}$$
$$- (\psi_1+\psi_2)\Big[\lambda(n-1) + \lambda|_{M_1}(n_1-1) + \lambda|_{M_2}(n_2-1)\Big]$$
$$- \psi_2(\psi_1-\psi_2)\Big[\mu(n-1) + \mu|_{M_1}(n_1-1) + \mu|_{M_2}(n_2-1)\Big]$$
$$+ (\psi_1-\psi_2)\Big[n(n-1)\pi(H) + n_1(n_1-1)\pi(H_1) + n_2(n_2-1)\pi(H_2)\Big]. \quad (39)$$

The equality in (39) *is attained if and only if $\mathcal{M}$ is a mixed totally geodesic isometric immersion and meets the condition* (17).

Corollary 10. *Let $\mathcal{M}$ be an n-dimensional warped product θ-slant submanifold of an m-dimensional locally golden product space form $(\overline{\mathcal{M}} = \mathcal{M}_1(c_1) \times \mathcal{M}_2(c_2), g, \phi)$ with semi-symmetric metric connections. Then,*

$$n_2\frac{\Delta f}{f} \le \frac{n^2(n-2)}{2(n-1)}||\mathcal{H}||^2$$
$$+ \frac{1}{40}(c_1+c_2)\Big\{10n_1n_2 - 4\big[tr^2\phi + tr\phi - cos^2\theta(trT)\big]\Big\}$$
$$\pm \frac{1}{8\sqrt{5}}(c_1-c_2)\{4n_1n_2 - 4tr\phi\}$$
$$- 2\Big[\lambda(n-1) + \lambda|_{M_1}(n_1-1) + \lambda|_{M_2}(n_2-1)\Big]. \quad (40)$$

The equality in (40) *is attained if and only if $\mathcal{M}$ is a mixed totally geodesic isometric immersion and meets the condition* (17).

Corollary 11. *Let $\mathcal{M}$ be an n-dimensional warped product θ-slant submanifold of an m-dimensional locally golden product space form $(\overline{\mathcal{M}} = \mathcal{M}_1(c_1) \times \mathcal{M}_2(c_2), g, \phi)$ with semi-symmetric non-metric connections. Then,*

$$n_2\frac{\Delta f}{f} \le \frac{n^2(n-2)}{2(n-1)}||\mathcal{H}||^2$$

$$+\frac{1}{40}(c_1+c_2)\Big\{10n_1n_2-4\big[tr^2\phi+tr\phi-cos^2\theta(trT)\big]\Big\}$$
$$\pm\frac{1}{8\sqrt{5}}(c_1-c_2)\{4n_1n_2-4tr\phi\}$$
$$-\Big[\lambda(n-1)+\lambda|_{M_1}(n_1-1)+\lambda|_{M_2}(n_2-1)\Big]$$
$$+\Big[n(n-1)\pi(H)+n_1(n_1-1)\pi(H_1)+n_2(n_2-1)\pi(H_2)\Big]. \quad (41)$$

The equality in (41) *is attained if and only if* $\mathcal{M}$ *is a mixed totally geodesic isometric immersion and meets the condition* (17).

Remark 2. *We can obtain results similar to the results (9), (10), and (11) for copper, silver, nickel, bronze, subtle, etc., by proving specific values to p and q.*

Remark 3. *We can also obtain the results (9), (10), and (11) for particular classes of the* θ*-slant submanifolds, i.e., invariant and anti-invariant submanifolds by providing particular values* $\theta = 0$ *and* $\theta = \frac{\pi}{2}$*, respectively.*

5. Conclusions

We have the following conclusions from our findings in this article:

1. We delved into the realm of geometric inequalities, with a particular focus on Chen's inequality. Our investigation revolved around its application to assess the square norm of the mean curvature vector and the warping function of warped product slant submanifolds. Within the framework of locally metallic product space forms with quarter-symmetric metric connection, we successfully established this geometric inequality and explored its implications.
2. By examining the conditions under which equality is achieved within the inequality, we gained valuable insights into the intricacies of warped product slant submanifolds. Our findings shed light on the underlying geometric properties and the relationships between the mean curvature vector, the warping function, and the ambient space.
3. The implications of our research extend beyond the theoretical realm. The established geometric inequality and its equality conditions provide a powerful tool for studying and characterizing warped product slant submanifolds in locally metallic product space forms. This has potential applications in various fields, such as differential geometry, mathematical physics, and even in applied sciences where understanding the geometric properties of submanifolds is crucial.

Overall, our study contributes to the existing body of knowledge by providing a deeper understanding of Chen's inequality and its significance in the context of warped product slant submanifolds. We hope that our findings will inspire further research and stimulate new avenues of exploration in the fascinating field of geometric inequalities and their applications.

Future Work

The following could be future research topics for the study of Chen's inequality in the context of warped product slant submanifolds within locally metallic product space forms with quarter-symmetric metric connections:

1. Further studies may involve extending Chen's inequality to other classes of geometric spaces or submanifolds. One possible approach to this would be to examine whether it can be applied to other kinds of submanifolds, including minimum submanifolds, hypersurfaces, Lagrangian submanifolds, etc., and to examine the implications in those situations.
2. Further investigation into the characteristics and properties of warped product slant submanifolds is possible. This might involve creating additional geometric inequali-

ties unique to this class of submanifolds, as well as analyzing the behavior of warping functions and mean curvature vectors in various dimensions and situations.
3. Beyond the quarter-symmetric metric connection, different kinds of metric connections can be taken into consideration to advance the research. Analyzing Chen's inequality in relation to other metric connections may yield insightful comparisons.
4. Subsequent investigations may utilize computational or numerical techniques to verify and investigate the outcomes derived from analytical procedures. In order to investigate the behavior of mean curvature vectors and warping functions and to confirm the accuracy and applicability of Chen's inequality in real-world situations, this can include running numerical experiments or simulations.
5. Interdisciplinary research can be facilitated by working with scientists in adjacent domains like mathematical physics, geometric analysis, or differential geometry. Collaboration with specialists in other fields can result in fresh insights, alternative uses, and a better understanding of Chen's inequality's significance.

Future research can advance geometric inequalities, our knowledge of warped product slant submanifolds, and the field of differential geometry as a whole by exploring these directions.

Author Contributions: Conceptualization, Y.L., M.A., M.A.K., I.A.-D. and M.Z.Y.; methodology, Y.L., M.A., M.A.K., I.A.-D. and M.Z.Y.; investigation, Y.L., M.A., M.A.K., I.A.-D. and M.Z.Y.; writing—original draft preparation, Y.L., M.A., M.A.K., I.A.-D. and M.Z.Y.; writing—review and editing, Y.L., M.A., M.A.K., I.A.-D. and M.Z.Y. All authors have read and agreed to the published version of the manuscript.

Funding: This work was funded by Imam Mohammad Ibn Saud Islamic University (IMSIU) (grant number IMSIU-RP23078).

Data Availability Statement: Data are contained within the article.

Acknowledgments: This work was supported and funded by the Deanship of Scientific Research at Imam Mohammad Ibn Saud Islamic University (IMSIU) (grant number IMSIU-RP23078).

Conflicts of Interest: The authors declare no conflicts of interest in this paper.

References

1. Habermann, L.; Jost, J. Metrics on Riemann surfaces and the geometry of moduli spaces. In *Geometric Theory of Singular Phenomena in Partial Differential Equations (Cortona, 1995)*; Cambridge University Press: Cambridge, UK, 1998; Volume XXXVIII, pp. 53–70.
2. Madsen, I.; Weiss, M. The stable moduli space of Riemann surfaces: Mumford's conjecture. *Ann. Math.* **2007**, *165*, 843–941. [CrossRef]
3. Antón-Sancho, Á. F_4 and $PS_p(8,\mathbb{C})$-Higgd pairs understood as fixed points of the moduli space of E_6-Higgs bundles over a compact Riemannian surface. *Open Math.* **2022**, *20*, 1723–1733. [CrossRef]
4. Antón-Sancho, Á. Fixed Points of Automorphisms of the Vector Bundle Moduli Space Over a Compact Riemann Surface. *Mediterr. J. Math.* **2024**, *21*, 20. [CrossRef]
5. Antón-Sancho, Á. Fixed points of principal E_6-bundles over a compact algebraic curve. *Quaest. Math.* **2024**, *47*, 501–513. [CrossRef]
6. Gothen, P.; Zúñiga-Rojas, R. Stratifications on the moduli space of Higgs bundles. *Port. Math.* **2017**, *74*, 127–148. [CrossRef]
7. Narasimhan, M.S.; Ramanan, S. Moduli of vector bundles on a compact Riemann surface. *Ann. Math.* **1969**, *89*, 19–51. [CrossRef]
8. Newstead, P.E. Characteristic classes of stable bundles of rank 2 over an algebraic curve. *Trans. Am. Math. Soc.* **1972**, *169*, 337–345. [CrossRef]
9. Bishop, R.L.; O'Neill, B. Manifolds of negative curvature. *Trans. Am. Math. Soc.* **1967**, *145*, 1–45. [CrossRef]
10. Beem, J.K.; Ehrlich, P.E.; Powell, T.G. Warped product manifolds in relativity. In *Selected Studies: Physics-Astrophysics, Mathematics, History of Science*; North-Holland: New York, NY, USA, 1982.
11. O'Neill, B. *Semi-Riemannian Geometry with Applications to Relativity*; Academic Press: New York, NY, USA, 1983.
12. Kılıc, E.; Tripathi, M.M.; Gülbahar, M. Chen–Ricci inequalities for submanifolds of Riemannian and Kaehlerian product manifolds. In *Annales Polonici Mathematici*; Instytut Matematyczny Polskiej Akademii Nauk: Warsaw, Poland, 2016; Volume 116, pp. 37–56.
13. Gülbahar, M.; Trıpathı, M.M.; Kılıç, E. Inequalities involving k-Chen invariants for submanifolds of Riemannian product manifolds. *Commun. Fac. Sci. Univ. Ank. Ser. Math. Stat.* **2019**, *68*, 466–483. [CrossRef]
14. Gülbahar, M.; Erkan, E.; Düzgör, M. Ricci Curvatures on Hypersurfaces of Almost Product-like Statistical Manifolds. *J. Eng. Technol. Appl. Sci.* **2024**, *9*, 33–46. [CrossRef]

15. Chen, B.Y. On isometric minimal immersions from warped products into real space forms. *Proc. Edinb. Math. Soc.* **2002**, *45*, 579–587. [CrossRef]
16. Chen, B.Y. Geometry of warped products as riemannian submanifolds and related problem. *Soochow J. Math.* **2002**, *28*, 125–157.
17. Chen, B.Y. Geometry of warped product submanifolds: A survey. *J. Adv. Math. Stud.* **2013**, *6*, 1–43.
18. Chen, B.Y. CR-warped products in complex projective spaces with compact holomorphic factor. *Monatsh. Math.* **2004**, *141*, 177–186. [CrossRef]
19. Uddin, S.; Chen, B.Y.; Al-Solamy, F.R. Warped product bi-slant immersions in Kaehler manifolds. *Mediterr. J. Math.* **2017**, *14*, 1–11. [CrossRef]
20. Munteanu, M.I. Warped product contact CR-submanifolds of Sasakian space forms. *Publ. Math. Debr.* **2005**, *66*, 75–120. [CrossRef]
21. Mustafa, A.; De, A.; Uddin, S. Characterization of warped product submanifolds in Kenmotsu manifolds. *Balkan. J. Geom. Appl.* **2015**, *20*, 86–97.
22. Uddin, S.; Al-Solamy, F.R. Warped product pseudo-slant immersions in Sasakian manifolds. *Publ. Math. Debr.* **2017**, *91*, 331–348. [CrossRef]
23. Qu, Q.; Wang, Y. Multiple warped products with a quarter-symmetric connection. *J. Meth. Anal. Appl.* **2015**, *431*, 955–987. [CrossRef]
24. Wang, Y. Chen inequalities for submanifolds of complex space forms and Sasakian space forms with quarter-symmetric connection. *Int. J. Geom. Methods Mod. Phys.* **2019**, *16*. [CrossRef]
25. Goldberg, S.I.; Yano, K. Polynomial structures on manifolds. *Kodai Math. Semin. Rep.* **1970**, *2*, 199–218. [CrossRef]
26. Bahadir, O.; Uddin, S. Slant submanifolds of golden Riemannian manifols. *J. Math. Ext.* **2019**, *13*, 1–10.
27. Hretcanu, C.E.; Crasmareanu, M. Metallic structures on Riemannian manifolds. *Rev. Un. Mat. Argent.* **2013**, *54*, 15–27.
28. Blaga, A.M.; Hretcanu, C.E. Invariant, anti-invariant and slant-submanifols of metallic Riemannian manifold. *Novi Sad J. Math.* **2018**, *48*, 57–82. [CrossRef]
29. Yano, K.; Kon, M. *Structures on Manifolds: Series in Pure Mathematics*; World Scientific: Singapore, 1984.

axioms

MDPI

Article

Geometry of Torsion Gerbes and Flat Twisted Vector Bundles

Byungdo Park

Department of Mathematics Education, Chungbuk National University, Cheongju 28644, Republic of Korea; byungdo@chungbuk.ac.kr

Abstract: Gerbes and higher gerbes are geometric cocycles representing higher degree cohomology classes, and are attracting considerable interest in differential geometry and mathematical physics. We prove that a 2-gerbe has a torsion Dixmier–Douady class if and only if the gerbe has locally constant cocycle data. As an application, we give an alternative description of flat twisted vector bundles in terms of locally constant transition maps. These results generalize to n-gerbes for $n = 1$ and $n \geq 3$, providing insights into the structure of higher gerbes and their applications to the geometry of twisted vector bundles.

Keywords: gerbe; 2-gerbe; smooth Deligne cohomology; Dixmier–Douady class; twisted vector bundles

MSC: primary 53C08; secondary 55R65; 55N05

1. Introduction

In modern differential geometry, the study of higher categorical structures has led to significant advancements in our understanding of manifolds and their invariants. Gerbes and higher gerbes, as geometric realizations of such structures, play a crucial role in this landscape, connecting diverse areas such as algebraic topology, complex geometry, and mathematical physics.

$U(1)$-gerbes are geometric objects representing degree 3 integral cohomology classes, just as line bundles represent degree 2 integral cohomology classes. Gerbes were originally introduced by Giraud [1], and began to be used more often in the context of algebraic topology and differential geometry after Brylinski [2]. In particular, Murray [3] conceived and constructed an explicit and geometric model of a gerbe, called a bundle gerbe, as opposed to a description as a certain *sheaf of groupoids*. This model by Murray has been further developed by several authors. Most notably, Stevenson has developed a geometric model of a 2-gerbe, and the 2-stack structure of gerbes was considered [4,5], which was further studied by Waldorf [6], and equivariant refinements were studied in [7,8]. Gerbes and higher gerbes have been applied to several problems in mathematics and physics. For example, twisted K-theory and Ramond–Ramond field classifications [9–12], local formulas for 2d Wess–Zumino (WZ) action [13] and its Feynman amplitude interpreted as a bundle gerbe holonomy [14,15], geometric string structures [16], and even topological insulators [17–19].

As mentioned above, there are several models for higher gerbes with connection. To list a few, there are bundle n-gerbes with connection, sheaves of higher groupoids, and a map into a classifying ∞-stack $\mathbb{B}^{n+2}_{\nabla}$. However, one of the most classical and elementary models would be the Deligne cocycle model, consisting of Čech cocycles and local differential form data. Indeed, the Deligne complex is the natural home for studying differential geometric cocycles such as line bundles with connections and (higher) gerbes with connection.

This article is a brief technical report on differential geometry of torsion gerbes. Namely, we prove that a necessary and sufficient condition for the Dixmier–Douady class of a 2-gerbe to be torsion is that its cocycle data consist of locally constant maps, and its proof essentially generalizes for the case of n-gerbes with $n = 1$ or $n \geq 3$. The idea comes

Citation: Park, B. Geometry of Torsion Gerbes and Flat Twisted Vector Bundles. *Axioms* **2024**, *13*, 504. https://doi.org/10.3390/axioms13080504

Academic Editor: Mića Stanković

Received: 29 June 2024
Revised: 22 July 2024
Accepted: 24 July 2024
Published: 26 July 2024

from a well-known fact on flat vector bundles, i.e., a necessary and sufficient condition for a vector bundle to admit a flat connection is that the Čech-cocycle data of the underlying vector bundle consists of locally constant maps. Using our results on torsion 2-gerbes, we also prove a generalization of this fact to flat twisted vector bundles.

As is well-known, a gerbe being torsion or not is crucial in studying geometric cocycles of twisted K-theory. Indeed, if a geometric cocycle admits a nontorsion twist, it has to be an infinite dimensional construction (see [9,12]). Therefore, we expect that our results will be useful in studying finite-dimensional constructions such as twisted vector bundles or bundle gerbe modules with finite-dimensional fibers.

This paper is organized as follows. In Section 2, we review the $U(1)$-gerbe with connections and its higher analogues. This section also serves the purpose of setting up notations and terminologies we will be using throughout this paper. In Section 3, we prove that a 2-gerbe is torsion if and only if its cocycle data consists of locally constant functions. In Section 4, we apply our main theorem to prove a twisted analogue of a classical fact that a vector bundle is flat if and only if there exist local trivializations whose transition maps are locally constant.

2. Preliminaries

In this section, we review (higher) gerbes with connection. Throughout this paper, all of our manifolds are smooth manifolds, and all of our maps are smooth maps, unless specified otherwise. In particular, X always denotes a manifold. By gerbes, we will always mean $U(1)$-gerbes. We will use the notation $U_{i_1\cdots i_n}$ to denote an n-fold intersection $U_{i_1} \cap \cdots \cap U_{i_n}$. If an open cover is locally finite and every n-fold intersection is contractible for all $n \in \mathbb{Z}^+$, we will call it a good cover. On a smooth manifold, a good cover always exists. A Čech cocycle $\zeta = (\zeta_{i_1\cdots i_n})$ is said to be *completely normalized* if $\zeta_{i_1\cdots i_n} \equiv 1$ whenever there is a repeated index, and $\zeta_{\sigma(i_1)\cdots\sigma(i_n)} = (\zeta_{i_1\cdots i_n})^{\mathrm{sign}(\sigma)}$ for any $\sigma \in \mathfrak{S}_n$, where $\mathfrak{S}_n$ is the symmetric group on n letters.

2.1. gerbes with connection

In this subsection, we shall review a Čech cocycle description of a gerbe with connections. See Gawędzki and Reis [15] and Hitchin [20] for a broader account.

Definition 1. *Let X be a manifold and $\mathcal{U} := \{U_i\}_{i\in\Lambda}$ an open cover of X. A **gerbe** over X subordinate to $\mathcal{U}$ is a $U(1)$-valued completely normalized Čech 2-cocycle $\{\lambda_{kji}\} \in \check{Z}^2(\mathcal{U}, U(1))$. A **connection** on a gerbe $\{\lambda_{kji}\}$ on $\mathcal{U}$ is a pair $(\{A_{ji}\}, \{B_i\})$ consisting of a family of differential 1-forms $\{A_{ji} \in \Omega^1(U_{ij}; \sqrt{-1}\mathbb{R})\}_{i,j\in\Lambda}$, and a family of differential 2-forms $\{B_i \in \Omega^2(U_i; \sqrt{-1}\mathbb{R})\}_{i\in\Lambda}$, satisfying the following relations:*

- $\lambda_{kji}\lambda_{lji}^{-1}\lambda_{lki}\lambda_{lkj}^{-1} = 1;$
- $d\log\lambda_{kji} = A_{ji} + A_{ik} + A_{kj};$
- $B_j - B_i = dA_{ji}.$

*From $dB_i = dB_j$ for all $i, j \in \Lambda$, the family of exact 3-forms $\{dB_i\}_{i\in\Lambda}$ defines a global closed differential 3-form H. The differential form H is called the **curvature** of the gerbe, or the **Neveu–Schwarz 3-form**.*

A gerbe with connections on $\mathcal{U}$ is therefore a Deligne cocycle of degree 2. Notice that our total differential is $D = d + (-1)^q\delta$ on $\check{C}^p(\mathcal{U}, \Omega^q)$. Throughout the rest of this paper, $\widehat{\lambda} = (\{\lambda_{kji}\}, \{A_{ji}\}, \{B_i\})$ always denotes a gerbe with connections defined on an open cover $\mathcal{U} = \{U_i\}_{i\in\Lambda}$ of X, and H denotes the 3-curvature form of $\widehat{\lambda}$.

Definition 2. *Two gerbes with connections $\widehat{\lambda}$ and $\widehat{\lambda}'$ are **isomorphic** if $\widehat{\lambda}'$ is obtained by adding a total degree 2 Deligne coboundary to $\widehat{\lambda}$, i.e., $\widehat{\lambda}' = \widehat{\lambda} + D\widehat{\mu}$ for some $\widehat{\mu} \in \check{C}^1(\mathcal{U}, \Omega^0) \oplus \check{C}^0(\mathcal{U}, \Omega^1)$.*

Remark 1. *Let* $\{\lambda_{kji}\} \in \check{Z}^2(\mathcal{U}, U(1))$ *be a gerbe, and* $\delta : \check{H}^2(\mathcal{U}, U(1)) \to H^3(X; 2\pi i\mathbb{Z})$ *be the connecting map. The image in* $H^3_{dR}(X; \sqrt{-1}\mathbb{R})$ *of the cohomology class* $\delta([\lambda]) \in H^3(X; 2\pi i\mathbb{Z})$ *coincides with the cohomology class of* $H \in H^3_{dR}(X; \sqrt{-1}\mathbb{R})$ *(see Brylinski ([2] p. 175) Corollary 4.2.8.). Here, the cohomology class* $\delta([\lambda])$ *is a topological invariant of a gerbe, called the* Dixmier–Douady class.

2.2. Higher gerbes with connection

In the previous subsection, we have seen that a gerbe with connections is a degree 2 Deligne cocycle. It is possible to generalize it to higher degrees for a cocycle definition of an n-gerbe with connections. Compare Stevenson [4,5] and Gajer [21].

Definition 3. *Let* X *be a manifold, and* $\mathcal{U} := \{U_i\}_{i\in\Lambda}$ *be an open cover of* X*. An* ***n*-gerbe** *over* X *subordinate to* $\mathcal{U}$ *is a* $U(1)$*-valued completely normalized Čech* $(n+1)$*-cocycle* $\{\lambda_{i_{n+2}\cdots i_1}\} \in \check{Z}^{n+1}(\mathcal{U}, U(1))$*. A* **connection** *on an n-gerbe* $\{\lambda_{i_{n+2}\cdots i_1}\}$ *on* $\mathcal{U}$ *is an* $(n+1)$*-tuple* $(\{A^{(1)}_{i_{n+1}\cdots i_1}\}, \{A^{(2)}_{i_n\cdots i_1}\}, \cdots, \{A^{(n+1)}_{i_1}\})$*, consisting of a family of differential k-forms* $\{A^{(k)}_{i_{n+k-2}\cdots i_1} \in \Omega^k(U_{i_{n+k-2}\cdots i_1}; \sqrt{-1}\mathbb{R})\}_{i_{n+k-2},\cdots,i_1\in\Lambda}$*, satisfying that the* $(n+2)$*-tuple* $\widehat{\lambda} = (\lambda, A^{(1)}, \cdots, A^{(n+1)})$ *is a degree* $(n+1)$*-Deligne cocycle, i.e.,* $D\widehat{\lambda} = 0$*. The differential* $(n+1)$*-forms* $\{A^{(n+1)}_i\}$ *defined on each open set satisfy* $dA^{(n+1)}_i = dA^{(n+1)}_j$ *for all* $i, j \in \Lambda$*; the family of exact* $(n+2)$*-forms* $\{dA^{(n+1)}_i\}_{i\in\Lambda}$ *defines a global closed differential* $(n+2)$*-form* $\mathcal{H}$*. The differential form* $\mathcal{H}$ *is called the* ***curvature*** *of the n-gerbe.*

Definition 4. *Two n-gerbes with connection* $\widehat{\lambda}$ *and* $\widehat{\lambda}'$ *are* **isomorphic** *if* $\widehat{\lambda}'$ *is obtained by adding a total degree* $n+1$ *Deligne coboundary to* $\widehat{\lambda}$*, i.e.,* $\widehat{\lambda}' = \widehat{\lambda} + D\widehat{\mu}$ *for some* $\widehat{\mu} \in \check{C}^n(\mathcal{U}, \Omega^0) \oplus \check{C}^{n-1}(\mathcal{U}, \Omega^1) \oplus \cdots \oplus \check{C}^0(\mathcal{U}, \Omega^n)$

Similarly for gerbes, an n-gerbe $\lambda \in \check{Z}^{n+1}(\mathcal{U}, U(1))$ has an higher analogue of the Dixmier–Douady class in $H^{n+2}(X; 2\pi i\mathbb{Z})$ as its topological invariant. Its image in $H^{n+2}_{\mathrm{dR}}(X; \sqrt{-1}\mathbb{R})$ coincides with the curvature $\mathcal{H}$ of n-gerbe (Cf. Stevenson [4], Chapter 11).

Remark 2. *For later use, we give explicit formula of the cocycle condition for a 2-gerbe with connection* $(\{\lambda_{lkji}\}, \{A_{kji}\}, \{B_{ji}\}, \{C_i\})$*.*

C1. $\lambda_{kji}\lambda^{-1}_{lji}\lambda_{lki}\lambda^{-1}_{lkj} = 1$;
C2. $d\log\lambda_{lkji} = A_{kji} - A_{lji} + A_{lki} - A_{lkj}$;
C3. $dA_{kji} = -B_{ji} + B_{ki} - B_{kj}$;
C4. $dB_{ji} = C_i - C_j$.

3. Main Theorems

In this section, we shall state and prove our main theorems on a necessary and sufficient condition for a 2-gerbe having a torsion Dixmier–Douady class. We state and prove the sufficiency and then the necessity.

Theorem 1. *Let* X *be a manifold,* $\mathcal{U} = \{U_i\}_{i\in\Lambda}$ *be an open cover of* X*, and* $\lambda = \{\lambda_{lkji}\}$ *be a 2-gerbe on* X*. If each* λ_{lkji} *is a locally constant map, then this 2-gerbe determines a torsion class* $\delta([\lambda])$ *in* $H^4(X; 2\pi i\mathbb{Z})$*.*

Proof. Suppose that $(\{A_{kji}\}, \{B_{ji}\}, \{C_i\})$ is a connection on the given 2-gerbe λ. Since λ_{lkji} are locally constant maps, it follows that $A_{kji} - A_{lji} + A_{lki} - A_{lkj} = \lambda^{-1}_{lkji}d\lambda_{lkji} = 0$. Accordingly, we could have chosen a connection with $A_{kji} \equiv 0$, $B_{ji} \equiv 0$, and $C_i := \zeta|_{U_i}$ for some $\zeta \in \Omega^3(X; \sqrt{-1}\mathbb{R})$, since the quadruple $(\{\lambda_{kji}\}, \{0\}, \{0\}, \{\zeta|_{U_i}\})$ satisfies the cocycle conditions **C1** to **C4** in Remark 2. Moreover, since the curvature 4-form of this 2-gerbe with connections is exact, it follows that $\delta([\lambda]) \otimes \mathbb{R} = [d\zeta] = 0$, i.e., $\delta([\lambda])$ is a torsion class in $H^4(X, 2\pi i\mathbb{Z})$. Here, $\delta : \check{H}^3(\mathcal{U}, U(1)) \to H^4(X; 2\pi i\mathbb{Z})$ is the connecting map. □

Proceeding similarly as in the above proof, a similar theorem also holds for n-gerbes for $n = 1$ or ≥ 3, as stated in the following corollary.

Corollary 1. *Let X be a manifold, $\mathcal{U} = \{U_i\}_{i\in\Lambda}$ be an open cover of X, and $\lambda = \{\lambda_{i_{n+2}\cdots i_1}\}$ be an n-gerbe on X. If each $\lambda_{i_{n+2}\cdots i_1}$ is a locally constant map, then this n-gerbe determines a torsion class $\delta([\lambda])$ in $H^{n+2}(X; 2\pi i\mathbb{Z})$.*

Theorem 2. *Let X and λ be as above. Suppose that the 2-gerbe λ is defined on a good cover $\mathcal{U} = \{U_i\}_{i\in\Lambda}$, and also that λ determines a torsion class $\delta([\lambda])$ in $H^4(X; 2\pi i\mathbb{Z})$. Then, given any connection $(\{A_{kji}\}, \{B_{ji}\}, \{C_i\})$ on this 2-gerbe, there exists a 2-gerbe with connection $(\widetilde{\lambda}, \widetilde{A}, \widetilde{B}, \widetilde{C})$ that has an underlying 2-gerbe consisting of a family of locally constant maps $\widetilde{\lambda}_{lkji} : U_{ijkl} \to U(1)$, such that the difference between (λ, A, B, C) and $(\widetilde{\lambda}, \widetilde{A}, \widetilde{B}, \widetilde{C})$ is a Deligne coboundary of degree 3.*

Proof. Suppose a 2-gerbe λ determines a torsion class $\delta([\lambda])$ in $H^4(X; 2\pi i\mathbb{Z})$. We first choose an arbitrary connection $(\{A_{kji}\}, \{B_{ji}\}, \{C_i\})$ on the 2-gerbe λ. For the curvature $\mathcal{H}$ of the 2-gerbe, $\delta([\lambda]) \otimes \mathbb{R} = [\mathcal{H}]$ is satisfied, and since the 2-gerbe is a torsion, $[\mathcal{H}]$ has a representative $d\zeta$, where ζ is a differential 3-form on X. Now, from $dC_i = \mathcal{H}|_{U_i} = d\zeta|_{U_i}$, we have $d(\zeta|_{U_i} - C_i) = 0$, and since U_i is contractible, by Poincaré's Lemma, $\zeta|_{U_i} - C_i = d\Pi_i$ for some $\Pi_i \in \Omega^2(U_i; i\mathbb{R})$. We define

$$\widetilde{C}_i := C_i + d\Pi_i = \zeta|_{U_i}.$$

Applying **C4**, we see that $d(B_{ji} + \Pi_i - \Pi_j) = 0$. Again by Poincaré's Lemma, there exists $\xi_{ji} \in \Omega^1(U_{ij}; \sqrt{-1}\mathbb{R})$, such that

$$B_{ji} + \Pi_i - \Pi_j = d\xi_{ji}. \tag{1}$$

We set

$$\widetilde{B}_{ji} := B_{ji} + \Pi_i - \Pi_j - d\xi_{ji} = 0.$$

Applying **C3** and Equation (1), we have $d(A_{kji} + \xi_{ji} - \xi_{ki} + \xi_{kj}) = 0$. Again, there exists $\chi_{kji} \in \Omega^0(U_{ijk}; U(1))$ such that $A_{kji} + \xi_{ji} - \xi_{ki} + \xi_{kj} = d\log\chi_{kji}$, so we define

$$\widetilde{A}_{kji} := A_{kji} + \xi_{ji} - \xi_{ki} + \xi_{kj} - d\log\chi_{kji} = 0$$
$$\widetilde{\lambda}_{lkji} := \lambda_{lkji}\chi_{kji}^{-1}\chi_{lji}\chi_{lki}^{-1}\chi_{lkj}.$$

It can be readily seen that $\widehat{\widetilde{\lambda}} = (\{\widetilde{\lambda}_{lkji}\}, \{\widetilde{A}_{kji}\}, \{\widetilde{B}_{ji}\}, \{\widetilde{C}_i\})$ satisfies conditions from **C1** to **C4**, where $\widetilde{A}_{kji} \equiv 0 \equiv \widetilde{B}_{ji}$ and $\widetilde{C}_i$ is a restriction of a global 3-form ζ to U_i. The 2-gerbe cocycles being locally constant follows from the cocycle condition **C2** for $\widehat{\widetilde{\lambda}}$. In addition, $\widehat{\lambda} = (\{\lambda_{lkji}\}, \{A_{kji}\}, \{B_{ji}\}, \{C_i\})$ satisfies $\widehat{\widetilde{\lambda}} = \widehat{\lambda} + D\widehat{\chi}$ where $\widehat{\chi} = (\{\chi_{kji}^{-1}\}, \{-\xi_{ji}\}, \{\Pi_i\})$. □

Proceeding similarly as in the above proof, a similar theorem also holds for torsion n-gerbes for $n = 1$ or ≥ 3, as stated in the following corollary.

Corollary 2. *Let X and λ be as above. Suppose that the n-gerbe λ is defined on a good cover $\mathcal{U} = \{U_i\}_{i\in\Lambda}$, and also that λ determines a torsion class $\delta([\lambda])$ in $H^{n+2}(X; 2\pi i\mathbb{Z})$. Then, given any connection $(\{A^{(1)}_{i_{n+1}\cdots i_1}\}, \{A^{(2)}_{i_n\cdots i_1}\}, \cdots, \{A^{(n+1)}_{i_1}\})$ on this n-gerbe, there exists an n-gerbe with connection $(\widetilde{\lambda}, \widetilde{A}^{(1)}, \cdots, \widetilde{A}^{(n+1)})$ that has an underlying n-gerbe consisting of a family of locally constant maps $\widetilde{\lambda}_{i_{n+2}\cdots i_1} : U_{i_{n+2}\cdots i_1} \to U(1)$, such that the difference between $(\lambda, A^{(1)}, \cdots, A^{(n+1)})$ and $(\widetilde{\lambda}, \widetilde{A}^{(1)}, \cdots, \widetilde{A}^{(n+1)})$ is a Deligne coboundary of degree $n+1$.*

4. Application: Flatness of Twisted Vector Bundle

In this section, we briefly review what a twisted vector bundle with connections is. After that, we recall an alternative characterization of a flat vector bundle via locally constant transition maps. We apply Corollary 2 to state and prove its twisted analogue.

Definition 5. *Let $\mathcal{U} = \{U_i\}_{i\in\Lambda}$ be an open cover of X, and λ be a $U(1)$-valued completely normalized Čech 2-cocycle. A* ***λ-twisted vector bundle*** *E of rank n over X consists of a family of product bundles $\{U_i \times \mathbb{C}^n : U_i \in \mathcal{U}\}_{i\in\Lambda}$ together with transition maps*

$$g_{ji} : U_{ij} \to U(n)$$

satisfying

$$g_{ii} = \mathbf{1}, \quad g_{ji} = g_{ij}^{-1}, \quad g_{kj}g_{ji} = g_{ki}\lambda_{kji}.$$

The gerbe λ in this definition is also called a *twist*. A λ-twisted vector bundle is *smooth* if all transition maps and gerbe cocycle data are smooth maps. We shall write a λ-twisted vector bundle E over X of rank n as a triple $(\mathcal{U}, \{g_{ji}\}, \{\lambda_{kji}\})$.

Definition 6. *Let $\widehat{\lambda} = (\{\lambda_{kji}\}, \{A_{ji}\}, \{B_i\})$ be a gerbe with connections, and $E = (\mathcal{U}, \{g_{ji}\}, \{\lambda_{kji}\})$ be a smooth λ-twisted vector bundle of rank n. A* ***connection*** *on E compatible with $\widehat{\lambda}$ is a family $\Gamma = \{\Gamma_i \in \Omega^1(U_i; \mathfrak{u}(n))\}_{i\in\Lambda}$ satisfying*

$$\Gamma_i - g_{ji}^{-1}\Gamma_j g_{ji} - g_{ji}^{-1}dg_{ji} = -A_{ji} \cdot \mathbf{1}, \tag{2}$$

where $A_{ji} \in \Omega^1(U_{ij}; i\mathbb{R})$. Here, $\mathfrak{u}(n)$ denotes the Lie algebra of $U(n)$, and $\mathbf{1}$ the identity matrix.

It is easy to see that Equation (2) is compatible with the cocycle condition of gerbes with connection, i.e., $\delta(A)_{kji} \cdot \mathbf{1} = d\log\lambda_{kji} \cdot \mathbf{1}$. A standard argument using partitions of unity shows that, for any λ-twisted vector bundle E, there exists a connection on E compatible with $\widehat{\lambda}$.

Definition 7. *Let $\widehat{\lambda} = (\{\lambda_{kji}\}, \{A_{ji}\}, \{B_i\})$ be as above, and (E, Γ) be a λ-twisted vector bundle $(\mathcal{U}, \{g_{ji}\}, \{\lambda_{kji}\})$ of rank n with a connection Γ compatible with $\widehat{\lambda}$. The* ***curvature form*** *of Γ is the family $R = \{R_i \in \Omega^2(U_i; \mathfrak{u}(n))\}_{i\in\Lambda}$, where $R_i := d\Gamma_i + \Gamma_i \wedge \Gamma_i$.*

The following proposition is a well-known characterization of a flat vector bundle.

Proposition 1. *If a vector bundle E over X admits a flat connection ∇, then there exists a cocycle consisting of locally constant transition maps. Conversely, if a cocycle (g_{ji}) of a vector bundle E over X defined on an open cover $\mathcal{U} = \{U_i\}_{i\in\Lambda}$ consists of locally constant maps, then E admits a flat connection.*

Proof. Since ∇ is a flat connection, there exists a locally trivial open cover $\mathcal{U} = \{U_i\}_{i\in\Lambda}$ such that the connection form ω_i on U_i is identically zero. Let $\{g_{ji}\}$ be a cocycle of the vector bundle E over X defined on the open cover $\mathcal{U}$. Connection forms satisfy the following gauge transformation formula:

$$\omega_i = g_{ji}^{-1}\omega_j g_{ji} + g_{ji}^{-1}dg_{ji}.$$

It follows that $dg_{ji} = 0$, and hence each g_{ji} is a locally constant map. Conversely, if each g_{ji} is locally constant, then $dg_{ji} = 0$. So, we can take $\omega \equiv 0$ for each $i \in \Lambda$. □

A λ-twisted vector bundle admits only torsion twists. By Corollary 2, a torsion gerbe with connections is always isomorphic to a gerbe with connection $\widehat{\lambda} = (\{\lambda_{kji}\}, \{A_{ji}\}, \{B_i\})$

where all λ_{kji} are locally constant, $A_{ji} \equiv 0$, and $B_i = \zeta|_{U_i}$ for a globally defined differential form $\zeta \in \Omega^2(X; i\mathbb{R})$.

Theorem 3. *Let $\widehat{\lambda} = (\{\lambda_{kji}\}, \{A_{ji}\}, \{B_i\})$ be a gerbe with connections, provided that every λ_{kji} is locally constant, and $A_{ji} = 0$ for all $i, j \in \Lambda$. $E = (\mathcal{U}, \{g_{ji}\}, \{\lambda_{kji}\})$ is a λ-twisted vector bundle that admits a connection $\Gamma = \{\Gamma_i\}_{i\in\Lambda}$ compatible with the connection of $\widehat{\lambda}$ such that $R_i \equiv 0$ for each $i \in \Lambda$, if and only if each g_{ji} is locally constant.*

Proof. Suppose a λ-twisted vector bundle with connection (E, Γ) is flat, i.e., $R_i \equiv 0$. Then, over each $U_i \in \mathcal{U}$, it admits a parallel framing such that the connection form $\Gamma_i \equiv 0$. By Equation (2), we obtain $dg_{ji} = 0$ and, hence, g_{ji} is locally constant. Suppose each g_{ji} is locally constant. The family $\Gamma_{i i\in\Lambda}$ with $\Gamma_i \equiv 0$ is a connection on E. The corresponding curvature form $R_i \equiv 0$. □

5. Discussion

In this paper, we have investigated the differential geometry of torsion gerbes, focusing on providing a necessary and sufficient condition for the Dixmier–Douady class of a 2-gerbe to be torsion. Our primary result demonstrates that a 2-gerbe is torsion if and only if its cocycle data consists of locally constant functions. This insight extends to n-gerbes for $n = 1$ and $n \geq 3$, offering a generalized perspective on the structure of higher gerbes.

We drew upon the well-established understanding of flat vector bundles, wherein the existence of a flat connection is characterized by locally constant Čech cocycles. This analogy underscored the significance of locally constant cocycle data in the context of gerbes. We extended this result for the case of flat twisted vector bundles, thereby broadening the applicability of our findings.

In summary, this paper contributes to the deeper understanding of the geometry and topology of torsion gerbes and their higher analogues, offering new tools and perspectives for future research in both mathematics and theoretical physics. For example, our results can be applied to investigating the role of locally constant cocycle data in the differential geometry of twisted vector bundles over orbifolds, and more general stratified spaces.

Funding: This research received no funding.

Institutional Review Board Statement: Not applicable.

Informed Consent Statement: Not applicable.

Data Availability Statement: No new data were used or created in this study. Data sharing is not applicable to this article.

Acknowledgments: We thank Mahmoud Zeinalian for giving motivations to write this paper.

Conflicts of Interest: The author declares no conflicts of interest.

References

1. Giraud, J. *Cohomologie Non Abélienne*; Die Grundlehren der mathematischen Wissenschaften, Band 179; Springer: Berlin/Heidelberg, Germany; New York, NY, USA, 1971; p. ix + 467.
2. Brylinski, J.-L. *Loop Spaces, Characteristic Classes and Geometric Quantization*; Progress in Mathematics; Birkhäuser Boston, Inc.: Boston, MA, USA, 1993; Volume 107, p. xvi + 300. [CrossRef]
3. Murray, M.K. Bundle gerbes. *J. Lond. Math. Soc.* **1996**, *54*, 403–416. [CrossRef]
4. Stevenson, D. The Geometry of Bundle Gerbes. Ph.D. Thesis, The University of Adelaide, Adelaide, Australia, 2000.
5. Stevenson, D. Bundle 2-gerbes. *Proc. Lond. Math. Soc.* **2004**, *88*, 405–435. [CrossRef]
6. Waldorf, K. More morphisms between bundle gerbes. *Theory Appl. Categ.* **2007**, *18*, 240–273.
7. Murray, M.K.; Roberts, D.M.; Stevenson, D.; Vozzo, R.F. Equivariant bundle gerbes. *Adv. Theory Math. Phys.* **2017**, *21*, 921–975. [CrossRef]
8. Park, B.; Redden, C. A classification of equivariant gerbe connections. *Commun. Contemp. Math.* **2019**, *21*, 1850001. [CrossRef]
9. Bouwknegt, P.; Carey, A.L.; Mathai, V.; Murray, M.K.; Stevenson, D. Twisted K-theory and K-theory of bundle gerbes. *Commun. Math. Phys.* **2002**, *228*, 17–45. [CrossRef]

10. Mathai, V.; Stevenson, D. Chern character in twisted *K*-theory: Equivariant and holomorphic cases. *Commun. Math. Phys.* **2003**, *236*, 161–186. [CrossRef]
11. Park, B. Geometric models of twisted differential *K*-theory I. *J. Homotopy Relat. Struct.* **2018**, *13*, 143–167. [CrossRef]
12. Gorokhovsky, A.; Lott, J. A Hilbert bundle description of differential *K*-theory. *Adv. Math.* **2018**, *328*, 661–712. [CrossRef]
13. Gawędzki, K. Topological actions in two-dimensional quantum field theories. In *Nonperturbative Quantum Field Theory (Cargèse, 1987)*; Nato Science Series B; Plenum: New York, NY, USA, 1988; Volume 185, pp. 101–141.
14. Carey, A.L.; Mickelsson, J.; Murray, M.K. Bundle gerbes applied to quantum field theory. *Rev. Math. Phys.* **2000**, *12*, 65–90. [CrossRef]
15. Gawędzki, K.; Reis, N. WZW branes and gerbes. *Rev. Math. Phys.* **2002**, *14*, 1281–1334. [CrossRef]
16. Waldorf, K. String connections and Chern-Simons theory. *Trans. Am. Math. Soc.* **2013**, *365*, 4393–4432. [CrossRef]
17. Gawędzki, K. Bundle gerbes for topological insulators. In *Advanced School on Topological Quantum Field Theory*; Mathematical Institute of the Polish Academy of Sciences: Warsaw, Poland, 2018; Volume 114, pp. 145–180.
18. Gawędzki, K. Square root of gerbe holonomy and invariants of time-reversal-symmetric topological insulators. *J. Geom. Phys.* **2017**, *120*, 169–191. [CrossRef]
19. Mathai, V.; Thiang, G.C. Differential topology of semimetals. *Commun. Math. Phys.* **2017**, *355*, 561–602. [CrossRef]
20. Hitchin, N. Lectures on special Lagrangian submanifolds. In *Winter School on Mirror Symmetry, Vector Bundles and Lagrangian Submanifolds (Cambridge, MA, 1999)*; AMS/IP Studies in Advanced Mathematics; American Mathematical Society: Providence, RI, USA, 2001; Volume 23, pp. 151–182. [CrossRef]
21. Gajer, P. Geometry of Deligne cohomology. *Invent. Math.* **1997**, *127*, 155–207. [CrossRef]

axioms

Article

Quasi-Canonical Biholomorphically Projective Mappings of Generalized Riemannian Space in the Eisenhart Sense

Vladislava M. Milenković [1] and Mića S. Stanković [2,*]

[1] Faculty of Technology, University of Niš, Bulevar oslobodjenja 124, 16000 Leskovac, Serbia; vanja.dunja91@gmail.com or vladislava@tf.ni.ac.rs
[2] Faculty of Sciences and Mathematics, University of Niš, Višegradska 33, 18000 Niš, Serbia
* Correspondence: mica.stankovic@pmf.edu.rs

Abstract: In this paper, quasi-canonical biholomorphically projective and equitorsion quasi-canonical biholomorphically projective mappings are defined. Some relations between the corresponding curvature tensors of the generalized Riemannian spaces GR_N and $G\overline{R}_N$ are obtained. At the end, the invariant geometric object of an equitorsion quasi-canonical biholomorphically projective mapping is found.

Keywords: quasi-canonicalbiholomorphically projective mapping; equitorsion quasi-canonical biholomorphically projective mapping; curvature tensor; invariant geometric object

MSC: Primary 53B05; Secondary 53B20; 53C15

Citation: Milenković, V.M.; Stanković, M.S. Quasi-Canonical Biholomorphically Projective Mappings of Generalized Riemannian Space in the Eisenhart Sense. *Axioms* **2024**, *13*, 528. https://doi.org/10.3390/axioms13080528

Academic Editor: Anna Maria Fino

Received: 27 June 2024
Revised: 30 July 2024
Accepted: 31 July 2024
Published: 3 August 2024

1. Introduction and Preliminaries

Differentiable manifolds GR_N with a non-symmetric metric tensor, GA_N with a non-symmetric affine connection and their mappings were, and still are, the subject of interest of many scientists [1–14]. The use of a non-symmetric basic tensor and non-symmetric connection became especially relevant after the appearance of the works of A. Einstein related to creating the unified field theory, where the symmetric part of the basic tensor is related to gravitation, and the antisymmetric one to electromagnetism. We can say that, after A. Einstein [15,16], the main steps were made by L. P. Eisenhart [17,18].

Geometric mappings are interesting, both theoretically and practically. Geodesic and almost geodesic lines play an important role in geometry and physics [1,2,6,19]. The movement of many types of mechanical systems, as well as bodies or particles in gravitational or electromagnetic fields, in continual constant surroundings, is often conducted in paths, which can be looked upon as geodesic lines of Riemannian or affine connected spaces, which are defined by the energetic regime along which the process takes place. So, for example, two Riemannian spaces, which admit reciprocal geodesic mapping, describe processes which are unfolded by an equivalent exterior load and equal orbit, but different energetic regimes. In this case, one of these processes can be modeled by another. During recent years, many papers have been devoted to the theory of holomorphically projective mappings; let us mention J. Mikeš, S.M. Minčić, M.S. Stanković, Lj. S. Velimirović, M. Lj. Zlatanović, etc. [7,12–14,19]. This paper is a natural continuation of the research published in paper [20] in which biholomorphically projective mappings were studied, and they can be observed as a kind of generalization of holomorphically projective mappings.

A generalized Riemannian space GR_N in the sense of Eisenhart's definition [18] is a differentiable N-dimensional manifold, equipped with a non-symmetric metric tensor g_{ij}. The connection coefficients of the space GR_N are the generalized Cristoffel's symbols of the second kind [19]:

$$\Gamma^i_{jk} = g^{\underline{ip}}\Gamma_{p.jk}, \tag{1}$$

where $||g^{\underline{ij}}|| = ||g_{\underline{ij}}||^{-1}$, $g_{\underline{ij}} = \frac{1}{2}(g_{ij} + g_{ji})$ and

$$\Gamma_{i.jk} = \frac{1}{2}(g_{ji,k} - g_{jk,i} + g_{ik,j}),$$

where, for example, $g_{ij,k} = \frac{\partial g_{ij}}{\partial x^k}$. We suppose that $\det ||g_{ij}|| \neq 0$, $\det ||g_{\underline{ij}}|| \neq 0$. In the general case, the connection coefficients are not symmetric, i.e., $\Gamma^i_{jk} \neq \Gamma^i_{kj}$, and they can be represented as the sum of the symmetric and antisymmetric parts

$$\Gamma^i_{jk} = S^i_{jk} + T^i_{jk}, \tag{2}$$

where the symmetric and antisymmetric part of Γ^i_{jk} are given by the formulas

$$\Gamma^i_{\underline{jk}} = \frac{1}{2}(\Gamma^i_{jk} + \Gamma^i_{kj}) = S^i_{jk}, \quad \Gamma^i_{\underset{\vee}{jk}} = \frac{1}{2}(\Gamma^i_{jk} - \Gamma^i_{kj}) = T^i_{jk}. \tag{3}$$

The magnitude T^i_{jk} is the torsion tensor of the space GR_N.

In a generalized Riemannian space, one can define four kinds of covariant derivatives [9]. For example, for a tensor a^i_j in GR_N we have

$$\begin{aligned}
a^i_{j\underset{1}{|}m} &= a^i_{j,m} + \Gamma^i_{pm} a^p_j - \Gamma^p_{jm} a^i_p, \\
a^i_{j\underset{2}{|}m} &= a^i_{j,m} + \Gamma^i_{mp} a^p_j - \Gamma^p_{mj} a^i_p, \\
a^i_{j\underset{3}{|}m} &= a^i_{j,m} + \Gamma^i_{pm} a^p_j - \Gamma^p_{mj} a^i_p, \\
a^i_{j\underset{4}{|}m} &= a^i_{j,m} + \Gamma^i_{mp} a^p_j - \Gamma^p_{jm} a^i_p,
\end{aligned} \tag{4}$$

where $\underset{\theta}{|}$ $(\theta = 1, 2, 3, 4)$ denotes a covariant derivative of the kind θ and $a^i_{j,m} = \dfrac{\partial a^i_j}{\partial x^m}$.

In the case of the space GR_N, we have twelve curvature tensors, and S. M. Minčić proved that there are five independent ones. In this paper, we will consider the following five independent curvature tensors [9]:

$$\begin{aligned}
\underset{1}{R}{}^i_{jmn} &= \Gamma^i_{jm,n} - \Gamma^i_{jn,m} + \Gamma^p_{jm}\Gamma^i_{pn} - \Gamma^p_{jn}\Gamma^i_{pm}, \\
\underset{2}{R}{}^i_{jmn} &= \Gamma^i_{mj,n} - \Gamma^i_{nj,m} + \Gamma^p_{mj}\Gamma^i_{np} - \Gamma^p_{nj}\Gamma^i_{mp}, \\
\underset{3}{R}{}^i_{jmn} &= \Gamma^i_{jm,n} - \Gamma^i_{nj,m} + \Gamma^p_{jm}\Gamma^i_{np} - \Gamma^p_{nj}\Gamma^i_{pm} + \Gamma^p_{mn}(\Gamma^i_{pj} - \Gamma^i_{jp}), \\
\underset{4}{R}{}^i_{jmn} &= \Gamma^i_{jm,n} - \Gamma^i_{nj,m} + \Gamma^p_{jm}\Gamma^i_{np} - \Gamma^p_{nj}\Gamma^i_{pm} + \Gamma^p_{nm}(\Gamma^i_{pj} - \Gamma^i_{jp}), \\
\underset{5}{R}{}^i_{jmn} &= \frac{1}{2}(\Gamma^i_{jm,n} + \Gamma^i_{mj,n} - \Gamma^i_{jn,m} - \Gamma^i_{nj,m} \\
&\quad + \Gamma^p_{jm}\Gamma^i_{pn} - \Gamma^p_{jn}\Gamma^i_{mp} + \Gamma^p_{mj}\Gamma^i_{pn} - \Gamma^p_{nj}\Gamma^i_{pm}).
\end{aligned} \tag{5}$$

Let GR_N and $G\overline{R}_N$ be two generalized Riemannian spaces. We will observe these spaces in the common system of coordinates defined by the mapping $f : GR_N \to G\overline{R}_N$. If Γ^h_{ij} and $\overline{\Gamma}^h_{ij}$ are connection coefficients of the spaces GR_N and $G\overline{R}_N$, respectively, then

$$P^h_{ij} = \overline{\Gamma}^h_{ij} - \Gamma^h_{ij} \tag{6}$$

is the deformation tensor of the connection for a mapping f.

The relations between the corresponding curvature tensors of the spaces GR_N and $G\overline{R}_N$ are obtained in [19] as follows:

$$\begin{aligned}
\underset{1}{\overline{R}}{}^i_{jmn} &= \underset{1}{R}{}^i_{jmn} + P^i_{jm\underset{1}{|}n} - P^i_{jn\underset{1}{|}m} + P^p_{jm}P^i_{pn} - P^p_{jn}P^i_{pm} + 2T^p_{mn}P^i_{jp},\\
\underset{2}{\overline{R}}{}^i_{jmn} &= \underset{2}{R}{}^i_{jmn} + P^i_{mj\underset{2}{|}n} - P^i_{nj\underset{2}{|}m} + P^p_{mj}P^i_{np} - P^p_{nj}P^i_{mp} + 2T^p_{nm}P^i_{pj},\\
\underset{3}{\overline{R}}{}^i_{jmn} &= \underset{3}{R}{}^i_{jmn} + P^i_{jm\underset{2}{|}n} - P^i_{nj\underset{1}{|}m} + P^p_{jm}P^i_{np} - P^p_{nj}P^i_{pm} + 2P^p_{nm}(T^i_{pj} + \underset{\vee}{P}{}^i_{pj}),\\
\underset{4}{\overline{R}}{}^i_{jmn} &= \underset{4}{R}{}^i_{jmn} + P^i_{jm\underset{2}{|}n} - P^i_{nj\underset{1}{|}m} + P^p_{jm}P^i_{np} - P^p_{nj}P^i_{pm} + 2P^p_{mn}(T^i_{pj} + \underset{\vee}{P}{}^i_{pj}),\\
\underset{5}{\overline{R}}{}^i_{jmn} &= \underset{5}{R}{}^i_{jmn} + \frac{1}{2}(P^i_{jm\underset{3}{|}n} - P^i_{jn\underset{4}{|}m} + P^i_{mj\underset{4}{|}n} - P^i_{nj\underset{3}{|}m}\\
&\quad + P^p_{jm}P^i_{pn} - P^p_{jn}P^i_{mp} + P^p_{mj}P^i_{np} - P^p_{nj}P^i_{pm}),
\end{aligned} \tag{7}$$

where P^h_{ij} is a deformation tensor for a mapping f, $\underset{\vee}{P}{}^h_{ij}$ is its antisymmetric part and T^h_{ij} is a torsion tensor.

2. Quasi-Canonical Biholomorphically Projective Mappings

In paper [20], we define biholomorphically projective mappings between two generalized Riemannian spaces GR_N and $G\overline{R}_N$ with almost complex structures that are equal in a common system of coordinates defined by the mapping $f : GR_N \to G\overline{R}_N$. We have considered a generalized Riemannian space GR_N with a non-symmetric metric tensor g_{ij} and almost complex structure F^h_i such that $F^h_i \neq a\delta^h_i$, where a is scalar invariant, and we have defined the biholomorphically projective curve of the kind θ $(\theta = 1,2)$ and the biholomorphically projective mapping of the kind θ $(\theta = 1,2)$.

Definition 1 ([20]). *In the space GR_N, a curve l given in parametric form*

$$x^i = x^i(t),\ (i = 1, ..., N),$$

is said to be biholomorphically projective of the kind θ $(\theta = 1,2)$ if it satisfies the following equation:

$$\lambda^h{}_{\underset{\theta}{|}p}(t)\lambda^p(t) = a(t)\lambda^h(t) + b(t)F^h_p\lambda^p(t) + c(t)\overset{2}{F}{}^h_p\lambda^p(t),$$

where a, b and c are functions of parameter t, $\lambda^i = \dfrac{dx^i}{dt}$ and $\overset{2}{F}{}^h_p = F^h_q F^q_p$.

Definition 2 ([20]). *A diffeomorphism $f : GR_N \to G\overline{R}_N$ is a biholomorphically projective mapping of the kind θ $(\theta = 1,2)$ if biholomorphically projective curves of the kind θ $(\theta = 1,2)$ of the space GR_N are mapped to the biholomorphically projective curves of the kind θ of the space $G\overline{R}_N$.*

Since it holds [20]

$$\lambda^h{}_{\underset{1}{|}p}\lambda^p = \frac{d\lambda^h}{dt} + \Gamma^h_{pq}\lambda^p\lambda^q = \lambda^h{}_{\underset{2}{|}p}\lambda^p,$$

we conclude that the biholomorphically projective curves of the first kind and the biholomorphically projective curves of the second kind match, so we will simply call them the biholomorphically projective curves. Therefore, the biholomorphically projective curves of the spaces GR_N and $G\overline{R}_N$, respectively, satisfy relations [20]

$$\frac{d\lambda^h}{dt} + \Gamma^h_{pq}\lambda^p\lambda^q = a\lambda^h + bF^h_p\lambda^p + c\overset{2}{F}{}^h_p\lambda^p, \tag{8}$$

$$\frac{d\lambda^h}{dt} + \overline{\Gamma}^h_{pq}\lambda^p\lambda^q = \overline{a}\lambda^h + \overline{b}F^h_p\lambda^p + \overline{c}\overset{2}{F}{}^h_p\lambda^p, \tag{9}$$

where a, b, c, $\overline{a}$, $\overline{b}$ and $\overline{c}$ are functions of parameter t, $\lambda^i = \frac{dx^i}{dt}$, and Γ^h_{pq} and $\overline{\Gamma}^h_{pq}$ are connection coeficients of the spaces GR_N and $G\overline{R}_N$, respectively, $\overset{2}{F}{}^h_p = F^h_q F^q_p$.

From Equations (8) and (9) we obtain [20]

$$(\overline{\Gamma}^h_{pq} - \Gamma^h_{pq})\lambda^p\lambda^q = \psi\lambda^h + \sigma F^h_p\lambda^p + \tau\overset{2}{F}{}^h_p\lambda^p,$$

where we denote $\psi = \overline{a} - a, \sigma = \overline{b} - b, \tau = \overline{c} - c$. We can set $\psi = \psi_p\lambda^p, \sigma = \sigma_p\lambda^p, \tau = \tau_p\lambda^p$. Now, we have [20]

$$(\overline{\Gamma}^h_{pq} - \Gamma^h_{pq} - \psi_p\delta^h_q - \sigma_p F^h_q - \tau_p\overset{2}{F}{}^h_q)\lambda^p\lambda^q = 0.$$

From this we conclude that the following relation is satisfied [20]:

$$\overline{\Gamma}^h_{ij} = \Gamma^h_{ij} + \psi_{(i}\delta^h_{j)} + \sigma_{(i}F^h_{j)} + \tau_{(i}\overset{2}{F}{}^h_{j)} + \xi^h_{ij}, \tag{10}$$

and the deformation tensor has the form

$$P^h_{ij} = \psi_{(i}\delta^h_{j)} + \sigma_{(i}F^h_{j)} + \tau_{(i}\overset{2}{F}{}^h_{j)} + \xi^h_{ij}, \tag{11}$$

where (ij) is a symmetrization without division by indices i, j; ψ_i, σ_i and τ_i are vectors; $\overset{2}{F}{}^h_p = F^h_q F^q_p$ and ξ^h_{ij} is an antisymmetric tensor.

Inspired by the form of the deformation tensor (11), we will define a new type of mapping. Let GR_N and $G\overline{R}_N$ be two generalized Riemannian spaces with almost complex structures F^h_i and $\overline{F}^h_i$, respectively, where $F^h_i = \overline{F}^h_i$ in the common system of coordinates defined by the mapping $f : GR_N \to G\overline{R}_N$, and assume that it holds that $F^h_i \neq a\delta^h_i$, where a is scalar invariant.

The mapping $f : GR_N \to G\overline{R}_N$ is a quasi-canonical biholomorphically projective mapping if in the common coordinate system the connection coefficients Γ^h_{ij} and $\overline{\Gamma}^h_{ij}$ satisfy the relation

$$\overline{\Gamma}^h_{ij} = \Gamma^h_{ij} + \psi_{(i}\delta^h_{j)} + \tau_{(i}\overset{2}{F}{}^h_{j)} + \xi^h_{ij}, \tag{12}$$

where (ij) is a symmetrization without division by indices i, j; ψ_i and τ_i are vectors; $\overset{2}{F}{}^h_p = F^h_q F^q_p$ and ξ^h_{ij} is an antisymmetric tensor.

Let P^h_{ij} be a deformation tensor with respect to the quasi-canonical biholomorphically projective mapping $f : GR_N \to G\overline{R}_N$. Then, from (6) and (12), we have

$$P^h_{ij} = \psi_{(i}\delta^h_{j)} + \tau_{(i}\overset{2}{F}{}^h_{j)} + \xi^h_{ij}. \tag{13}$$

3. Some Relations between Curvature Tensors

In this section, we will find the relations between the corresponding curvature tensors of the spaces GR_N and $G\overline{R}_N$.

According to relations (5), (7) and (13), for the curvature tensor of the first kind we have

$$
\begin{aligned}
\underset{1}{\overline{R}}{}^i_{jmn} &= \underset{1}{R}{}^i_{jmn} + \underset{1}{\psi}_{j<n}\delta^i_{m>} + \underset{1}{\psi}_{<m|n>}\delta^i_j + \psi_p\xi^p_{j<m}\delta^i_{n>} + 2\psi_j\xi^i_{mn} \\
&+ \tau_j\tau_{<n}\overset{4}{F}{}^i_{m>} + \tau_m\overset{2}{F}{}^i_{j\underset{1}{|}n} - \tau_n\overset{2}{F}{}^i_{j\underset{1}{|}m} + \tau_{j\underset{1}{|}<n}\overset{2}{F}{}^i_{m>} + \tau_j\overset{2}{F}{}^i_{<m\underset{1}{|}n>} + \tau_{<m\underset{1}{|}n>}\overset{2}{F}{}^i_j \\
&+ \tau_p\tau_j\overset{2}{F}{}^p_{<m}\overset{2}{F}{}^i_{n>} + \tau_p\tau_{<m}\overset{2}{F}{}^i_{n>}\overset{2}{F}{}^p_j + \xi^p_{jm}\xi^i_{pn} - \xi^p_{jn}\xi^i_{pm} + \xi^i_{j<m\underset{1}{|}n>} + \tau_{(p}\overset{2}{F}{}^i_{n)}\xi^p_{jm} \\
&+ \tau_{(j}\overset{2}{F}{}^p_{m)}\xi^i_{pn} - \tau_{(j}\overset{2}{F}{}^p_{n)}\xi^i_{pm} - \tau_{(p}\overset{2}{F}{}^i_{m)}\xi^p_{jn} + 2T^p_{mn}(\psi_{(j}\delta^i_{p)} + \tau_{(j}\overset{2}{F}{}^i_{p)} + \xi^i_{jp}),
\end{aligned}
\tag{14}
$$

where (ij) is a symmetrization without division , $< ij >$ is an antisymmetrization without division by indices i, j and

$$
\begin{aligned}
&\overset{2}{F}{}^h_j = F^h_pF^p_j,\ \overset{3}{F}{}^h_j = F^h_pF^p_qF^q_j,\ \overset{4}{F}{}^h_j = F^h_pF^p_qF^q_rF^r_j, \\
&\underset{1}{\psi}_{jn} = \psi_{j\underset{1}{|}n} - \psi_j\psi_n - \psi_p\tau_{(j}\overset{2}{F}{}^p_{n)}.
\end{aligned}
\tag{15}
$$

Based on the facts given above, we have obtained the following statement.

Theorem 1. *A quasi-canonical biholomorphically projective relation between the curvature tensors of the first kind of the generalized Riemannian spaces GR_N and $G\overline{R}_N$ is given by Formula (14), where T^h_{ij} is the torsion tensor and the notation is the same as in (15).*

From relations (5), (7) and (13), for the curvature tensor of the second kind, we obtain the following:

$$
\begin{aligned}
\underset{2}{\overline{R}}{}^i_{jmn} &= \underset{2}{R}{}^i_{jmn} + \underset{2}{\psi}_{j<n}\delta^i_{m>} + \underset{2}{\psi}_{<m|n>}\delta^i_j - \psi_p(\xi^p_{nj}\delta^i_m - \xi^p_{mj}\delta^i_n) \\
&+ 2\psi_j\xi^i_{nm} + \tau_j\tau_{<n}\overset{4}{F}{}^i_{m>} + \tau_m\overset{2}{F}{}^i_{j\underset{2}{|}n} - \tau_n\overset{2}{F}{}^i_{j\underset{2}{|}m} + \tau_{j\underset{2}{|}<n}\overset{2}{F}{}^i_{m>} \\
&+ \tau_j\overset{2}{F}{}^i_{<m\underset{2}{|}n>} + \tau_{<m\underset{2}{|}n>}\overset{2}{F}{}^i_j + \tau_p\tau_j\overset{2}{F}{}^p_{<m}\overset{2}{F}{}^i_{n>} + \tau_p\tau_{<m}\overset{2}{F}{}^i_{n>}\overset{2}{F}{}^p_j \\
&+ \tau_{(p}\overset{2}{F}{}^i_{n)}\xi^p_{mj} + \tau_{(j}\overset{2}{F}{}^p_{m)}\xi^i_{np} - \tau_{(j}\overset{2}{F}{}^p_{n)}\xi^i_{mp} - \tau_{(p}\overset{2}{F}{}^i_{m)}\xi^p_{nj} \\
&+ \xi^p_{mj}\xi^i_{np} - \xi^p_{nj}\xi^i_{mp} + \xi^i_{mj\underset{2}{|}n} - \xi^i_{nj\underset{2}{|}m} + 2T^p_{mn}(\psi_{(j}\delta^i_{p)} + \tau_{(j}\overset{2}{F}{}^i_{p)} + \xi^i_{pj}),
\end{aligned}
\tag{16}
$$

where $\overset{2}{F}{}^h_j$, $\overset{4}{F}{}^h_j$, are determined by Formula (15) and

$$
\underset{2}{\psi}_{jn} = \psi_{j\underset{2}{|}n} - \psi_j\psi_n - \psi_p\tau_{(j}\overset{2}{F}{}^p_{n)}.
\tag{17}
$$

Therefore, the following theorem is valid.

Theorem 2. *A quasi-canonical biholomorphically projective relation between the curvature tensors of the second kind of the generalized Riemannian spaces GR_N and $G\overline{R}_N$ is given by Formula (16), where T^h_{ij} is the torsion tensor and the notation is the same as in (15) and (17).*

Considering relations (5), (7) and (13), for the curvature tensor of the third kind we have the following:

$$\begin{aligned}
\underset{3}{\overline{R}}{}^i_{jmn} &= \underset{3}{R}{}^i_{jmn} + \psi_{\underset{2}{j}n}\delta^i_m - \psi_{\underset{1}{j}m}\delta^i_n + (\psi_{m\underset{2}{|}n} - \psi_{n\underset{1}{|}m})\delta^i_j + 2\psi_j\xi^i_{nm} \\
&- \psi_p(\xi^p_{nj}\delta^i_m - \xi^p_{mj}\delta^i_n) + \tau_j\tau_{<n}\overset{4}{F}{}^i_{m>} + \tau_m\overset{2}{F}{}^i_{j\underset{2}{|}n} - \tau_n\overset{2}{F}{}^i_{j\underset{1}{|}m} \\
&+ \tau_{j\underset{2}{|}n}\overset{2}{F}{}^i_m - \tau_{j\underset{1}{|}m}\overset{2}{F}{}^i_n + \tau_j(\overset{2}{F}{}^i_{m\underset{2}{|}n} - \overset{2}{F}{}^i_{n\underset{1}{|}m}) + (\tau_{m\underset{2}{|}n} - \tau_{n\underset{1}{|}m})\overset{2}{F}{}^i_j \\
&+ \tau_p\tau_j\overset{2}{F}{}^p_{<m}\overset{2}{F}{}^i_{n>} + \tau_p\tau_{<m}\overset{2}{F}{}^i_{n>}\overset{2}{F}{}^p_j + \xi^p_{jm}\xi^i_{np} - \xi^p_{nj}\xi^i_{mp} - \xi^i_{jm\underset{2}{|}n} - \xi^i_{nj\underset{1}{|}m} \\
&+ \tau_{(p}\overset{2}{F}{}^i_{n)}\xi^p_{mj} + \tau_{(j}\overset{2}{F}{}^p_{m)}\xi^i_{np} - \tau_{(j}\overset{2}{F}{}^p_{n)}\xi^i_{pm} - \tau_{(p}\overset{2}{F}{}^i_{m)}\xi^p_{nj} \\
&+ 2(\psi_{(n}\delta^p_{m)} + \tau_{(n}\overset{2}{F}{}^p_{m)} + \xi^p_{nm})(T^i_{pj} + \xi^i_{pj}),
\end{aligned} \tag{18}$$

where the notation is the same as in (15) and (17).

In this way, the following theorem is proven.

Theorem 3. *A quasi-canonical biholomorphically projective relation between the curvature tensors of the third kind of the generalized Riemannian spaces* GR_N *and* $G\overline{R}_N$ *is given by Formula* (18)*, where* T^h_{ij} *is the torsion tensor and the notation is the same as in* (15) *and* (17).

Using relations (5), (7) and (13), for a curvature tensor of the fourth kind we obtain the following:

$$\begin{aligned}
\underset{4}{\overline{R}}{}^i_{jmn} &= \underset{4}{R}{}^i_{jmn} + \psi_{\underset{2}{j}n}\delta^i_m - \psi_{\underset{1}{j}m}\delta^i_n + (\psi_{m\underset{2}{|}n} - \psi_{n\underset{1}{|}m})\delta^i_j + 2\psi_j\xi^i_{nm} \\
&- \psi_p(\xi^p_{nj}\delta^i_m - \xi^p_{mj}\delta^i_n) + \tau_j\tau_{<n}\overset{4}{F}{}^i_{m>} + \tau_m\overset{2}{F}{}^i_{j\underset{2}{|}n} - \tau_n\overset{2}{F}{}^i_{j\underset{1}{|}m} \\
&+ \tau_j(\overset{2}{F}{}^i_{m\underset{2}{|}n} - \overset{2}{F}{}^i_{n\underset{1}{|}m}) + \tau_{j\underset{2}{|}n}\overset{2}{F}{}^i_m - \tau_{j\underset{1}{|}m}\overset{2}{F}{}^i_n + (\tau_{m\underset{2}{|}n} - \tau_{n\underset{1}{|}m})\overset{2}{F}{}^i_j \\
&+ \tau_p\tau_j\overset{2}{F}{}^p_{<m}\overset{2}{F}{}^i_{n>} + \tau_p\tau_{<m}\overset{2}{F}{}^i_{n>}\overset{2}{F}{}^p_j + \xi^p_{jm}\xi^i_{np} - \xi^p_{nj}\xi^i_{mp} + \xi^i_{jm\underset{2}{|}n} - \xi^i_{nj\underset{1}{|}m} \\
&+ \tau_{(p}\overset{2}{F}{}^i_{n)}\xi^p_{mj} + \tau_{(j}\overset{2}{F}{}^p_{m)}\xi^i_{np} - \tau_{(j}\overset{2}{F}{}^p_{n)}\xi^i_{pm} - \tau_{(p}\overset{2}{F}{}^i_{m)}\xi^p_{nj} \\
&+ 2(\psi_{(n}\delta^p_{m)} + \tau_{(n}\overset{2}{F}{}^p_{m)} + \xi^p_{mn})(T^i_{pj} + \xi^i_{pj}),
\end{aligned} \tag{19}$$

where the notation is the same as in (15) and (17). This proves the next statement.

Theorem 4. *A quasi-canonical biholomorphically projective relation between the curvature tensors of the fourth kind of the generalized Riemannian spaces* GR_N *and* $G\overline{R}_N$ *is given by Formula* (19)*, where* T^h_{ij} *is the torsion tensor and the notation is the same as in* (15) *and* (17).

Considering relations (5), (7) and (13), for the curvature tensor of the fifth kind we have the following:

$$
\begin{aligned}
\underset{5}{\overline{R}}{}^{i}{}_{jmn} &= \underset{5}{R}{}^{i}{}_{jmn} + 2\psi_j T^i_{mn} + 2\psi_n T^i_{mj} + \frac{1}{2}(\underset{3}{\psi}_{jn} + \underset{4}{\psi}_{jn})\delta^i_m \\
&+ \frac{1}{2}(\psi_{<m\underset{3}{|}n>} + \psi_{<m\underset{4}{|}n>})\delta^i_j - \frac{1}{2}(\underset{3}{\psi}_{jm} + \underset{4}{\psi}_{jm})\delta^i_n \\
&+ \frac{1}{2}\overset{2}{F}{}^i_j(\tau_{<m\underset{3}{|}n>} + \tau_{<m\underset{4}{|}n>}) + \frac{1}{2}\overset{2}{F}{}^i_m(\tau_{j\underset{3}{|}n} + \tau_{j\underset{4}{|}n}) - \frac{1}{2}\overset{2}{F}{}^i_n(\tau_{j\underset{3}{|}m} + \tau_{j\underset{4}{|}m}) \\
&- \frac{1}{2}\tau_n(\overset{2}{F}{}^i_{j\underset{3}{|}m} + \overset{2}{F}{}^i_{j\underset{4}{|}m}) + \frac{1}{2}\tau_m(\overset{2}{F}{}^i_{j\underset{3}{|}n} + \overset{2}{F}{}^i_{j\underset{4}{|}n}) + \frac{1}{2}\tau_j(\overset{2}{F}{}^i{}_{<m\underset{3}{|}n>} + \overset{2}{F}{}^i{}_{<m\underset{4}{|}n>}) \\
&+ \tau_p\tau_j\overset{2}{F}{}^p_{<m}\overset{2}{F}{}^i_{n>} + \tau_p\tau_{<m}\overset{2}{F}{}^i_{n>}\overset{2}{F}{}^p_j + \tau_j\tau_{<n}\overset{4}{F}{}^i_{m>} + \xi^p_{jm}\xi^i_{pn} - \xi^p_{jn}\xi^i_{mp} \\
&+ \frac{1}{2}(\xi^i_{jm\underset{3}{|}n} - \xi^i_{nj\underset{3}{|}m} - \xi^i_{jn\underset{4}{|}m} + \xi^i_{mj\underset{4}{|}n}),
\end{aligned}
\tag{20}
$$

where the notation is the same as in (15) and

$$
\begin{aligned}
\underset{3}{\psi}_{jn} &= \psi_{j\underset{3}{|}n} - \psi_j\psi_n + \psi_p\tau_{(j}\overset{2}{F}{}^p_{n)}, \\
\underset{4}{\psi}_{jn} &= \psi_{j\underset{4}{|}n} - \psi_j\psi_n + \psi_p\tau_{(j}\overset{2}{F}{}^p_{n)}.
\end{aligned}
\tag{21}
$$

Based on the facts given above, we have proved the next theorem related to curvature tensors of the fifth kind.

Theorem 5. *A quasi-canonical biholomorphically projective relation between the curvature tensors of the fifth kind of the generalized Riemannian spaces GR_N and $G\overline{R}_N$ is given by Formula (20), where T^h_{ij} is the torsion tensor and the notation is the same as in (15) and (21).*

4. Equitorsion Quasi-Canonical Biholomorphically Projective Mapping

The mapping $f : GR_N \to G\overline{R}_N$ is an equitorsion quasi-canonical biholomorphically projective mapping,if the torsion tensors of the spaces GR_N and $G\overline{R}_N$ are equal in a common coordinate system after the mapping f. In this case, based on (6) and (13), we conclude that

$$\xi^h_{ij} = 0. \tag{22}$$

Then, relation (13) becomes

$$P^h_{ij} = \psi_{(i}\delta^h_{j)} + \tau_{(i}\overset{2}{F}{}^h_{j)}. \tag{23}$$

Considering (22), from (14), we obtain the following:

$$
\begin{aligned}
\underset{1}{\overline{R}}{}^{i}{}_{jmn} &= \underset{1}{R}{}^{i}{}_{jmn} + \underset{1}{\psi}_{j<n}\delta^i_{m>} + \psi_{<m\underset{1}{|}n>}\delta^i_j + \tau_j\tau_{<n}\overset{4}{F}{}^i_{m>} \\
&+ \tau_m\overset{2}{F}{}^i_{j\underset{1}{|}n} - \tau_n\overset{2}{F}{}^i_{j\underset{1}{|}m} + \tau_{j\underset{1}{|}<n}\overset{2}{F}{}^i_{m>} + \tau_j\overset{2}{F}{}^i{}_{<m\underset{1}{|}n>} + \tau_{<m\underset{1}{|}n>}\overset{2}{F}{}^i_j \\
&+ \tau_p\tau_j\overset{2}{F}{}^p_{<m}\overset{2}{F}{}^i_{n>} + \tau_p\tau_{<m}\overset{2}{F}{}^i_{n>}\overset{2}{F}{}^p_j + 2T^p_{mn}(\psi_{(j}\delta^i_{p)} + \tau_{(j}\overset{2}{F}{}^i_{p)}),
\end{aligned}
\tag{24}
$$

Hence, the next theorem holds.

Theorem 6. *An equitorsion quasi-canonical biholomorphically projective relation between the curvature tensors of the first kind of the generalized Riemannian spaces GR_N and $G\overline{R}_N$ is given by Formula (24), where T^h_{ij} is the torsion tensor and the notation is the same as in (15).*

The relation between the curvature tensors of the second kind (16), after applying relation (22), becomes the following:

$$\begin{aligned}\underset{2}{\overline{R}}{}^{i}{}_{jmn} &= \underset{2}{R}{}^{i}{}_{jmn} + \underset{2}{\psi}_{j<n}\delta^{i}_{m>} + \psi_{<m\underset{2}{|}n>}\delta^{i}_{j} + \tau_j\tau_{<n}\overset{4}{F}{}^{i}_{m>} \\ &+ \tau_m\overset{2}{F}{}^{i}_{j\underset{2}{|}n} - \tau_n\overset{2}{F}{}^{i}_{j\underset{2}{|}m} + \tau_{j\underset{2}{|}<n}\overset{2}{F}{}^{i}_{m>} + \tau_j\overset{2}{F}{}^{i}_{<m\underset{2}{|}n>} + \tau_{<m\underset{2}{|}n>}\overset{2}{F}{}^{i}_{j} \\ &+ \tau_p\tau_j\overset{2}{F}{}^{p}_{<m}\overset{2}{F}{}^{i}_{n>} + \tau_p\tau_{<m}\overset{2}{F}{}^{i}_{n>}\overset{2}{F}{}^{p}_{j} + 2T^{p}_{mn}(\psi_{(j}\delta^{i}_{p)} + \tau_{(j}\overset{2}{F}{}^{i}_{p)} + \xi^{i}_{pj}).\end{aligned} \tag{25}$$

In this way, the following theorem is proven.

Theorem 7. *An equitorsion quasi-canonical biholomorphically projective relation between the curvature tensors of the second kind of the generalized Riemannian spaces GR_N and $G\overline{R}_N$ is given by Formula (25), where T^h_{ij} is the torsion tensor.*

The relation between the curvature tensors of the third kind (16), with respect to (22), becomes the following:

$$\begin{aligned}\underset{3}{\overline{R}}{}^{i}{}_{jmn} &= \underset{3}{R}{}^{i}{}_{jmn} + \underset{2}{\psi}_{jn}\delta^{i}_{m} - \underset{1}{\psi}_{jm}\delta^{i}_{n} + (\psi_{m\underset{2}{|}n} - \psi_{n\underset{1}{|}m})\delta^{i}_{j} + \tau_j\tau_{<n}\overset{4}{F}{}^{i}_{m>} \\ &+ \tau_m\overset{2}{F}{}^{i}_{j\underset{2}{|}n} - \tau_n\overset{2}{F}{}^{i}_{j\underset{1}{|}m} + \tau_{j\underset{2}{|}n}\overset{2}{F}{}^{i}_{m} - \tau_{j\underset{1}{|}m}\overset{2}{F}{}^{i}_{n} + \tau_j(\overset{2}{F}{}^{i}_{m\underset{2}{|}n} - \overset{2}{F}{}^{i}_{n\underset{1}{|}m}) \\ &+ (\tau_{m\underset{2}{|}n} - \tau_{n\underset{1}{|}m})\overset{2}{F}{}^{i}_{j} + \tau_p\tau_j\overset{2}{F}{}^{p}_{<m}\overset{2}{F}{}^{i}_{n>} + \tau_p\tau_{<m}\overset{2}{F}{}^{i}_{n>}\overset{2}{F}{}^{p}_{j} \\ &+ 2(\psi_{(n}\delta^{p}_{m)} + \tau_{(n}\overset{2}{F}{}^{p}_{m)})T^{i}_{pj},\end{aligned} \tag{26}$$

and we may formulate the following theorem.

Theorem 8. *An equitorsion quasi-canonical biholomorphically projective relation between the curvature tensors of the third kind of the generalized Riemannian spaces GR_N and $G\overline{R}_N$ is given by Formula (26), where T^h_{ij} is the torsion tensor and the notation is the same as in (15) and (17).*

In particular, from relations (19) and (22), we have

$$\begin{aligned}\underset{4}{\overline{R}}{}^{i}{}_{jmn} &= \underset{4}{R}{}^{i}{}_{jmn} + \underset{2}{\psi}_{jn}\delta^{i}_{m} - \underset{1}{\psi}_{jm}\delta^{i}_{n} + (\psi_{m\underset{2}{|}n} - \psi_{n\underset{1}{|}m})\delta^{i}_{j} + \tau_j\tau_{<n}\overset{4}{F}{}^{i}_{m>} \\ &+ \tau_m\overset{2}{F}{}^{i}_{j\underset{2}{|}n} - \tau_n\overset{2}{F}{}^{i}_{j\underset{1}{|}m} + \tau_j(\overset{2}{F}{}^{i}_{m\underset{2}{|}n} - \overset{2}{F}{}^{i}_{n\underset{1}{|}m}) + \tau_{j\underset{2}{|}n}\overset{2}{F}{}^{i}_{m} - \tau_{j\underset{1}{|}m}\overset{2}{F}{}^{i}_{n} \\ &+ (\tau_{m\underset{2}{|}n} - \tau_{n\underset{1}{|}m})\overset{2}{F}{}^{i}_{j} + \tau_p\tau_j\overset{2}{F}{}^{p}_{<m}\overset{2}{F}{}^{i}_{n>} + \tau_p\tau_{<m}\overset{2}{F}{}^{i}_{n>}\overset{2}{F}{}^{p}_{j} \\ &+ 2(\psi_{(n}\delta^{p}_{m)} + \tau_{(n}\overset{2}{F}{}^{p}_{m)})T^{i}_{pj}.\end{aligned} \tag{27}$$

Therefore,the next theorem holds.

Theorem 9. *An equitorsion quasi-canonical biholomorphically projective relation between the curvature tensors of the fourth kind of the generalized Riemannian spaces GR_N and $G\overline{R}_N$ is given by Formula (27), where T^h_{ij} is the torsion tensor and the notation is the same as in (15) and (17).*

Analogously, from (20), with respect to (22), we obtain the following:

$$\begin{aligned}
\overline{\underset{5}{R}}{}^{i}{}_{jmn} &= \underset{5}{R}{}^{i}{}_{jmn} + 2\psi_j T^i_{mn} + 2\psi_n T^i_{mj} + \frac{1}{2}(\underset{3}{\psi}_{jn} + \underset{4}{\psi}_{jn})\delta^i_m \\
&+ \frac{1}{2}(\underset{3}{\psi}_{<m|n>} + \underset{4}{\psi}_{<m|n>})\delta^i_j - \frac{1}{2}(\underset{3}{\psi}_{jm} + \underset{4}{\psi}_{jm})\delta^i_n \\
&+ \frac{1}{2}\overset{2}{F}{}^i_j(\underset{3}{\tau}_{<m|n>} + \underset{4}{\tau}_{<m|n>}) + \frac{1}{2}\overset{2}{F}{}^i_m(\underset{3}{\tau}_{j|n} + \underset{4}{\tau}_{j|n}) - \frac{1}{2}\overset{2}{F}{}^i_n(\underset{3}{\tau}_{j|m} + \underset{4}{\tau}_{j|m}) \\
&- \frac{1}{2}\tau_n(\overset{2}{F}{}^i_{j\underset{3}{|}m} + \overset{2}{F}{}^i_{j\underset{4}{|}m}) + \frac{1}{2}\tau_m(\overset{2}{F}{}^i_{j\underset{3}{|}n} + \overset{2}{F}{}^i_{j\underset{4}{|}n}) + \frac{1}{2}\tau_j(\overset{2}{F}{}^i_{<m\underset{3}{|}n>} + \overset{2}{F}{}^i_{<m\underset{4}{|}n>}) \\
&+ \tau_p\tau_j \overset{2}{F}{}^p_{<m}\overset{2}{F}{}^i_{n>} + \tau_p\tau_{<m}\overset{2}{F}{}^i_{n>}\overset{2}{F}{}^p_j + \tau_j\tau_{<n}\overset{4}{F}{}^i_{m>},
\end{aligned} \tag{28}$$

i.e., the following theorem is valid:

Theorem 10. *An equitorsion quasi-canonical biholomorphically projective relation between the curvature tensors of the fifth kind of the generalized Riemannian spaces GR_N and $G\overline{R}_N$ is given by Formula* (28), *where T^h_{ij} is the torsion tensor and the notation is the same as in* (15) *and* (21).

5. Invariant Geometric Objects of Quasi-Canonical Biholomorphically Projective Mappings

In this section, we will obtain an invariant geometric object of an equitorsion quasi-canonical biholomorphically projective mapping. In relation to that, in relation (23), let us set

$$\tau_i = \psi_p F^p_i.$$

Then, we have

$$\overline{\Gamma}^h_{ij} - \Gamma^h_{ij} = \psi_{(i}\delta^h_{j)} + \psi_p F^p_{(i}\overset{2}{F}{}^h_{j)}. \tag{29}$$

Contractingby indices h and i in (29), assuming that it is valid that

$$Tr(F^2) = 0, \text{ i.e., } \overset{2}{F}{}^p_p = F^p_q F^q_p = 0 \tag{30}$$

and

$$\overset{3}{F}{}^h_j = F^h_p F^p_q F^q_j = e\delta^h_j \ \ (e = \pm 1, 0), \tag{31}$$

we obtain

$$\psi_j = \frac{1}{N+1+e}(\overline{\Gamma}^p_{\underline{pj}} - \Gamma^p_{\underline{pj}}). \tag{32}$$

Substituting (32) in (29) we have

$$\begin{aligned}
&\overline{\Gamma}^h_{ij} - \frac{1}{N+1+e}\left(\overline{\Gamma}^p_{\underline{pi}}\delta^h_j + \overline{\Gamma}^p_{\underline{pj}}\delta^h_i + \overline{\Gamma}^q_{\underline{qp}}F^p_{(i}\overset{2}{F}{}^h_{j)}\right) \\
&= \Gamma^h_{ij} - \frac{1}{N+1+e}\left(\Gamma^p_{\underline{pi}}\delta^h_j + \Gamma^p_{\underline{pj}}\delta^h_i + \Gamma^q_{\underline{qp}}F^p_{(i}\overset{2}{F}{}^h_{j)}\right).
\end{aligned} \tag{33}$$

If we denote

$$\mathcal{QCT}^h_{ij} = \Gamma^h_{ij} - \frac{1}{N+1+e}\left(\Gamma^p_{\underline{pi}}\delta^h_j + \Gamma^p_{\underline{pj}}\delta^h_i + \Gamma^q_{\underline{qp}}F^p_{(i}\overset{2}{F}{}^h_{j)}\right), \tag{34}$$

relation (33) can be presented in the form

$$\overline{\mathcal{QCT}}{}^h_{ij} = \mathcal{QCT}^h_{ij}, \tag{35}$$

where $\overline{QCT}^h_{ij}$ is an object of the space $G\overline{R}_N$. The magnitude QCT^h_{ij} is called a *Thomas equitorsion quasi-canonical biholomorphically projective parameter*and it is not a tensor.

Accordingly, we conclude that the following assertion is valid.

Theorem 11. *The geometric object QCT^h_{ij} given by Equation (34) is an invariant of the equitorsion quasi-canonical biholomorphically projective mapping $f : GR_N \to G\overline{R}_N$, provided that relations (30) and (31) are valid.*

6. Discussion

This paper is a continuation of the research discussed in paper [20]. The form of the deformation tensor of a biholomorphically projective mapping allows us to define new types of mappings. Here, we have defined quasi-canonical biholomorphically projective mappings and equitorsion quasi-canonical biholomorphically projective mappings. Also, we obtained some relations between the corresponding curvature tensors of the generalized Riemannian spaces GR_N and $G\overline{R}_N$ and we found an invariant geometric object of an equitorsion quasi-canonical biholomorphically projective mapping which is of the Thomas type. Apart from the mapping defined in this paper, it is possible to consider some other types of mapping, which will be the subject of our further research. Also, the goal of further research will be to find new invariant geometric objects. The findings of this paper also motivate us to answer the following questions: (i) Are there any interpretations from a physical point of view? (ii) What is the geometrical significance?

Author Contributions: Both authors have equally contributed to this work. Both authors wrote, read, and approved the final manuscript. All authors have read and agreed to the published version of the manuscript.

Funding: V.M.M. acknowledges the grant of the Ministry of Science, Technological Development and Innovation of the Republic of Serbia 451-03-65/2024-03/200133 for carrying out this research. M.S.S. acknowledges the grant of the Ministry of Science, Technological Development and Innovation of the Republic of Serbia 451-03-65/2024-03/200124 for carrying out this research.

Data Availability Statement: The original contributions presented in the study are included in the article, further inquiries can be directed to the corresponding author.

Acknowledgments: The authors would like to thank the referees for their valuable comments which helped to improve the manuscript.

Conflicts of Interest: The authors declare no conflicts of interest.

References

1. Berezovski, V.E.; Mikeš, J. On the classification of almost geodesic mapings of affine connection spaces. *Acta Univ. Palacki. Olomouc. Fac. Rer. Nat. Math.* **1996**, *35*, 21–24.
2. Berezovski, V.E.; Mikeš, J. On special almost geodesic mappings of type π_1 of spaces with affine connection. *Acta Univ. Palacki. Olomouc Fac. Rer. Nat. Math.* **2004**, *43*, 21–26.
3. Berezovski, V.E.; Mikeš, J. Almost geodesic mappings of spaces with affine connection. *J. Math. Sci.* **2015**, *207*, 389–409. [CrossRef]
4. Ivanov, S.; Zlatanović, L.M. Connection on Non-Symmetric (Generalized) Riemanian Manifold and Gravity. *Class. Quantum Gravity* **2016**, *33*, 075016. [CrossRef]
5. Konovenko, N.G.; Kurbatova, I.N.; Cventuh, E. $2F$-planar mappings of pseudo-Riemannian spaces with f-structure. *Proc. Int. Geom. Cent.* **2018**, *11*, 39–51. (In Ukrainian)
6. Mikeš, J. Geodesic mappings of affine-connected and Riemannian spaces. *J. Math. Sci. N. Y.* **1996**, *78*, 311–333. [CrossRef]
7. Mikeš, J. Holomorphically projective mappings and their generalizations. *J. Math. Sci. N. Y.* **1998**, *89*, 1334–1353. [CrossRef]
8. Mikeš, J.; Pokorná, O.; Starko, G. *On almost Geodesic Mappings $\pi_2(e)$ onto Riemannian Spaces*; Rendiconti del Circolo Matematico di Palermo, Serie II; Circolo Matematico di Palermo: Palermo, Italy, 2004; (Suppl. 72), pp. 151–157.
9. Minčić, S.M. *Independent Curvature Tensors and Pseudotensors of Spaces with Non-Symmetric Affine Connexion*; Differential Geometry, Colloquia Mathematica Societatis János Bolayai: Budapest, Hungary, 1979; Volume 31, pp. 445–460.
10. Minčić, S.M.; Stanković, M.S. *Equitorsion Geodesic Mappings of Generalized Rimannian Spaces*; Publications de L'Institut Mathéma-tique: Belgrade, Serbia, 1997; Volume 61, pp. 97–104.
11. Prvanović, M. Four curvature tensors of non-symmetric affine connexion. In Proceedings of the Conference "150 Years of Lobachevski Geometry" (Kazan 1976), Moscow, Russia, 1977 ; pp. 199–205. (In Russian)

12. Stanković, M.S.; Minčić, S.M.; Velimirović, L.S. On holomorphically projective mappings of generalized Kählerian spaces. *Mat. Vesn.* **2002**, *54*, 195–202.
13. Stanković, M.S.; Minčić, S.M.; Velimirović, L.S. On equitorsion holomorphically projective mappings of generalized Kählerian spaces. *Czechoslov. Math. J.* **2004**, *54*, 701–715. [CrossRef]
14. Stanković, M.S.; Zlatanović, M.L.; Velimirović, L.S. Equitorsion holomorphically projective mappings of generalized Kählerian space of the first kind. *Czechoslov. Math. J.* **2010**, *60*, 635–653. [CrossRef]
15. Einstein, A. Generalization of the relativistic theory of gravitation. *Ann. Math.* **1945**, *46*, 576–584. [CrossRef]
16. Einstein, A. *Relativistic Theory of the Non-Symmetric Field*, 5th ed.; Appendix II in the Book "The Meaning of Relativity"; Princeton University Press: Princeton, NY, USA, 1955; Volume 49.
17. Eisenhart, L.P. *Non-Riemannian Geometry*; American Mathematical Society: New York, NY, USA, 1927.
18. Eisenhart, L.P. Generalized Riemannian spaces I. *Proc. Natl. Acad. Sci. USA* **1951**, *37*, 311–315. [CrossRef] [PubMed]
19. Minčić, S.M.; Stanković, M.S.; Velimirović, L.S. *Generalized Riemannian Spaces and Spaces of Non-Symmetric Affine Connection*; Monography; University of Niš, Faculty of Science and Mathematics: Niš, Serbia, 2013.
20. Milenković, V.M.; Zlatanović, L.M. Biholomorphically projective mappings of generalized Riemannian space in the Eisenhart sense. *Quaest. Math.* **2021**, *45*, 979–991. [CrossRef]

axioms

MDPI

Article

Ensuring Topological Data-Structure Preservation under Autoencoder Compression Due to Latent Space Regularization in Gauss–Legendre Nodes

Chethan Krishnamurthy Ramanaik [1,*,†], Anna Willmann [2,†], Juan-Esteban Suarez Cardona [2], Pia Hanfeld [2], Nico Hoffmann [3] and Michael Hecht [2,4,*]

1 Forschungsinstitut CODE, University of the Bundeswehr Munich, 85579 Neubiberg, Germany
2 CASUS—Center for Advanced Systems Understanding, Helmholtz-Zentrum Dresden-Rossendorf e.V. (HZDR), 01328 Dresden, Germany; a.willmann@hzdr.de (A.W.); j.suarez-cardona@hzdr.de (J.-E.S.C.); p.hanfeld@hzdr.de (P.H.)
3 SAXONY.ai, 01217 Dresden, Germany
4 Mathematical Institute, University of Wrocław, 50-384 Wrocław, Poland
* Correspondence: chethan.krishnamurthy@unibw.de (C.K.R.); m.hecht@hzdr.de (M.H.)
† These authors contributed equally to this work.

Abstract: We formulate a data-independent latent space regularization constraint for general unsupervised autoencoders. The regularization relies on sampling the autoencoder Jacobian at Legendre nodes, which are the centers of the Gauss–Legendre quadrature. Revisiting this classic allows us to prove that regularized autoencoders ensure a one-to-one re-embedding of the initial data manifold into its latent representation. Demonstrations show that previously proposed regularization strategies, such as contractive autoencoding, cause topological defects even in simple examples, as do convolutional-based (variational) autoencoders. In contrast, topological preservation is ensured by standard multilayer perceptron neural networks when regularized using our approach. This observation extends from the classic FashionMNIST dataset to (low-resolution) MRI brain scans, suggesting that reliable low-dimensional representations of complex high-dimensional datasets can be achieved using this regularization technique.

Keywords: autoencoder; regularization; data manifold learning

MSC: 53A07; 57R40; 53C22

Citation: Ramanaik, C.K.; Willmann, A.; Suarez Cardona, J.-E.; Hanfeld, P.; Hoffmann, N.; Hecht, M. Ensuring Topological Data-Structure Preservation under Autoencoder Compression Due to Latent Space Regularization in Gauss–Legendre Nodes. *Axioms* **2024**, *13*, 535. https://doi.org/10.3390/axioms13080535

Academic Editors: Mića Stanković and Stefano De Marchi

Received: 3 June 2024
Revised: 5 July 2024
Accepted: 27 July 2024
Published: 7 August 2024

1. Introduction

Systematic analysis and post-processing of high-dimensional and high-throughput datasets [1,2] is a current computational challenge across disciplines such as neuroscience [3–5], plasma physics [6–8], and cell biology and medicine [9–12]. In the machine learning (ML) community, autoencoders (AEs) are commonly considered the central tool for learning a low-dimensional *one-to-one representation* of high-dimensional datasets. These representations serve as a baseline for feature selection and classification tasks, which are prevalent in bio-medicine [13–17].

AEs can be considered as a *non-linear extension* of classic *principal component analysis* (PCA) [18–20]. Comparisons for linear problems are provided in [21]. While addressing the non-linear case, AEs face the challenge of preserving the topological data structure under AE compression.

To state the problem: We mathematically formalize AEs as pairs of continuously differentiable maps (φ, ν), $\varphi : \Omega_{m_2} \longrightarrow \Omega_{m_1}$, $\nu : \Omega_{m_1} \longrightarrow \Omega_{m_2}$, $0 < m_1 < m_2 \in \mathbb{N}$, defined on bounded domains $\Omega_{m_1} \subseteq \mathbb{R}^{m_1}$ and $\Omega_{m_2} \subseteq \mathbb{R}^{m_2}$. Commonly, φ is termed the *encoder*, and ν the *decoder*. We assume that the data $D \subseteq \mathcal{D}$ is sampled from a regular or even smooth data manifold $\mathcal{D} \subseteq \Omega_{m_2}$, with $\dim \mathcal{D} = m_0 \leq m_1$.

We seek to find proper AEs (φ, ν) yielding homeomorphic latent representations $\varphi(D) = \mathcal{D}' \cong \mathcal{D}$. In other words, the restrictions $\varphi_{|\mathcal{D}} : \mathcal{D} \longrightarrow \mathcal{D}'$ and $\nu_{|\mathcal{D}'} : \mathcal{D}' \longrightarrow \mathcal{D}$ of the encoder and decoder result in one-to-one maps, being inverse to each other:

$$\mathcal{D} \cong \mathcal{D}' = \varphi(\mathcal{D}) \subseteq \Omega_{m_1}, \quad \nu(\varphi(x)) = x, \text{ for all } x \in \mathcal{D}. \tag{1}$$

While the second condition in Equation (1) is usually realized by minimization of a reconstruction loss, this is insufficient for guaranteeing the one-to-one representation $\mathcal{D} \cong \mathcal{D}'$.

To realize AEs matching both requirements in Equation (1), we strengthen the condition by requiring the decoder to be an embedding of the whole latent domain $\Omega_{m_1} \supset \mathcal{D}$, including $\nu(\Omega_{m_1}) \supset \mathcal{D}$ in its interior. See Figure 1 for an illustration. We mathematically prove and empirically demonstrate this *latent regularization strategy* to deliver regularized AEs (AR-REG), satisfying Equation (1).

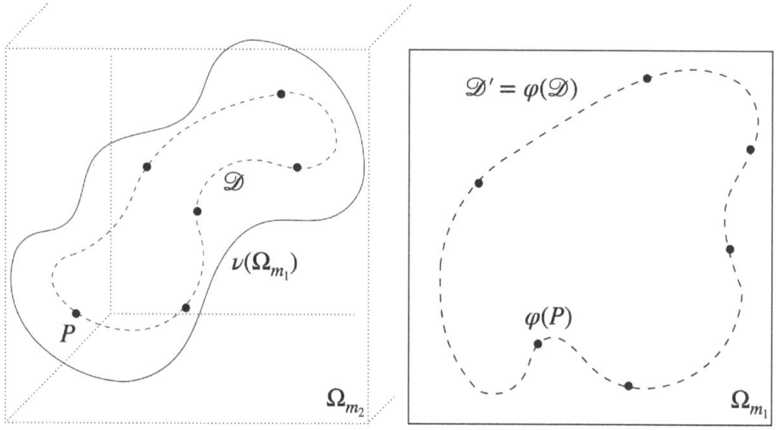

Figure 1. Illustration of the latent representation $\mathcal{D}' = \varphi(\mathcal{D}) \subseteq \Omega_{m_1}$ of the data manifold $\mathcal{D} \subseteq \Omega_{m_2}$, $\dim \mathcal{D} = m_0 < m_1 < m_2 \in \mathbb{N}$ given by the autoencoder (φ, ν). The decoder is a one-to-one mapping of the hypercube Ω_{m_1} to its image $\nu(\Omega_{m_1}) \supset \mathcal{D}$, including $\mathcal{D}$ in its interior and consequently guaranteeing Equation (1).

Our investigations are motivated by recent results of Hansen et al. [22–24], complementing other contributions [25–27] that investigate instabilities of machine learning methods from a general mathematical perspective.

1.1. *The Inherent Instability of Inverse Problems*

The *instability phenomenon of inverse problems* states that, in general, one cannot guarantee solutions of inverse problems to be stable. An excellent introduction to the topic is given in [22] with deeper treatments and discussions in [23,24].

In our setup, these insights translate to the fact that, in general, the *local Lipschitz constant*

$$L_\varepsilon(\nu, y) = \sup_{0 < \|y' - y\| < \varepsilon} \frac{\|\nu(y') - \nu(y)\|}{\|y' - y\|}, \quad \varepsilon > 0$$

of the decoder $\nu : \Omega_{m_1} \longrightarrow \Omega_{m_2}$ at some latent code $y \in \Omega_1$ might be unbounded. Consequently, small perturbations $y' \approx y$ of the latent code can result in large differences of the reconstruction $\|\nu(y') - \nu(y)\| \gg 0$. This fact generally applies and can only be avoided if an additional control on the null space of the Jacobian of the encoder $\ker J(\varphi(x))$ is given. Providing this control is the essence of our contribution.

1.2. Contribution

Avoiding the aforementioned instability, requires the null space of the Jacobian of the encoder $\ker J(\varphi)$ to be perpendicular to the tangent space of $\mathcal{D}$

$$\ker J(\varphi) \perp T\mathcal{D}\,. \tag{2}$$

In fact, due to the *inverse function theorem*, see, e.g., [28,29], the conditions Equations (1) and (2) are equivalent. In Figure 1, $\ker J(\varphi)$ is illustrated to be perpendicular to the image of the whole latent domain Ω_{m_1}

$$\ker J(\varphi) \perp T\nu(\Omega_{m_1})\,, \quad \nu(\Omega_{m_1}) \supseteq \mathcal{D}\,,$$

being sufficient for guaranteeing Equation (2), and consequently, Equation (1).

While several state-of-the-art AE regularization techniques are commonly established, none of them specifically formulates this necessary mathematical requirement in Equation (2). Consequently, we are not aware of any regularization approach that can theoretically guarantee the regularized AE to preserve the topological data-structure, as we do in Theorem 1. Our computational contributions split into:

(C1) For realising a *latent space regularized* AE (AE-REG) we introduce the L^2-regularization loss

$$\mathcal{L}_{\mathrm{reg}}(\varphi,\nu) = \|J(\varphi\circ\nu) - I\|^2_{L^2(\Omega_{m_1})} \tag{3}$$

and mathematically prove the AE-REG to satisfy condition Equation (1), Theorem 1, when being trained due to this additional regularization.

(C2) To approximate $\mathcal{L}_{\mathrm{reg}}(\varphi,\nu)$ we revisit the classic Gauss–Legendre quadratures (cubatures) [30–34], only requiring sub-sampling of $J(\nu\circ\varphi)(p_\alpha)$, $p_\alpha \in P_{m,n}$ on a Legendre grid of sufficient high resolution $1 \ll |P_{m,n}|$ in order to execute the regularization. While the data-independent latent Legendre nodes $P_{m,n} \subseteq \Omega_{m_1}$ are contained in the smaller dimensional latent space, regularization of high resolution can be efficiently realised.

(C3) Based on our prior work [35–37], and [38–41], we complement the regularization through a *hybridisation approach* combining autoencoders with *multivariate Chebyshev-polynomial-regression*. The resulting Hybrid AE is acting on the polynomial coefficient space, given by pre-encoding the training data due to high-quality regression.

We want to emphasize that the proposed regularization is data-independent in the sense that it does not require any prior knowledge of the data manifold, its embedding, or any parametrization of $\mathcal{D}$. Moreover, while being integrated into the loss function, the regularization is independent of the AE architecture and can be applied to any AE realizations, such as convolutional or variational AEs. Our results show that already regularized MLP-based AEs perform superior to these alternatives.

As we demonstrate, the regularization yields the desired re-embedding, enhances the autoencoder's reconstruction quality, and increases robustness under noise perturbations.

1.3. Related Work—Regularization of Autoencoders

A multitude of supervised learning schemes, addressing representation learning tasks, are surveyed in [42,43]. Self-supervised autoencoders rest on *inductive bias learning* techniques [44,45] in combination with vectorized autoencoders [46,47]. However, the mathematical requirements, Equations (1) and (2) were not considered in these strategies at all. Consequently, one-to-one representations might only be realized due to a well-chosen inductive bias regularization for rich datasets [9].

This article focus on regularization techniques of purely unsupervised AEs. We want to mention the following prominent approaches:

(R1) *Contractive AEs* (ContraAE) [48,49] are based on an ambient Jacobian regularization loss

$$\mathcal{L}^*_{\mathrm{reg}}(\varphi,\nu) = \|J(\nu \circ \varphi) - I\|^2_{L^2(\Omega_{m_2})} \tag{4}$$

formulated in the ambient domain. This makes contraAEs arbitrarily contractive in perpendicular directions $(T\mathcal{D})^{\perp}$ of $T\mathcal{D}$. However, this is insufficient to guarantee Equation (1). In addition, the regularization is data dependent, resting on the training dataset, and is computationally costly due to the large Jacobian $J \in \mathbb{R}^{m_2 \times m_2}$, $m_2 \gg m_1 \geq 1$. Several experiments in Section 5 demonstrate contraAE failing to deliver topologically preserved representations.

(R2) *Variational AEs* (VAE), along with extensions like β-VAE, consist of *stochastic encoders and decoders* and are commonly used for density estimation and generative modelling of complex distributions based on minimisation of the Evidence Lower Bound (ELBO) [50,51]. The variational latent space distribution induces an implicit regularization, which is complemented by [52,53] due to a l_1-sparsity constraint of the decoder Jacobian.
However, as the contraAE-constraint, this regularization is computationally costly and insufficient for guaranteeing a one-to-one encoding, which is reflected in the degenerated representations appearing in Section 5.

(R3) *Convolutional AEs* (CNN-AE) are known to deliver one-to-one representations for a generic setup theoretically [54]. However, the implicit convolutions seems to prevent clear separation of tangent $T\mathcal{D}$ and perpendicular direction $(T\mathcal{D})^{\perp}$ of the data manifold $\mathcal{D}$, resulting in topological defects already for simple examples, see Section 5.

2. Mathematical Concepts

We provide the mathematical notions on which our approach rests, starting by fixing the notation.

2.1. Notation

We consider neural networks (NNs) $\nu(\cdot, w)$ of fixed architecture Ξ_{m_1,m_2}, specifying number and depth of the hidden layers, the choice of piece-wise smooth activation functions $\sigma(x)$, e.g., ReLU or sin, with input dimension m_1 and output dimension m_2. Further, $\Upsilon_{\Xi_{m_1,m_2}}$ denotes the parameter space of the weights and bias $w = (v,b) \in W = V \times B \subseteq \mathbb{R}^K$, $K \in \mathbb{N}$, see, e.g., [55,56].

We denote with $\Omega_m = (-1,1)^m$ the m-dimensional open *standard hypercube*, with $\|\cdot\|$ the standard Euclidean norm on $\mathbb{R}^m$ and with $\|\cdot\|_p$, $1 \leq p \leq \infty$ the l_p-norm. $\Pi_{m,n} = \mathrm{span}\{x^\alpha\}_{\|\alpha\|_\infty \leq n}$ denotes the $\mathbb{R}$-*vector space of all real polynomials* in m variables spanned by all monomials $x^\alpha = \prod_{i=1}^m x_i^{\alpha_i}$ of *maximum degree* $n \in \mathbb{N}$ and $A_{m,n} = \{\alpha \in \mathbb{N}^m : \|\alpha\|_\infty = \max_{i=1,\dots,m}\{|\alpha_i|\} \leq n\}$ the corresponding multi-index set. For an excellent overview on functional analysis we recommend [57–59]. Here, we consider the Hilbert space $L^2(\Omega_m, \mathbb{R})$ of all *Lebesgue measurable* functions $f : \Omega_m \longrightarrow \mathbb{R}$ with finite L^2-norm $\|f\|^2_{L^2(\Omega_m)} < \infty$ induced by the inner product

$$< f, g >_{L^2(\Omega_m)} = \int_{\Omega_m} f \cdot g \, d\Omega_m \,, \; f, g \in L^2(\Omega, \mathbb{R})\,. \tag{5}$$

Moreover, $C^k(\Omega_m, \mathbb{R})$, $k \in \mathbb{N} \cup \{\infty\}$ denotes the *Banach spaces* of continuous functions being k-times continuously differentiable, equipped with the norm

$$\|f\|_{C^k(\Omega_m)} = \sum_{i=0}^{k} \sup_{x \in \Omega_m} |D^\alpha f(x)| \,, \|\alpha\|_1 \leq k\,.$$

2.2. Orthogonal Polynomials and Gauss–Legendre Cubatures

We follow [30–33,60] for recapturing: Let $m, n \in \mathbb{N}$ and $P_{m,n} = \oplus_{i=1}^{m} \mathrm{Leg}_n \subseteq \Omega_m$ be the m-dimensional Legendre grids, where $\mathrm{Leg}_n = \{p_0, \dots, p_n\}$ are the $n+1$ *Legendre nodes* given by the roots of the *Legendre polynomials* of degree $n+2$. We denote $p_\alpha = (p_{\alpha_1}, \dots, p_{\alpha_m}) \in P_{A_{m,n}}$, $\alpha \in A_{m,n}$. The Lagrange polynomials $L_\alpha \in \Pi_{A_{m,n}}$, defined by $L_\alpha(p_\beta) = \delta_{\alpha,\beta}$, $\forall\, \alpha, \beta \in A_{m,n}$, where $\delta_{\cdot,\cdot}$ denotes the *Kronecker delta*, are given by

$$L_\alpha = \prod_{i=1}^{m} l_{\alpha_i,i}\,, \quad l_{j,i} = \prod_{j \neq i, j=0}^{m} \frac{x_i - p_j}{p_i - p_j}\,. \tag{6}$$

Indeed, the L_α are an orthogonal L^2-basis of $\Pi_{m,n}$,

$$\langle L_\alpha, L_\beta \rangle_{L^2(\Omega_m)} = \int_{\Omega_m} L_\alpha(x) L_\beta(x) d\Omega_m = w_\alpha \delta_{\alpha,\beta}\,, \tag{7}$$

where the *Gauss–Legendre cubature weight* $w_\alpha = \|L_\alpha\|^2_{L^2(\Omega_m)}$ can be computed numerically. Consequently, for any polynomial $Q \in \Pi_{m,2n+1}$ of degree $2n+1$ the following cubature rule applies:

$$\int_{\Omega_m} Q(x) d\Omega_m = \sum_{\alpha \in A_{m,n}} w_\alpha Q(p_\alpha)\,. \tag{8}$$

Thanks to $|P_{m,n}| = (n+1)^m \ll (2n+1)^m$ this makes *Gauss–Legendre integration* a very powerful scheme, yielding

$$\langle Q_1, Q_2 \rangle_{L^2(\Omega_m)} = \sum_{\alpha \in A_{m,n}} Q_1(p_\alpha) Q_2(p_\alpha) w_\alpha\,, \tag{9}$$

for all $Q_1, Q_2 \in \Pi_{m,n}$.

In light of this fact, we propose the following AE regularization method.

3. Legendre-Latent-Space Regularization for Autoencoders

The regularization is formulated from the perspective of classic differential geometry, see, e.g., [28,61–63]. As introduced in Equation (1), we assume that the training data $D_{\text{train}} \subseteq \mathcal{D} \subseteq \mathbb{R}^{m_2}$ is sampled from a regular data manifold. We formalise the notion of autoencoders:

Definition 1 (autoencoders and data manifolds). *Let* $1 \leq m_0 \leq m_1 \leq m_2 \in \mathbb{N}$, $\mathcal{D} \subseteq \Omega_{m_2}$ *be a (data) manifold of dimension* $\dim \mathcal{D} = m_0$. *Given continuously differentiable maps* $\varphi : \Omega_{m_2} \longrightarrow \Omega_{m_1}$, $\nu : \Omega_{m_1} \longrightarrow \Omega_{m_2}$ *such that:*

(i) *ν is a right-inverse of φ on $\mathcal{D}$, i.e, $\nu(\varphi(x)) = x$ for all $x \in \mathcal{D}$.*
(ii) *φ is a left-inverse of ν, i.e, $\varphi(\nu(y)) = y$ for all $y \in \Omega_{m_1}$*

Then we call the pair (φ, ν) a proper autoencoder *with respect to $\mathcal{D}$.*

Given a proper AE (φ, ν), φ yields a low dimensional homeomorphic re-embedding of $\mathcal{D} \cong \mathcal{D}' = \varphi(\mathcal{D}) \subseteq \mathbb{R}^{m_1}$ as demanded in Equation (1) and illustrated in Figure 1, fulfilling the stability requirement of Equation (2).

We formulate the following losses for deriving proper AEs:

Definition 2 (regularization loss). *Let $\mathcal{D} \subseteq \Omega_{m_2}$ be a C^1-data manifold of dimension $\dim \mathcal{D} = m_0 < m_1 < m_2$ and $\emptyset \neq D_{\text{train}} \subseteq \mathcal{D}$ be a finite training dataset. For NNs $\varphi(\cdot, u) \in \Xi_{m_2,m_1}$, $\nu(\cdot, w) \in \Xi_{m_1,m_2}$ with weights $(u, w) \in \Upsilon_{\Xi_{m_2,m_1}} \times \Upsilon_{\Xi_{m_1,m_2}}$, we define the loss*

$$\mathcal{L}_{D_{\text{train}},n} : \Upsilon_{\Xi_{m_2,m_1}} \times \Upsilon_{\Xi_{m_1,m_2}} \longrightarrow \mathbb{R}^+, \quad \mathcal{L}_{D_{\text{train}},n}(u, w) = \mathcal{L}_0(D_{\text{train}}, u, w) + \lambda \mathcal{L}_1(u, w, n)\,,$$

where $\lambda > 0$ *is a hyper-parameter and*

$$\mathcal{L}_0(D_{\text{train}}, u, w) = \sum_{x \in D_{\text{train}}} \|x - \nu(\varphi(x,u),w)\|^2 \tag{10}$$

$$\mathcal{L}_1(u,w,n) = \sum_{\alpha \in A_{m_1,n}} \| I - J\left(\varphi(\nu(p_\alpha, w), u)\right)\|^2, \tag{11}$$

with $I \in \mathbb{R}^{m_1 \times m_1}$ *denoting the identity matrix,* $p_\alpha \in P_{m_1,n}$ *be the Legendre nodes, and* $J(\varphi(\nu(p_\alpha, w)) \in \mathbb{R}^{m_1 \times m_1}$ *the Jacobian.*

We show that the AEs with vanishing loss result to be proper AEs, Defintion 1.

Theorem 1 (Main Theorem). *Let the assumptions of Definition 2 be satisfied, and* $\varphi(\cdot, u_n) \in \Xi_{m_2,m_1}$, $\nu(\cdot, w_n) \in \Xi_{m_1,m_2}$ *be sequences of continuously differentiable NNs satisfying:*

(i) *The loss converges* $\mathcal{L}_{D_{\text{train}},n}(u_n, w_n) \xrightarrow[n\to\infty]{} 0.$

(ii) *The weight sequences converge*

$$\lim_{n\to\infty}(u_n, w_n) = (u_\infty, w_\infty) \in \Upsilon_{\Xi_{m_2,m_1}} \times \Upsilon_{\Xi_{m_1,m_2}}.$$

(iii) *The decoder satisfies* $\nu(\Omega_{m_1}, w_n) \supseteq \mathcal{D}$, $\forall n \geq n_0 \in \mathbb{N}$ *for some* $n_0 \geq 1$.

Then $(\varphi(\cdot, w_n), \nu(\cdot, u_n)) \xrightarrow[n\to\infty]{} (\varphi(\cdot, w_\infty), \nu(\cdot, u_\infty))$ *uniformly converges to a proper autoencoder with respect to* $\mathcal{D}$.

Proof. The proof follows by combining several facts: First, the *inverse function theorem* [29] implies that any map $\rho \in C^1(\Omega_m, \Omega_m)$ satisfying

$$J(\rho(x)) = I, \ \forall x \in \Omega_m, \text{ and } \rho(x_0) = x_0, \tag{12}$$

for some $x_0 \in \Omega_m$ is given by the identity, i.e., $\rho(x) = x$, $\forall x \in \Omega_m$.

Secondly, the Stone–Weierstrass theorem [64,65] states that any continuous map $\rho \in C^0(\Omega_m, \Omega_m)$, with coordinate functions $\rho(x) = (\rho_1(x), \ldots, \rho_m(x))$ can be uniformly approximated by a polynomial map $Q_\rho^n(x) = (Q_{\rho,1}^n(x), \ldots, Q_{\rho,m}^n(x))$, $Q_{\rho,i}^n(x) \in \Pi_{m,n}$, $1 \leq i \leq m$, i.e, $\|\rho - Q_\rho^n\|_{C^0(\Omega_m)} \xrightarrow[n\to\infty]{} 0$.

Thirdly, while the NNs $\varphi(\cdot, w)$, $\nu(\cdot, u)$ depend continuously on the weights u, w, the convergence in *(ii)* is uniform. Consequently, the convergence $\mathcal{L}_{D_{\text{train}},n}(u_n, w_n) \xrightarrow[n\to\infty]{} 0$ of the loss implies that any sequence of polynomial approximations $Q_\rho^n(x)$ of the map $\rho(\cdot) = \varphi(\nu(\cdot, w_\infty), u_\infty)$ satisfies

$$\sum_{\alpha \in A_{m_1,n}} \| I - J\left(Q_\rho^n(p_\alpha)\right)\|^2 = 0$$

in the limit for $n \to \infty$. Hence, Equation (12) holds in the limit for $n \to \infty$ and consequently $\varphi(\nu(y, w_\infty), u_\infty) = Q_\rho^\infty(y) = y$ for all $y \in \Omega_{m_1}$ yielding requirement *(ii)* of Definition 1.

Given that assumption *(iii)* is satisfied, in completion, requirement *(i)* of Definition 1 holds, finishing the proof. □

Apart from ensuring topological maintenance, one seeks for high-quality reconstructions. We propose a novel hybridization approach, delivering both.

4. Hybridization of Autoencoders due to Polynomial Regression

The hybridisation approach rests on deriving *Chebyshev Polynomial Surrogate Models* $Q_{\Theta,d}$ fitting the initial training data $d \in D_{\text{train}} \subseteq \Omega_{m_2}$. For the sake of simplicity, we motivate the setup in case of images:

Let $d = (d_{ij})_{1 \le i,j \le r} \in \mathbb{R}^{r \times r}$ be the intensity values of an image on an equidistant pixel grid $G_{r \times r} = (g_{ij})_{1 \le i,j \le r} \subseteq \Omega_2$ of resolution $r \times r$, $r \in \mathbb{N}$. We seek for a polynomial

$$Q_\Theta : \Omega_2 \longrightarrow \mathbb{R}, \quad Q_\Theta \in \Pi_{2,n},$$

such that evaluating Q_Θ, $\Theta = (\theta_\alpha)_{\alpha \in A_{2,n}} \in \mathbb{R}^{|A_{2,n}|}$ on $G_{r \times r}$ approximates d, i.e., $Q_\Theta(g_{ij}) \approx d_{ij}$ for all $1 \le i,j \le r$. We model Q_Θ in terms of *Chebyshev polynomials of first kind* well known to provide excellent approximation properties [33,35]:

$$Q_\Theta(x_1, x_2) = \sum_{\alpha \in A_{2,n}} \theta_\alpha T_{\alpha_1}(x_1) T_{\alpha_2}(x_2). \tag{13}$$

The expansion is computed due to standard least-square fits:

$$\Theta_d = \operatorname{argmin}_{C \in \mathbb{R}^{|A_{2,n}|}} \|RC - d\|^2, \tag{14}$$

where $R = (T_\alpha(g_{ij}))_{1 \le i,j, \le n, \alpha \in A_{2,n}} \in \mathbb{R}^{r^2 \times |A_{2,n}|}$, $T_\alpha = T_{\alpha_1} \cdot T_{\alpha_2}$ denotes the regression matrix.

Given that each image (training point) $d \in D_{\text{train}}$ can be approximated with the same polynomial degree $n \in \mathbb{N}$, we posterior train an autoencoder (φ, ν), only acting on the polynomial coefficient space $\varphi : \mathbb{R}^{|A_{2,n}|} \longrightarrow \Omega_{m_1}$, $\nu \Omega_{m_1} \longrightarrow \mathbb{R}^{|A_{2,n}|}$ by exchanging the loss in Equation (10) due to

$$\mathcal{L}_0^*(D_{\text{train}}, u, w) = \sum_{d \in D_{\text{train}}} \|d - R \cdot \nu(\varphi(\Theta_d, u), w)\|^2 \tag{15}$$

In contrast to the regularization loss in Definition 2, here, pre-encoding the training data due to polynomial regression decreases the input dimension $m_2 \in \mathbb{N}$ of the (NN) encoder $\varphi : \Omega_{m_2} \longrightarrow \Omega_{m_1}$. In practice, this enables to reach low dimensional latent dimension by increasing the reconstruction quality, as we demonstrate in the next section.

5. Numerical Experiments

We executed experiments, designed to validate our theoretical results, on HEMERA a NVIDIA V100 cluster at HZDR. Complete code benchmark sets and supplements are available at https://github.com/casus/autoencoder-regularisation, accessed on 2 June 2024. The following AEs were applied:

- **(B1)** *Multilayer perceptron autoencoder (MLP-AE)*: Feed forward NNs with activation functions $\sigma(x) = \sin(x)$.
- **(B2)** *Convolutional autoencoder (CNN-AE)*: Standard convolutional neural networks (CNNs) with activation functions $\sigma(x) = \sin(x)$, as discussed in **(R3)**.
- **(B3)** *Variational autoencoder*: MLP based (MLP-VAE) and CNN based (CNN-VAE) as in [50,51], discussed in **(R2)**.
- **(B4)** *Contractive autoencoder (ContraAE)*: MLP based with with activation functions $\sigma(x) = \sin(x)$ as in [48,49], discussed in **(R1)**.
- **(B5)** *regularized autoencoder (AE-REG)*: MLP based, as in **(B1)**, trained with respect to the regularization loss from Definition 2.
- **(B6)** *Hybridised AE (Hybrid AE-REG)*: MLP based, as in **(B1)**, trained with respect to the modified loss in Definition 2 due to Equation (15).

The choice of activation functions $\sigma(x) = \sin(x)$ yields a natural way for normalizing the latent encoding to Ω_m and performed best compared to trials with ReLU, ELU or $\sigma(x) = \tanh(x)$. The regularization of AE-REG and Hybrid AE-REG is realized due to sub-sampling batches from the Legendre grid $P_{m,n}$ for each iteration and computing the autoencoder Jacobians due to automatic differentiation [66].

5.1. Topological Data-Structure Preservation

Inspired by Figure 1, we start by validating Theorem 1 for known data manifold topologies.

Experiment 1 (Cycle reconstructions in dimension 15). *We consider the unit circle* $S^1 \subseteq \mathbb{R}^2$, *a uniform random matrix* $A \in \mathbb{R}^{15,2}$ *with entries in* $[-2,2]$ *and the data manifold* $\mathcal{D} = \{Ax : x \in S^1 \subseteq \mathbb{R}^2\}$, *being an ellipse embedded along some 2-dimensional hyperplane* $H_A = \{Ax : x \in \mathbb{R}^2\} \subseteq \mathbb{R}^{15}$. *Due to Bezout's Theorem [67,68], a 3-points sample uniquely determines a circle in the 2-dimensional plane. Therefore, we executed the AEs for this minimal case of a set of random samples* $|D_{\text{train}}| = 3$, $D_{\text{train}} \subseteq \mathcal{D}$ *as training set.*

MLP-AE, MLP-VAE, and AE-REG consists of 2 hidden linear layers (in the encoder and decoder), each of length 6. The degree of the Legendre grid $P_{m,n}$ used for the regularization of AE-REG was set to $n = 21$, Definition 2. CNN-AE and CNN-VAE consists of 2 hidden convolutional layers with kernel size 3, stride of 2 in the first hidden layer and 1 in the second, and 5 filters per layer. The resulting parameter spaces $\Upsilon_{\Xi_{15,2}}$ are all of similar size: $|\Upsilon_{\Xi_{15,2}}| \sim 400$. All AEs were trained with the Adam optimizer [69].

Representative results out of 6 repetitions are shown in Figure 2. Only AE-REG delivers a feasible 2D re-embedding, while all other AEs cause overlappings or cycle-crossings. More examples are given in the supplements; whereas AE-REG delivers similar reconstructions for all other trials while the other AEs fail in most of the cases.

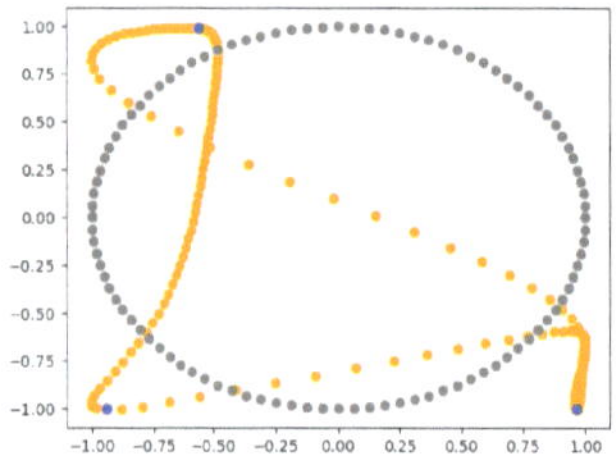

(**a**) CNN-AE

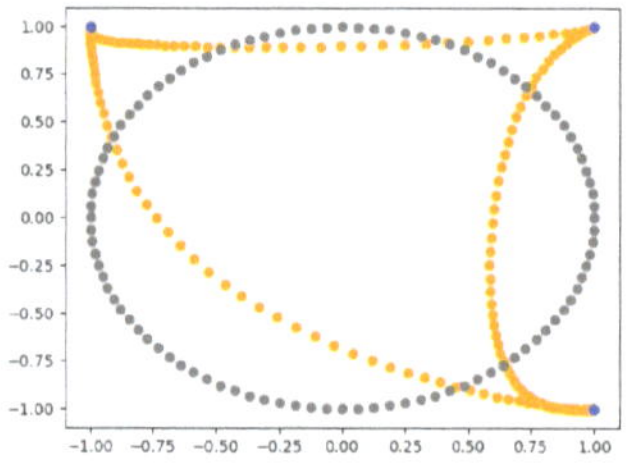

(**b**) CNN-VAE

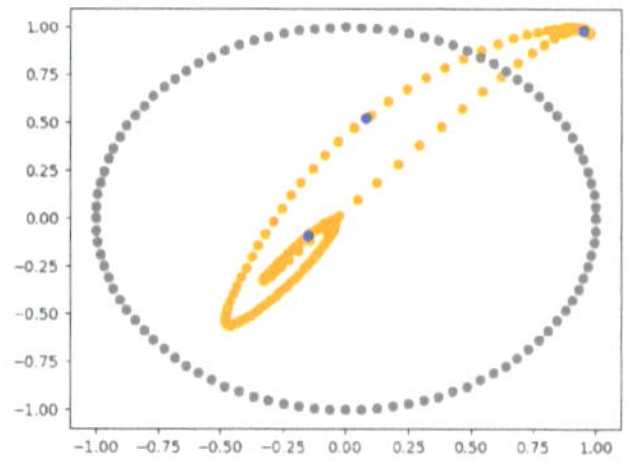

(**c**) ContraAE

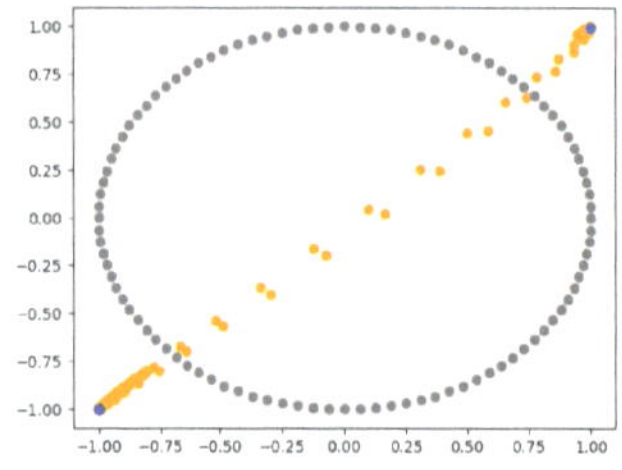

(**d**) MLP-VAE

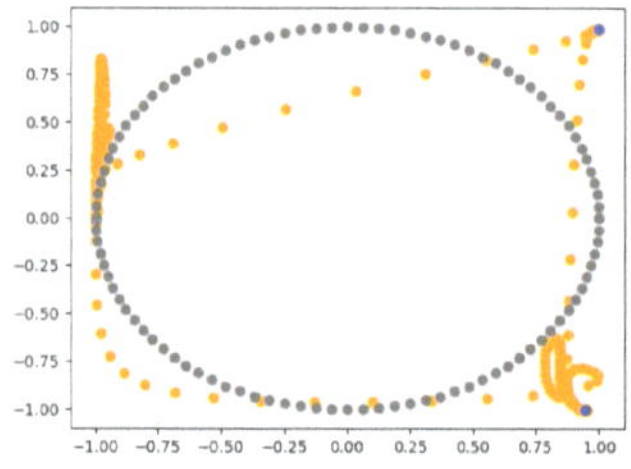

(**e**) MLP-AE

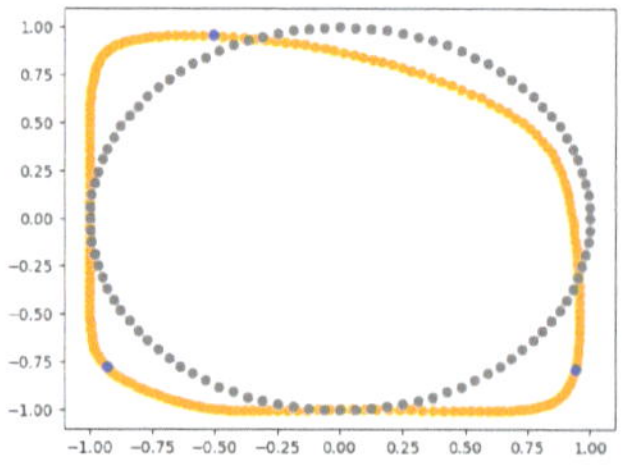

(**f**) AE-REG

Figure 2. Circle reconstruction using various autoencoder models.

Linking back to our initial discussions of ContraAE **(R1)**: The results show that the ambient domain regularization formulated for the ContraAE, is insufficient for guaranteeing a one-to-one encoding. Similarily, CNN-based AEs cause self-intersecting points. As initially discussed in **(R3)**, CNNs are invertible for a generic setup [54], but seem to fail sharply separating tangent $T\mathcal{D}$ and perpendicular direction $T\mathcal{D}^{\perp}$ of the data manifold $\mathcal{D}$.

We demonstrate the impact of the regularization to not belonging to an edge case by considering the following scenario:

Experiment 2 (Torus reconstruction). *Following the experimental design of Experiment 1 we generate challenging tori embeddings of a squeezed torus with radii* $0 < r, R, r = 0.7\ R = 2.0$ *in dimension* $m = 15$ *and dimension* $m = 1024$ *due to multiplication with random matrices* $A \in [-1, 1]^{m \times 3}$. *We randomly sample 50 training points* $|D_{\text{train}}| = 50$, $D_{\text{train}} \subseteq \mathcal{D}$ *and seek for their 3D re-embedding due to the AEs. We choose a Legendre grid* $P_{m,n}$ *of degree* $n = 21$.

A show-case is given in Figure 3, visualized by a dense set of 2000 test points. As in Experiment 1 only AE-REG is capable for re-embedding the torus in a feasible way. MLP-VAE, CNN-AE and CNN-VAE flatten the torus, ContraAE and MPL-AE cause self-intersections. Similar results occur for the high-dimensional case $m = 1024$, see the supplements. Summarizing the results suggests that without regularization AE-compression does not preserve the data topology. We continue our evaluation to give further evidence on this expectation.

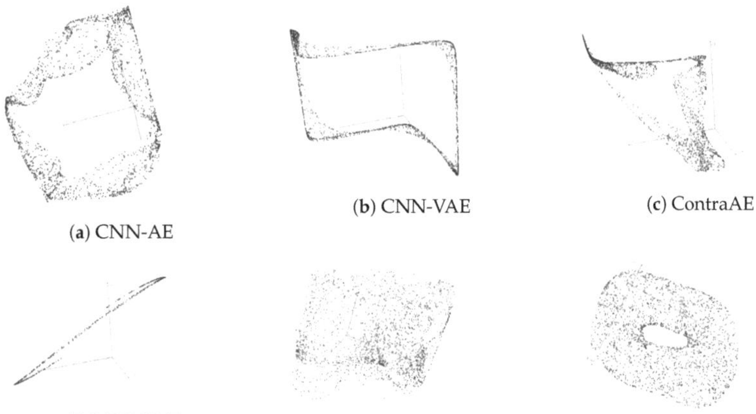

(**a**) CNN-AE (**b**) CNN-VAE (**c**) ContraAE

(**d**) MLP-VAE (**e**) MLP-AE (**f**) AE-REG

Figure 3. Torus reconstruction using various autoencoder models, dim = 15.

5.2. Autoencoder Compression for FashionMNIST

We continue by benchmarking on the the classic FashionMNIST dataset [70].

Experiment 3 (FashionMNIST compression). *The 70,000 FashionMNIST images separated into 10 fashion classes (T-shirts, shoes, etc.) being of* $32 \times 32 = 1024$*-pixel resolution (ambient domain dimension). For providing a challenging competition, we reduced the dataset to 24,000 uniformly sub-sampled images and trained the AEs for* 40% *training data and complementary test data, respectively. Here, we consider latent dimensions* $m = 4, 10$. *Results of further runs for* $m = 2, 4, 6, 8, 10$ *are given in the supplements.*

MLP-AE, MLP-VAE, AE-REG and Hybrid AE-REG consists of 3 hidden layers, each of length 100. The degree of the Legendre grid $P_{m,n}$ used for the regularization of AE-REG was set to $n = 21$, Definition 2. CNN-AE and CNN-VAE consists of 3 convolutional layers with kernel size 3, stride of 2. The resulting parameter spaces $\Upsilon_{\Xi_{15,2}}$ of all AEs are of similar size. Further details of the architectures are reported in the supplements.

We evaluated the reconstruction quality with respect to peak signal-to-noise ratios (PSNR) for perturbed test data due 0%, 10%, 20%, 50% of Gaussian noise encoded to latent dimension $m = 10$, and plot them in Figure 4. The choice $m = 10$, here, reflects the number of FashionMNIST-classes.

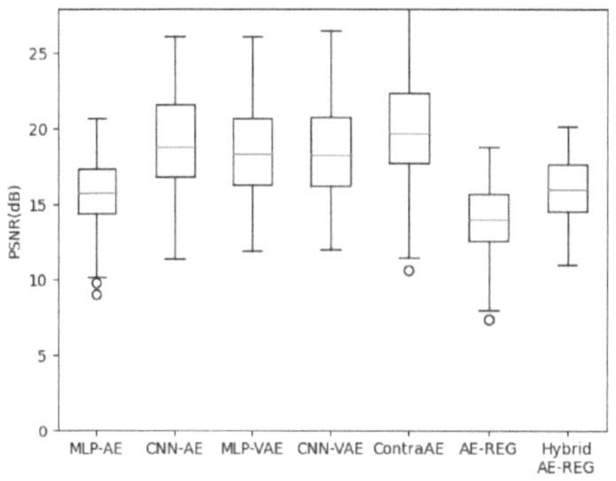

(**a**) Without Noise

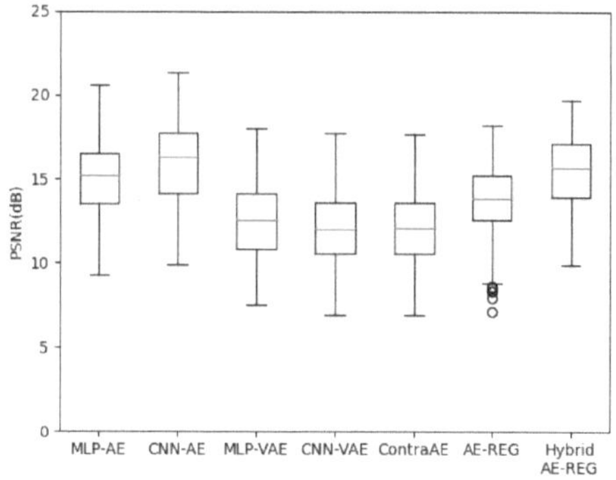

(**b**) 10% Gaussian Noise

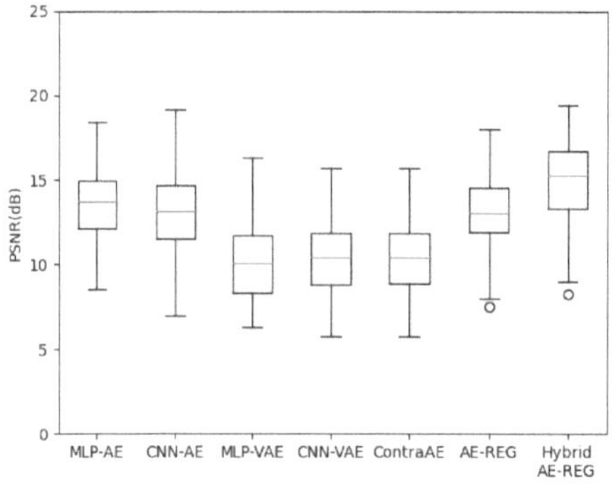

(**c**) 20% Gaussian Noise

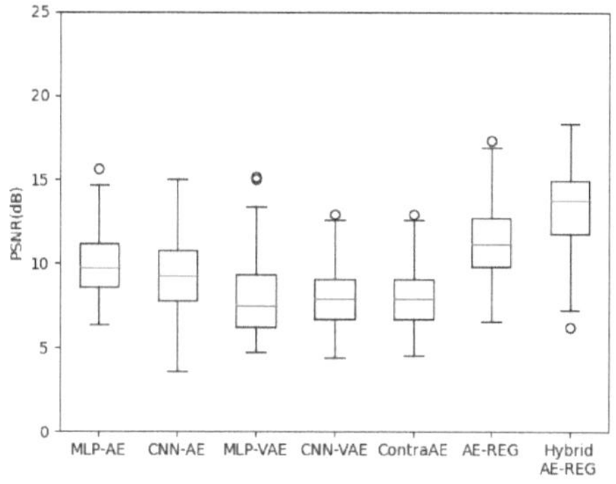

(**d**) 50% Gaussian Noise

Figure 4. FashionMNIST reconstruction with varying levels of Gaussian noise, latent dimension dim = 10.

While Hybrid AE-REG performs compatible to MLP-AE and worse than the other AEs in the non-perturbed case, its superiority appears already for perturbations with 10% of Gaussian noise and exceeds the reached reconstruction quality of all other AEs for 20% Gaussian noise or more. We want to stress that Hybrid AE-REG maintains its reconstruction quality throughout the noise perturbations (up to 70%, see the supplements). This outstanding appearance of robustness gives strong evidence on the impact of the regularization and well-designed pre-encoding technique due to the hybridization with polynomial regression. Analogue results appear when measuring the reconstruction quality with respect to the structural similarity index measure (SSIM), given in the supplements.

In Figure 5, show cases of the reconstructions are illustrated, including additional vertical and horizontal flip perturbations. Apart from AE-REG and Hybrid AE-REG (rows (7) and (8)), all other AEs flip the FashionMNIST label-class for reconstructions of images with 20% or 50% of Gaussian noise. Flipping the label-class is the analogue to topological defects as cycle crossings appeared for the non-regularized AEs in Experiment 1, indicating again that the latent representation of the FashionMNIST dataset given due to the non-regularized AEs does not maintain structural information.

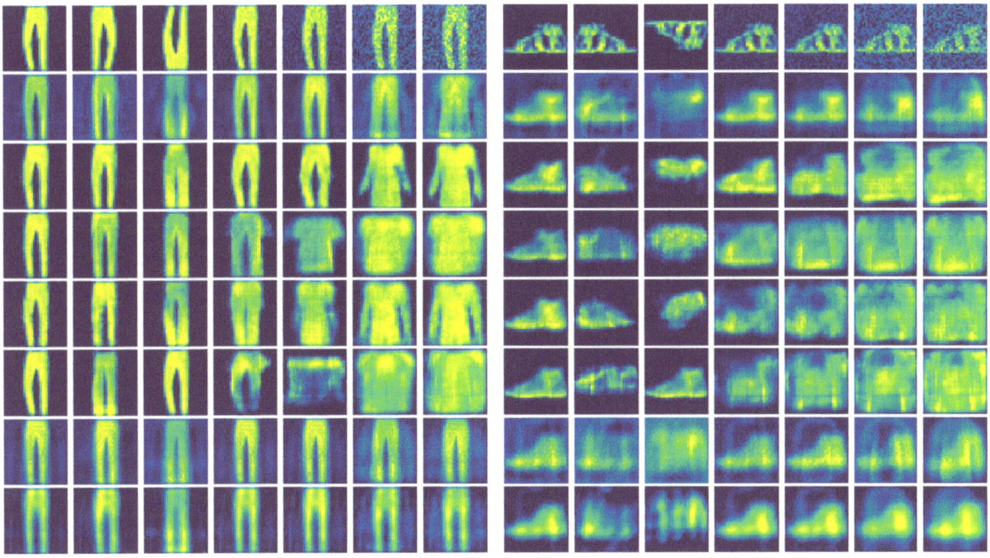

Figure 5. Two show cases of FashionMNIST reconstruction for latent dimension $m = 10$. First row shows the input image with vertical, horizontal flips, and 0%, 10%, 20%, 50%, 70% of Gaussian noise. Rows beneath show the results of (2) MLAP-AE, (3) CNN-AE, (4) MLP-VAE, (5) CNN-VAE, (6) ContraAE, (7) AE-REG, and (8) Hybrid AE-REG.

While visualization of the FashionMNIST data manifold is not possible, we decided to investigate its structure by computing geodesics. Figure 6 provides show cases of decoded latent-geodesics $\nu(\gamma)$ with respect to latent dimension $m = 4$, connecting two AE-latent codes of the encoded test data that has been initially perturbed by 50% Gaussian noise before encoding. The latent-geodesics γ have to connect the endpoints along the curved encoded data manifold $\mathcal{D}' = \varphi(\mathcal{D})$ without forbidden short-cuts through $\mathcal{D}'$. That is why the geodesics are computed as shortest paths for an early Vietoris–Rips filtration [71] that contains the endpoints in one common connected component. More examples are given in the supplements.

Apart from CNN-AE and AE-REG, all other geodesics contain latent codes of images belonging to another FashionMNIST-class, while for Hybrid AE-REG this happens just once. We interpret these appearances as forbidden short-cuts of $\nu(\gamma)$ through $\mathcal{D}$, caused by topological artefacts in $\mathcal{D}' = \varphi(\mathcal{D})$.

AE-REG delivers a smoother transition between the endpoints than CNN-AE, suggesting that though the CNN-AE geodesic is shorter, the regularized AEs preserve the topology with higher resolution.

5.3. Autoencoder Compression for Low-Resolution MRI Brain Scans

For evaluating the potential impact of the hybridisation and regularization technique to more realistic high-dimensional problems, we conducted the following experiment.

Experiment 4 (MRI compression). *We consider the MRI brain scans dataset from Open Access Series of Imaging Studies (OASIS) [72]. We extract two-dimensional slices from the three-dimensional MRI images, resulting in* 60,000 *images of resolution* 91×91*-pixels. We follow Experiment 3 by splitting the dataset into* 40% *training images and complementary test images and compare the AE compression for latent dimension* $m = 40$. *Results for latent dimension* $m = 10, 15, 20, 40, 60, 70$ *and* 5%, 20%, 40%, 80% *training data are given in the supplements, as well as further details on the specifications.*

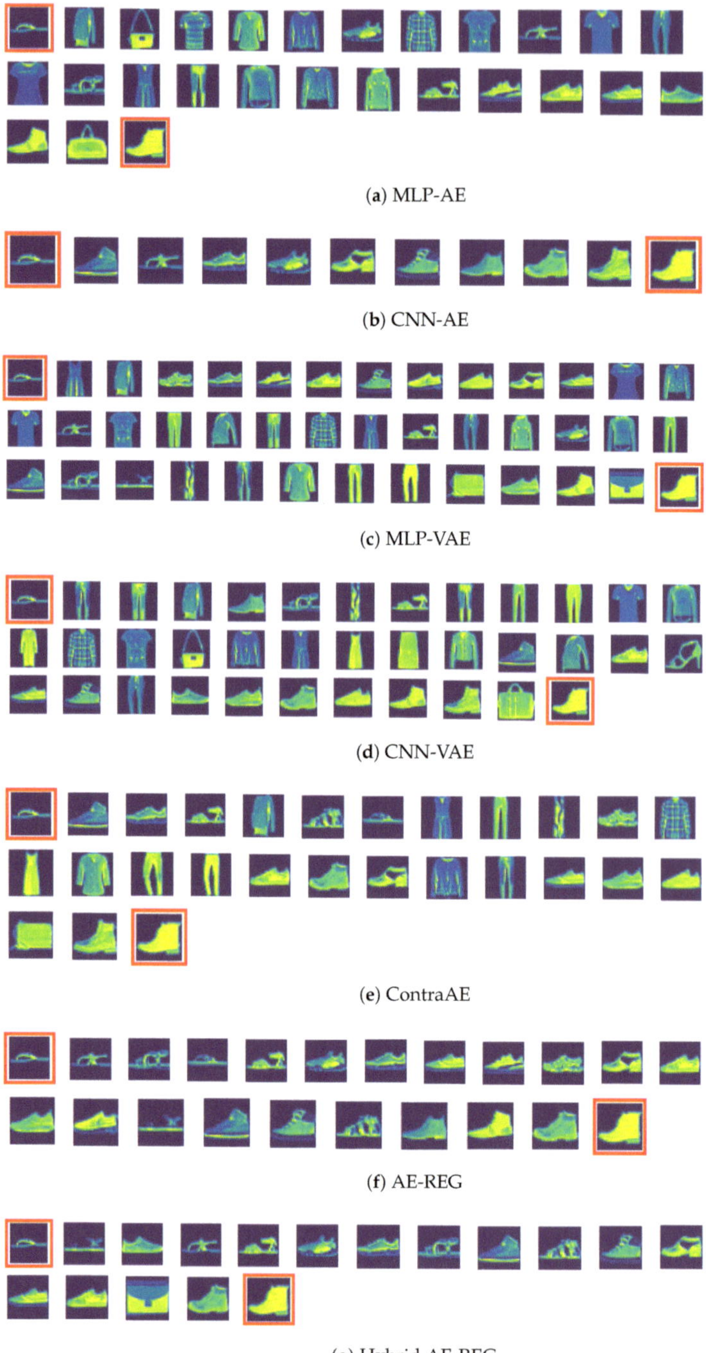

Figure 6. FashionMNIST geodesics in latent dimension dim = 4.

We keep the architecture setup of the AEs, but increase the NN sizes to 5 hidden layers each consisting of 1000 neurons. Reconstructions measured by PSNR are evaluated in Figure 7. Analogous results appear for SSIM, see the supplements.

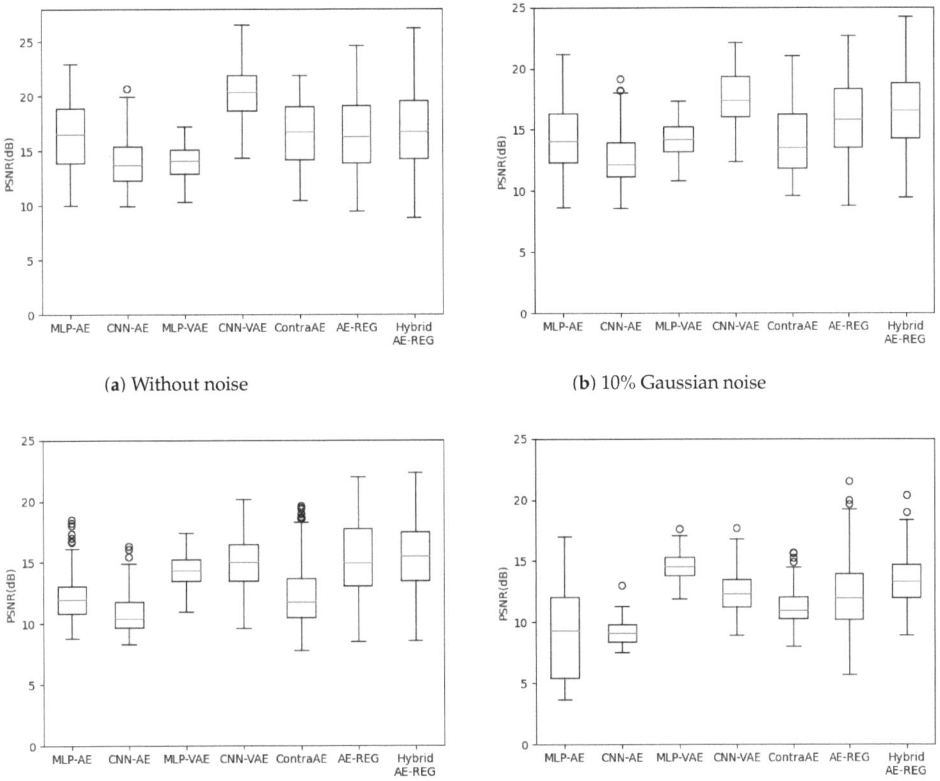

(**a**) Without noise (**b**) 10% Gaussian noise

(**c**) 20% Gaussian noise (**d**) 50% Gaussian noise

Figure 7. MRI reconstruction, latent dimension dim = 40.

As in Experiment 3, we observe that AE-REG and Hybrid AE-REG perform compatible or slightly worse than the other AEs in the unperturbed scenario, but show their superiority over the other AEs for 10% Gaussian noise, or 20% for CNN-VAE. Hybrid AE-REG specifically maintains its reconstruction quality under noise perturbations up to 20% (maintains stable for 50%). The performance increase compared to the un-regularized MLP-AE becomes evident and validates again that a strong robustness is achieved due to the regularization.

A show case is given in Figure 8. Apart from Hybrid AE-REG (row (8)), all AEs show artefacts when reconstructing perturbed images. CNN-VAE (row (4)) and AE-REG (row (7)) perform compatible and maintain stability up to 20% Gaussian noise perturbation.

In Figure 9, examples of geodesics are visualized, being computed analogously as in Experiment 3 for the encoded images once without noise and once by adding 10% Gaussian noise before encoding. The AE-REG geodesic consists of similar slices, including one differing slice for 10% Gaussian noise perturbation. CNN-VAE delivers a shorter path; however, it includes a strongly differing slice, which is kept for 10% of Gaussian noise. CNN-AE provides a feasible geodesic in the unperturbed case; however, it becomes unstable in the perturbed case.

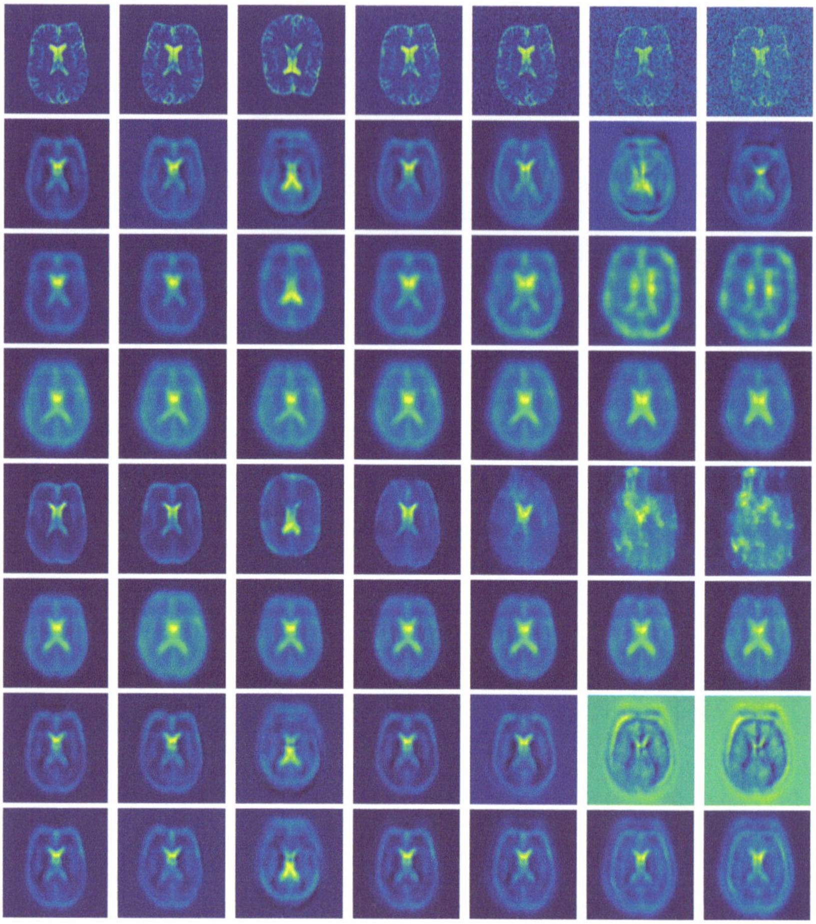

Figure 8. MRI show case. First row shows the input image with vertical, horizontal flips, and 0%, 10%, 20%, 50%, 70% of Gaussian noise. Rows beneath show the results of (2) MLAP-AE, (3) CNN-AE, (4) MLP-VAE, (5) CNN-VAE, (6) ContraAE, (7) AE-REG, and (8) Hybrid AE-REG.

We interpret the difference of the AE-REG to CNN-VAE and CNN-AE as an indicator for delivering consistent latent representations on a higher resolution. While the CNN-AE and AE-REG geodesics indicate that one may trust the encoded latent representations, the CNN-AE encoding may not be suitable for reliable post-processing, such as classification tasks. More showcases are given in the supplements, showing similar unstable behaviour of the other AEs.

Summarizing, the results validate once more regularization and hybridization to deliver reliable AEs that are capable for compressing datasets to low-dimensional latent spaces by preserving their topology. A feasible approach to extend the hybridization technique to images or datasets of high resolution is one of the aspects we discuss in our concluding thoughts.

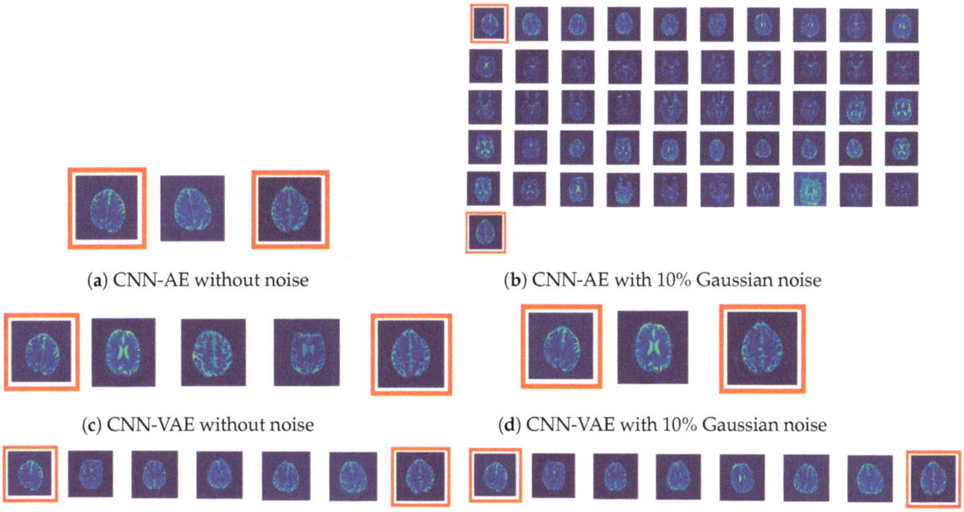

(**a**) CNN-AE without noise (**b**) CNN-AE with 10% Gaussian noise

(**c**) CNN-VAE without noise (**d**) CNN-VAE with 10% Gaussian noise

(**e**) AE-REG without noise (**f**) AE-REG with 10% Gaussian noise

Figure 9. MRI geodesics for latent dimension dim $= 40$ with various levels of Gaussian noise.

6. Conclusions

We delivered a mathematical theory for addressing encoding tasks of datasets being sampled from smooth data manifolds. Our insights were condensed in an efficiently realizable regularization constraint, resting on sampling the encoder Jacobian in Legendre nodes, located in the latent space. We have proven the regularization to guarantee a re-embedding of the data manifold under mild assumptions on the dataset.

We want to stress that the regularization is not limited to specific NN architectures, but already strongly impacts the performance of simple MLPs. Combinations with initially discussed vectorised AEs [44,45] might extend and improve high-dimensional data analysis as in [9]. When combined with the proposed polynomial regression, the hybridised AEs increase strongly in reconstruction quality. For addressing images of high resolution or multi-dimensional datasets, dim ≥ 3, we propose to apply our recent extension of these regression methods [35].

In summary, the regularized AEs performed far better than the considered alternatives, especially with regard to maintaining the topological structure of the initial dataset. The present computations of geodesics provides a tool for analysing the latent space geometry encoded by the regularized AEs and contributes towards explainability of reliable feature selections, as initially emphasised [13–17].

While structural preservation is substantial for consistent post-analysis, we believe that the proposed regularization technique can deliver new reliable insights across disciplines and may even enable corrections or refinements of prior deduced correlations.

Author Contributions: Conceptualization, M.H.; methodology, C.K.R., J.-E.S.C. and M.H.; software, C.K.R., P.H. and A.W.; validation, C.K.R., P.H., J.-E.S.C. and A.W.; formal analysis, J.-E.S.C. and M.H.; investigation, C.K.R. and A.W.; resources N.H.; data curation, N.H.; writing—original draft preparation, C.K.R. and M.H.; writing—review and editing, M.H.; visualization, C.K.R. and A.W.; supervision, N.H. and M.H. All authors have read and agreed to the published version of the manuscript.

Funding: This work was partially funded by the Center of Advanced Systems Understanding (CASUS), financed by Germany's Federal Ministry of Education and Research (BMBF) and by the Saxon Ministry for Science, Culture and Tourism (SMWK) with tax funds on the basis of the budget approved by the Saxon State Parliament.

Data Availability Statement: Complete code benchmark sets and supplements are available at https://github.com/casus/autoencoder-regularisation, accessed on 2 June 2024.

Acknowledgments: We express our gratitude to Ivo F. Sbalzarini, Giovanni Volpe, Loic Royer, and Artur Yamimovich for their insightful discussions on autoencoders and their significance in machine learning applications.

Conflicts of Interest: Author Nico Hoffmann was employed by the company SAXONY.ai. The remaining authors declare that the research was conducted in the absence of any commercial or financial relationships that could be construed as a potential conflict of interest.

References

1. Pepperkok, R.; Ellenberg, J. High-throughput fluorescence microscopy for systems biology. *Nat. Rev. Mol. Cell Biol.* **2006**, *7*, 690–696. [CrossRef] [PubMed]
2. Perlman, Z.E.; Slack, M.D.; Feng, Y.; Mitchison, T.J.; Wu, L.F.; Altschuler, S.J. Multidimensional drug profiling by automated microscopy. *Science* **2004**, *306*, 1194–1198. [CrossRef] [PubMed]
3. Vogt, N. Machine learning in neuroscience. *Nat. Methods* **2018**, *15*, 33. [CrossRef]
4. Carlson, T.; Goddard, E.; Kaplan, D.M.; Klein, C.; Ritchie, J.B. Ghosts in machine learning for cognitive neuroscience: Moving from data to theory. *NeuroImage* **2018**, *180*, 88–100. [CrossRef] [PubMed]
5. Zhang, F.; Cetin Karayumak, S.; Hoffmann, N.; Rathi, Y.; Golby, A.J.; O'Donnell, L.J. Deep white matter analysis (DeepWMA): Fast and consistent tractography segmentation. *Med. Image Anal.* **2020**, *65*, 101761. [CrossRef] [PubMed]
6. Karniadakis, G.E.; Kevrekidis, I.G.; Lu, L.; Perdikaris, P.; Wang, S.; Yang, L. Physics-informed machine learning. *Nat. Rev. Phys.* **2021**, *3*, 422–440. [CrossRef]
7. Rodriguez-Nieva, J.F.; Scheurer, M.S. Identifying topological order through unsupervised machine learning. *Nat. Phys.* **2019**, *15*, 790–795. [CrossRef]
8. Willmann, A.; Stiller, P.; Debus, A.; Irman, A.; Pausch, R.; Chang, Y.Y.; Bussmann, M.; Hoffmann, N. Data-Driven Shadowgraph Simulation of a 3D Object. *arXiv* **2021**, arXiv:2106.00317.
9. Kobayashi, H.; Cheveralls, K.C.; Leonetti, M.D.; Royer, L.A. Self-supervised deep learning encodes high-resolution features of protein subcellular localization. *Nat. Methods* **2022**, *19*, 995–1003. [CrossRef]
10. Chandrasekaran, S.N.; Ceulemans, H.; Boyd, J.D.; Carpenter, A.E. Image-based profiling for drug discovery: Due for a machine-learning upgrade? *Nat. Rev. Drug Discov.* **2021**, *20*, 145–159. [CrossRef]
11. Anitei, M.; Chenna, R.; Czupalla, C.; Esner, M.; Christ, S.; Lenhard, S.; Korn, K.; Meyenhofer, F.; Bickle, M.; Zerial, M.; et al. A high-throughput siRNA screen identifies genes that regulate mannose 6-phosphate receptor trafficking. *J. Cell Sci.* **2014**, *127*, 5079–5092. [CrossRef]
12. Nikitina, K.; Segeletz, S.; Kuhn, M.; Kalaidzidis, Y.; Zerial, M. Basic Phenotypes of Endocytic System Recognized by Independent Phenotypes Analysis of a High-throughput Genomic Screen. In Proceedings of the 2019 3rd International Conference on Computational Biology and Bioinformatics, Nagoya, Japan, 17–19 October 2019; pp. 69–75.
13. Ronneberger, O.; Fischer, P.; Brox, T. U-net: Convolutional networks for biomedical image segmentation. In *Proceedings of the Medical Image Computing and Computer-Assisted Intervention–MICCAI 2015: 18th International Conference, Munich, Germany, 5–9 October 2015*; Proceedings, Part III 18; Springer: Berlin/Heidelberg, Germany, 2015; pp. 234–241.
14. Galimov, E.; Yakimovich, A. A tandem segmentation-classification approach for the localization of morphological predictors of C. elegans lifespan and motility. *Aging* **2022**, *14*, 1665. [CrossRef]
15. Yakimovich, A.; Huttunen, M.; Samolej, J.; Clough, B.; Yoshida, N.; Mostowy, S.; Frickel, E.M.; Mercer, J. Mimicry embedding facilitates advanced neural network training for image-based pathogen detection. *Msphere* **2020**, *5*, e00836-20. [CrossRef] [PubMed]
16. Fisch, D.; Yakimovich, A.; Clough, B.; Mercer, J.; Frickel, E.M. Image-Based Quantitation of Host Cell–Toxoplasma gondii Interplay Using HRMAn: A Host Response to Microbe Analysis Pipeline. In *Toxoplasma gondii: Methods and Protocols*; Humana: New York, NY, USA, 2020; pp. 411–433.
17. Andriasyan, V.; Yakimovich, A.; Petkidis, A.; Georgi, F.; Witte, R.; Puntener, D.; Greber, U.F. Microscopy deep learning predicts virus infections and reveals mechanics of lytic-infected cells. *Iscience* **2021**, *24*, 102543. [CrossRef] [PubMed]
18. Sánchez, J.; Mardia, K.; Kent, J.; Bibby, J. *Multivariate Analysis*; Academic Press: London, UK; New York, NY, USA; Toronto, ON, Canada; Sydney, Australia; San Francisco, CA, USA, 1979.
19. Dunteman, G.H. Basic concepts of principal components analysis. In *Principal Components Analysis*; SAGE Publications Ltd.: London, UK, 1989; pp. 15–22.
20. Krzanowski, W. *Principles of Multivariate Analysis*; OUP Oxford: Oxford, UK, 2000; Volume 23.
21. Rolinek, M.; Zietlow, D.; Martius, G. Variational Autoencoders Pursue PCA Directions (by Accident). In Proceedings of the 2019 IEEE/CVF Conference on Computer Vision and Pattern Recognition (CVPR), Long Beach, CA, USA, 15–20 June 2019; Volume 2019, pp. 12398–12407. [CrossRef]
22. Antun, V.; Gottschling, N.M.; Hansen, A.C.; Adcock, B. Deep learning in scientific computing: Understanding the instability mystery. *SIAM News* **2021**, *54*, 3–5.

23. Gottschling, N.M.; Antun, V.; Adcock, B.; Hansen, A.C. The troublesome kernel: Why deep learning for inverse problems is typically unstable. *arXiv* **2020**, arXiv:2001.01258.
24. Antun, V.; Renna, F.; Poon, C.; Adcock, B.; Hansen, A.C. On instabilities of deep learning in image reconstruction and the potential costs of AI. *Proc. Natl. Acad. Sci. USA* **2020**, *117*, 30088–30095. [CrossRef]
25. Chen, H.; Zhang, H.; Si, S.; Li, Y.; Boning, D.; Hsieh, C.J. Robustness Verification of Tree-based Models. In *Proceedings of the Advances in Neural Information Processing Systems*; Wallach, H., Larochelle, H., Beygelzimer, A., d'Alché-Buc, F., Fox, E., Garnett, R., Eds.; Curran Associates, Inc.: Red Hook, NY, USA, 2019; Volume 32.
26. Galhotra, S.; Brun, Y.; Meliou, A. Fairness testing: Testing software for discrimination. In Proceedings of the 2017 11th Joint Meeting on Foundations of Software Engineering, Paderborn, Germany, 4–8 September 2017; pp. 498–510.
27. Mazzucato, D.; Urban, C. Reduced products of abstract domains for fairness certification of neural networks. In *Proceedings of the Static Analysis: 28th International Symposium, SAS 2021, Chicago, IL, USA, 17–19 October 2021*; Proceedings 28; Springer: Berlin/Heidelberg, Germany, 2021; pp. 308–322.
28. Lang, S. *Differential Manifolds*; Springer: Berlin/Heidelberg, Germany, 1985.
29. Krantz, S.G.; Parks, H.R. The implicit function theorem. Modern Birkhäuser Classics. In *History, Theory, and Applications*, 2003rd ed.; Birkhäuser/Springer: New York, NY, USA, 2013; Volume 163, p. xiv.
30. Stroud, A. *Approximate Calculation of Multiple Integrals: Prentice-Hall Series in Automatic Computation*; Prentice-Hall: Englewood, NJ, USA, 1971.
31. Stroud, A.; Secrest, D. *Gaussian Quadrature Formulas*; Prentice-Hall: Englewood, NJ, USA, 2011.
32. Trefethen, L.N. Multivariate polynomial approximation in the hypercube. *Proc. Am. Math. Soc.* **2017**, *145*, 4837–4844. [CrossRef]
33. Trefethen, L.N. *Approximation Theory and Approximation Practice*; SIAM: Philadelphia, PA, USA, 2019; Volume 164.
34. Sobolev, S.L.; Vaskevich, V. *The Theory of Cubature Formulas*; Springer Science & Business Media: Berlin/Heidelberg, Germany, 1997; Volume 415.
35. Veettil, S.K.T.; Zheng, Y.; Acosta, U.H.; Wicaksono, D.; Hecht, M. Multivariate Polynomial Regression of Euclidean Degree Extends the Stability for Fast Approximations of Trefethen Functions. *arXiv* **2022**, arXiv:2212.11706.
36. Suarez Cardona, J.E.; Hofmann, P.A.; Hecht, M. Learning Partial Differential Equations by Spectral Approximates of General Sobolev Spaces. *arXiv* **2023**, arXiv:2301.04887.
37. Suarez Cardona, J.E.; Hecht, M. Replacing Automatic Differentiation by Sobolev Cubatures fastens Physics Informed Neural Nets and strengthens their Approximation Power. *arXiv* **2022**, arXiv:2211.15443.
38. Hecht, M.; Cheeseman, B.L.; Hoffmann, K.B.; Sbalzarini, I.F. A Quadratic-Time Algorithm for General Multivariate Polynomial Interpolation. *arXiv* **2017**, arXiv:1710.10846.
39. Hecht, M.; Hoffmann, K.B.; Cheeseman, B.L.; Sbalzarini, I.F. Multivariate Newton Interpolation. *arXiv* **2018**, arXiv:1812.04256.
40. Hecht, M.; Gonciarz, K.; Michelfeit, J.; Sivkin, V.; Sbalzarini, I.F. Multivariate Interpolation in Unisolvent Nodes–Lifting the Curse of Dimensionality. *arXiv* **2020**, arXiv:2010.10824.
41. Hecht, M.; Sbalzarini, I.F. Fast Interpolation and Fourier Transform in High-Dimensional Spaces. In *Proceedings of the Intelligent Computing*; Arai, K., Kapoor, S., Bhatia, R., Eds.; Advances in Intelligent Systems and Computing; Springer International Publishing: Cham, Switzerland, 2018; Volume 857, pp. 53–75.
42. Sindhu Meena, K.; Suriya, S. A Survey on Supervised and Unsupervised Learning Techniques. In *Proceedings of the International Conference on Artificial Intelligence, Smart Grid and Smart City Applications*; Kumar, L.A., Jayashree, L.S., Manimegalai, R., Eds.; Springer International Publishing: Cham, Switzerland, 2020; pp. 627–644.
43. Chao, G.; Luo, Y.; Ding, W. Recent Advances in Supervised Dimension Reduction: A Survey. *Mach. Learn. Knowl. Extr.* **2019**, *1*, 341–358. [CrossRef]
44. Mitchell, T.M. *The Need for Biases in Learning Generalizations*; Citeseer: Berkeley, CA, USA, 1980.
45. Gordon, D.F.; Desjardins, M. Evaluation and selection of biases in machine learning. *Mach. Learn.* **1995**, *20*, 5–22. [CrossRef]
46. Wu, H.; Flierl, M. Vector quantization-based regularization for autoencoders. In Proceedings of the AAAI Conference on Artificial Intelligence, New York, NY, USA, 7–12 February 2020; Volume 34, pp. 6380–6387.
47. Van Den Oord, A.; Vinyals, O.; Kavukcuoglu, K. Neural discrete representation learning. *Adv. Neural Inf. Process. Syst.* **2017**, *30*, 6306–6315.
48. Rifai, S.; Mesnil, G.; Vincent, P.; Muller, X.; Bengio, Y.; Dauphin, Y.; Glorot, X. Higher order contractive auto-encoder. In Proceedings of the Joint European Conference on Machine Learning and Knowledge Discovery in Databases, Athens, Greece, 5–9 September 2011; Springer: Berlin/Heidelberg, Germany, 2011; pp. 645–660.
49. Rifai, S.; Vincent, P.; Muller, X.; Glorot, X.; Bengio, Y. Contractive auto-encoders: Explicit invariance during feature extraction. In Proceedings of the 28th International Conference on International Conference on Machine Learning (ICML), Bellevue, WA, USA, 28 June–2 July 2011.
50. Kingma, D.P.; Welling, M. Auto-encoding variational bayes. *arXiv* **2013**, arXiv:1312.6114.
51. Burgess, C.P.; Higgins, I.; Pal, A.; Matthey, L.; Watters, N.; Desjardins, G.; Lerchner, A. Understanding disentangling in β-VAE. *arXiv* **2018**, arXiv:1804.03599.
52. Kumar, A.; Poole, B. On implicit regularization in β-VAEs. In Proceedings of the 37th International Conference on Machine Learning (ICML 2020), Vienna, Austria, 12–18 July 2020; pp. 5436–5446.

53. Rhodes, T.; Lee, D. Local Disentanglement in Variational Auto-Encoders Using Jacobian L_1 Regularization. *Adv. Neural Inf. Process. Syst.* **2021**, *34*, 22708–22719.
54. Gilbert, A.C.; Zhang, Y.; Lee, K.; Zhang, Y.; Lee, H. Towards understanding the invertibility of convolutional neural networks. In Proceedings of the 26th International Joint Conference on Artificial Intelligence, Melbourne, Australia, 19–25 August 2017; pp. 1703–1710.
55. Anthony, M.; Bartlett, P.L. *Neural Network Learning: Theoretical Foundations*; Cambridge University Press: Cambridge, MA, USA, 2009.
56. Goodfellow, I.; Bengio, Y.; Courville, A. *Deep Learning*; MIT Press: Cambridge, MA, USA, 2016.
57. Adams, R.A.; Fournier, J.J. *Sobolev Spaces*; Academic Press: Cambridge, MA, USA, 2003; Volume 140.
58. Brezis, H. *Functional Analysis, Sobolev Spaces and Partial Differential Equations*; Springer: Berlin/Heidelberg, Germany, 2011; Volume 2.
59. Pedersen, M. *Functional Analysis in Applied Mathematics and Engineering*; CRC Press: Boca Raton, FL, USA, 2018.
60. Gautschi, W. *Numerical Analysis*; Springer Science & Business Media: Berlin/Heidelberg, Germany, 2011.
61. Chen, W.; Chern, S.S.; Lam, K.S. *Lectures on Differential Geometry*; World Scientific Publishing Company: Singapore, 1999; Volume 1.
62. Taubes, C.H. *Differential Geometry: Bundles, Connections, Metrics and Curvature*; OUP Oxford: Oxford, UK, 2011; Volume 23.
63. Do Carmo, M.P. *Differential Geometry of Curves and Surfaces: Revised and Updated*, 2nd ed.; Courier Dover Publications: Mineola, NY, USA, 2016.
64. Weierstrass, K. Über die analytische Darstellbarkeit sogenannter willkürlicher Funktionen einer reellen Veränderlichen. *Sitzungsberichte K. Preußischen Akad. Wiss. Berl.* **1885**, *2*, 633–639.
65. De Branges, L. The Stone-Weierstrass Theorem. *Proc. Am. Math. Soc.* **1959**, *10*, 822–824. [CrossRef]
66. Baydin, A.G.; Pearlmutter, B.A.; Radul, A.A.; Siskind, J.M. Automatic differentiation in machine learning: A survey. *J. Mach. Learn. Res.* **2018**, *18*, 1–43.
67. Bézout, E. *Théorie Générale des Équations Algébriques*; de l'imprimerie de Ph.-D. Pierres: Paris, France, 1779.
68. Fulton, W. *Algebraic Curves (Mathematics Lecture Note Series)*; The Benjamin/Cummings Publishing Co., Inc.: Menlo Park, CA, USA, 1974.
69. Kingma, D.P.; Ba, J. Adam: A method for stochastic optimization. *arXiv* **2014**, arXiv:1412.6980.
70. Xiao, H.; Rasul, K.; Vollgraf, R. Fashion-MNIST: A Novel Image Dataset for Benchmarking Machine Learning Algorithms. *arXiv* **2017**, arXiv:1708.07747.
71. Moor, M.; Horn, M.; Rieck, B.; Borgwardt, K. Topological autoencoders. In Proceedings of the International Conference on Machine Learning, Virtual, 13–18 July 2020; pp. 7045–7054.
72. Marcus, D.S.; Wang, T.H.; Parker, J.; Csernansky, J.G.; Morris, J.C.; Buckner, R.L. Open Access Series of Imaging Studies (OASIS): Cross-sectional MRI data in young, middle aged, nondemented, and demented older adults. *J. Cogn. Neurosci.* **2007**, *19*, 1498–1507. [CrossRef]

MDPI AG
Grosspeteranlage 5
4052 Basel
Switzerland
Tel.: +41 61 683 77 34

Axioms Editorial Office
E-mail: axioms@mdpi.com
www.mdpi.com/journal/axioms

www.ingramcontent.com/pod-product-compliance
Lightning Source LLC
LaVergne TN
LVHW072325090526
838202LV00019B/2354
* 9 7 8 3 7 2 5 8 2 3 0 1 7 *